The Scientific Life

The Scientific Life

A MORAL HISTORY OF A LATE MODERN VOCATION

STEVEN SHAPIN

The University of Chicago Press   Chicago and London

STEVEN SHAPIN is the Franklin L. Ford Professor of the History of Science at Harvard University. He is the author of *A Social History of Truth*, *The Scientific Revolution*, and, with Simon Schaffer, *Leviathan and the Air-Pump*. He is also a frequent contributor to the *London Review of Books* and has written for the *New Yorker*.

The University of Chicago Press, Chicago 60637
The University of Chicago Press, Ltd., London
© 2008 by The University of Chicago
All rights reserved. Published 2008
Printed in the United States of America

17 16 15 14 13 12 11 10 09 08   1 2 3 4 5

ISBN-13: 978-0-226-75024-8 (cloth)
ISBN-10: 0-226-75024-8 (cloth)

Library of Congress Cataloging-in-Publication Data

Shapin, Steven.
    The scientific life : a moral history of a late modern vocation / Steven Shapin.
        p.   cm.
    Includes bibliographical references and index.
    ISBN-13: 978-0-226-75024-8 (cloth : alk. paper)
    ISBN-10: 0-226-75024-8 (pbk.)
    1. Scientists—Moral and ethical aspects.   2. Science—Economic aspects.   I. Title.
    Q180.55.M67S53   2008
    174′.95—dc22   2008004588

♾ The paper used in this publication meets the minimum
requirements of the American National Standard for Information
Sciences—Permanence of Paper for Printed Library Materials, ANSI Z39.48-1992.

For

BARBARA ROSENKRANTZ

&

NAT MARSHALL

Vocation

designation or destination to a particular state
or profession; a calling by the will of God; summons; call;
inducement; employment; calling; occupation; trade

Nuttall's Standard Dictionary, 1913

a summons or strong inclination to a particular state or
course of action; *especially*: a divine call to the religious life;
an entry into the priesthood or a religious order;
the work in which a person is regularly employed

Merriam-Webster's Online Dictionary

One's ordinary occupation, business, or profession

Oxford English Dictionary, 1989

Contents

Acknowledgments ix   Preface xiii

*

1 · Knowledge and Virtue: 1
The Way We Live Now

2 · From Calling to Job: 21
*Nature, Truth, Method, and Vocation from the
Seventeenth to the Nineteenth Centuries*

3 · The Moral Equivalence of the Scientist: 47
A History of the Very Idea

4 · Who Is the Industrial Scientist? 93
The View from the Tower

5 · Who Is the Industrial Scientist? 127
The View from the Managers

6 · The Scientist and the Civic Virtues: 165
The Moral Life of Organized Science

7 · The Scientific Entrepreneur: 209
Money, Motives, and the Place of Virtue

8 · Visions of the Future: 269
Uncertainty and Virtue in the World of
High-Tech and Venture Capital

The Way We Live Now: 305
Epilogue

*

Notes 315

Bibliography 401   Index 441

Acknowledgments

My respect for my academic colleagues, and my appreciation of the many demands on their time, are so great that I have not inflicted the completed manuscript of this book on any of them. Rebecca Herzig of Bates College was kind enough to read large portions of it about two years before it was done, and her criticisms were enormously perceptive and constructive—more helpful than I deserve. Chris Phillips in my department at Harvard worked with me in the last stages, helping me assemble the book and, especially, trying to shorten it, and I owe him a very great debt. Neither Rebecca nor Chris, of course, bears the slightest responsibility for remaining flaws.

I want especially to mention my former student Charles Thorpe, who was an essential interlocutor during the early stages of writing and thinking and who assisted in the ethnographic parts of chapter 7. The intermittent gestures to his work on Oppenheimer do not adequately reflect the extent of his contribution. I also wish to acknowledge Mark Peter Jones, a former student in the Sociology Department of the University of California, San Diego, whose doctoral thesis on entrepreneurial biotech is a rich and resonant source for anyone interested in late modern technoscience and entrepreneurial science, and who made a large number of critical comments, both stylistic and substantive, after the manuscript had been submitted, many of which I tried to act upon.

The number of people with whom I have talked over many years about the substance of various parts of the book is very large, and most of them did not know how helpful they had been or, indeed, what project they were assisting. These include (in no special order) Paul Rabinow, Peter Buck, Adrian Johns, David Kaiser, Andy Lakoff, David Edgerton, Robert Kohler, Yaron Ezrahi, Daniel S. Greenberg, Hélène Mialet, Rebecca Lemov, Naomi Oreskes, Richard Yeo, Soraya de Chadarevian, Bill Clark, Grischa Metlay, Heather Douglas, Geert Somsen, Jeff Sturchio, Charis Thompson, Cyrus Mody, and Kasper Risbjerg Eskildsen. The readers for the University of Chicago Press were extraordinarily careful and understanding, even, and especially, when they had to be critical. They immeasurably improved what still must be a very imperfect book.

The idea for this book originated partly during a happy year at the Center for Advanced Study in the Behavioral Sciences at Stanford. I did not know at the time that this was the project that would ultimately emerge, but the free and easy intellectual environment of the Center allowed such a strange project to take root in my mind. Research for chapter 7 was partly supported by grant P-00-02 from the University of California Industry-University Cooperative Research Program's Economic Impact Research Program in 2002–2003, and I thank all the scientists and engineers who allowed me to talk with them in this connection, but who, owing to conditions of my grant, I was not able to name. I greatly regret that, and, more to the point, so too do many of my interviewees, none of whom expressed any objections to being directly quoted. Chapters 4 and 5 use some much modified material from my "Who Is the Industrial Scientist? Commentary from Academic Sociology and from the Shop-Floor in the United States, ca. 1900–ca. 1970," in *The Science-Industry Nexus: History, Policy, Implications*, Nobel Symposium 123, eds. Karl Grandin, Nina Wormbs, and Sven Widmalm (Canton, MA: Science History Publications, 2004), pp. 337–363, and chapter 2 uses a small portion of material previously published as "The Man of Science," in *The Cambridge History of Science, Vol. 3: Early Modern Science*, eds. Lorraine Daston and Katharine Park (Cambridge: Cambridge University Press, 2006), pp. 179–191, and "The Image of the Man of Science," in *The Cambridge History of Science, Vol. 4: Eighteenth-Century Science*, ed. Roy Porter (Cambridge: Cambridge University Press, 2003), pp. 159–183. The last stages of assembling this book were made much easier by the dedication and unfailing good humor of Ellen Guarente. I thank my copyeditor Michael Koplow for his extraordinary skills and my editor Christie Henry, without whose encouragement this book would either have taken a very different form or not come into being at all.

My wife Abigail Barrow has permanently ensured my humility already by not reading one book dedicated to her, so I have no need for an additional dose, but she assisted this book in more ways than even she knows, and there is a legitimate sense in which this is more her book than mine.

Preface

I have written two books about aspects of seventeenth-century science that ended with brief speculations about "the way we live now."[1] Some of my historian-friends were puzzled about these gestures; others disapproved, taking this as further proof that my commitment to the purity and particularity of history was wanting. They were right.

It wasn't that I had spent almost twenty years in an interdisciplinary science studies unit, then almost fifteen in a sociology department. Neither experience much affected my disciplinary allegiance, just because I never properly had one. There was, for me, no choice but to embrace the quotidian particularity of the past and to write about it in as much detail as I could manage and readers might bear. At the same time, I take for granted three things that many historians seem to find, in some degree, incompatible: (1) that historians *should* commit themselves to writing about the past, as it really was, and that the institutional intention of history writing *must* embrace such a commitment; (2) that we *inevitably* write about the past as an expression of present concerns, and that we have no choice in this matter; and (3) that we *can* write about the past to find out about how it came to be that we live as we now do, and, indeed, for giving better descriptions of the way we live now.[2] I do not think that there *is* a conflict between these aims or that the "presentness" of any account of the past has to be denied for historical writing to have whatever credibility it possesses.

The political philosopher Michael Oakeshott reckoned that concern with the particularities of the past just defines the institutional intention of the historian's role: "The historian is interested in the deadness of the past, and in its dissimilarity from the present. What attracts his eye and fires his enthusiasm is diversity. He has a preternatural sensibility for the minute and detailed differences which distinguish one situation from another, one man from another, one age from another. The modern instance does not attract him, for he knows that similarities appear only when details are neglected." Yet Oakeshott also recognized something he called "a practical past": "Here the past is thought of as merely that which preceded the present, that from which the present has grown or developed, and the significance of the past is taken to lie in the fact that it has been influential in deciding the present or future fortunes of men." This "practical past," he declared, belongs not to the historian but to the social scientist.[3] Two different sorts of people, occupying two different academic roles, had to undertake these distinct types of inquiry. Oakeshott was quite right about the diversity of the ways in which the past may be apprehended, and in this respect his views about the nature of history are widely shared. But I can see no rationale—other than the contingent and ever-changing ways in which some academic disciplines decide to manage their territorial claims—why concern with this "practical past" should not be acknowledged as *history*. So the concluding gestures of my previous books were, to me, no mere rhetorical flourishes. Wanting a better understanding of "the way we live now" was indeed a motivation for writing those detailed books about passages of early modern science. Some years ago, I came to feel that it was time to act more systematically upon those sensibilities, so here I *start* with a sketch of some issues involved in describing aspects of how we live now, specifically how we think about the most powerful forms of knowledge and about those who make and manipulate that knowledge.

There are some obvious ways in which this is a very different book from those I have undertaken before. Like some of my other work, it is about the relationship between knowledge and the virtues of people—a relationship that has taken many different forms over the course of time but which has no historical "origin." Here, I decide to pick up the story in the early modern period: the radically novel configurations of people, practices, and institutions that flourish in present-day technoscience have their deep histories, and I want to encourage readers to think about what is indeed now radically new and what is a reconfiguration of the deep past. But the core of the book starts around the beginning of the twentieth century and it has no historical terminus at all. The last substantive chapter is about aspects of present-day practices for making the *future*, or, at least, about how certain people come to have enough confidence in envisaged futures to act in the present on pictures of possible worlds-to-come.

I refer intermittently to the sort of society and culture I want to understand as "late modernity." I do not intend anything very precise by this usage. I mean to pick out the time-span of roughly 1900 to the present, and late modernity seems as good a term as any for that purpose, especially as "early modernity" is a recognized historians' designation for the period roughly 1550–1730, leaving (if one prefers) the Enlightenment and the Industrial Revolution for the unmodified "modern."[4] Social theorists have appropriated "late modernity," along with "high modernity" (Anthony Giddens), "reflexive modernity" (Giddens, Ulrich Beck, and Scott Lash), "liquid modernity" (Zygmunt Bauman), and (of course) postmodernity for more programmatically ambitious definitional agendas. I very much like the theorists' intermittent attention to the accelerating institutional, intellectual, and moral *uncertainties* of the present and recent past, and this book will have a lot to say about the relationship between uncertainty and the personal dimensions of institutional action.[5]

The present book deals not with English gentleman-amateurs but very substantially with American industrial scientists, entrepreneurs, venture capitalists, and Organization Men: research managers at electrical and photographic firms, team-playing organic chemists, Southern Californian investors in high-tech companies, engineering professors trying to develop and sell intellectual property and to get ahead in their academic careers. My heroes are not, in the main, and in the usual sense of the word, heroic; if what they do changes the world—and it does—then most of their world-changing actions have a mundane character; and, throughout the twentieth century, and into the present, many external commentators seem to find their motives ignoble. It is often said that they are *ordinary people*. Sometimes that ordinariness is celebrated by the scientists and technologists themselves; sometimes the imputation of ordinariness is objected to by those deputing themselves as Defenders of Science: "The old notion of the scientist as hero," the evolutionary psychologist Steven Pinker complains, "has been replaced by the idea of scientists as amoral nerds at best."[6] We *know* these people, or think we do, and it is not their historical otherness that has to be insisted on but their *familiarity* and that of the world they, and we, inhabit together. Yet it is a familiarity that has to be recovered and recuperated from some widely circulated academic narratives. There are many ill-founded stories about what both their otherness and their ordinariness consist in.

This is overwhelmingly, though not exclusively, an American story. The cultural and social backgrounds described in chapters 2 and 3 necessarily draw in European material, since twentieth-century America is, of course, the historical legatee of many world traditions. With chapter 4, the scene shifts to the United States and substantially remains there. I very occasionally draw in material from other developed societies, since, after all, the late modern world is at least partly "globalized": Americans knew about and responded to developments occurring elsewhere; people and ideas moved back and forth. In the area of industrial science, for example, American patterns in the early twentieth century were importantly shaped by Germany, while by the end of the century the rest of the world was desperately trying to understand, and to package for import, institutional and cultural configurations thought to have their origins and natural homes in the United States. I have no interest here in arguing that the American experience is central to, or uniquely causative of, what is happening in the rest of the world, but there is so much worldwide curiosity about—emulation or abhorrence of—the American patterns I describe that it is scarcely necessary for me to do so. One of my purposes is to stir up curiosity about how Americans think about knowledge and knowers, how they organize their institutional practices on the bases of that thinking. And insofar as that curiosity is acquitted, a window is opened onto one of the major sites in which more widely distributed features of the late modern condition have been made. What is scientific *knowledge* considered to be in late modern America? What kind of *people* are the bearers of that knowledge thought to be? And, most importantly, what relations obtain between the *authority* of knowledge and the *character* of knowers? How does people-knowledge figure in late modern science and technology? Whether or not the rest of the world will come closely to resemble America in these respects must be an open question. The practical response to that question should, however, depend upon getting as detailed an account as we can about what happened, and is happening, in American science and technology.

For all the differences in time period and cultural settings, there are many ways in which this book is a direct outgrowth of my prior studies of knowledge and society in early modern England. Here the links are not empirical but thematic. Late modern American industrial scientists and venture capitalists are not English gentlemen; Southern Californian public universities are not seventeenth-century (or even twenty-first-century)

Cambridge colleges; networking entrepreneurs form very different social configurations from conversational polite society in a Restoration London drawing-room. Yet change the focus of engagement and many of the same predicaments and practices come into view. Knowledge of things still depends upon knowledge of people. The world of the face-to-face and the familiar still figures in making and warranting knowledge. The late modern expert still retains some characteristics of the early modern virtuoso. Trust in familiar people still has not been replaced by the apparatuses of surveillance, control, and institutional discipline. This is a book about some centrally important social and intellectual configurations of late modernity that continue to resemble those of "the world we have lost" but whose resemblances have become largely invisible in many academic accounts of "the way we live now."

⋆ I ⋆

Knowledge and Virtue

THE WAY WE LIVE NOW

> We are here partly interested in the origin of precisely the irrational
> element which lies in this, as in every conception of a calling.
>
> Max Weber, *The Protestant Ethic and the Spirit of Capitalism*

THE PERSONAL AND THE IMPERSONAL:
AN ESSENTIAL TENSION

This book has two purposes: the first is to show how and why people
and their virtues matter to the making and the authority of late modern
bodies of technical knowledge; the second is to give some account of why
that now may seem an odd, even perverse, claim. What does it mean to say
that people matter? I mean that we cannot understand how various scientific and technological knowledges are made, and made authoritative,
without appreciating the roles of familiarity, trust, and the recognition
of personal virtues. And the reason such a claim may seem perverse is that
both these knowledges and the means by which they are produced are
widely accounted *impersonal*—having nothing to do with personal characteristics and patterns of familiarity and enjoying their special authority
through *being understood* to have no such dependencies. That is why this
book has two tasks—to establish the claim about personal virtues and to
account for its apparent oddness—and I shall try to weave them together,
to keep them both in play and in their natural tension. It is only through
sensitivity to both the personal and the attributed impersonal—to the
role of familiar people in the making and maintenance of scientific and
technological knowledges and to the evident strangeness of asserting such
a role—that we can appreciate some key features of how we now live and
know: what our technoscientific knowledge is, how it is produced, how
its cultural authority is secured.

In generic terms, this book belongs not to the sociology of scientific
knowledge but to cultural history and to a broadly sociological interest in

forms of behavior within institutions. So when I say that I am concerned with the authority of science, or even the authority of scientific knowledge, I do not deal here with why groups of scientists might prefer theory *A* over theory *B*, or even with the wider credibility of specific claims, such as those concerning the safety or danger of radioactive isotopes. Rather, I mean to direct attention to the conditions in which what is taken to be science is or is not considered credible, in which those responsible for that knowledge are or are not thought to be reliable sources, in which their way of life in making scientific knowledge is or is not reckoned to be one that conduces to the reliability of that knowledge. This is, so to speak, the *external* referent of allusions to scientific authority, and it points to beliefs about science circulating in the general culture and, to some extent, within the scientific community itself on occasions when its members choose to reflect upon their identity and authority. But there is also an *internal* dimension to this project. While widely diffused beliefs about science and scientists are of considerable interest in their own right, I want also to understand how a range of such beliefs match up with what can be learned about how science is made, how scientists think of themselves and the institutions they inhabit, how scientists are thought of, dealt with, and managed by members of the institutions they inhabit and those with which they come in contact. There are pervasive mismatches between aspects of the external and internal accounts, and the interpretation of these fault lines is a theme that runs through the book. And among the most telling of these mismatches is that between the attributed impersonality of late modern science-making and the rich repertoires of affect-saturated familiarity that one uncovers when looking closely at quotidian institutional practices. I identify my objects of interest here as belonging to *technoscience*. As a term of academic art, this has been closely associated with the work of Bruno Latour, whose rejection of conventional distinctions between what belongs to science and what to society, politics, or the economy employed the notion of technoscience "to describe all the elements tied to the scientific contents no matter how dirty, unexpected, or foreign they may seem."[1] In the present context, the usage is both more diffuse and more mundane. In the beginning of my period of interest—in the early part of the twentieth century—distinctions between science and technology, between the institutions in which each was done, and between the motives and personal characteristics of the scientist and the technologist were widely held, were accounted relatively clear, and were often consequential in institutional and cultural action. At the end of the period covered by this book—the late twentieth century and the present—such distinctions continue to circulate in much external commentary on science and technology, and on universities and industry, but they have been greatly eroded *within* the worlds of science and technology and within the institutions where science and technology happen. "Technoscience" here is just a term that allows me historically to follow natural knowledge and its embodiment in material artifacts without taking a position on what is science and what technology.

To argue for the importance, even the centrality, of the personal dimension in late modern technoscience is directly to confront a sensibility that defines almost all academic, and probably much lay, thought about late modern culture. Isn't the regime of trust, familiarity, and personal virtue precisely "the world we have lost"?[2] What is modernity, and even more its "late" version, but the subjugation of subjectivity to objectivity, the personal to the methodically mechanical, the individual to the institutional, the contingent and the spontaneous to the rule of rule? It is widely said that we now trust in impersonal criteria, not in people; in rationally organized and regulated institutions rather than in charismatic leaders. This is the sort of thing Max Weber meant when he pointed to the "separation of business from the household, which completely dominates modern economic life," and which was the spatial manifestation of familiarity's decline.[3] As late moderns, it is claimed, we are not able to call upon the resources of familiarity in addressing social and intellectual problems, nor would it be considered legitimate to do so. People are accounted weak; rules and institutions are accounted strong. People are arbitrary and malleable; rules and institutions are bound by stable criteria. This was the sentiment informing Weber's account of *charismatic authority* and its characteristic gesture: "It has been written . . . , but *I* say unto you." Charisma meets needs "that go beyond those of everyday routine," where routinization and bureaucratic organization are the stamp of the modern: "In radical contrast to bureaucratic organization, charisma knows no formal or regulated appointment or dismissal, no career, advancement or salary, no supervisory or appeals body, no local or purely technical jurisdiction, and no permanent institutions in the manner of bureaucratic agencies."[4] So, insofar as late modernity's technoscientific experts are almost wholly professionalized, organized, and regularly remunerated, we are tempted to talk straightforwardly about the "waning of charisma" and the consequent "diminishing importance of individual action."[5] Charisma, personal authority, and familiarity rule where rules do not, and that is why their absence or diminution can be talked about as among the markers of modernity.

There are, however, reasons to reject, or severely to qualify, much of this academic common sense about late modern realities. First, while the irrelevance of the personal in scientific knowledge-making has been vigorously asserted at least since the seventeenth century, familiar people and their virtues have *always* been pertinent to the making, maintenance, transmission, and authority of knowledge. Whose integrity and competence can one trust? Much early modern rhetoric cautioning against reliance on people as knowledge-sources should be read as arguments against specific *modes* of authority and not against trusting familiar people in general. The elimination of the personal has been repeatedly announced and celebrated. Late modernity is not alone in this. I have made this kind of argument in previous books and need not repeat it here. Whatever is true about knowledge-making in general should be true about particular historical moments of knowledge-making.[6] Second, when considering the supposed ejection of the personal from late modern technoscience, the endemic problem of theoretical metonymy presents itself—taking the part for the whole, an account of certain aspects of late modernity for its range of quotidian realities, what is considered the essence or leading edge of change for the ways things *are*. Paul Rabinow has observed that "in the sphere of meaning, the mark of modernity is fracture and pluralism . . . Modernity is the principle of de-magification, not its colonial triumph."[7] Metonymic bias afflicts practically all influential social and cultural theories—almost necessarily so insofar as theoretical representations are, at most, abridgments of the social realties they purport to describe. Take, for example, the move from noting those aspects of present-day society that are "globalized" to describing ours *as* "a globalized society"; from rightly remarking on the *importance* of quantification in late modernity to describing number as our *privileged* way of remedying problems of bias, interest, and mistrust; from drawing attention to reflexive *bits* of our culture to describing our culture *as* reflexive. Theoretical metonymy *is* a problem to those wanting a more filled-in picture, rather than a pencil sketch, of the way we live now. But dissatisfaction with stories about depersonalization and demoralization amounts to more than that.

One could say that the related resources of personal virtue, familiarity, and charisma are *neglected aspects* of late modernity and that they *survive* in more vigor than some theorists allow. In itself, that might be a useful thing to say, and material in this book *can* be enlisted in its support. But my argument is not confined to claims about survival: I try to establish that the characteristics and virtues of familiar people now matter *more* than they have for very many years and that this mattering concentrates in just those intellectual and institutional configurations from which the most consequential *changes* of late modernity emerge. We are not here talking about premodern survivals or vestiges but about accelerating late modern realities. The closer you get to the heart of technoscience, and the closer you get to the scenes in which technoscientific futures are made, the greater is the acknowledged role of the personal, the familiar, and even the charismatic. Much of this book shows how it is that personal virtue, familiarity, and charisma feature in such characteristically late modern configurations as the industrial research laboratory and the entrepreneurial network. That is this book's central contention, and almost everything leads up to it and is enlisted to support it.

But just because this way of looking at things goes so much against the grain of academic commentary, the book has to take on another task: it has to give some sort of historical account of why that commentary takes the form it does. And, after all, the argument in this book is not entirely lacking in support among the theorists. It was in connection with Zygmunt Bauman's attempt to describe "postmodernity" that he remarked, "Human passions used to be considered too errant and fickle, and the task to make human cohabitation secure too serious, to entrust the fate of human coexistence to moral capacities of human persons. What we come to understand now is that fate can be entrusted to little else."[8] I will try to show in some detail why Bauman was quite right. It is what I call the intense and accelerating *normative uncertainties* of late modernity that draw upon, stress, and mobilize these supposedly premodern resources, uncertainties that reach their highest pitch in many of the scenes in which new scientific knowledges and new technological artifacts are made. Late modernity proliferates uncertainties; radical uncertainties mark the venues from which technoscientific futures emerge; and it is in the quotidian management of those uncertainties that the personal, the familiar, and the charismatic flourish. As the social theorist Stephen Turner puts it, "Weber never imagined that what the future held was a new age of charisma . . . , a new age in which the extraordinary is ordinary, in which changes in values and attitudes led by the example and personal force of publicly acclaimed personalities is a characteristic feature of the culture."[9]

The late modern condition has been repeatedly idealized, deplored, and celebrated: it remains still to be adequately *described*. Describing late modernity is evidently a far more daunting task than theorizing it—so difficult that I aim only to offer accounts of a very few features of it, albeit features that I consider perspicuous—especially revealing of late modern textures of technical knowledge, order, and authority. I narrow down my interests even more; I want to describe *who truth-speakers are* in late modernity: what kinds of people, with what kind of attributed and acted-upon characteristics, are the bearers of our most potent forms of knowledge. Nor am I alone in thinking that a description of late modernity's truth-speakers is a good way to understand much more about the way we live now: Rabinow talks about his project in similar terms, and, likewise, identifies his work with Weberian questions and sensibilities. Who *are* the people who bear late modern technical knowledges, and what is the (radically changing) nature of the institutional configurations in which they work and from which their knowledge emerges? What moral warrants stand behind their claims to knowledge and their authority to realize consequent technologies? Who, as Rabinow puts it, "has the authority—and the responsibility—to represent experience and knowledge" in late modernity?[10] And how do we cope with a situation in which "those authorized to speak the truth require vast sums of money to practice their sciences and thereby to produce those truths on which we so firmly believe ourselves to be dependent"?[11]

THE INVISIBLE SCIENTIST

Why are questions about the identity of late modernity's truth-speakers perspicuous? After all, the widely stipulated moral ordinariness of technical experts, and the institutional ordinariness of their jobs, as well as the supposed impersonality, transparency, and efficacy of Scientific Method, count as reasons why personal characteristics *do not matter* to the constitution, authority, and status of technical knowledges. These, too, are late modern sensibilities that must be described and appreciated before they can be qualified or criticized. The voices insisting on the impersonality of late modern technoscientific knowledges, and knowledge-making practices, are many, eloquent, and insistent. Writing about the same time as Weber, the American economist and social theorist Thorstein Veblen offhandedly remarked on the elimination of "the personal equation" in modern times, and especially in science and industry. Like Weber, Veblen observed that "no effort is spared to eliminate all bias of personality from the technique of the results of science or scholarship."[12] What the physiologist Claude Bernard wrote fifty years before was now in the process of passing into a cultural commonplace: "Art is I, Science is We," and so the impersonality of the means of scientific production and the absence of a personal mark on the product were reliable and visible signs of its authenticity. No one man's opinion represents scientific truth: "The revolution which the experimental method has effected in the sciences is this: it has put a scientific criterion in the place of personal authority." As science folds the particular into the general, "the names of promoters of science disappear little by little, and the further science advances, the more it takes an impersonal form and detaches itself from the past."[13] In 1890, the French philosopher Ernest Renan wrote that "[the scientist's] goal is not to be read, but to insert one stone in the great edifice . . . the life of the scientist can be summarized in two or three results, whose expression will occupy but a few lines or disappear completely in more advanced formulations."[14]

Over the past several centuries there has been much disagreement about exactly where in the map of culture to draw the line between the relevance and irrelevance of the personal. Did impersonality mark out natural science in particular or did it distinguish science as well as other forms of abstract and rational thought from work considered imaginative? But by the mid-nineteenth century there was general agreement that this was a crucial boundary; that its placement had to do with the relative roles of Reason, Method, Emotion, Imagination, and Genius; and that it concerned the nature and status of knowledge and the means by which different sorts of knowledges were produced. Immanuel Kant said that genius did not properly belong to the natural sciences since, unlike the imaginative arts, everything in scientific productions can be *learned*, and can, therefore, be imitated by essentially anyone.[15] In America, Ralph Waldo Emerson traced the transparency of creative persons in general to the circumstances and psychic requirements of the work they did:

> Great geniuses have the shortest biographies. Their cousins can tell you nothing about them. They lived in their writings, and so their house and street life was trivial and commonplace. If you would know their tastes and complexions, the most admiring of their readers most resembles them. Plato especially has no external biography. If he had lover, wife, or children, we hear nothing of them. He ground them all into paint. As a good chimney burns its smoke, so a philosopher converts the value of all his fortunes into his intellectual performances.[16]

And, just after the death of the greatest genius of twentieth-century science, Roland Barthes dwelt on the mythic significance of Einstein's *brain*, not as a part signifying the whole but the whole itself. It is the material brain, the equation-producing "machine of genius," that tells you all you need to know about the man.[17] Intermittently, there are still attempts to circumscribe the role of the personal in writing about literary figures—John Updike once remarked that "the main question about literary biography is, surely, Why do we need it at all?"[18] —but a reviewer of a biography of Descartes quickly deflected Updike's sentiment onto the biography of *philosophers*: "The philosopher lives his or her life . . . in the quiet seclusion of the mind—and it is because of what he writes or thinks there that he interests us. Everything else, what Updike calls the 'sensational stuff,' is merely secondary or extraneous, and certainly irrelevant to an evaluation of the work."[19]

While the institutionalized expression of such an idea may be a mark of the late modern condition, some of those who gave voice to it in the nineteenth century differed about whether or not the irrelevance of the personal in intellectual productions was anything specially to do with present times. Some reckoned that it was, but that one should resist any splitting of intellectual product from knowers' moral mode of life. In 1859, Søren Kierkegaard insisted, against what he took to be an insidiously growing and characteristically contemporary assumption to the contrary, "that authorship is and ought to be a serious calling implying an appropriate mode of personal existence." People have now lost sight of that fact and the moral imperatives that underpin it. If you go back to Antiquity, you will see what it meant to be an author: "But in our age, which reckons as wisdom that which is truly the mystery of unrighteousness, viz. that one need not inquire about the communicator, but only about the communication, the objective only—in our age what is an author? An author is merely an *x*, even when his name is signed, something quite impersonal." In this circumstance, Kierkegaard concluded, resides the "demoralization of the modern state."[20] In 1845, the Scottish politician and man of letters Henry Brougham described the specific biographical intractability of subjects whose lives were given over to the pursuit of abstract or universal truths: "When the studies of a philosopher"—and in this context Brougham included the natural philosopher—"and especially of a mathematician, have been described, his discoveries recorded, and his writings considered, his history has been written. There is little else to say of such a man: his private life is generally uninteresting and unvaried."[21] In the same spirit, T. H. Huxley later wrote about an apocryphal Babylonian philosopher: "Happily Zadig is in the position of a great many other philosophers. What he was like when he was in the flesh, indeed whether he existed at all, are matters of no great consequence. What we care about in a light is that it shows the way, not whether it is lamp or candle, tallow or wax."[22] And, as chapter 3 shows, present-day sentiments of this general sort have attained the status of how-could-it-be-otherwise matters of fact. They are a resource allowing us to move about on the map of knowledge, to distribute value, and to sort out differences of value and authority between intellectual modes.

During the Second World War, the English zoologist John R. Baker vigorously defended the idea of scientific genius against its socialist detractors, and even compared the temperament of the creative scientist to that of the musician and artist. But he insisted nevertheless on "an obvious distinction" between the scientific and the artistic author: "If Mozart had not composed that immortal work of genius, the overture to 'Le Nozze di Figaro,' no one else would ever have done so; but if Kekulé had not lived, structural formulae and the benzene ring would not have remained for ever hidden: someone else would eventually have dreamed the same dreams."[23] Derek de Solla Price's once influential *Little Science, Big Science* gives an example of how this sentiment had become matter of course by the 1960s. Reflecting on sociological studies of "multiple discoveries," Price noted that scientists are interchangeable in a way that creative artists are not. Personal qualities matter little in science because Method is the main engine of discovery and a unitary external Reality is the stable object of scientific knowledge. Like Bernard, Price noted that while art is "intensely personal" science is "social," since "the scientist needs recognition by his peers." It is the social nature of science that cancels out personal identity and renders it uninteresting and irrelevant. And the very object of scientific knowledge tells against the consequence of authorial identity.[24] It is just very much harder, as a matter of late modern cultural fact, that Roth/Zuckerman should escape personal responsibility for Portnoy/Carnovsky than to hold James Watson personally to account for the double-helical structure of DNA. Nor do artists *want* to escape personal responsibility if they should say, with Oscar Wilde, that "I have put my genius into my life; I have put only my talent into my works."[25] We know what they mean, and no scientist could credibly say the same.

IS AND OUGHT: KNOWLEDGE AND THE KNOWER

Late modernity is supposedly marked by the extension of impersonal means of control to ever new domains, ultimately bringing all of social life under the sway of impersonal reason. It is claimed that we know how to rationally plan our present and even our future. Late modernity is accounted the triumph of the bureaucrats and the planners, and, by natural extension, of science understood in bureaucratic and planning terms: not just science as knowledge of nature, but science as knowledge of ourselves as natural objects, and, finally, of science as knowledge of science itself—the rational closure of the reflexive circle. This was what Alfred North Whitehead meant when he announced in *Science and the Modern World* that "the greatest invention of the nineteenth century was the invention of the method of invention."[26] The full expression of the rule of rule over spontaneity is found in the confidence that the production of truth can be not just rationally organized but effectively planned. This is not a confidence that belongs specifically to any one of the great ideological cleavages of the twentieth century. Late capitalism has endorsed it as enthusiastically as State socialism. And, as I shall show, much the same can be said about late modern *rebellion* against this idea: both conservatives and liberals have found the idea of a rational method of invention equally bizarre; hardheaded venture capitalists vigorously embrace "the personal equation" in investing in radically uncertain futures; many academic scientists—one might think as a matter of course—have rejected the idea that their inquiries can be effectively planned; chapter 5 will show that some of the most eloquent opponents of the idea of rigorous planning have been industrial research managers. But those theorists who identify late modernity with the disappearance of "the personal equation" from science and industry have, seemingly, won the official academic argument. It is just this widely shared view of the disappearance or dissolution of the personal in late modern technoscience and its institutions that seems to inform some of our most confident characterizations of "the way we live now."

One of the keystones of official late modern culture is the distinction between the domains of the "is" and the "ought," and its alleged institutionalization in a range of social and cultural practices. The transit from the descriptive to the normative was not just the "naturalistic fallacy" marked by philosophers from David Hume to Henry Sidgwick to G. E. Moore; it was also ontologically unsustainable and politically inadvisable.[27] During the early years of the twentieth century, recognition of this fallacy became a cultural commonplace, disengaged from any particular philosopher's instantiation. In 1905, the French mathematician Henri Poincaré, even while drawing an analogy between the search for scientific and moral truth, acknowledged "that it may seem that I am misusing words, that I combine thus under the same name two things having nothing in common; that scientific truth, which is demonstrated, can in no way be likened to moral truth, which is felt." Ethics and science, Poincaré noted, "have their own domains"; "they can never conflict since they can never meet. There can no more be immoral science than there can be scientific morals."[28] Ludwig Wittgenstein ended his 1929 lecture on ethics by identifying the distinction between ethical and scientific inquiries: "Ethics so far as it springs from a desire to say something about the ultimate meaning of life, the absolute good, the absolute valuable, can be no science. What it says does not add to our knowledge in any sense."[29] Veblen insisted that science "knows nothing of policy or utility, of better or worse."[30] If there were such people as "moral experts" in modern society, they were not to be found in the laboratory or speaking from a scientific podium. And this was just what Weber argued in "Science as a Vocation." The scientist—the German *der Wissenschaftler* including the social scientist, of course—had neither the moral competence nor the moral right to use the lecture-room or the learned journal to pronounce on *what ought to be done*. The "demagification of the world" definitively shattered the early modern liaison between the role of the scientist and the role of the priest and moralist. No one—"aside from certain big children who are indeed found in the natural sciences"—still believes that science is either a way to God or a key to moral action.[31] Weber here endorsed Tolstoy's sentiments about the fact of science-as-it-then-was: "Science is meaningless because it gives no answer to our question, the only question important for us: 'What shall we do and how shall we live?'"[32] It followed that the scientist's vocation *morally* required the active renunciation of any special moral make up or claims to any special moral authority. Weber said it *about* natural science, and natural scientists—of greater and lesser talents—said it repeatedly from *within* the heart of science.[33]

The greatest physicist of the twentieth century endorsed the impossibility of moving from "is" to "ought":

> All scientific statements and laws have one characteristic in common: they are "true or false" (adequate or inadequate). Roughly speaking, our reaction to them is "yes" or "no." The scientific way of thinking has a further characteristic. The concepts which it uses to build up its coherent systems are not expressing emotions. For the scientist, there is only "being," but no wishing, no valuing, no good, no evil, no goal. As long as we remain within the realm of science proper, we can never meet with a sentence of the type: "Thou shalt not lie."[34]

"Knowledge of what *is* does not open the door directly to what *should be*."[35] Although Einstein famously said that "the man of science is a poor philosopher," he clearly knew all about the naturalistic fallacy.[36] The scientist's authority, Einstein stipulated, was of a quite special and limited sort—and here Einstein tracked Weber's own formulation—smacking of "the Puritan's restraint."[37] Again and again, the most publicly moralistic of modern scientists—and the one whose moral stature was most publicly recognized—insisted upon the natural scientist's *lack* of moral authority: science could not endow its practitioners with any such authority.

Immediately after the Second World War, James Bryant Conant's presidential charge to his Harvard colleagues to design the curricular forms of *General Education for a Free Society*—the famous "Red Book"—produced a nice and confident distinction between what natural science could and could not do. The difference between the natural sciences and the humanities is just that "the former describe, analyze, and explain; the latter appraise, judge, and criticize. In the first, a statement is judged as true or false; in the second, a result is judged as good or bad. The natural sciences do not take it on themselves to evaluate the worth of what they describe."[38] A little later, Yale physicist (and part-time Kantian philosopher) Henry Margenau reacted with distaste to the suggestion of his philosophical colleague F. S. C. Northrop that the ideological differences of the postwar world could and should be resolved by getting straight on "the methods of scientific verification in natural science." Margenau vigorously insisted that Northrop had misunderstood the nature of science: "Science is not equipped with devices capable of rendering ethical judgments. While it may tell you how one may kill most efficiently, it will not—in my opinion, it will never—tell you whether it is right to kill."[39] A Bell Labs physicist used topical terms to ridicule Northrop's faith in the ethical capacities of science: "Let us define a communist as a person who opposes private ownership of a farm or factory, and a democrat as a person who favors private ownership. Does Northrop mean to say that if we physicists were to come to agreement among ourselves on the basic doctrines of theoretical physics, then either the communist would give up his opposition to private ownership or else the democrat become opposed to it?"[40]

In the 1960s, the theoretical physicist Richard Feynman, positively celebrating the "social irresponsibility" of the atomic scientist, rediscovered the naturalistic fallacy for himself: "The principle that observation is the judge imposes a severe limitation to the kind of questions that can be answered [by the scientist]. They are limited to questions that you can put this way: 'if I do this, what will happen?' Questions like, 'should I do this?' and 'what is the value of this?' are not of the same kind . . . That is the step the scientist cannot take . . . As far as I know in the gathering of scientific evidence, there doesn't seem to be anywhere, anything that says whether the Golden Rule is a good one or not."[41] Later, while Feynman was writing about the absolute necessity of internal scientific integrity — the honesty that's needed to guard against finding what you want to find — he carefully distinguished this sort of integrity from that expected in ordinary life: "I am not trying to tell you what to do about cheating on your wife, or fooling your girlfriend, or something like that" — the kind of personal conduct about which he often publicly boasted — "when you're not trying to be a scientist, but just trying to be an ordinary human being. We'll leave those problems up to you and your rabbi."[42] There were just no grounds in the nature of science — properly understood — or in the makeup of the scientist — properly understood — to expect expertise in the natural order to translate into virtue in the moral order. These types of expertise, and of legitimate authority, were not fungible. But it was left to several of Weber's later followers, especially in America, to draw out the full implications of this sentiment for understandings of what kind of figure the scientist was: a quite ordinary sort of person. We are still exploring the implications of this insistence for the cultural authority of scientific knowledge.

THE ANTINOMIES WE LIVE WITH

Insofar as expressions of this sort are routine formulations in and around late modern natural scientific culture, one cannot simply say that they are wrong. I shall describe in some detail the elaboration of a twentieth-century academic culture that either vigorously asserted or accepted as a matter of course what I will call the "de-moralization" of society's technical experts.[43] That culture is real, pervasive, and consequential. No account of late modernity can or should ignore it. However, I show that the presumption of de-moralization coexists in late modernity both with contrary sentiments and with massive evidence about technoscientific practices that points to different conclusions altogether. Accordingly, the description of "the way we live now" cannot be unitary, simple, or tidy. I need to say how it is that personal virtue *still matters* to the making and warranting of late modern technoscience *and* I have to give some account of why it is so widely said that it does *not matter*. It would be convenient to be able to tell a story of linear transition from one discrete sensibility to another: from a sacred to a secular world, from trust-in-familiar-people to anonymous trust in impersonal standards and faceless institutions; from virtue to institutional control as a solution to problems of credibility and authority. It would be handy to say that before one historical period we lived one way and thereafter another way. There is a market for stories of this sort: they are easy to tell and easy to remember, and they are, therefore, well suited to the academic contexts in which they are, in the main, transmitted from one generation to the next. And they get our blood up when we want to *complain* about "the way we live now" and to feel nostalgic for "the world we have lost." Yet I do not believe that late modern practical realities are adequately described by any such stories.

In the place of these stories, I want to offer something that is much less tidy, but which, I hope, is more faithful to quotidian aspects of "the way we live now." Part of my story is, conveniently, about historical *change*. For example, I will trace the historical trajectory from the early to the late modern that delivered to us the contemporary commonplace about the moral ordinariness of scientists; I will document changing conceptions of what scientific knowledge is about and how it is made; and I will describe radical changes in the institutional circumstances in which scientists worked. Far from denying the realities of historical change, this book aims to provide a richer picture of what some of those real and substantial changes consisted of. For all that, the late modern realities I want to talk about are best described in "pointillist" terms. "The way we live now" is subject to radically different accounts, according to the "we" whose testimony we listen to, and according to the occasions in which we organize our affairs and talk about how we do so. So late modern ways of talking about knowledge and the knower have undergone enormous changes, but the changes are not absolute. Late modernity must be characterized through heterogeneity, through multiple occasions and circumstances, and through the elaboration of different cultural sectors that privilege different ways of talking about how we live. I will describe a few of these complexities, and I will give some account of the terms in which they come to inhabit "the same" culture.

THE ARGUMENT SUMMARIZED

The next two chapters chart the social and cultural transition from science as a calling to science as a job. That transition was never complete, yet trends in that direction obsessed commentators in the nineteenth and early twentieth centuries. It was part of whatever might be meant by the secularizing and modernizing process. Chapter 2 traces a lineage from early modern conceptions of the natural philosopher as a "priest of nature" through the period of Weber's denial that the scientist, lacking moral superiority, was in any position to pronounce on public moral matters. I show that changing stipulations about the character of the scientist were enfolded in changing conceptions of what scientific knowledge was about, in changing notions of how the scientist went about securing knowledge, and in changing associations between the pursuit of natural knowledge and the structures that produced wealth and that projected power. Chapter 3 takes the story further into the twentieth century and through the Second World War, which so spectacularly mobilized, organized, and recognized the secular power of science. Here the central topic is the history of a specific idea and the conditions in which articulations of that idea became a cultural commonplace. The idea is that of the moral ordinariness of the scientist, and I describe the conditions of its emergence, notably including the rise of organized Big Science. At the same time, and responding to many of the same processes that rendered the scientist's moral ordinariness a commonplace, I point out that it was a commonplace with a vigorous opposition: there were many commentators who did not accept as a matter of fact that the scientist was morally equivalent to anyone else and who reckoned that such stipulations were an index of something gone *wrong*—both *in* science and *with* the late modern order.

The twentieth-century integration of science with the civic structures that projected power and created wealth had far-reaching consequences for the appreciation of who the scientist was and what vouched for the authority of scientific knowledge. Chapter 4 shows how central that integration was to topics addressed in academic social science and other external cultural commentary from about the middle of the twentieth century. Academic sociologists both described and warned against the violations of the special values of science that were occurring through its industrialization and organization; more practically oriented social scientists and management experts responded to industrial and governmental concerns by getting to work on the supposed *problems* presented by scientists' character to the increasing fact of organization and by organization to the integrity of scientific knowledge. The scientific community was said to be endowed with a special set of structural virtues—e.g., universalism, disinterestedness, anti-authoritarianism—and these virtues were necessary for making objective knowledge. And yet the attributed virtues were deeply problematic. Objective scientific knowledge, and its material consequences, were desperately needed by the emerging Cold War State, but, at the same time, the expression of scientific virtues was widely thought to pose pressing practical problems—both for the effective organization of science in industry and government laboratories and for national security. I show how central these worries were to strands of American social science, cultural commentary, and management theorizing by the middle of the twentieth century, emerging most strongly in the 1950s and 1960s. Views for and against the moral ordinariness of science stand astride some of the major fault lines of American culture in the middle third of the twentieth century.

For all the academic insistence on the organized—and, especially, the industrialized—scientist as devoid of the proper virtues, it is not self-evident that external commentary adequately described the quotidian realities of technoscience in these new settings, and sections towards the end of chapter 4 offer some reasons why matters appeared as they did to so many external commentators. But what does the new life of organized science look like when described not by outside observers but by *participants*? How did those directly concerned with managing the work of organized scientists talk about them and their virtues? Did they recognize any link between the nature of the investigative enterprise and the moral constitution of investigators? How did notions of virtue figure in industrial knowledge-making? Chapter 5 introduces the notion of *normative uncertainty* attaching in some degree to any research enterprise; it shows that this uncertainty was widely appreciated by those in day-to-day charge of industrial research; and it indicates how the result of this recognition was a practical attribution of personal virtue to the scientist. De-moralization does *not* adequately describe the life of the industrial scientist.

One specific aspect of organized Big Science was the object of much external, and some internal, criticism. This was the fact, and the celebration, of *teamwork*. The collective conduct of science was seen to stand in tension with one of the most enduring and pervasive sensibilities about the nature of the intellectual life, whether sacred or secular: Truth was more solitary than social; inquiry conducted collectively was likely to result in outcomes that were at best mediocre and at worst pathological. And in mid-twentieth-century America nowhere was that sensibility given more eloquent and influential voice than in William H. Whyte's *The Organization Man*, central chapters of which dealt with teamwork and its allegedly disastrous consequences in industrial science. Chapter 6 documents these sentiments about organized science and sets them in the context of American Cold War culture. Yet it also looks closely at what team science was like in practical industrial settings and it describes an economy of civic virtue that flourished in such settings. Virtue is not so easily expunged from late modern technoscience, and it is better to see how the relationship between virtue and the pursuit of knowledge has been reconfigured than to assume it has been dispensed with.

As the twentieth century progressed, a life in science began for the first time to hold out the possibility of a comfortable way of living. That is one thing included in saying that science was becoming a job and no longer a calling. This is, presumably, what Weber was gesturing at when, in the conclusion to *The Protestant Ethic*, he announced that the idea of calling had become hollowed out, and nowhere more than in the United States: "The idea of duty in one's calling prowls about in our lives like the ghost of dead religious beliefs." Since we late moderns cannot directly relate calling "to the highest spiritual and cultural values," the individual just ceases to justify it at all and it becomes "a purely mundane passion." Perhaps now, perhaps in the world to come, expert practitioners will just be (Weber here gesturing at Nietzsche) "'specialists without spirit, sensualists without heart.'"[44] Yet even at the middle of the century it was widely assumed that doing science was not, and could not be, a road to riches. No one doing science could possibly be doing it for the sake of accumulating wealth. In the 1970s, and increasingly in the last decades of the past century and the early years of this one, that changed. If not for all areas of science, then certainly for some, the possibility emerged of becoming rich through doing science. Where does the figure of the scientist-entrepreneur fit on the map of virtue? What is thought to motivate the scientist-entrepreneur? Much external commentary, again, portrays the scientist-entrepreneur, and, more generally, the scientist choosing to move from the academy to industry, as following a money motive. Chapter 7 aims to shift the discussion from celebration and accusation to description: how do scientists make their decisions about where to do their work? how do they think about universities and industry as places to do that work, and what institutional virtues and vices do they attribute to each? Much of this chapter derives from interviews and conversations. I wanted to retrieve from the frontlines of present-day technoscientific knowledge-making something of what it *feels like* to those trying to make a career, to make knowledge, and to make sense of the increasingly uncertain institutional worlds they inhabit. The last substantive chapter remains within the world of the present, yet it deals with the means used to make technoscientific *futures*. Chapter 8 seeks to understand how venture capitalists make decisions about investable futures. It describes the moral fabric of the worlds in which they engage with technoscientific entrepreneurs to judge what technologies, what markets, and what people are most likely to produce profitable outcomes.

These worlds are at the cutting edge of late modernity, and they are widely accounted among the most instrumental and ruthlessly calculative segments of our culture. They are also the worlds that confront uncertainty in some of its most radical manifestations, and, because of that radical uncertainty, judgment takes a specially personal form. Decisions about what ventures to support turn out to be highly personal: judgments of business opportunities and technologies proceed importantly through judgments of familiar people and their virtues. People matter.

It is the normal fate of books to be misunderstood, or at least to evoke understandings in readers at some angle from those the author intended. That can't be helped, but there is one set of misunderstandings I think I can predict and that I want briefly to address in advance. Nothing in this book should be read as a *celebration* of late modern American culture, of late modern technoscience, or, specifically, of industrial or entrepreneurial science. That is simply because it is meant as a work of description and interpretation rather than advocacy. For similar reasons, nothing here should be taken as a *criticism* of late modern American universities or of academic science, and nothing here supports an idea that there are, after all, no differences between the academy and industry, or that there is nothing about the ideal of disinterested inquiry that is worth defending. Some things *are* criticized in this book. Certain *stories* about the "essential nature of science," the "essential nature of the scientist," the "essential natures" of such institutions as the university and industry are, indeed, compared to concrete realities and found to be problematic, even as I take these stories seriously as consequential cultural tropes. (Here, as elsewhere, the assumption that the historian must choose between "rhetoric" and "reality" should be rejected, while it still remains sensible and important to ask *which* stories—tropes and rhetorical specifications—hold up best when juxtaposed to the patterns of quotidian institutional life.)

It is almost certain that this book's main readers will be academics and, more specifically, historians and social scientists. Given that, a misunderstanding that especially concerns me is one that may flow from what I have to say in later chapters about entrepreneurial science and about late modern relationships between academia and industry. I say there that many standard contrasts between late modern academic and industrial science are poorly founded. The picture of the university as an Ivory Tower of unconstrained scientific inquiry and industry as a regimented, de-moralized, and mercenary Iron Cage did not describe early twentieth-century realities very well and describes early twenty-first-century realities even worse. Universities, and academic science, have changed; industry and corporate science have changed too. More generally, institutional realities—both academic and corporate—have always been so heterogeneous that the contrasts that have the greatest grip on our minds are not those between the range of mundane institutional realities but between ideal types. And it is probably best to treat ideal-typical contrasts largely as apologetic resources in ideological battles, some of which, indeed, are fought not *between* academia and industry but *within* sectors of academia. It may suit some apologetic purposes to spin nostalgic stories about a past-that-never-was and to take ideal-typifications as adequate descriptions of quotidian institutional realities, but if these stories and typifications cannot stand up to close historical and sociological scrutiny, then they defeat the very values they purport to defend.

That said, I see no reason to deny my attachment—both emotional and practical—to the regulative ideal of disinterested inquiry, to such freedom of inquiry as our current institutional arrangements make possible, and to the virtues that—as I will argue—*remain* linked to, and embedded within, the life of inquiry. After all, I like to think that this book is the outcome of such inquiry and that the academic environment I inhabit has made it possible. When I worked in Britain, I lived through some of the Thatcherite depredations of the British university system, and then moved to a distinguished American public university whose chancellor, a physicist who had spent virtually the whole of his previous career at AT&T's Bell Labs, announced that "as scholars, we should not seek knowledge for its own sake."[45] I am one of many scholars alarmed at threats to such spaces of free inquiry as continue to exist in late modern American universities—threats from political interference, threats from commercializing imperatives, and, not at all least, internal threats from excesses in disciplinary professionalization and consequent rigidities and orthodoxies. However, I differ from many of my colleagues in rejecting the notion that there is some essential and necessary difference between academia and industry with respect to the possibilities of inquiry. The late modern research university is a mongrel, the result of a set of historical contingencies. It is not now possible, and perhaps it never was possible, to identify a set of values to which all its members subscribe or conditions of work that describe what all its faculty do.[46] And perhaps we might also acknowledge that industry is no longer, if it ever was, the sort of institution that *necessarily* represents a threat to the virtues associated with the life of inquiry. Those virtues can thrive in industry and they can be compromised in academia. If that is the case, and if we are indeed committed to such virtues, then we need not defend them by defending *any particular institution*, still less the institution that goes under the name of a modern research university. We can and should recognize a variety of institutional environments in which the virtues can flourish, make them visible as such, and show the good that they do in whatever institution is prudent enough to recognize and encourage them.[47] I am not saying that the life of inquiry, in whatever institutional environment it is authentically found, *should* be a vocation and *should* embrace the attendant virtues; I am saying that it inevitably *does*, and, that being the case, the question of why that circumstance is often not recognized acquires salience. I am not being perverse or paradoxical when I suggest that the "managerial ethos" increasingly being imposed on universities is a *misrepresentation* of the practices of much innovative industry and that universities ought to welcome, rather than resist, many points of comparison with how sectors of innovative industry actually do manage creative people. At the end of the day, however, I am with Max Weber in doubting the legitimacy of scholarly moralizing. So, having said the necessary minimum about possible misunderstandings, I get on with the task of *describing* the scientific vocation and its changes in late modern America.

From Calling to Job

NATURE, TRUTH, METHOD, AND
VOCATION FROM THE SEVENTEENTH
TO THE NINETEENTH CENTURIES

If you recall Swammerdam's statement, "Here I bring you the proof of
God's providence in the anatomy of a louse," you will see what the scientific
worker . . . conceived to be his task: to show the path to God.

Max Weber, *Science as a Vocation*

MORAL EQUIVALENCE AND THE DISCIPLINES

Writing during the Second World War, with the existence of both liberal science and liberal society under threat, the American sociologist Robert K. Merton (1910–2003) announced that there was nothing special about scientists as people: "A passion for knowledge, idle curiosity, altruistic concern with the benefit to humanity, and a host of other special motives have been attributed to the scientist. The quest for distinctive motives appears to have been misdirected." There is, he said, "no satisfactory evidence" that scientists are "recruited from the ranks of those who exhibit an unusual degree of moral integrity" or that the objectivity of scientific knowledge proceeds from "the personal qualities of scientists."[1] Merton's insistence on what I call the "moral equivalence" of scientists is now a commonplace, but it was not a commonplace at the time he gave voice to it, and he was good enough a historian to appreciate aspects of its novelty. This chapter describes how, why, to what extent, and with what consequences late modernity's most powerful knowers came to be portrayed as ordinary people. It is a story that bears upon the authority of technoscientific knowledge and of the institutions that house that knowledge.

One reason that Merton insisted upon moral equivalence has local academic significance and should be treated straightaway. Merton was a sociologist, writing in a Harvard academic setting in which the social sciences represented a relatively novel way of thinking. He was arguing for the legitimacy, coherence, and interest of social-structural frameworks for explaining cultural conduct.[2] For any sociologist, and most especially for those concerned with highly valued forms of knowledge, an argument had to be made against *individualism*—the sufficiency of unique individuals, their dispositions and capacities, as explanations of intellectual outcomes. The idea of *genius* was one such obstacle, and, because Merton did not take up a sociological account of scientific *knowledge*, he had no need to tackle that limit on sociological accounting. But the adherence of scientists to a quite special code of conduct definitely was a proper object for sociologists' scrutiny, and this is the explicitly described context in which Merton made his skeptical remarks about scientists' special motives. Sociologists were being reassured that they *could* give adequate and interesting accounts of many features of science in terms of concepts appropriate to their discipline—not personal motives or individual constitutions, but the community's "norms" and "ethos," internalized by individuals, but articulated, enforced, and belonging to the scientific community. As Merton insisted elsewhere—reiterating a caution by Durkheim—sociologists should not conflate "institutional" and "motivational" levels of analysis. Their explanatory resources ultimately belonged to the realm of social structure, not to the dispositions of individuals.[3] If anything was special about scientists' conduct, its cause was to be sought in their institutional environment. Merton was confidently conjecturing that, as a *matter of fact*, a psychological inventory of scientists—present and past—would not turn up either constitutional or motivational differences between them and other relevant types of person, and he was presenting that claim as a *matter of course* against the background of apparent contemporary presumptions to the contrary. So Merton's stipulation of moral equivalence was, among other things, a tactic in building an academic discipline—justifying its procedures and bounding it from other disciplines (notably psychology) and from what was taken as the matter-of-fact individualism of the common culture.[4]

Merton was here attempting to talk sociologically about structures and processes right at the heart of the late modern condition. And in so doing he became a surprising precursor of Foucault's celebrated identification of the typically post–World War II figure of the "specific intellectual," the descendant of the philosophers who used to speak transcendent, eternal, and universal truth to power, but whose role was now defined by providing particular expert services to power.[5] Among the major obstacles to a properly sociological story about modernity were any residual sentiments about the special motivational constitution of the modern scientist or about the personal moral basis of the scientist's cultural authority. So the moral equivalence of the scientist was a key feature of academic discipline formation, but it was also a site in which some of the tensions of tectonic cultural change became visible.

Insofar as academic sociology traded in structural items, then, it was obliged to offer arguments against the sufficiency or the pertinence of individualistic items. Yet, under this description, social science was not the only academic discipline so placed. The philosopher Karl Popper was in whole-hearted agreement with Merton's sentiments. His 1950 essay on "The Sociology of Knowledge," while not mentioning Merton by name, noted that scientific objectivity has the character of a "social institution": "What we call 'scientific objectivity' is not the product of the individual scientist's impartiality, but a product of the social or public character of scientific method; and the individual scientist's impartiality is, so far as it exists, not the source but rather the result of this socially or institutionally organized objectivity of science." In Popper's view, what was fundamentally wrong with then-current sociology of knowledge was just its *individualism*, its "naïve" presumption that scientific objectivity had to be grounded in "the individual scientist's impartiality or objectivity": "We are all suffering under our own system of prejudices . . . , and scientists are no exception to this rule, even though they may have purged themselves from some of the prejudices in their particular field."[6]

THE NATURE OF NATURE AND THE NATURE OF KNOWLEDGE

Merton insisted on moral equivalence against the background of what he acknowledged as widespread sentiments to the contrary. His tone was correcting what the early moderns called "vulgar errors," whether these "errors" were found in other academic disciplines or in the common culture. In the late 1930s and early 1940s, expressions of moral equivalence were rare. By the end of Merton's life, such "errors" had been substantially eliminated from the academy. It has now become an official commonplace, seemingly in need of no special evidence or argument, that scientists are morally no different from anyone else, and, more generally, that the "personal equation" has been eliminated from the scenes in which powerful technoscientific knowledge is produced. So, it is well to remind ourselves of the scheme of things in which the moral superiority of those who spoke Truth about Nature was *itself* a cultural commonplace. What did such presumptions look like? How were they sustained? And what did they say about knowledge, the knower, and the known?

From the early modern period through much of the nineteenth and even early twentieth century, there were three major bases for conceiving of the natural philosopher, or scientist, as morally superior to other sorts of people.[7] The first was a conception of the *referent* of scientific knowledge: what kind of entity did you know about when you knew about Nature? The second concerned views about the character or quality of scientific knowledge and the methods by which that knowledge was secured. And the third flowed from appreciations of what sorts of people, and in what circumstances, pursued scientific knowledge. Knowing about Nature considered as Divine Creation is quite a different enterprise from knowing about nature as a chance concatenation of atoms. The first sort of Nature has the capacity to uplift, and, indeed, the possession of authentic natural knowledge can be signaled by its moral effects on knowers: Aristotle considered this to be the case, so did members of the great natural theological tradition that spanned the period from the seventeenth to the mid-nineteenth century.[8] English Restoration practitioners liked to consider themselves "priests of nature," the contemplation of God's Second Book rendering them pious.[9] The eighteenth-century Unitarian chemist Joseph Priestley wrote that "a Philosopher ought to be something greater, and better than another man." If the man of science was not already virtuous, then the "contemplation of the works of God should give a sublimity to his virtue, should expand his benevolence, extinguish everything mean, base, and selfish in [his] nature."[10] And one finds much the same sort of causal argument in John Herschel's *Preliminary Discourse* of 1830: "The observation of the calm, energetic regularity of nature, the immense scale of her operations, and the certainty with which her ends are attained, tends, irresistibly, to tranquilize and re-assure the mind, and render it less accessible to repining, selfish, and turbulent emotions."[11]

These sensibilities did not disappear from the culture, or even specifically from academic culture, with the naturalism of Darwin's *Origin of Species*. In 1916, Sir Richard Gregory, physicist and the editor of *Nature* magazine, articulated views of the sanctity of science, proceeding from the sanctity of its object, which differed little from those expressed by Herschel, Priestley, or even Boyle. The study of Nature elevates those who pursue it: "The conviction that devotion to the study of Nature exalts the Creator gives courage and power to those who possess it; it is the Divine afflatus which inspires and enables the highest work in science."[12] Given Nature so conceived as an object of inquiry, one might legitimately expect those who studied it to be *better* than other people. This is the *referent* of scientific knowledge that can make sense of the notion—pervasive between Antiquity and the mid-nineteenth century—that the *point* of philosophy—and not just natural philosophy—might be a better manner of living in the world: moral, not just material, utility.[13]

And that is one reason why Weber pronounced the "de-magification of the world" and the death of natural theology in the same essay that rejected any notion that the scientist had the kind of authority that entitled him to pronounce on *what ought to be done*. The secularization of inquiry, announced by the Scientific Naturalists inspired by Darwin, meant that the *object* of scientific inquiry—Nature—was to be relocated from the sacred to the secular domain. (And its "case" was to be lowered accordingly.)[14] If you were a Scientific Naturalist, or if you subscribed to their assumptions, then knowing about nature was no longer like knowing a divinely written book, but like knowing how a car engine worked. The automobile mechanic—like the natural scientist—can be regarded as an expert, even a highly valued, powerful, and well-remunerated expert, but nothing uplifting—or at least nothing particularly uplifting—is now officially associated with the scientist's object of inquiry: no morals, no lessons, and no special authority to pronounce on what ought to be done. Weber asked himself what it could possibly mean in 1918 to talk about "science as a way 'to God.'" He answered his own question. Any such talk must be nonsense: "Science, this specifically irreligious power? That science today is irreligious no one will doubt in his innermost being, even if he will not admit it to himself."[15] The severance of a long-standing tradition causally linking scientific inquiry to personal and public morality is indexed by a remark attributed to the Scottish socialist physiologist J. B. S. Haldane in the 1920s. Haldane was in the company of some English theologians, who asked him what he could conclude about the nature of the Creator from a study of His Creation. Haldane's response was: "an inordinate fondness for beetles."[16] The flippancy of the manner is more telling than the substance of the inference.

Knowledge about God's Creation was a different thing, with different bearings upon the character of the knower, than knowledge of a demagified world. And here the *character* and *quality* of natural knowledge is hard to separate from its referent. How to describe scientific knowledge vis-à-vis what scientific knowledge is *about*? In early modern practice, one robust response would be that natural philosophers were engaged in matching their knowledge to that of the Creator, insofar as that goal was not considered impious. As God had created Nature, God's knowledge was perfect, and, to the extent that human beings were permitted to find out God's creative secrets, then the task of the natural philosopher was the imitation of God and the quality of the philosopher's knowledge was distinguished from that of the common people. So a characteristic trope of natural philosophy during the Scientific Revolution was an evaluative contrast between genuine philosophical knowledge of what lay behind sensory appearances and the superficial, sense-based knowledge of "the vulgar." Galileo's discussion of the Copernican system, for example, insisted that Scriptural reference to the movement of the Sun was an intentional authorial adaptation to the superficial knowledge of the vulgar: "It is sufficiently obvious that to attribute motion to the sun and rest to the earth was . . . necessary lest the shallow minds of the common people should become confused, obstinate, and contumnacious."[17] And Newton had to stipulate the precise ways in which he used notions like time, space, motion, etc., because "the common people conceive those quantities under no other notions but from the relation they bear to sensible objects. And thence arise certain prejudices."[18] As the early modern natural philosopher imitated God, so he at the same time marked a distinction between the quality and character of his knowledge and that of the common people. That three-way relationship between natural philosophy, metaphysics, and talk of God's works bound natural knowledge to religion and to the moral discourses enfolded in Christian religion. "And thus much concerning God," Newton wrote in the General Scholium to the *Mathematical Principles of Natural Philosophy*, "to discourse of whom from the appearances of things, does certainly belong to Natural Philosophy."[19] It is, indeed, another way of describing what was involved in the culture of natural theology—proving God's existence and attributes from the evidence of His Creation. Accordingly, whatever happened to the career of religiously based moral authority was bound up with notions of what natural knowledge was *about*.

However, early modern natural knowledge is not describable simply as natural philosophy, nor were all practitioners of natural philosophy agreed about what it might mean to seek the Truth behind appearances. The uplift and personal virtue traditionally said to be evoked by inquiry was causally attached to some instantiations—but not others—of what we continue anachronistically to call "the scientific role." It is not accidental that Priestley specified that it was the *philosopher* who might be better than another man: there were other designations of an intellectual role available to him. As Peter Dear and others have shown, early modern natural philosophy (considered as the search for universal and certain Truths about the underlying physical order of nature) was (until Newton, and even after) quite a different exercise from mathematics (considered as the search for operative regularities independent of any particular physical reality that might give rise to them). It is the natural *philosopher* who was, so to speak, in the Truth or the Reality business, not the mathematician, the physician, or the engineer.[20] Newton's enterprise of writing the *mathematical* principles of natural philosophy was, therefore, a far more difficult and unstable fusion than has been widely appreciated. Different conceptions both of the referents of knowledge and of the quality and character of knowledge were involved in the practices of early mathematics and of natural philosophy. Mathematics gave certainty, but at the cost of correspondence to Creation; natural philosophy gave correspondence, but sacrificed certainty. From the perspective of this book, however, the salient point is that these practices differed in their conceptions of truth, in their relationships to religion, and in their capacity for personal moral uplift.[21]

THE TRUTH BUSINESS AND ITS CAREER

Towards the end of the nineteenth century and early twentieth century a variety of related philosophical terms of art were developed to describe formal philosophies of science that, in one way or another, rejected the idea of scientific Truth as correspondence to God's reality, to the ultimate reality that was supposed to lie behind appearances. Each of them insisted on the distinction between the practice of science and the practice of metaphysics:

- *operationalism:* the meaning of a proposition consists of the operations involved in proving or applying it;
- *instrumentalism:* scientific concepts and theories are just useful tools that allow one to explain and predict, but need not be assessed by their truth-as-correspondence-to-reality;
- *phenomenalism:* science can and should be disengaged from any talk of what lies beyond or behind appearances—scientific knowledge is grounded not in "reality" but in sensations;
- *positivism:* metaphysical speculations are scientifically illegitimate, and sense-data are the only proper objects of knowledge and criteria for judging it;
- *conventionalism:* scientific theories are conventional claims to be assessed by their simplicity and utility and not by their truth-as-correspondence;

- *pragmatism:* when metaphysics comes up, change the subject, and insist instead on the intelligibility and propriety of truth considered simply as what works;
- *probabilism:* familiar in science since the seventeenth century, but now increasingly stressed to distinguish the legitimately modest quality of scientific certainty (about theories) with the vaulting ambition of dogmatists, speculative philosophers, and theologians; and, finally,
- *falsificationism:* best known through Karl Popper's claim in *The Logic of Scientific Discovery* (1934) that scientific generalizations can never be verified but only falsified, and that, therefore, legitimate scientific method can never establish the Truth of theories.

Academic philosophers make fine distinctions between such positions, but, from the point of view of cultural history, it is important to note what they have in common: each aims to sever the links that bound early modern natural philosophy to religion by way of metaphysics and notions of God's Truth. Just as the Scientific Naturalism of the late nineteenth century *lowered the case* of "nature," so all of these characterizations of the quality and character of scientific knowledge *lowered the case* of "truth." And some, indeed, quite explicitly identified the metaphysical tendencies of religious discourse as an intellectual pathology, to be cured by deflationary conceptions of proper scientific knowledge. More and more, early modern insistence on the radical differences between profound philosophical knowledge and superficial vulgar knowledge was explicitly set aside. When T. H. Huxley insisted that science was "nothing but *trained and organised common sense,*" differing from ordinary cultural practices only in regimens of expert training and in the social forms in which it happened, he was speaking for very many fellow-scientists and presaging what soon became the scientists' common sense about science.[22] American-style deflationary conceptions explicitly democratized the nature of scientific knowledge and, therefore, the character of the scientist. And in so doing, they inevitably redefined the identity of the divine. In the 1890s, American scientists acknowledged that more limited, modest, and provisional conceptions of scientific knowledge might give nonscientists a "sense of dissatisfaction and incompleteness ... The results of scientific study ... may appear vague, indefinite, incompetent to satisfy the loftier yearnings of the soul of man for something utterly true, immutably real." But they should resist the lure of metaphysical snake-oil salesmen, for these deflationary notions of scientific knowledge were, after all, "noble, inspiring, consolatory ... presenting aims which are at once practical, humanitarian and spiritually uplifting."[23] Pragmatism gave philosophical grip to those more diffuse sentiments. In 1904, William James wrote eloquently about the proper usage of notions of cultural nobility: "In this real world of sweat and dirt, it seems to me that when a view of things is 'noble,' that ought to count as a presumption against its truth, and as a philosophic disqualification. The prince of darkness may be a gentleman, as we are told he is, but whatever the God of earth and heaven is, he can surely be no gentleman."[24]

It would be quite wrong to speak of the simple *replacement* of the metaphysical ambitions of the early modern natural philosopher by the operationalist epistemology of the modern scientist. Both ideas of what it was to do science, and both views of the character of scientific knowledge, not only coexisted through the nineteenth century but continue to do so today.[25] Yet there can be no doubt that the place of metaphysical ambitions in science and of correspondence theories of Truth have been declining in significance among *scientists*—if not among academic philosophers of science—for more than a century. Among the proponents of these anti-metaphysical conceptions of scientific knowledge in the late nineteenth and early twentieth century were the scientists Ernst Mach, Pierre Duhem, W. K. Clifford, John Tyndall, T. H. Huxley, Henri Poincaré, and Karl Pearson, and only the "Anglo-Saxon heresy" that excludes human scientists from the title would omit Auguste Comte, Herbert Spencer, and William James. Scientists themselves were, so to speak, getting out of the Truth and Reality business and affiliating themselves with more modest and more active conceptions of what their knowledge was *about*.[26] In 1899, the physicist Henry Rowland, making no allusions to pragmatism or to any other formal philosophy of science, explicitly contrasted the scientific with the "vulgar" or "ordinary crude" mind. But here the "vulgar" supposedly believed in a version of scientific Truth that schooled practitioners did not: the scientist alone properly appreciated that "there is no such thing as absolute truth and absolute falsehood."[27] The American biologist David Starr Jordan blandly noted that "man can come in contact with no ultimate truth of any sort,"[28] and a few years earlier, E. W. Scripture—an experimental psychologist at Yale—announced that science had definitively freed itself of the dead hand of speculative philosophy. If the philosopher was the person you turned to for answers to general questions, then the sciences now had a better answer—a *collection* of specialists bringing their expertise to bear on any intelligibly framed general question: "Philosophy has no relation to the sciences," and scientists can now "dispense with the philosopher." Cut loose from its advisory function to the scientific

community, deductive philosophy now had no *point*: "The day of philosophical systems is past."[29] Why not let Scientific Method take over the
tasks traditionally assigned to religion and moral philosophy? A turn-of-
the-century American botanist remarked that "one might term science an
intellectual religion and not go wide of the mark."[30]

Certainly, by the 1920s, many scientists were propagating such views
in public, and addressing themselves *to* the public, without the excursus
into formal philosophizing found in the work of such predecessors as
Mach, Duhem, and Poincaré. So, in the context of this book, the significance of such accounts of the nature of scientific knowledge was not
solely or mainly their presence and status in academic philosophy; it was
their prevalence in scientists' *own* publicly circulated stories about what it
was they were doing and what they reckoned ought to be believed about
genuine scientific knowledge. These accounts increasingly represented
the way that scientists themselves tended to think about their work,
whatever conflict there might now be with outdated notions of Truth or
Reality.[31] Such tendencies internal to science were only augmented by the
later "crisis" in knowledge precipitated by quantum theory. By the 1920s,
Albert Einstein was reminding the general reader that "it is difficult even
to attach a precise meaning to the term 'scientific truth,'" its semantics
varying radically according to context.[32] And in the post–World War II
period, an eminent research director at Bell Labs insisted that public trust
in science was dependent upon realizing that there were limits to the
notions of scientific "truth" and "certainty": "Scientific findings, scientific
facts, are usually thought of as symbols of certainty. But people must realize that these findings are certain only with respect to a particular frame of
reference."[33] C. P. Snow surely spoke for most scientists when he bump-
tiously stipulated that "by *truth*, I don't intend anything complicated . . .
I am using the word as a scientist uses it. We all know that the philosophical examination of the concept of empirical truth gets us into some
curious complexities, but most scientists really don't care."[34] Over the
first several decades of the twentieth century, the disengagement of
scientists from classically absolutist and universalistic notions of Truth
became commonplace, and chapter 3 will document some of the cultural
settings in which this disengagement occurred. The laity might imagine
that the scientists had simply inherited the cultural authority of the
priests, but many scientists circulated views of scientific knowledge very
different from priestly notions of Truth. The scientist was properly to be
understood not on the model of the philosopher but on the model of the
engineer and technician. The active was to replace the contemplative;

technique (an attempt to control) was to replace speculation (an attempt to understand and to tell the Truth about the world).

Pragmatist philosophers frequently gestured at theirs as a new philosophy of science and even specified that they were attempting to replace a discredited theory of truth with a new and better one, an illegitimate metaphysics with a defensible and minimalist ontology, but an Italian follower of James gave the game away in 1907 when he wrote in a popular American periodical that pragmatism "is really *less a philosophy than a method of doing without philosophy*."[35] The trajectory of anti-metaphysical notions of scientific knowledge closely tracked that of anti-clericalism and secularism: some nineteenth-century Scientific Naturalists made those relations quite clear, and the Viennese-style logical empiricism that emerged from the 1920s made the equation between metaphysics and religious nonsense a centerpiece of the commendation of a "scientific philosophy."[36] Writing after the Second World War about the rise of this scientific philosophy, Hans Reichenbach spelled out in just what way it was *mistaken* to think of the scientist on the model of the priest and scientific knowledge on the model of religious certainties:

> The overestimation of the reliability of scientific results is not restricted to the philosopher; it has become a general feature of modern times . . . The belief that science has the answer to all questions—that if somebody is in need of technical information, or is ill, or is troubled by some psychological problem, he merely has to ask the scientist in order to obtain an answer—is so widespread that science has taken over a social function which originally was satisfied by religion: the function of offering ultimate security. The belief in science has replaced, in large measure, the belief in God . . . No wonder the mathematical scientist appeared as a sort of little god, whose teachings had to be accepted as exempt from all doubt. All the dangers of theology, its dogmatism and its control of thought through the guaranty of certainty, reappear in a philosophy that regards science as infallible.[37]

If the public did indeed continue to think of the scientist along the lines of the priest, that, according to Reichenbach and many others, was nothing to do with how scientists themselves rightly understood who they were and what they legitimately knew. The culture evidently was still populated by two quite incompatible modes of thinking about knowledge and the knower.

When Claude Bernard wrote in 1865 that "Art is I; Science is We," he meant to show how Method ensured the objectivity of scientific knowledge by disciplining or dissolving the role of the personal in its making. There was no need to know much about the person who made scientific discoveries, not—in Bernard's formulation—because there was nothing special about him but because whatever idiosyncratic character he had was dissolved in the collectivity, and therefore anonymity, of the scientific voice.[38] Methodical discipline of the personal and the contingent was, of course, an ambition at least as old as the Scientific Revolution, when Francis Bacon analyzed the "distempers of knowledge" introduced by the role of the Idols—the distorting effects of language, convention, interest, and personal bias—and suggested an inductive method as therapy, and when Descartes commended high rationalism as a solution to crises in the foundations of belief. Faith in Method grew even as incompatible versions of what such a Method might be proliferated. Yet one key feature all early modern Methods had in common was a belief that their principles could be formalized, written down, transmitted with ease from one person to another, and implemented by each person so as to yield reliable knowledge. For Method to fulfill such expectations, it would have to be as unlike spontaneously varying and uncontrollable human nature as possible. It would have to be invariant and impersonal in its operation.

It remains legitimate to draw attention to the seventeenth century as the time when various accounts of rational Method were articulated in attempts to discipline bias, arbitrariness, interest, and the idiosyncratic. But none of those formal prescriptions of Method was securely institutionalized, and all of them coexisted in a culture in which contrary accounts were understood and valued. Notions of the quite special individual not only *called to* the study of Nature but divinely *inspired* in his inquiries were common cultural currency in the seventeenth century, and later. Newton's remark that if he had seen further than others it was because "he stood on the shoulders of giants" has for many years been cited in support of a norm of humility, supposedly essential to the idea of science, and compatible with Bernard's sentiment. Yet, as historians have realized for some time, it was no such thing: Newton was probably insulting his physically stunted opponent Robert Hooke, but, more importantly, gesturing towards the so-called *prisca sapientia* (and related *prisca theologi*) traditions. Within this sensibility, there had anciently existed a vast body of primitive, pristine, and powerful knowledge that had been lost or submerged (in Christian idioms through Original Sin). That primitive knowledge had been secretly handed on through a sort of apostolic tradition, the lineage including Hermes Trismegistus (now believed to be mythic), Moses, Archimedes, and, in various versions, Socrates, Aristotle, Jesus, and

Aquinas. *These* were the "shoulders" on which Newton wished it to be understood he stood. The point was not modesty but colossal pride in who he was. The way Newton knew was not as other men knew.[39] In another idiom, Robert Boyle accounted for special success in discovering Nature's secrets through a providentialist conception of the knower. He did not doubt that "the favour of God does (much more than most men are aware of) vouchsafe to promote some men's proficiency in the study of nature." God guided certain philosophers' intuitions, "directing them to those happy and pregnant hints, which an ordinary skill and industry may so improve, as to do such things, and make such discoveries by virtue of them, as both others, and the person himself, whose knowledge is thus increased, would scarce have imagined to be possible." God chose to which individuals He would so reveal His mysteries.[40] Boyle's anonymous first publication claimed that extraordinary discoveries in physick were "rather inspired than acquired."[41] The natural philosopher was doing God's work, reading God's Book, and making discoveries by divine guidance. The ascription of scientific discovery to divine inspiration or providence was never unproblematic or uncontestable in the seventeenth century, but it was *intelligible*.

The idea of inspired scientific genius was partly secularized in the eighteenth and nineteenth centuries, but did not come close to disappearing—Romanticism was compatible with either a Christian or a pantheist version of inspired genius—nor is it absent from late modern culture.[42] Later chapters draw attention to twentieth- and twenty-first-century practitioners' skepticism about Method and their insistence on the continuing significance of the personal in the making of technical knowledge. Nevertheless, a range of cultural and institutional changes from the eighteenth through the twentieth centuries gave the notion of Scientific Method much wider cultural distribution and lay plausibility than it previously had. And it was these changes that created part of the cultural credibility of notions of scientific impersonality and of the moral equivalence of the scientist.[43] While, for many twentieth-century commentators, Einstein continued to signify both the morality and the spontaneity of genius, by the 1950s Roland Barthes could coherently gesture at a contrary imagination when he talked about Einstein as "a genius so lacking in magic that one speaks about his thought as of a functional labour analogous to the mechanical making of sausages, the grinding of corn or the crushing of ore: he used to produce thought, continuously, as a mill makes flour."[44]

One pertinent circumstance was the accumulating integration of scientific expertise into the organized structures of State power and, later, of commerce. State use of expertise, again, was no new thing in the eighteenth century—it was, of course, a notable feature of Antiquity—but a series of changes from the late seventeenth through the nineteenth centuries strongly affected the structure of scientific careers and the perception of what kind of thing science was. These included the further centralization of the nation-state and the role of technical experts in that process; the increasing dependence of warfare on technique; and the global extension of State power through colonialism and imperialism. While the State had always drawn upon scientific expertise in an ad hoc fashion, by the eighteenth century it increasingly gave systematic institutional form to the mobilization of such expertise. And in the course of this mobilization, collective methodical discipline became a more common feature of the settings in which scientific experts worked. The projection of an idea of science as methodically disciplined and intellectually mundane was not, however, peculiar to science in a State or commercial setting.[45] Stress on mundane methodical discipline was a notable feature of emerging skepticism about the role of "genius" during the Industrial Revolution. In the year that Darwin published the *Origin of Species*, the Victorian arch-apostle of hard work and application, Samuel Smiles, made clear what he thought about the relative significance of genius versus disciplined application in making scientific knowledge. There *is* such a thing as genius, but its role has been systematically exaggerated: "fortune is usually on the side of the industrious"; what you need is "common sense, attention, application, and perseverance." "Drudgery" is the price of success. So science is on a cognitive level with all sorts of other practical activities.[46] Indeed, when the eugenicist and Scientific Naturalist Francis Galton surveyed the characteristics of English men of science in the 1870s and asked them whether they thought they possessed any special talent, his half-cousin Charles Darwin replied: "None except for business, as evinced by keeping accounts, replies to correspondence, and investing money very well. Very methodical in all my habits."[47] From the mid-nineteenth century on, it was widely insisted that scientific knowledge was made not by acts of heroic individual genius, or even through acts of imagination, but by deploying the bourgeois virtues of industrializing society.[48]

WHEN SCIENCE WAS A CALLING

Early modern students of nature conducted their inquiries in a variety of institutional settings and occupied a variety of social roles. Some were remunerated to conduct their inquiries, but not many. There were a handful of State astronomers; Robert Hooke was paid to carry out experiments for the Royal Society of London; at the very end of the seventeenth century and the early years of the eighteenth century Edmond Halley was engaged by the Royal Navy to undertake voyages for astronomical, geophysical, and cartographic purposes; and there were a small number of court chemists, mathematicians, and miscellaneous philosophers. That is to say, the number of people paid by the State or its affiliated institutions to find out new knowledge of nature in the seventeenth century could probably be counted on one's fingers and toes. This inculdes professors. The university professor was engaged to be a custodian of knowledge and to transmit it to the next generation. The physician and surgeon were remunerated to keep people healthy and to treat them when they were ill. The cleric was responsible for being a mouthpiece for God's words; for living a blameless, if not holy, life; and for ensuring the moral conduct of his community. All of the people occupying these roles might do scientific research (as we now put it), but doing it was not their *business*. The early modern Speaker of Truth about Nature was, almost without exception, not a professional but an amateur. He was understood to do it not because it was his job—though, indeed, he had related responsibilities that involved technical expertise—but because, in some irreducible sense, he wanted to do it, or even because he was called to do it.

In no early modern social situation was that free commitment to inquiry as consequential as it was for the gentleman—figures such as the Honourable Robert Boyle (a son of the earl of Cork) or René Descartes (a scion of a legal and medical family aspiring to, and ultimately attaining, the status of *noblesse de robe*). The pursuit of natural knowledge was not a recognized aspect of the gentlemanly role, and serious scholarship—of any sort—might conflict with polite expectations. Descartes' father was famously said to have been ashamed of René, alone among his three boys, "a son stupid enough to have had himself bound in calf." Scholarship ran the risk of appearing pedantic; authorship smacked of illegitimate fame-seeking.[49] Commitments of this sort, therefore, had to be specially explained and justified. One way of justifying such commitments was to link them to conceptions of Christian gentility, links that were especially strong in Protestant cultures but which were also found in Roman Catholic settings. The gentleman studying Nature could be understood, as I have indicated, to read God's Second Book, and was therefore engaged in a kind of religious activity.[50] And, of course, very many men of science in the early modern period were themselves in holy orders—Copernicus, Marin Mersenne, Pierre Gassendi, John Wilkins, John Ray, and Stephen

Hales are a few of the more notable instances—and thus the circumstances of their social role helped forge an embodied link between priestly and scientific vocations. Professors too could share in the moral authority of the clerically controlled institutions from which they spoke. Lay practitioners well understood religious justifications for the pursuit of natural knowledge, and they knew how to link the moral purposes of their inquiries to those of the priestly vocation. Godly subject matter might make for godly scholars, even if their occupational role was secular. This was the major way in which the culture of natural theology sustained an understanding of the man of science as virtuous beyond the normal run of scholars. Yet eighteenth-century cultures that were *not* powerfully marked by natural theology also produced portrayals of the man of science as specially or uniquely virtuous.

The *éloges* presented in commemoration of recently deceased members of the Paris Academy of Sciences are the eighteenth century's most highly developed and influential portraits of the virtuous man of science. While a natural theological idiom was not especially strong in that setting, other resources were available to display the superior virtue of the man of science. Many of the more than two hundred *éloges* composed from 1699 to 1791 by Bernard le Bovier de Fontenelle—and his successors Jean-Jacques Dortous de Mairan, Jean-Paul Grandjean de Fouchy, and the marquis de Condorcet—drew upon Stoic and Plutarchan tropes to establish both the special moral qualities possessed by those drawn to science and the additional virtues that a life dedicated to scientific Truth encouraged in its devotees.[51] Like many of Plutarch's Greek and Roman heroes, Fontenelle's eighteenth-century men of science were described as embodiments of Stoic fortitude and self-denial. The life of science held out few prospects of material reward and little hope of fame, honor, or the applause of the polite and political worlds. The dedication to Truth drawing men to such a life was made manifest by the neglect of self and of material self-interest, and by a cool disregard for public favor and approval. Such power as men of science came to possess was not vaingloriously sought for but thrust upon them by patrons often wanting the material goods sometimes understood to derive from scientific knowledge. Sincerity, candor, tranquility, and contentment were naturally instilled in men who lived for the love of Nature's Truth. In Scotland, Adam Smith was much impressed with Fontenelle's *éloges*, and his *Theory of Moral Sentiments* drew an inference from the disengagement and integrity of the scientific life to the communal virtue of those who lived it:

Mathematicians and Natural Philosophers, from their independency upon the public opinion, have little temptation to form themselves into factions and cabals, either for the support of their own reputation, or for the depression of that of their rivals. They are almost always men of the most amiable simplicity of manners, who live in good harmony with one another, are the friends of one another's reputation, enter into no intrigue in order to secure the public applause, but are pleased when their works are approved of, without being either much vexed or very angry when they are neglected.

Things were different, Smith suggested, with "poets, or with those who value themselves upon what is called fine writing," who lived for, and on, public applause and who were therefore prone to faction and back-biting.[52] By the 1770s, these sentiments were supplemented by Condorcet's Renaissance-humanist preferences for a life of action and civic benevolence. The man of science, in Condorcet's picture, had the capacity to benefit the public realm both materially and spiritually. Condorcet's *éloge* of Benjamin Franklin accordingly celebrated both Franklin's technological ingenuity and the political reformism that was reckoned to flow from the very nature of modern scientific inquiry. Science would at once produce technological change and encourage those mental and moral attributes that would naturalize rational industrial society.[53] The *Encyclopédie*'s character of the ideal "philosopher" celebrated his integrity and free action by identifying him as an *honnête homme*, polite, civil, and autonomous. No slave to system or to dogma, he serves Truth alone. Unlike the pedants and ascetics of the past, this philosopher "knows how to divide his time between solitude and social intercourse . . . He looks on civil society as a divinity on earth." His concern is with the benefit and good order of civil society. Living in society, not apart from it, the philosopher was to be accounted one of its most valuable members.[54]

Insofar as the man of science was a special form of the scholar, the social circumstances affecting the scholarly life bore on him as well. In the 1784 essay "What Is Enlightenment?" Kant recognized that the "freedom to make public use of one's reason" was everywhere circumscribed by the obligations people owed to the institutions within which they worked. The military man owed allegiance to his superior officers; the lawyer and government official to their relevant hierarchies; the cleric to church dogma. Call that the "private use of reason" and accept that such private uses may everywhere be compromised. But the "public use of reason" is that which a scholar performs before the reading public, and in that he can and must be free: as a "scholar he has complete freedom, even the calling, to communicate to the public all his carefully tested and well meaning thoughts."[55] So, while Kant was here concerned with the different acts a person might perform within the same role or institution, he was also addressing the role and circumstances in which different people found themselves. Who was the person who devoted himself to such scholarly pursuits in general and to science in particular? The image of the selfless man of science, offering much to society and neither receiving nor expecting to receive much in return, was given credibility by some recognized social circumstances affecting scientific work. In the eighteenth century, as in the seventeenth, a decision to pursue many forms of scientific learning might well be taken against plausible calculations of material self-interest, and often against strong parental desires or directions. For those lacking independent means, the professions of law, religion, and medicine were understood to assure an honest and legitimate living. Very many eighteenth-century men of science chose their calling against their fathers' encouragement towards a career at the bar or in the church; in maturity others managed to combine scientific research with at least nominal legal, administrative, or clerical careers; and many others managed the much easier combination of science and medicine. But social respectability was only dubiously associated with the calling of the practical mathematician or engineer, and it was difficult to envisage clear remunerative and polite career prospects for the physicist, the geographer, the naturalist, or, to a lesser extent, the astronomer. In 1830, Charles Babbage's *Reflections on the Decline of Science in England* dwelt extensively on the financial dissuasives to a career in science compared to the security of the established professions: "In England, those who have hitherto pursued science, have in general no very reasonable grounds of complaint; they knew, or should have known, that there was no demand for it, that it led to little honour, and to less profit."[56] To Babbage, these circumstances appeared solely as *problems*, no longer to be celebrated as indices of intellectual integrity.

If you were battling to rise from the lower orders—for example, the electrician Stephen Gray, the chemist John Dalton, or the geologist William Smith—a career as scientific lecturer, author, or technical consultant might have both its material and social attractions, and if you possessed independent means freeing you from material concerns—as did, for example, the naturalists the comte de Buffon, the earl of Bute, or Sir Joseph Banks; the physicist Henry Cavendish; and the geological chemist Sir James Hall—you could afford to adopt an insouciant attitude towards remuneration, towards orthodox notions of cultural respectability, and even towards scientific authorship and the public assertion of property in intellectual goods.[57] But for many in middling social circumstances—from younger sons of the aristocracy to the offspring of the professional and mercantile classes—scientific inquiry would have to be combined with an adequately remunerated professional or public role. There were many such possible hybrid forms of life in the eighteenth century beyond those attached to the universities and the learned professions: Antoine Laurent Lavoisier famously served as a "tax-farmer"; Leibniz and Johann Wolfgang von Goethe as government officials; Charles Augustin Coulomb worked as a military and civil engineer; and the young Alexander von Humboldt as both a diplomat and a supervisor of mining. For those of intermediate social standing, a decision to devote oneself solely or mainly to scientific scholarship might be understood—against this background—as testimony to a particular selfless and whole-hearted kind of dedication. What could account for a commitment to science other than a genuine calling?[58]

The intercalation of scientific expertise into the structures of power and profit continued and gathered pace into the nineteenth century, but the historical differences here are matters of degree. Insofar as the State was concerned, it had always needed a wide range of technical expertise. Antiquity knew all about the roles of, for example, the mathematically competent military engineer who could design fortifications, the astronomer who made calendars, and the physicians and surgeons who could advise on diet or cut for the stone. The man of science as remuneratively engaged "civic expert" was not a novel entity in the seventeenth and eighteenth centuries. Nor were cultural appreciations of his role particularly linked to the rhetoric of utility that, from the seventeenth century, picked out the special capacity of some Methodologically modernized versions of natural science to contribute to useful outcomes and to augment State power. The point here does not hinge on the hoary debate over the relations between scientific *theory* and technical utility; rather it concerns the roles and the historical appreciations of scientifically knowing *people*. And what the eighteenth and nineteenth centuries witnessed was a vast expansion in the numbers of scientifically trained people matter-of-factly employed as civic experts in many commercial, military, and governmental settings. This state of affairs did not depend upon the acceptance of a causal and direct connection between abstract knowledge and useful outcomes. What the State wanted, and what it increasingly could secure from scientifically trained practitioners, was not natural philosophy but instrumental expertise, not knowledge but knowledge-power, not Truth but competence in predicting and controlling. And this accelerating expansion in the links between the State, commerce, and natural knowledge had crucial bearings on appreciations of the identity of both the man of science and scientific knowledge.[59]

Enough had been achieved in this direction by the end of the nineteenth century that it could generate a self-conscious intellectual opposition. In 1874, a letter writer to the *New York Times* noted the current criticism of John Tyndall's scientific naturalism, but insisted that science had only itself to blame: "The constant straining for material progress which goes on in the world has its inevitable effect upon scientists, who, in spite of their science, are only men." They confine themselves to the material domain and neglect the spiritual, and, in this neglect, it was said that they generate their own opposition.[60] In 1885, the Victorian socialist, mystic, and sexual reformer Edward Carpenter thought that modern science was wholly legitimate as long as it represented its theories and generalizations merely as transitional means towards achieving practical ends: "For practical results and brief predictions [modern science] affords a quantity of useful generalisations—shorthand notes and conventional symbols and pocket summaries of phenomena—which bear about the same relation to the actual world that a map does to the country it represents." And this was the technical expertise that so effectively allowed the State to extend its power and commerce to expand its possibilities for the creation of wealth. But when science reckoned either that its representations captured reality or that it could set its claims against those of emotion and feeling, then it misdescribed its own nature and became an illegitimate form of dogma, as pernicious as the religious superstitions it thought to supplant.[61] Tolstoy found Carpenter's formulations appealing, and his essay "Modern Science," which influenced Max Weber, was prepared as a preface to a Russian translation of a collection of Carpenter's essays. The scientists of today think, Tolstoy wrote, that "our science is the most important activity in the world, and we men of science are the most important and necessary people in the world." Tolstoy didn't approve; like Carpenter, he thought that science ought to renounce the "experimental method" and the is/ought divide—but he recognized these conceptions of science as consequential facts about the authority of technical expertise, now divorced from virtue.[62]

There is no need to carry on this story about social roles much closer to this book's major focus on the twentieth century. The point of principle has been established: through the eighteenth and nineteenth centuries the pursuit of science was increasingly integrated into the structures of power and profit. Insofar as that integration was recognized in the culture, appreciation of the nature of science and the character of the man of science might undergo significant changes—changes in conceptions of what scientific knowledge was *about*, in the *methods* used to produce it, in the *personal characteristics* of those who pursued scientific knowledge. The *vita contemplativa* as the circumstance for the pursuit of natural knowledge was being replaced by the *vita activa*, the re-situation of the man of science from the cloistered domain of contemplation to the civic sphere, the transition from the sacred to the social virtues, and thus the fulfillment of Bacon's dream. For Ralph Waldo Emerson the figure of the "scholar" represented "man thinking," but the ideal of the new "American Scholar" was that he should no longer be "a recluse, a valetudinarian" but now a man fitted for action, whose knowledge derived from action and who was positioned, through thought, to act vigorously in the world: "There can be no scholar without the heroic mind. The preamble of thought, the transition through which it passes from the unconscious to the conscious, is action."[63] This is just what the biologist and peace activist David Starr Jordan said at the end of the nineteenth century: "The chief value of nature study in character building is that, like life itself, it deals in realities." Scientific research "if it be genuine is essentially doing." And just because science could be conceived as "doing," scientific knowledge could be both power *and* civic virtue: "Wisdom is knowing what it is best to do next. Virtue is doing it. Doing right becomes habit if it is pursued long enough. It becomes a 'second nature' or a higher heredity."[64] In 1924, the founder of American technical industrial consultancy, the Boston chemist Arthur D. Little, drew contemporary lessons from the life of Benjamin Franklin as an ideal synthesis of *vitæ activa* and *contemplativa*: "His remarkable career should refute forever the fallacy which, unfortunately, is still current, that the man of science is temperamentally unfitted for the practical business of life . . . Science was made for life and life is more than science." If you really needed to refute that still-current "fallacy," then look at America's modern Franklins—in the corporate laboratories of General Electric, General Bakelite, and Bell Telephone.[65] And so this account of the secularization of science and the normalization of the scientific career now veers close to the sorts of stories about the Making of Modernity that this book means also to qualify, to circumscribe, and, finally, to challenge.

Some initial qualifications have to be made. First, the integration of science into structures of power and profit was never more than partial in the nineteenth century. The figure of the man of science as an amateur, conducting inquiry without expectation of a remunerated career, did *not* disappear. The most famous scientist of the century was a gentleman-amateur: Charles Darwin was never *employed* to produce scientific knowledge, nor was the knowledge he then produced designed to be of use to contemporary structures of power and profit. On the *Beagle*, Darwin was the captain's guest, on the Admiralty books for food, although he paid an extra sum for the captain's table. Far from receiving remuneration, the total cost of the voyage to Darwin's father was £1200.[66] In Britain alone, the list of amateur-scientists in the late eighteenth and nineteenth century includes some of the most influential figures in all the sciences. And, while the cleric-scientist did become a much rarer figure in the nineteenth century, the "founder" of genetics, Gregor Mendel, was a monk, and several British geologists were in holy orders in addition to holding academic posts. Much talk of "the professionalization of science" during the nineteenth century has been too crude. The increasing integration of science into the State and commerce did not simply turn "science" into a "career." Both the notion of "science" and that of "career" need to be spelled out. What both the State and the institutions of commerce wanted was *expertise*, and embodied expertise is what they bought in and supported. This is just what Charles Babbage complained of in his 1830 *Reflections on the Decline of Science*. He meant there to distinguish the different *sorts* of inquiry and their different standings vis-à-vis social favor and State support. And he was offering not a disengaged description of the state of science in France or Germany, but an *argument*, appealing to national competition and national self-interest, why the British State ought to underwrite certain conceptions of the scientific career.[67] It was a distinction that was later institutionalized: Babbage contrasted what he called "the more difficult and abstract sciences" with those "connected with objects of profit." He did not think that these latter were in "decline" or that they required imminent State action. Useful inventions received their reward, either directly, from patents and profit, or indirectly, through imminent expectations of profit, but "all abstract truth is entirely excluded from reward under this system. It is only the application of principles to common life which can be thus rewarded." And here he offered one of the earlier forms of the Golden Goose argument in favor of supporting the "abstract" or, as we now say, "pure" sciences, just because their ultimate benefit to the State will be incalculably, and unpredictably, great: "If, therefore, it is important to the country that abstract principles should be applied to practical use, it is clear that it is also important that encouragement should be held out to the few who are capable of adding to the number of those truths on which such applications are founded."[68]

It was this sensibility that led in the late eighteenth century to Benjamin Franklin's response to a question about the utility of his research — "Of what use is a newborn babe?" — and by mid-century, to Faraday's famous response to Gladstone (then Chancellor of the Exchequer), when asked of what use was his work on electromagnetism: "I don't know, but I'm sure that one day you will tax it."[69] By the end of the century, and especially in utilitarian America, men of science were celebrating the civic worth of almost any sort of scientific inquiry — "the foundation of industrial advance was laid by workers in pure science" — many of them totally abandoning purely cultural justifications.[70] Babbage wanted cultural keywords used correctly: "science" ought properly to designate those abstract, disengaged, and disinterested inquiries that both required subvention and justified such subvention by their ultimate utility. Pure science ought to be supported and made into a career, but now the arguments in favor of that career structure had been substantially shifted — away from both sacred and aristocratic justifications and towards an indefinitely *postdated utilitarianism*.[71] For all that, Babbage's usages and arguments had only limited consequences. In nineteenth-century Britain, the notion of "abstract" scientific inquiry as a career, supported by the State, was no more than a vision, however vigorously it was lobbied for by late Victorian Scientific Naturalists. In the United States, Tocqueville observed a deep democratically rooted disinclination to support, so to speak, "science for its own sake":

> In America the purely practical part of science is admirably understood, and careful attention is paid to the theoretical portion which is immediately requisite to application . . . In this respect the Americans carry to excess a tendency that is, I think, discernible, though in a less degree, among all democratic nations . . . In aristocratic ages science is more particularly called upon to furnish gratification to the mind; in democracies, to the body.[72]

There were legal obstacles too: the United States Constitution was understood to bar a Federal government role in supporting such a thing. Why should citizens pay the bill for intellectual self-gratification?[73] Through the nineteenth century, and well into the twentieth, a range of distinctions and stipulated causal relations between "the scientist" and "the engineer," and between "pure" and "applied science," emerged as argumentative tactics — both as evaluative justifications for social subvention of "abstract inquiries" and as descriptions of their respective characters and bases of authority. In the 1890s, even Americans keen to make the Golden Goose argument wanted it understood that those capable of abstract inquiries belonged to a different intellectual order than the engineer: "The quality of mind that discovers the laws of nature is of a higher order than that which makes application of them."[74] Necessarily, there were fewer minds with abstract abilities, and the laws of supply and demand *should* ensure their greater reward. In the 1920s, an eminent French physiologist wrote—in a wildly popular book translated into English—that "probably the chief characteristic of true savants—whether they be archæologists, mathematicians, chemists, astronomers, or physicists—is that they do not endeavour to apply their work in practice. They are not concerned with the application of theory." Neither the engineer nor the physician was a savant: "To construct a ship, or to save a patient, is to act rather than think."[75] And so their respective characters belong to the active versus the contemplative worlds.

Accordingly, by the end of the nineteenth century, the transition from science as a calling to science as a job had at most just begun.[76] Changes in knowledge, in method, and in the character of the knower were indeed taking place, but the cultural assemblages referring to these things were neither coherent nor stable. Agitation for the establishment of a normal scientific career was not the same thing as its institutionalized reality. In the 1870s, Francis Galton described the secular, innate (and usually inherited) taste that the man of science had for the enterprise, and the obstacles such an innate taste encountered in the choice of science as a profession: those making that choice "must do so in spite of the fact that it is more unremunerative than any other pursuit." Much has indeed changed, Galton acknowledged, from the time when the current crop were boys, when the clergy controlled education and "crushed the inquiring spirit." But there was now much hope that new emerging professional opportunities "will, even in our days, give rise to the establishment of a sort of scientific priesthood, throughout the kingdom, whose high duties would have reference to the health and well-being of the nation in its broadest sense, and whose emoluments and social position would be made commensurate with the importance and variety of their functions."[77]

But not just yet, neither in Britain nor in America. In the 1880s, an eminent American physician addressed an audience of Washington savants in terms that would have been recognizable to the early moderns: "The man of science, as defined by his eulogists, is the *beau idéal* of a philosopher, a man whose life is dedicated to the advancement of knowledge for its own sake, and not for the sake of money or fame, or of professional position or advancement. He undertakes scientific investigations exclusively or mainly because he loves the work itself, and not with any reference to the probable utility of the results." There was, the writer admitted, an emerging problem about how much money the scientist could make without compromising his primary commitment to truth, and here the answer was satisfyingly concrete: "There are some reasons for thinking that the maximum limit is about $5,000 per annum . . . The more they demonstrate their indifference to mere pecuniary considerations, the more creditable it is to them; so much all are agreed on."[78] An appreciation that a life in science was *not* a fit way of making a bourgeois living persisted in America well into the twentieth century, and, while that circumstance was widely regretted, it could also be called upon to establish the integrity of the scientist's vocation. In the mid-1920s, a physicist observed that the practitioner's role was neither financially rewarding nor socially valued, and that "this is just as true to-day as it was in the days of the dependent philosopher slaves of Greece or the roaming impecunious scholars of the Medieval and Dark ages." And that made it all the more remarkable that recruits still kept "coming up to take the place of each one who is called away to higher realms of truth by death." There could be no better sign that science was a calling and not a job.[79] In late Victorian Britain and America, the emerging practice of scientific expert witnessing put enormous public pressure on the idea of the scientist's virtue. Some spokesmen for the scientific community applauded the practice, notably arguing that the adversarial structure in which experts disagreed would advance the search for truth and ensure *communal* virtue. A leading chemist announced that "it was not dishonourable for a scientist to earn his living." Others, including the editor of *Nature*, thought that the spectacle of scientists hiring themselves out to vested interests would erode public perceptions that *any* scientist was constitutionally incorruptible, insisting, as historian Christopher Hamlin writes, that "the search for knowledge was a far higher calling than any activity whose main aim was making a living."[80]

In 1918, Thorstein Veblen, while excoriating American universities for their administrators' pursuit of the materially useful, nevertheless pointed to what he saw as a "long-term idealistic drift" towards the recognition of disinterested inquiry and of the university as its proper and only home: "This profitless quest of knowledge has come to be the highest and ulterior aim of modern culture."[81] And, most tellingly for present purposes, Max Weber articulated a vigorously anti-Methodical view of scientific discovery—"Ideas come to us when they please, not when it pleases us"—and rejected any idea that utilitarian science could possibly be a vocation. Recognition of the value of knowledge "for its own sake" was just essential "if the quest for such knowledge is to be a 'vocation.'" True, science did contribute to "the technology of controlling life by calculating external objects as well as man's activities. Well, you will say, that, after all, amounts to no more than the greengrocer," and Weber agreed that the scientist's vocation could not be justified on the same grounds as the shopkeeper's. The authentic man of science, in Weber's view, was someone whose immense moral authority came from a legitimate understanding of what scientific knowledge was about and what it was not about, from his mastery of understanding a de-magified world, and, above all, from a proper sense of *identity*. The scientist was not to be mistaken for a priest, nor his knowledge for priestly knowledge:

> It is not the gift of grace of seers and prophets dispensing sacred values and revelations, nor does it partake of the contemplation of sages and philosophers about the meaning of the universe. This, to be sure, is the inescapable condition of our historical situation. We cannot evade it so long as we remain true to ourselves.[82]

So at the beginning of the twentieth century the identity of the scientist was radically unstable. To be a scientist was *still* something of a calling but it was *becoming* something of a job; it was still associated with the idea of social disengagement but increasingly recognized as a source of civically valued power and wealth; it was still associated with a notion of special personal virtue but it was on the cusp of moral ordinariness. Chapter 3 takes up the story of how the idea of the morally ordinary scientist became a twentieth-century cultural institution and what work that idea did in the culture.

The Moral Equivalence
of the Scientist

A HISTORY OF THE VERY IDEA

The American's conception of the teacher who faces him is: he
sells me his knowledge and his methods for my father's money, just
as the greengrocer sells my mother cabbage. And that is all.

Max Weber, *Science as a Vocation*

"SCIENTISTS ARE HUMAN TOO"

The previous chapter started by suggesting that Merton's 1942 insistence
upon the scientist's moral equivalence had the character of an argument
against persisting "vulgar error." The knowing sociologist felt obliged to
address still well-entrenched presumptions to the contrary, and had to
show that alternatives to them were possible, intelligible, and disciplinar-
ily prudent. That is to say, during the first half of the twentieth century,
stipulations of moral equivalence were frankly *reactionary*—they were
explicitly reacting against "what everybody believed" or, at least, had
recently believed. The chapter then described how presumptions of the
moral superiority of those speaking Nature's Truth were institutionalized
in seventeenth- and eighteenth-century cultures, offering an account of
what these institutions were, how they worked, and how alternatives to
them began to appear. But when did the sort of matter-of-fact dismissals
represented in Merton's remark first appear? When I began to consider
that question, I thought it possible that the late 1930s and early 1940s
were the first period in which moral equivalence was culturally intelligible, and I could not locate any stipulation before Merton's that was strictly
similar—in sense, tone, and manner. That initial guess proved factually
wrong, but not seriously so. The early twentieth century *was* the period
when presumptions of moral equivalence began significantly to populate
both American and Western European culture; such presumptions themselves became commonplace after the Second World War; and the United
States was the setting where these presumptions had their greatest grip
and resonance. This chapter tracks the idea of moral equivalence back

and forth in history, using Merton's wartime essay as a fulcrum. I want to point out some of the cultural contexts in which the presumption of moral equivalence figured, and I want to identify *continuing* sources of resistance to the idea through the twentieth century. Moral equivalence *is* a late modern commonplace, but it is a commonplace whose legitimacy is still *not* universally acknowledged, and which is invoked in quite specific contexts of use. Accordingly, this chapter has four parts: (1) it gives an account of moral equivalence sentiments as expressed in and around the 1930s; (2) it traces a genealogy of those sentiments back into the nineteenth century; (3) it describes beliefs in moral superiority persisting into the setting in which Merton wrote, and beyond; and (4) it follows these presumptions of moral equivalence into the post–World War II period, showing how they were invoked in a variety of cultural exercises, including the description and justification of research as a *job* in the era of Big Science, industrialized science, and what came to be called the military-industrial-academic complex.

The assertion that science has a "human face," that scientists "are human too," however much that state of affairs might now be presented as a matter of course, was not common in the 1930s, in the United States or elsewhere in Western culture. Apart from Merton's, there were not many such stipulations. The great American cynic H. L. Mencken foreshadowed Merton when, in 1918, he observed that "the value the world sets upon motives is often grossly unjust and inaccurate." And the example he used to make his point was that of the scientist, to whom great altruism was often, and wrongly, attributed. In fact, Mencken insisted, what motivated the scientist was a somewhat more intense version of quite ordinary curiosity: "His prototype is not the liberator releasing slaves, the good Samaritan lifting up the fallen, but a dog sniffing tremendously at an infinite series of rat-holes."[1] In 1936, the English physicist William George offhandedly remarked that "no very detailed knowledge of research workers is necessary to discover that they remain human always. They are not a special kind of human being."[2] Merton footnoted another source of such sentiments in his 1942 essay, a now scarcely known 1938 book intriguingly called *Scientists Are Human* by the Scottish expatriate psychologist David Lindsay Watson, with a foreword by John Dewey.[3] Merton had already written an irritated, but generally friendly, review of the book several years before in *Isis*, warning of its "eccentricity" but applauding its recognition of "the social forces" that "produce scientific habits of mind."[4] He had also conducted an odd correspondence with Watson, in which Watson tried to enlist Merton's assistance in getting a much longer version of his manuscript published.[5] Watson wanted it understood that scientists share "passions and self-deceptions . . . with the rest of mankind": "The pharisees are firmly established in our modern scientific institutions." These are the undeniable facts of the modern condition, Watson judged, but very few scientists had yet brought themselves round unflinchingly to look the facts in the face.[6] So Watson was insisting, against what he took to be still-dominant sentiment in the scientific community, on the moral equivalence of science-as-it-now-actually-was. There was something quite positive about that state of affairs, something that was in the nature of science. The making of scientific knowledge, as Michael Polanyi was soon to observe, was a "personal" matter: its "human side" was not a marginal factor; it was constitutive of the very idea of science.[7] But, as the "pharisee" remark makes clear, the moral equivalence of science was a circumstance brought about by current institutional arrangements and current conceptions of the nature of knowledge. And, according to Watson, those conceptions were ultimately disastrous.

Watson's insistence on moral equivalence was, therefore, part of his *criticism* of the way things were going. The "human side" of science was constitutive, and this meant that scientists ought to be *better* motivated than the current run of humankind; they ought to be inspired, not technicians of mechanical method; they ought to have the integrity of the freebooter, not the comfortable security of the placeholder. The emerging professionalization of the scientific occupation meant that scientists were now jobholders, in fact, if not in ideal theory, even in academia. And the knowledge produced by ordinary people was itself ordinary stuff: "Their truths are therefore the truths adequate for this sort of life." Ultimately, the fault lay in the institutionalization of notions of mechanical method sketched in the previous chapter. And, insofar as the culture comes to credit a mechanically methodical conception of scientific knowledge, "in the lay mind, the qualities of both science and the scientist [will be] inferred to be those of the uninspired hod-carrier."[8]

As a general matter, scientists' pre–World War II statements of moral equivalence functioned as *criticisms* of "the way we live now." Of these, Albert Einstein's were the most consequential. In one context, a personal disavowal of extraordinariness worked simply as an expression of good manners, and, perhaps, also of personal annoyance at the moral and intellectual expectations that his celebrity had set up in the public culture: "It strikes me as unfair, and even in bad taste," Einstein wrote in 1921, "to select a few individuals for boundless admiration, attributing superhuman powers of mind and character to them." It was as much as to say "I am but a man."[9] But other expressions of moral equivalence had more general reference. In the same year that Weber delivered his lecture on *Science as a Vocation*, Einstein observed that "in the temple of science are many mansions, and various indeed are they that dwell therein and the motives that have led them thither." Some make science "their own special sport," taking to it "out of a joyful sense of superior intellectual power." Increasing numbers of others "are to be found in the temple who have offered the products of their brains on this altar for purely utilitarian purposes." For such men, in Einstein's opinion, "any sphere of human activity will do, if it comes to a point; whether they become engineers, officers, tradesmen, or scientists depends on circumstances." So, as a late modern matter of fact, all kinds of people, drawn for all sorts of motives, were doing science. Still, it was a matter of fact that Einstein regretted. The true man of science was someone who, so to speak, could not help it. He was drawn to science by the fineness of his personality: "A finely tempered nature longs to escape from personal life into the world of objective perception and thought." Exceptional people; exceptional motives; and exceptional modes of knowing.[10]

Marxist scientists were certainly keen to stress their status as "workers," and, thus, might have been specially inclined towards the cognitive and moral leveling that so often accompanied socialist political commitments among "brain workers." In these connections, the term "scientist" was frequently replaced with "research worker" or "scientific worker." The British National Association of Scientific Workers was founded in 1918, changing its name in 1927 to the Association of Scientific Workers.[11] In 1925, the association produced a report, summarized for a U.S. audience in *Science*, which specified how properly to view the motives of such workers: "The motives of research workers are, generally speaking, as mixed and as commonplace as those of their neighbors. It is well to recognize this fact, and to discard the illusion that the research worker necessarily pursues a lofty course inspired by an ideal superior to that which moves the remainder of mankind." Of course, it was conceded, research workers wanted to pursue knowledge "for its own sake," but that motive was insufficient to sustain the worker through the laborious research process, and mundane concerns for reputation and financial reward were both routine and necessary to the scientific occupation.[12] As early as 1929, in a text shaped at least as much by Freudian as Marxist sensibilities, J. D. Bernal, who had been a socialist since undergraduate days, wrote that "the mere observation of scientists should be sufficient at the present to show that . . . in every respect, save their work, they resemble their non-scientific brothers."[13]

After the war, arguing for further integration of science into the economic and political institutions of the State, Bernal made it plain that "scientists are just one kind of worker—as necessary as, but not more necessary than, any other."[14]

Other twentieth-century stipulations of moral equivalence are harder to categorize. In 1928, the *New York Times* reported on controversies then raging in French science: "Scientists themselves admit that laymen have an exaggerated idea of the 'scientific spirit.' They are ready enough to assert that in the realm of science the motives of objective truth, of freedom from personal prejudice and vainglory, are more potent than in any other field of human activity. But scientists are human, and . . . the history of science is by no means free from passionate controversies in which such human weaknesses as envy and spite have manifested themselves." The ideal says that scientists should not be attached to their theories, but the reality says that they are; they ideal says that they ought to be unconcerned for glory, but the reality says they are as much concerned as anyone else.[15] A few years later, A. V. Hill, a Nobel-winning Cambridge physiologist and vigorous anti-Nazi activist (though disavowing Marxist allegiances), insisted on scientists' moral ordinariness, addressing both the essential humanity and universality of science:

> It is true that many distinguished scientists have been men of great general capacity; a man of such capacity is likely to be distinguished at any task he undertakes. The converse, however, is certainly not true; many of the most important contributors to science have been extreme specialists—rather dull dogs: others have been dreamers, poets, artists, rather than men of broad understanding. Their view on general topics may be entertaining, but they demand no special attention.[16]

It was insisted, again, that expertise was not fungible. There was no more reason to impute any special ideological creed or moral makeup to scientists as a group than to any other brain-workers. The residual virtue of "tolerance" was, of course, all the virtue that Hill needed in his attack on the Nazification of German science. Otherwise, he now found it odd that people should still think of scientists as in any way special. He told of his own surprise when, as a young man, he met a physiologist who was the author of a paper he particularly admired and found that he was just a regular sort of person. And that gave him his text to lecture on "the humanity of science": "one of his chief purposes" was to make sure it was understood "that the scientist at work is a human being like the rest of us."[17]

So, in the "between the wars" period, there was a variety of cultural settings in which specifications of moral equivalence were expressed and a variety of purposes to which these specifications were put: the justification of academic disciplines and their characteristic methods; the criticism of what were taken to be contemporary tendencies in the institutionalization of science; the ideologically driven stipulation of the scientist's class-character; the moral exculpation of certain categories of intellectuals from aspersions attached particularly to them; and, of course, a range of other idiosyncratic purposes in which such stipulations might be recruited, as it were, accidentally—not a major point to be made, but as a passing remark, drawing upon a pool of iconoclastic sentiment elsewhere in the culture. But such remarks in this period *were* indeed iconoclastic, just in the sense that they all *reacted* self-consciously against what was taken still to be prevailing sentiment. When you said that the scientist was nobody special, you *knew* that "fashion"—vulgar or otherwise—was against you. A later section of this chapter will describe what that contemporary fashion in such matters looked like, but I need now to trace the idea of moral equivalence back as far as I can in the culture. Does it indeed have a nineteenth-century genealogy? Or is its expression an authentic mark of the institutional and cultural changes occurring in the early to middle parts of the twentieth century?

VIRTUE, BIOGRAPHY, AND PHILOSOPHY: 1830–1920

When and how did the moral equivalence of the scientist became a cultural institution—a standardized expression or sentiment, one that you could invoke in a taken-for-granted way, or even criticize as an attitude *assumed* to be a cultural given? Later in this chapter, I point to post–World War II developments, especially in America, as providing the prime conditions for this institutionalization. But what I cannot do with great confidence is to say when that expression or sentiment *originated*. Merton's stipulations about moral equivalence were, indeed, doing relatively new things in the culture, but sentiments of that general sort were *not* wholly unprecedented.

One notable antecedent arose from controversies over the proper posture of the scientific biographer. In 1831, the Scottish natural philosopher David Brewster wrote a life of Isaac Newton. Securing access to a cache of manuscript evidence that contained evidence of Newton's alchemical concerns and of his morally culpable conduct in the priority dispute with Leibniz over the invention of the calculus, Brewster nevertheless chose to pass over such evidence in silence: "The social character of Sir Isaac Newton was such as might have been expected from his intellectual attainments. He was modest, candid, and affable, and without any of the eccentricities of genius, suiting himself to every company, and speaking of himself and others in such a manner that he was never even suspected of vanity."[18] A later edition of Brewster's biography, appearing in 1855, vigorously repeated the earlier assessment of Newton's noble personal character, but went some way to admitting imperfections (especially in connection with the calculus priority dispute). And the mathematician Augustus de Morgan both applauded and further encouraged what he saw as the gradual emergence of a more honest form of scientific biography: "The scientific fame of Newton, the power which he established over his contemporaries, and his own general high character, gave birth to the desirable myth that his goodness was paralleled only by his intellect. That unvarying dignity of mind is the necessary concomitant of great power of thought, is a pleasant creed, but hardly attainable except by those whose love for their faith is insured by their capacity for believing what they like."[19]

There was now no longer any need for this kind of "myth," de Morgan announced: we can look scientific truth, and those that produce it, straight in the face. That is a leading characteristic of our times, and it is a characteristic we can and should applaud: "We live, not merely in sceptical days, which doubt of Troy and will none of Romulus, but in discriminating days, which insist on the distinction between intellect and morals." Modern biographers have a duty to historical Truth, while they still belong to an intellectual culture that has yet entirely to slough off an older obligation to celebrating the virtuous subject: "Though biography be no longer an act of worship, it is not yet a solemn and impartial judgment: we are in the intermediate stage, in which advocacy is the aim, and in which the biographer, when a thought is more candid than usual, avows that he is to *do his best* for his client."[20] De Morgan foresaw a better future, in which even that residual purpose would be shed, and in which the obligation to historical Truth could flourish without compromise: "The time will come when [Newton's] social weaknesses are only quoted in proof of the completeness with which a high feeling may rule the principal occupation of life, which has a much slighter power over the subordinate ones. Strange as it may seem, there have been lawyers who have been honest in their practice, and otherwise out of it: there have been physicians who have shown humanity and kindness, such as no fee could ever buy, at the bedside of the patient and nowhere else."[21]

Just as the culture is fragmented, so the biographer, and the biographer's readers, can accept as a matter of fact the fragmented nature of the biographer's subject. Excellences are parceled out among these fragments. Why ever should we expect that excellence of mind is necessarily accompanied by excellence of morals? The excellences of an integral self are neither to be found nor ought they to be demanded as an act of homage: "Let a flaw be a flaw, because it is a flaw: Newton is not the less Newton."[22] Knowledge is one thing; virtue another. Newton remains an authentic hero, but his heroism now properly centers on an *aspect* of his life. The self is sorted out into its *roles*, and the scientific life of the mind—whether private or public—is not just one of those roles, but the *only* role germane to the scientific biographer concerned with the evaluation of the subject's scientific achievements. While present-day scientific biographers now display a remarkable taste for "the private life" as a mode of deflation, denigration, or, more rarely, positive humanization, few have any response to de Morgan's question about what any of this should have to do with the status of scientific ideas. Honest biography *cannot*, according to this sensibility, bear upon the status and worth of knowledge.[23]

The practice of honest biography was also on the mind of Thomas Babington Macaulay, in his celebrated early Victorian *Edinburgh Review* essay on the life of Francis Bacon. Here he took to task the author of a Bacon biography for what Macaulay called "the delusion . . . under the influence of which a man ascribes every moral excellence to those who have left imperishable monuments of their genius . . . Plato is never sullen. Cervantes is never petulant. Demosthenes never comes unseasonably. Dante never stays too long." But that, according to Macaulay, is just one of the idols of *our* literary tribe: dead authors in general give us pleasure and cannot offend us, so we see them as virtuous. As a matter of general fact, we do *not* speak ill of the dead. We have few *occasions* to do so: much more pertinent—and more fun—to speak ill of the living. Macaulay had no opinion about whether scientists and philosophers on the whole actually tended towards virtue. His concern was not specifically with men of science but with pervasive attitudes towards "dead authors." But in general we are unlikely to be justified in our genuflections towards dead authors whose work we admire. They all had their flaws: "Nothing can be more certain than that such men have not always deserved to be regarded with respect or affection." The great Bacon was a man, like other men. Why should a dispassionate observer or biographer expect otherwise?[24]

Still another context for expressions of moral equivalence before the 1930s was provided by developments *within* philosophy and cultural theory. Chapter 2 traced changing notions of Truth and Method through the late nineteenth and early twentieth centuries, and those particular chickens came home to roost in Nietzsche's rejection of the Western philosophical tradition and how it had understood its objects. According to Nietzsche, that tradition not only *deserved* disrepute, it had largely *achieved* its own disrepute through the sheer ordinariness of those who produced philosophical knowledge: "On the whole, speaking generally, it may just have been the humanness, all-too-humanness of the modern philosophers themselves, in short, their contemptibleness, which has injured most radically the reverence for philosophy." The philosopher—and one assumes that Nietzsche meant to designate all who pretended to speak Truth—had inherited the public character of someone elevated, apart, and special, but, once inspected, that character proved to be shallow: "The philosopher has long been mistaken and confused by the multitude, either with the scientific man and ideal scholar, or with the religiously elevated, desensualised, desecularised visionary and God-intoxicated man; and even yet when one hears anybody praised, because he lives 'wisely,' or 'as a philosopher,' it hardly means anything more than 'prudently and apart.'" "What is the scientific man?" Nietzsche asked in *Beyond Good and Evil*, and he answered his own question: "a commonplace sort of man, with commonplace virtues."[25]

Nietzsche took from Ralph Waldo Emerson the notion that the genuine philosopher *ought* to be a hero, but he also insisted that the philosopher ought to make manifest the genuineness of thought in the healthy constitution of his *body*. The cure for effete, ascetic, and unhealthy Apollonian philosophy, obsessed with logical order and universals, was a vigorous and heroic Dionysian philosophical life of the body. In the past, "the philosophic spirit had, in order to be *possible* to any extent at all, to masquerade and disguise itself as one of the *previously fixed* types of the contemplative man, to disguise itself as priest, wizard, soothsayer, as a religious man generally: the *ascetic ideal* has for a long time served the philosopher as a superficial form, as a condition which enabled him to exist."[26] The ascetic ideal had yielded a desiccated and pathological form of knowledge: with its notions of "'pure reason,' 'absolute spirituality,' 'knowledge-in-itself.'" There was, in fact, "only a seeing from a perspective, only a 'knowing' from a perspective, and the *more* emotions we express over a thing, the *more* eyes, different eyes, we train on the same thing, the more complete will be our 'idea of that thing, our 'objectivity.'" So Nietzsche analyzed the distempers of philosophy through both the pathology of ascetic ideals and through the modern *unsustainability* of the idea of the philosopher as a morally heroic ascetic. What was needed was the Superman, healthy knowledge produced by a supremely healthy body.[27]

While Nietzsche had little to say about the particular goals and practices of the natural sciences, his rejection of universals and absolutes shows a radical deflation of notions of transcendent scientific Truth discussed in the preceding chapter. That is how it was taken by early twentieth-century cultural commentators disturbed by the joint Deaths of God and Truth. The epigraph to *The Treason of the Intellectuals* (1927) by the French rationalist philosopher Julien Benda (1867–1956) complained that "the world is suffering from a lack of faith in a transcendental truth." The role of intellectuals in bringing about that "lack of faith" was their "treason," their betrayal of the culture and of their proper role in it.[28] Benda defined the intellectuals (*les clercs*) and the laity through opposing drives. The laity were those whose "whole function consists essentially in the pursuit of material interests," and they belonged in the world of *meum et tuum*, in the civic sphere. The intellectuals, however, were "all those who seek their joy in the practice of an art or science or metaphysical speculation, in short in the possession of non-material advantages, and hence in a certain manner say: 'My kingdom is not of this world.' Indeed, throughout history, for more than two thousand years *until modern times*, I see an uninterrupted series of philosophers, men of religion, men of literature, artists, men of learning . . . whose influence, whose life, were in direct opposition to the realism of the multitudes." But the late modern order collapsed these distinctions: the knowledge produced by these mongrelized intellectuals disengaged itself from the universal and the absolute. No more transcendental and universal Truth, now only local truths, so that one hears of "German science" and "French science," "bourgeois truth" and "working-class truth." No more absolutes and universals, only pragmatic conceptions of what works to serve particular contingent interests.[29]

Knowledge and the social roles of the knowledge-producer have adapted themselves to each other. It was the success of expertise in selling itself to the State and to commerce that resulted in the selling of its soul: "One of the principal causes" of the end of the universal and the absolute "is that the modern world has made the 'clerk' into a citizen, subject to all the responsibilities of a citizen, and consequently to despise lay passions is far more difficult for him than for his predecessors . . . The 'clerk' is not only conquered, he is assimilated." Now the intellectuals too were playing the political game—serving not humanity but the nation-state that increasingly engaged their services, no longer speaking Truth to power but putting their manipulative expertise at the service of power. In the wake of the Great War, Benda wrote that "to-day the 'clerk' has made himself Minister of War." And that is the context in which Benda announced the moral and motivational equivalence of the modern intellectual: "The truth is that the 'clerks" have become as much laymen as the laymen themselves."[30] As Foucault would later put it, the modern *clerc* was on the way to making himself into a "specific intellectual," and, in the process, his de-moralization was his own doing.[31]

THE SCIENTIST IN SERVICE

The integration of the scientist into the structures of commerce and power represented a success: the perception that scientific inquiry was materially necessary gave science institutional security, and it was a major basis for the transformation of scientific inquiry into a significant occupation. At the same time, that integration was one circumstance giving rise to the notion of moral equivalence. Consider, for example, the views of George Schley, an executive with the big pharmaceutical company Eli Lilly. In 1937, urging that the commercially relevant outcomes of academic research be patented, Schley was fully aware of arguments against mixing distinct commercial and academic values, but if people wanted the increasingly acknowledged useful fruits of disinterested research, they had to be prepared to pay scientists the market rate for their work. The laborer was worthy of his hire. And, in the course of that argument, the Lilly executive insisted—against persisting contemporary presumptions to the contrary—that the scientist was motivationally no different from anyone else. If society, he said, was "willing to compensate the artisan with patents, why should it not similarly compensate the professor?" True, the university scientist did, indeed, "do much with no thought of advantage to himself": "But, by and large, he will do more, and tell society more, if there is the added incentive of personal advantage; and society will benefit. There is no cause for him to scorn personal advantage, or for him to permit others to scorn it for him. He is made of clay as other men . . . It is right for him to want things." Scientists, Schley insisted, had *always* been made of the same stuff as other men, and it was time for myth to be put aside.[32]

Despite the rapid integration of science into industrial and State institutions in the first decades of the twentieth century, the notion of moral equivalence still sat astride a cultural fault line: some, like Schley, celebrated that integration and insisted that it be recognized; others worried about it as a looming problem for the identity of the scientist and ultimately for the authority of scientific knowledge. Daniel Kevles's history of the American physics community notes that at the outbreak of the Great War many scientists were still insisting vigorously on the necessity of independence and disinterestedness, but there were consequences that flowed from this insistence: "The more American physicists committed themselves in research to meeting standards of productivity and merit that were internal to their profession, the more they disaffected the high-status, ex-cultivated Americans who had once responded to their claims of cultural and social leadership." This might even be a family affair; Irving Langmuir's brother thought he was prostituting himself by leaving academia to join General Electric's Research Laboratory: "You will betray your true self if you devote your life selfishly to private enterprises and personal acquisition. And the minute you allow yourself to deviate from the path of pure science, you will lose something in character."[33] Physicists had to find other constituencies, other ways of making themselves and their work appealing, and a combination of utility and good citizenship was that way.

Charles Sanders Peirce (1839–1914), who had worked for many years for the U.S. Coast and Geodetic Survey before becoming a philosopher, celebrated the uselessness of genuine research as a way of finding unique virtue among the scientists: "If a man occupies himself with investigating the truth of some question for some ulterior purpose, such as to make money, or to amend his life, or to benefit his fellows, he may be ever so much better than a scientific man, if you will . . . , but he is not a scientific man." Inutility of inquiry was a proximate cause of personal virtue: "A scientific man must be single-minded and sincere with himself. Otherwise, his love of truth will melt away, at once. He can, therefore, hardly be otherwise than an honest, fair-minded man . . . On the whole, scientific men have been the best of men. It is quite natural, therefore, that a young man who might develop into a scientific man should be a well-conducted person."[34] In 1885, the editor of *Popular Science Monthly* commented on a Congressional exposé of corruption in the Coast Survey, where Peirce was then employed, drawing a lesson about scientists' moral equivalence: "Like other men, they [scientists] are self-seeking, ambitious, and have their personal ends to gain. Can we assume that morally they are any better than their neighbors; or that, if they get possession of place and power, they will not use and pervert them to the promotion of their selfish objects?" No, the writer thought not, but, far from celebrating normalization, he hoped for better things and better characters to come: "It is to be hoped that in the future science will become so developed as to react upon character and give us men morally as well as intellectually superior; but we are far from any such happy

result yet." As matters stood, it was a mistake to presume that technical competence translated into political or moral competence. Scientists
should render unto Caesar what is Caesar's.[35]

Those embracing democratic values insisted that experts *should not* set
themselves above the people and they put a premium on moral equivalence. Democratic citizenship demanded it. In 1897, an astronomer employed at the U.S. Naval Observatory in Washington made the formula
clear. Sometimes the scientist was portrayed as a moral paragon; sometimes as "a harmless eccentric, a feeble specimen of manhood." But both
extremes were wrong; the present-day scientist was, and ought to be,
quite *normal*: "In general he differs from his fellows only in the possession
of some peculiar aptitude or talent for study or investigation in some department of science. He may be a good chemist, and shirk every duty of a
good citizen . . . In short, the manifestation of ability in scientific pursuits,
as in other walks of life, does not necessarily imply the possession of good
morals or the other qualities that make the good citizen." This astronomer
had enough experience of scientists in government service to know that
they too could succumb to the abuse of power and to the temptations of
jobbery: "The usual remark in such cases that, 'after all, scientific men are
only human,' is not sufficient excuse for any man whose first duty is to be
a good citizen."[36] Government scientists seemed especially attracted to
the idea of moral equivalence. In 1924, a U.S. Forest Service scientist remarked, almost as an aside, that "it is well to get away from the notion that
the research man must be a genius or an abnormal type. With occasional
exceptions he is merely an average person whose training and experience
have fitted him for research and perhaps have rendered him unfit for other
vocations." If you wanted good research men, then you had to pay them
and provide for them, and their families, the "social life" and "advantages"
to which they felt "entitled": "The research man is no longer a recluse
who shuts himself up in a laboratory and whose thoughts are centered
entirely on his own problems. He is made of the same clay and has much
the same needs and desires as other human beings. If he lives a life of self-
denial it is usually because of a small income rather than because of the
exactions of his calling." Even if he was willing to make personal sacrifices
for science, one could not expect him to sacrifice his family's interests. In
any case, if you gave research men a normal range of rewards, you will
make him "a better scientist and a better citizen."[37] Better pay buys both
better morals and better knowledge.

This sort of sentiment is not difficult to find in American cultural commentary from the 1880s to the 1930s. There was a set of persistent problems

that presented themselves in particularly acute form for American scientists in the period: first, how to assert both their utility to the mundane affairs of State and commerce; second, how to mobilize reasons why scientific study belonged to institutions of higher education concerned with both utility and self-cultivation; third, how to identify those special conditions of disengagement and relative autonomy that permitted the goose of pure science to lay its golden eggs; and, fourth, how to explain, and, if necessary, explain away, instances of bad behavior among scientists, especially such behavior as conflicted with ideal-typifications of what kind of superior people they were. Scientists were to be acknowledged as good citizens, their studies forming a desirable civic character. And yet, at the same time, they required a degree of freedom from the normal structures of accountability that bore upon other citizens and other recipients of public largesse.[38] These problems often pointed to differing portrayals of what sort of person the scientist was. Certainly, the scientist could be celebrated as a normal, healthy member of democratic society, sharing its typical motives and rewarded by the same sorts of things as any other good citizen. This portrayal was increasingly common in the period, especially popular among those sectors of the American scientific community wishing for further social support and further integration into the institutions of the State, education, and commerce. But it was never unopposed by very different portrayals. Purist scientists might continue to insist on moral and motivational distinction; critics of America's utilitarian culture, and of the place of expertise in corporate capitalism, might identify the moral ordinariness of the scientist in the course of cultural and social criticism.

For both lay and scientific Americans between the wars Sinclair Lewis's novel *Arrowsmith* (1925) represented the single most gripping, influential, and authoritative picture of what the scientist was like.[39] Generations of American scientists traced their conceptions of scientific research and their vocation for science to their youthful reading of *Arrowsmith*.[40] In fact, there was not one character type, but several conflicting ones, in the novel, as there were conflicting conceptions of the nature of the scientific calling. Martin Arrowsmith was the young man who might, or might not, have had a genuine calling for scientific research, but who was torn between pure science and medical practice, between an inquiry into the unknown and immediate service to humankind, between passionate asceticism and the bland pleasures of making a middle-American, middle-class sort of living. His various models included Rotarian Midwestern community physicians, pillars of the local community; romantically dashing public health doctors; and, decisively, the cosmopolitan pure scientist, Max Gottlieb. Between Gottlieb's charisma and the encouragement of Martin's women, Arrowsmith ultimately did decide he had a genuine vocation for research, and, in a final gesture rejecting the pull of society and its rewards, he separated himself from his metropolitan research institute and established his own laboratory in the equivalent of Thoreau's Walden.

However, while many readers drew from *Arrowsmith* an understanding of the scientist as morally special, others recognized Martin Arrowsmith's all-too-humanness, and, for them, the lesson was that the scientist-as-he-actually-was came equipped with the full range of human emotions and motives. Even the saintly Max Gottlieb, a German-Jewish emigré (hence Lewis's attempt at rendering a funny foreign accent) who had a family to support, had a brief fling as industrial scientist working for an ethical drug firm, at least initially purring with contentment at the salary, the assistants, and the instrumental resources: "This night, as he knelt, with the wrinkles softening in his drawn face, he meditated, 'I was asinine that I should ever scold the commercialists! This salesman fellow, he has his feet on the ground. How much more aut'entic the worst counter-jumper than frightened professors! Fine dieners [technicians]! Freedom! No teaching of imbeciles! Du Heiliger!'"[41] The *New Republic*'s review of *Arrowsmith* noted that "today [the scientist] sits in the seats of the mighty. He is the president of great universities, the chairman of semi-official governmental agencies, the trusted adviser of states and even corporations."[42] And *Science*'s review admired the novel's unprecedented *realism*: "The High Priests have taken off their false whiskers and given Mr. Average Citizen a peep at the ceremonies going on inside the Temples." All scientists, it said, will recognize "the wavering allegiance between Truth and Mammon." This tension is the "common property" of scientific community.[43] *Arrowsmith* was uniquely influential in distributing this new understanding of the scientist's nature, but other voices in the 1920s concurred. Popularizing the new physics, the chemist and Republican science writer Edwin Slosson said that the scientific revolutionaries were in fact as "clean-shaven, as youthful, and as jazzy as a foregathering of Rotarians . . . It must be admitted that the scientist of today is fully as much a man of the world as his brother, the businessman." The modern scientist yields nothing to the businessman in this-worldliness, for "it is [the scientist], indeed, who made the jazz age practicable; it was his researches into the properties of matter that gave us the automobile, the radio, and, one might add, the saxophone."[44]

For all its utility in projecting an understanding of the scientist's ordinariness, *Arrowsmith* also offered a vivid account of his identity that rejected the morally normal, utilitarian, and Rotarian portrayal. American culture in the Gilded Age and Progressive Era responded as much to the one as to the other. If Dr. Martin Arrowsmith represented a man struggling to know whether or not he had a vocation for research, his mentor Max Gottlieb stood for the pure, traditional ideal. A passionate atheist, Gottlieb translocated religious passions and motivational vocabulary into a materialistic scientific domain. His scientist's "prayer," which Martin learned from his master, formed the centerpiece of the novel's picture of professional vocation:

> God give me unclouded eyes and freedom from haste. God give me a quiet and relentless anger against all pretense and all pretentious work and all work left slack and unfinished. God give me a restlessness whereby I may neither sleep nor accept praise till my observed results equal my calculated results or in pious glee I discover and assault my error. God give me strength not to trust to God![45]

The crucial monologue by Max Gottlieb (= "Greatest God-Love") is a vigorous insistence on the *absence* of any moral or motivational equivalence between the scientist and the Rotarian:

> To be a scientist—it is not just a different job, so that a man should choose between being a scientist and being an explorer or a bond-salesman or a physician or a king or a farmer. It is a tangle of ver-y obscure emotions, like mysticism, or wanting to write poetry; it makes its victim all different from the good normal man. The normal man, he does not care much what he does except that he should eat and sleep and make love. But the scientist is intensely religious—he is so religious that he will not accept quarter-truths, because they are an insult to his faith . . . To be a scientist is like being a Goethe: it is born in you.[46]

Just months after the publication of *Arrowsmith*, and in the context of the Scopes evolution trial in Tennessee, a New York newspaper defiantly rejected as an "impudent conceit" the claims of such Christian fundamentalists as William Jennings Bryan that they had a monopoly of virtue and morality: "To contribute successfully to the progress of science requires more integrity of mind, more purity of heart, more unselfishness, more devotion, more unworldliness, than any other kind of human activity." There are unworthy scientists, but "among the men who are really doing the work of science a moral code exists and is followed which would put the rest of us to shame. The search for truth." That is the purpose for which God made reason.[47]

The saintly Gottlieb entered an American culture entirely familiar with the scientist-as-moral-hero. The manly, modest, and Truth-questing scientist of Henry Rowland's story was still in circulation. In 1922, the physicist and inventor Michael Pupin claimed that "the aims and aspirations and the life of American scientists have not changed" in the past half-century. It is "a life of saints and not of ordinary materialistic clay. Such a life cannot be attained without unceasing nursing of the spirit and unrelenting suppression of the flesh." Pupin was not alone in regarding scientists "as a new species of the human race."[48] The *New York Times* celebrated Robert Millikan's "fine and modest personality," and Millikan himself wrote about the scientist's virtues of "modesty, simplicity, straightforwardness, objectiveness, industry, honesty, human sympathy, altruism, reverence and a keen sense of social responsibility."[49] And a confident and eloquent assertion of the scientist's moral heroism came from one of America's leaders in industrial research, when Arthur D. Little not only celebrated the virtues of the scientist, but also produced one of the least qualified, and most confident, statements of the theological uses of science to be found in early twentieth-century America: "Theirs is a true vocation, a calling and election. It brings intellectual satisfactions more precious than fine gold. They live in a world where common things assume a beauty and a meaning veiled from other eyes; a world where revelation follows skillful questioning . . . The laboratory may be a temple as truly as the church."[50] By the 1930s, Hollywood was joining in the celebration of scientists' moral heroism: first, Ronald Colman in a mediocre John Ford realization of *Arrowsmith*, then Paul Muni's 1936 portrayal of Louis Pasteur, and in 1940 Edward G. Robinson as the selfless, Truth-driven, and humanitarian Paul Ehrlich.[51]

Four years before *Arrowsmith* appeared, America had become besotted with another cosmopolitan German Jew, the saintly Albert Einstein, who first visited the States in the spring of 1921, the same year in which he was awarded the Nobel Prize. As Kevles observes, "In the 1920s, by associating even the most abstruse theorist with the good works of technology and business, the popularizers helped make the pure scientist as such a highly respected figure of the decade. So, in his own special way, did Albert Einstein." Yet the American obsession with Einstein did not focus on the utility of his work or the ordinariness of his motives. Quite the opposite, for Einstein came to represent a pure form of ideal disengagement, humility, gentleness, and peaceableness. The *New York Times* celebrated his arrival by calling him "a poet in science" and describing his physically visible otherworldliness: "Under a high, broad forehead are large and luminous eyes, almost childlike in their simplicity and unworldliness."[52] Einstein represented the scientist as holy man. Kevles notes that American journalists avidly covering his doings "fastened upon Einstein's unassuming manner, the innocent timidity with which he would greet a drove of reporters . . . Merely by being in the same profession as Einstein, every physicist wore a halo of humility."[53] In August 1939, Einstein was persuaded by Leo Szilard, Eugene Wigner, and Edward Teller to sign a letter to President Roosevelt urging an immediate and massive effort to construct an atomic bomb. Six years later, when Einstein was given the news about Hiroshima, his response was concise and it was typical of the man: "Oy vey."[54]

World War II and its aftermath brought about massive changes in the social and cultural realities of American science, in understandings of what science was and who the scientist was. These changes were matters of degree, but they occurred on such a scale that they appeared to participants, as they do to later commentators, to bring about a state of affairs that had no substantial historical precedent or ancestry. World War I was sometimes called "the chemists' war," and World War II "the physicists' war," but, so far as the scale on which all sorts of science was mobilized and the cultural consequences for science are concerned, the neat symmetry ends with the rhetorical form.[55] Midway between the two wars, American physical scientists could plausibly say that the role of science in the 1914–1918 conflict was to have rendered war so destructive as now to be unthinkable. In 1930, the *New York Times* quoted Robert Millikan's reassurance that "science has . . . helped beyond all other agencies to make it the last war," and, as to those warning of the destructive potential of "sub-atomic energy, it is improbable that there is any appreciable amount that will ever be available for blowing-up purposes, however savage and selfish man might grow to be."[56]

"BLOOD ON THEIR HANDS"

The mobilization of American science during the Second World War—especially in the Manhattan Project and in the construction of radar, but spreading across much of the scientific landscape—propelled a generation of academic scientists into a world that was largely unfamiliar to them: the experience of large-scale organization; of teamwork; of interdisciplinary project-oriented research; of unlimited resources and severely limited time; of close contact with the sorts of people—especially the military and the commercial worlds—they had not known much about;

and, after the end of the war and the beginning of the Cold War, the experience—for some of them—of political power. During the war itself, mobilized scientists were generally too busy to reflect on what was happening, and, if they were not too busy, security considerations prevented any such public reflections. Subsequently, they struggled to make sense of their experiences. Some felt badly about what they had done; others said they experienced no guilt whatever. But all American scientists now enjoyed the fruits of wartime military labors in the form of vastly increased governmental and industrial funds; enhanced access to what C. P. Snow came to call the "corridors of power"; a hugely expanded job-market for academic, industrial, and government scientists; and heightened public respect for scientists' power. Scientists had never before possessed such authority, largesse, civic responsibility, and obligations. By free choice or not, some scientists now lived the *vita activa*, and, while there were still consequential worries about the extent to which they were indeed "normal citizens," they had never been more integrated into the civic sphere.

Scientists themselves expressed awareness of the watershed they had just traversed. J. Robert Oppenheimer, who was adamant in his insistence that the bomb be used on live targets, famously said that the physicists had now "known sin" and told an unimpressed President Truman that he felt that he had "blood on his hands."[57] The *Bulletin of the Atomic Scientists* swiftly provided a forum in which Manhattan Project scientists, and others concerned about the new realities, passionately debated the moral constitution and moral responsibilities of the scientist.[58] Scientists not involved in military work nevertheless said that Hiroshima had stained them all. The biochemist Erwin Chargaff, for example, wrote that the belief that the scientific profession as a whole "was a noble one . . . was certainly shattered in 1945."[59] If, as an atomic scientist, you felt guilt over Hiroshima, or even if you thought that others might wrongly insist on scientists' guilt, then moral equivalence might count as exculpation. Just after the publication of *The Two Cultures*, in which C. P. Snow so favorably contrasted the progressive and humanitarian scientist with the reactionary "literary intellectual," Snow insisted upon moral equivalence—or, as he put it, a "just perceptible" moral superiority—as a way of showing the public that there was no reason to be scared of scientists. "The rest of the world . . . is frightened of the scientists themselves and tends to think of them as radically different from other men. As an ex-scientist . . . I know that is nonsense." Scientists were "certainly" not morally or temperamentally "*worse* than other men."[60] And in 1956 Lee DuBridge, writing as president of Caltech, and concerned about the Cold War shortage of scientists and engineers, expressed anxiety about high school students' "nonsensical" belief that "'scientists are just technicians and makers of terrible weapons.'" Perhaps, DuBridge suggested, scientists should visit schools to show that they were ordinary human beings.[61] A piece in the *Bulletin of the Atomic Scientists* recovering the historical origins of the word "scientist" insisted that it "does not comprise ... a species, but includes such a variety that soon 'scientists' begin to appear as ordinary men." Properly speaking, "there is no such animal. There are only men, varying greatly in the extent to which they apply scientific method."[62]

Even those scientists who were *not* morally queasy about Los Alamos and Hiroshima—and there were very many who came to terms easily and enthusiastically with militarized science and its destructive uses— nevertheless recognized the changes the war had brought about both in scientists' quotidian conditions of existence and in their personal constitutions. Just after the bombs were dropped on Japan, a crude poem circulating in the University of Wisconsin's physics department noted with satisfaction the change these events had worked on the physicist's identity:

> The college professor, you think, is a dreamer,
> But see the shellacking he gave Hiroshima.
> The people that thought these fellows were wacky
> Were lucky they didn't live near Nagasaki.[63]

One of the "fathers" of the hydrogen bomb, the Los Alamos mathematician Stanislaw Ulam, wrote that Ivory Tower physicists "got their heads turned with the sudden realization of not only the practical but worldwide historical importance of their work—not to mention the more trivial but obvious matter of the enormous sums of money and physical facilities that surpassed anything in their previous experience." Perhaps, Ulam perceptively suggested, "this played a role in the personality change of some principals; with Oppenheimer, the director, it may have had a bearing on his subsequent activities, career, ideas, and role as a universal sage."[64] The mathematician, and mighty Cold Warrior, John von Neumann, became for many colleagues a paragon of the novel scientific character that Los Alamos and the Rad Lab had issued into existence. "Some of von Neumann's scientific admirers," Steve Heims writes, "have seen in him the 'new man,' the ideal type of future person, implied by his name."[65] At the most specific level, the actual building of the bomb projected new understandings of what kind of persons scientists were. Los Alamos itself—uneasily poised between scientific and military conceptions of its identity—was the largest technoscientific project that the world had ever seen, and its construction and operation were informed at every point by notions of what scientists were like and how they functioned. The military powers worried about what they took to be scientists' tendency towards both uncontrolled individualism and unconstrained collegiality. How could secrecy be maintained if scientists were more dedicated to Truth than to victory, more to the international Republic of Science than to the security of the United States of America, if they were—as sociologist Alvin Gouldner later put it—"cosmopolitans" rather than "locals"?[66] And would their much-talked-of moral sensibilities make them balk at either the construction or the use of weapons of mass destruction?[67]

The Cold War heightened and focused those concerns. The atomic scientists' supposed moralism, their internationalism, their reputation for political radicalism, their awkwardness in bureaucratic structures, and possibly the proportion of Jews among them as well, had worried the military and the security services during World War II, and those worries soon translated into the invigilating apparatus of Cold War America.[68] In 1948, Edward U. Condon, director of the National Bureau of Standards, expressed anxiety about what the emerging National Security State was thinking about the character of its scientists. Security considerations seemed to *presume* disloyalty in scientists: "There are those who seem to start with the assumption that a scientist is a peculiarly unstable sort of fellow with no sense of responsibility or capacity for living according to the rules. They seem to start from the false assumption that he is guilty of incapacity in this direction unless he can prove himself innocent."[69] It was true, Condon later wrote, that scientists had a tendency towards independent thinking and that this was widely interpreted as "just a kind of unruliness or bad-boy-ism" that had to be tolerated "in these eccentric fellows because they are the geese that lay the hydrogen bombs as well as many other great and good things," but it had to be appreciated that such independence was *not* a social pathology and was fundamental to the scientific life.[70] Similarly, the University of Pennsylvania botanist Conway Zirkle conceded "much adverse publicity" about scientists' collective disloyalty, but insisted that "the number of good scientists in the free world who are Communists or who follow the Communist line in science can be counted, perhaps on the fingers of one hand." This only makes sense, since scientists "do not like to be disciplined," and their independent natures mean that "the party would never trust honest scientists too far."[71] Although Condon insisted that the atomic scientists as a whole were distinguished for their *exceptional* discretion and loyalty, nevertheless his "main point" was that "scientists are not deserving of, nor should they get, any *better* treatment than the rest of our citizenry" with respect to security vetting. The proper presumption was that scientists were just as loyal as the average citizen—no more and no less.[72] Writing against the emerging baroque culture of secrecy, screening, and loyalty oaths, the Columbia University law professor Walter Gellhorn thought that "it is not too much to say that loyalty of scientists as a group has become a matter about which there is wide public concern." This was, however, a libel against a whole category of persons and there was no reason to presume that scientists were collectively less loyal than any other category of Americans.[73] Several years later, the director of Monsanto's atomic energy work was asked to account for America's worrying failure to attract enough talented young people in science and engineering. He speculated "that there has been so much talk about Scientist X in the headlines that there is a general feeling among the public that scientists and spies are practically the same thing; so again you turn the other way in some safe direction," to the study of English or the social sciences.[74] In the aftermath of the 1954 Oppenheimer hearings, Vannevar Bush worried that the witch-hunt against scientific traitors would make "young men hesitate to enter the scientific professions," so doing immense harm to America's ability to counter the Soviet threat.[75] At the same time, the American Association of Scientific Workers, resisting the imposition of loyalty oaths on scientists, complained of the pervasive tendency "to see in a scientist a 'potential atom spy.'"[76]

The class libel was damaging—to scientists and to the nation's security—and it needed to be refuted. Scientists' supposed internationalism, and the disloyalty flowing from it, figured consequentially in postwar Congressional hearings about how science should be supported and managed. In 1954, the astronomer Harlow Shapley told Congress that scientists "call ourselves American by citizenship, but our blood is cosmopolitan. The scientists should, as rapidly as possible, call themselves citizens of the world and not the citizens of individual countries."[77] Ten years later, *Life* magazine's group portrait, *The Scientist*, noted that the "very vocabulary of the scientist is international" and that "scientists have suffered considerably for their cosmopolitan convictions."[78] Some politicians accepted scientific cosmopolitanism as a matter of fact; others were alarmed and saw it as justification for a still more intrusive security apparatus. Lewis

M. Terman, a psychologist whose work was funded by the Office of Naval Research asked in 1955 "Are Scientists Different?" The nation needed to know the answer to that question because of the desperate scientist shortage and because of tensions building up between members of the scientific community and the government that required their services so urgently: "The scientist is looked upon by many as an object of suspicion, and he in turn is irked by the distrust he senses and by the restrictions government work imposes on him." There was *some* empirical support for the common imputation of scientific unsociability: "Nevertheless one must guard against overgeneralization. Actually all degrees of social adjustment and social understanding are found within each of the [occupational groups studied]. Everyone knows that some scientists are extremely adept in social perception and in social relations—sufficiently adept to become deans, college presidents or other administrative officials."[79] Much of this anxiety, and the accompanying efforts at normalizing the scientist, was occasioned by the exposure of atomic spies and by the Oppenheimer case. Oppenheimer presented himself as the embodied assemblage of such anxieties, and the Oppenheimer security hearings of 1954 provided a forum in which many of these worries were implicated, even if, for pragmatic reasons, they had to be expressed with great circumspection. While Oppenheimer had his security clearance withdrawn, the Gray Board nevertheless did not want it to be thought by the scientific community, or by the public, that scientists as a body were any more disloyal than anyone else. The country needed them, and scientists were to understand that *they* needed a secure America to do their work. Scientists, as scientists, could not avoid their civic responsibilities to the Cold War Order.

So the Gray Board's *Final Report* announced that "we know that scientists, with their unusual talents, are loyal citizens, and, for every pertinent purpose, normal human beings." It was probably meant as a warning as much as a description. Because of the new dependence of the State on science, the board rightly used the language of mutual obligation: "We must believe that they, the young and the old and all between, will understand that a responsible Government must make responsible decisions. If scientists should believe that such a decision in Government, however distasteful with respect to an individual, must be applicable to his whole profession, they misapprehend their own duties and obligations as citizens."[80] The State would continue to enlist scientific expertise, and to grant considerable autonomy and vast resources to those experts, but only on the condition that experts left whatever moral and political preferences they might have outside the doors to the corridors of power.

Better yet, they should learn that they had no entitlement to expressions of special moral authority or political judgment. Should they insist on such authority, they risked losing access to the enormous material support that was being offered to scientists in the postwar decades. After the Oppenheimer hearings, the board reminded the American scientific community of the boundaries that democratic society placed around technical expertise:

> A question can properly be raised about advice of specialists relating to moral, military and political issues, under circumstances which lend such advice an undue and in some cases decisive weight. Caution must be expressed with respect to judgments which go beyond areas of special and particular competence.[81]

Shortly afterwards, a commentator noted, "They did not care what [Oppenheimer's] moral scruples were. It was the fact that he had any at all which was derogatory."[82] Setting himself against increasingly popular postwar sensibilities that a scientifically reconstituted, and elaborately funded, sociology could cure social and political ills, Oppenheimer cautioned that "there are formidable differences between the problems of science and those of practice. The method of science cannot be directly adapted to the solution of problems in politics and in man's spiritual life."[83] But even Oppenheimer's self-denying ordinance did not go far enough for his critics. The is/ought distinction was to be institutionalized in the modern American scientific role as a condition of that role's political legitimacy.

THE LIMITS OF EXPERTISE

It was a contract that appealed to both parties: the political powers got expertise-on-tap without interference in their prerogatives; the scientists got money and a reconfigured, but still worthwhile, version of autonomy. It was that sort of stipulation that encouraged Foucault to identify Oppenheimer as the model of the "specific intellectual"—expertise stripped of special virtue, and, ideally, of the capacity for universal spokesmanship.[84] But that production of de-moralized and de-politicized expertise was a collective project: the powers required it and many technical experts enthusiastically collaborated in bringing it about. Writing after Hiroshima about scientists' and engineers' restricted sense of social responsibility, Robert Merton observed that it had "required an atomic bomb to shake many scientists loose from this tenaciously held doctrine," but he did not note how that same event pushed many other scientists into a vigorous disavowal of any such responsibility.[85] The scientist who did more than anyone to destroy Oppenheimer's career as an expert-on-tap for the National Security State argued that it was not an act of morality, but its opposite, for scientists in a democratic society to presume to take "ought" decisions into their hands. They had neither the moral capacity nor the political right. So, in the context of highly charged debates over the construction of a thermonuclear weapon, Edward Teller argued that

> the scientist is not responsible for the laws of nature. It is his job to find out how these laws operate. It is the scientist's job to find the ways in which these laws can serve the human will. However, it is *not* the scientist's job to determine whether a hydrogen bomb should be constructed, whether it should be used, or how it should be used. This responsibility rests with the American people and with their chosen representatives.[86]

Such sentiments did not belong solely to the American political right. On the opposite side of the political spectrum, the physicist, and liberal activist, Ralph Lapp thought it important to tell the public that "the society of scientists embraces a wide spectrum of personalities as does any professional grouping . . . Scientists as a group probably have no better sense of human values than any other group . . . To say that science seeks the truth does not endow scientists as a group with special wisdom of what is good for society." Moreover, the modern *disunity* of science—the lack of any core Method or concepts—was a further reason why one could not expect from scientists any coherent moral deliverances.[87] J. B. Conant returned to the presidency of Harvard from his war work with a commitment to giving the educated American public a more realistic sense of "the tactics and strategy of science"; it was in connection with these curricular initiatives that he wrote: "My own observations lead me to conclude that as human beings scientific investigators are statistically distributed over the whole spectrum of human folly and wisdom much as other men." Specialized scientific traditions, technical instrumentation, and, especially, the "given social environment" of a scientific life combined to ensure that "even an emotionally unstable person" might be "exact and impartial in his laboratory." But such a person possesses neither competence nor legitimate authority "once he closes the laboratory door behind him."[88] So those who reckoned that there was such a coherent and stable thing as "science" and its unique "Method" as well as those who thought such notions were nonsense *both* had occasion to specify the moral irrelevance of science and the moral ordinariness of its practitioners.

Anxieties that postwar, newly powerful scientists might constitute a "new priesthood," or "new Brahmins," and that science might be wrongly regarded as "a sacred cow" were widespread within the American scientific community. Some politicians worried about scientists as potentially disloyal and diabolical, but others continued to project an image of scientists as priest-magicians, and, while scientists enjoyed the benefits of that image, they were—rightly, as it turned out—concerned about the attendant expectations. An essay in the *Bulletin of the Atomic Scientists* about immediate postwar Congressional engagements with science noted that, after Hiroshima, "scientists became charismatic figures of a new era, if not a new world, in which science was the new religion and scientists the new prophets . . . Scientists appeared to [politicians] as superior beings who had gone far ahead of the rest of the human race in knowledge and power . . . Congressmen perceived scientists as being in touch with a supernatural world of mysterious and awesome forces whose terrible power they alone could control. Their exclusive knowledge set scientists apart and made them tower far above other men."[89] Late in 1945, the atomic physicist I. I. Rabi complained to Senator J. William Fulbright that scientists were being treated as a different class of human being, with different capacities and virtues, from whom the public had to be protected, and Fulbright agreed: "The reason for that is that you scientists scared us all to death with your atomic bomb and we are still very frightened about it."[90]

These "frightening" aspects of postwar science, together with the rhetoric of moral neutrality, were identified by some scientists as a cause of contemporary "antiscientific trends." "The climate of public opinion has changed," a Harvard geographer wrote, "from one in which scientists could bask in the sunshine of widespread admiration, respect, and even awe, to one in which storm clouds of suspicion, recrimination, and fear endanger" scientific progress. The root suspicion was that science was not only amoral in itself but that it was "largely responsible for the abandonment of moral principles and the destruction of ethical standards which have undoubtedly occurred in recent years." There was not a lot that the scientist—qua scientist—could do about this, and the answer was to find a storehouse of value in a domain that "transcends science."[91] An editor of *Harper's Magazine* agreed. Positioning himself as a nonscientist friend of science, disturbed by the anti-scientific tendencies of McCarthyite America, Eric Larrabee wrote that "the 'mad scientist' who is so pervasive a figure of modern folklore is not entirely the product of envy and ignorance. There is justice—poetic justice, if you like—in the popular view of the archetypical scientist as a warped and incomplete being, a man who has isolated one component of the universal experience and cultivated it to the exclusion of all others." The solution was to project an appreciation of non-arrogant science as just one form of truth among others, humanizing and normalizing the making of scientific knowledge. "I am relatively undisturbed," Larrabee concluded, "at the image of a world in which scientists would be indistinguishable from people, in which scientists would be men and women first and scientists second . . . The human condition is crowded with ambiguities, and all our acts have unintended consequences."[92] Writing during the Korean War, a botanist expressed anxiety about current urgings to impose a "moratorium" on further scientific progress until social values and wisdom—possibly aided by advances in academic social science—caught up: "Scientists are often charged with being sociologically irresponsible. They are criticized for giving society new knowledge and tools without guaranteeing that society will use them wisely. The charge is true, but the criticism is unfair." Society expects too much of scientists: it expects them to be simultaneously "investigators, inventors, social pastors, and spiritual guides." But they aren't: "They are citizens"; they "accept the morals of the society of which they are a part"; and "most of them do what is required of citizens in times of national emergency." It is the humanists who are in charge of "values"; let them take care of that side of things. Scientists are neither constituted nor equipped to do so.[93]

Just after the war, the chemist Anthony Standen's *Science Is a Sacred Cow* roundly condemned external "moralizers" and "evangelists" who were perpetuating an image of science as more mysterious than it was and scientists as more wonderful than they were. These moralizers spoke, as they thought, *for* science but not from *within* science. Real scientists knew better. True, scientific research was once "carried out by men of exceptional intelligence." But now "scientists are turned out by mass production in our universities, and they therefore include men of very ordinary, even mediocre, intellectual powers . . . We are having the wool pulled over our eyes if we let ourselves be convinced that scientists, taken as a group, are anything special in the way of brains. They are very ordinary professional men, and all they know is their own trade, just like all other professional men." The much-trumpeted tolerance and liberalism of the scientist was, again, just so much myth-making by people outside of science who didn't know what they were talking about: "[Scientists] easily absorb the preju-dices of those around them, and many of them are mildly reactionary, and have mild class feelings and race bias, in an unthinking sort of way."[94] At the outbreak of the Korean War, the scientific community was put in the awkward position of arguing for draft deferments for its young men—on grounds of vital national security concerns—while defending them-selves from the damaging charge of undemocratic elitism. Scientists, a spokesman argued, "expressly do not wish to be thought of as an elite corps above the 'common herd.'"[95] While worrying about declining num-bers of natural science graduates over the course of the Korean War, NSF director Alan T. Waterman stressed that "this does not mean that scien-tists should have special privileges and special treatment; nothing could be further from the thoughts of scientists, for the whole process of science is essentially and necessarily democratic."[96]

The critics and the guilt-wracked have had more than their share of historians' attention. They tended to be the better rhetoricians; their careers were often marked by drama, tragedy, or pathos; and their sen-timents appealed to the academy's humanists and social scientists, so many of whom were desperate to maintain alliances with that fraction of the increasingly prestigious natural scientific community who resisted abandoning the academic commons and the world of what Americans like to call "values." But, despite Edward Teller's worries about his col-leagues' misplaced moralism, and a resulting emptying-out of the postwar weapons laboratories, the Cold War State never had the slightest dif-ficulty finding suitable scientific talent to staff Los Alamos, Livermore, and the chemical and biological weapons installations. Several of the more influential, if less rhetorically gifted, atomic scientists enthusias-tically endorsed weapons work, either because they agreed with Teller that scientists had no moral right to refuse their democratically elected government's legitimate wishes, because they found the work interesting and well-paid, or because they considered the nation under imminent threat from Godless Communism, against which winning the arms race was the only conceivable defense. America's scientific left did not mo-nopolize moral arguments about, for example, whether or not scientists should build weapons of mass destruction.[97]

Yet even within the pervasive drift towards endorsing moral equiv-alence, some atomic scientists, and some of their cultural allies, wanted to rescue a morally special *something*. That residual moralism cannot be neglected, even while recognizing how modest it was and how close it came to the nullity of moral equivalence. In the mid-1950s, the influential English popularizer Jacob Bronowski circulated a strongly felt, but still limited, account of residual virtue, attached to all who lived the life of the mind and only in a somewhat stronger form to the life of science: "By the worldly standards of public life, all scholars in their work are of course oddly virtuous. They do not make wild claims, they do not cheat, they do not try to persuade at any cost, they appeal neither to prejudice nor to authority, they are often frank about their ignorance, their disputes are fairly decorous, they do not confuse what is being argued with race, politics, sex or age, they listen patiently to the young and to the old who both know everything. These are the general virtues of scholarship, and they are peculiarly the virtues of science."[98] The English physicist Edward Appleton humbly suggested that the scientific life, after all, had a certain capacity to inculcate humble virtues. He spoke about "the kinds of mental qualities and awareness that science requires in its followers." And he suggested "that the exercise of these skills has a value in itself which is ample justification of a scientific vocation. To go further might be claiming too much." In fact, Appleton was one of the last eminent scientists to insist on the largely lost sense of what it was to live the scientific life: "Our vocation can never be simply an occupation; it is, by its very nature, more than that—a dedication to an end."[99] More significantly for the Cold War American setting, just before his death, Oppenheimer drew a measured but sharp distinction between the vast technical knowledge that scientists possessed and the moral and political programs of action in which scientific knowledge was increasingly enlisted. Here again, Oppenheimer started by conceding the practical limitation of the is/ought distinction:

> Among the things of which we cannot talk without some ambiguity, and in which the objective structure of the sciences will play what is often a very minor part, but sometimes an essential one, are many questions which are not private, which are common questions, and public ones: the arts, the good life, the good society. There is to my view no reason why we [scientists] should come to these with a greater consensus or a greater sense of valid relevant experience than any other profession.

Expertise was not fungible, and it was a great, but common, mistake to suppose it was. It was important for the culture to understand that expertise did not transfer automatically, or even easily, from one domain to another. Only if the life of scientific inquiry was a good life could scientists even participate in moral debates.[100]

It was a message Oppenheimer had been preaching since his Fall from Political Grace in 1954, and it was consonant both with an intense inner humanism and with the embrace of de-moralized expertise forced on him by the security hearings: "Science," Oppenheimer modestly insisted, "is not all of the life of reason; it is a part of it." A life in science did indeed inspire some homely virtues—a degree of humility, tolerance, and integrity—but that is all: no moral heroes here. And so, Oppenheimer concluded, "In this field quite ordinary men, using what are in the last analysis only the tools which are generally available in our society, manage to unfold for themselves and all others who wish to learn, the rich story of one aspect of the physical world, and of man's experience."[101] Writing in the same year as Oppenheimer's security hearing, the French theoretician-theologian Jacques Ellul agreed, possibly more enthusiastically than Oppenheimer appreciated:

> We are forced to conclude that our scientists are incapable of any but the emptiest platitudes when they stray from their specialties. It makes one think back on the collection of mediocrities accumulated by Einstein when he spoke of God, the state, peace, and the meaning of life... Even J. Robert Oppenheimer, who seems receptive to a general culture, is not outside this judgment. His political and social declarations, for example, scarcely go beyond the level of those of the man on the street.

There was no point in asking about such people's motives or about *why* they did what they did: "The attitude of the scientists, at any rate, is clear. Technique exists because it is technique. The golden age will be because it will be. Any other answer is superfluous."[102]

Even as the postwar American scientific community vigorously projected into the culture a picture of moral and, indeed, cognitive ordinariness, circumstances made it pressing that the scientific life be even more appealing than it had ever been. Both the Cold War State and industry recognized themselves to be in urgent need of a vastly expanded pool of scientific expertise. Even scientists uncomfortable with some aspects of government intervention in science knew how to make that case, as when the biophysicist Detlev Bronk, writing in defense of scientific freedom and spontaneity, noted that "scientists are required by the thousands for the training and operation of our armed forces."[103] The best way of securing that mass of expertise was to pay market rates for it—both through increased salaries and through the enhanced support of the institutions that recruited and trained such experts. But it was also felt necessary to combat what were taken to be still-prevalent understandings of scientists' character. So leading spokespersons of government, industry, and the universities took it upon themselves to specify the ordinariness of the scientist and, therefore, the attractiveness of the scientific career to those who felt themselves to be neither geniuses nor morally special. Recruits to a scientific career should know this; they will arguably make better and happier scientists if they are disabused of misleading myth. During the early stages of the Korean War remobilization, a participant in a conference on recruitment to scientific careers was seriously worried about image problems, though one wonders what books he was reading: "Every time a scientist is portrayed in literature he is a harebrained fellow who looks queer and might act that way. Do you think the high school kids want to work hard and end up like that?"[104]

The nuclear chemist Glenn T. Seaborg—occupying an important public platform as chancellor of the University of California at Berkeley from 1958 to 1961 and chairman of the Atomic Energy Commission under Presidents Kennedy, Johnson, and Nixon—frequently lectured young and lay audiences on the attractions of a life in science. No young people should be put off a career in science by the idea that they were not intelligent enough: "There is plenty of room in scientific research for those who are not in the genius category." Maybe it's just a matter of "plain hard work," or, in the Edisonian formulation of genius, "One part inspiration to nine parts perspiration."[105] Nor should anyone believe that there was a temperament, or a moral constitution, specific to science. In an essay titled "The Scientist as a Human Being," Seaborg wrote that "old legends die hard, but none have been more persistent than the belief that the scientist is something more, or perhaps something less, than a human being." Contrary to legend, "For the most part, the scientist tends to look more and more like the rest of the population."[106] Much "flowery nonsense" has been put out about the high-minded motives of the scientist: "Within their specialties their natural intellectual capacities are greater than the average man's and their trained competence is certainly greater, but as human beings they are subject to the same shortcomings, the same wants, desires and drives as anyone else."[107]

Shortages of natural scientists and engineers in the 1950s and 1960s translated into a Federally supported agenda for much of the social sciences. The National Manpower Council at Columbia University, set up in 1951 with Ford Foundation money and with the enthusiastic support of the university's then-president Dwight David Eisenhower, recommended

"that foundations and universities encourage and support research designed to increase our understanding of educational and career choice
processes [and] of the factors facilitating the development of talent and
intellectual ability," and, indeed, these were the topics that engaged the
attentions of many Cold War social scientists, including sociologists of
science and organizations.[108] The next chapter describes pertinent research on creativity, innovation, and the circumstances of the organized
scientist, but government and foundation funds also flowed to researchers
concerned with young people's "images" of the scientist and the conditions in which they could be either attracted to, or repelled by, the idea of
taking up a scientific career. At the height of the Cold War, concern about
shortages of scientists and engineers required to "deal with the forces of
international Communism" sometimes drew attention to images of the
scientist—as something more or less than a normal human being—held
by children and adolescents. And these images, it was felt, formed an
obstacle to recruitment. That is one reason why the 1950s and 1960s witnessed an upsurge of interest in documenting, and seeking to change, such
attitudes.

In 1957 and 1958, a survey supported by the Rockefeller Foundation
distressingly found that 40% of the American public deemed scientists to
be "odd and peculiar people."[109] The American Association for the Advancement of Science had already encouraged work on this subject by the
distinguished anthropologist Margaret Mead.[110] Mead and her colleague
Rhoda Métraux found that, while the body of scientific knowledge was
highly valued—"responsible for progress," "necessary for the defense of
the country," "responsible for preserving more lives"—the image of the
scientist held by American high school students was "overwhelmingly
negative." He was held to be dedicated and brilliant, but an asocial, un-
derpaid drudge, who either worked in isolation or, if for a company, mechanically doing "as he was told" by his superiors. His loyalty was suspect,
and "he may even sell secrets to the enemy." Mead suggested that the mass
media should be encouraged to present a more "realistic" image of the scientist, happily "working in groups." Such an image would aid in attracting
children into scientific careers and also in voters' willingness to approve
funds for needed scientific facilities.[111] In 1957, an NSF-funded psychological study of scientific career choice worried about the persistence of
a picture of the scientist as "a paragon." The antidotes to this off-putting
idea were "investigations which are more scientific in their methods,"
and, indeed, these showed that "the scientist and his fellows exhibit a
rather wide range of intellectual abilities." More pertinently, commonly

imputed moral and constitutional differences could be refuted by rigorous psychological investigations: "In his personality traits and patterns the scientist appears less of a paragon," though, unfortunately, there was some support for the pervasive idea that the scientist "is somewhat poorly adjusted socially."[112]

Just after Sputnik went up, *Time* magazine expressed unease that "the layman's image of science as a gray, austere calling, suited only to eccentrics"—what the physicist Luis Alvarez called "the Einstein complex"—was hindering the mass recruitment necessary to ward off the Soviet challenge. It should be much more widely appreciated, *Time* insisted, that science was a great and pleasurable "adventure," for this is what normalized the scientist as an admired American man: "Asked what he is doing, the scientist is likely to reply, disconcertingly, that he is having 'fun'—a word that recurs again and again, along with 'adventure' when scientists talk about their work."[113] *Life* magazine too was enlisted into the enterprise of publicly normalizing the scientist: its lavishly illustrated 1964 special, *The Scientist*, cited academic psychological studies of the scientist's personality, and started out with a chapter called "Hero—and Human Being," addressed to what were taken as persistent, and nationally dysfunctional, stereotypes about constitutional specialness. Pride of place was given to the comments of an Australian radio astronomer: scientists "are just like anybody else. They have all their failings. Some are dedicated, some sharp as a whip, others dull as dishwater. I've known some of the great names of science, men who have done tremendous good for the world, and while I've known no scientist who's been in jail, I've known some who richly deserve to be."[114] An advisor to the NSF lamented the effect on recruitment of what he saw as a proliferation of negative stereotypes. There was a "highly competitive market" for brains, and so, as "repugnant" as the terminology might be to scientists, "selling" the scientific career meant that the "product" must be "tailor[ed] . . . for maximal appeal." Specifically, the understanding of science as belonging to the "contemplative" life had to be adjusted, since "the whole trend of our society, the values of our society, are in fact away from the contemplative, away from respect for and concern with the complex, away from a sense of calling, of dedication, of single-minded purpose." The figure of "the classical scientist," here identified with David Riesman's "inner-directed" person, was marked by "unswerving and selfless devotion to the quest for knowledge."[115] But this was the just the stereotype that was responsible for serious recruitment problems. That same shortage-driven impulse that helped normalize the notion of the scientific occupation was also accompanied by early gestures at gender normalization. Lee DuBridge was one of several scientists and policy makers during the Cold War who lamented that the physical sciences and engineering had "almost completely failed" to enlist women. "Psychologists tell us that there is, statistically, no essential difference between the kind of mental aptitudes found in men and in women. Why are there not just as many female engineers as male, thus doubling our potential supply?"[116]

BIG SCIENCE, NORMAL SCIENTISTS

The desire of the Cold War State was to provide the resources and incentives to make American science big, and Big Science is what was soon achieved. Big Science had remarkably few apologists, just because it had so little *need* of apologetic defense. Only a few of the atomic physicists who enjoyed guilt-free Federal or industrial largesse felt moved to respond to what they took as a counter-current of pathological nostalgia for a pre-war Golden Age of purity. The Berkeley physicist Luis Alvarez, however, noted that he spoke from experience in suspecting that distance lent false enchantment: "Some physicists," he wrote in 1975, "still long for the 'good old days' in which a single experimenter did everything in his own laboratory and published the results by himself . . . But I am also old enough to remember that the good old days were really not all that good: they were inefficient and lonely and frustrating, and often molded physicists into characters that required only a minimum of imagination to turn into the 'mad scientists' of the horror movies."[117] Also on the political right, the physicist Eugene Wigner would not rise to the bait when asked whether he agreed that Big Science had "damaged" science. You could do great things with expensive equipment, and this was "wonderful for science," though he felt obliged to admit that "the spirit of science has changed . . . The monastic spirit of science had an attraction for those of us who chose science as a monastic occupation . . . We were not interested in power."[118] But in the main scientists enjoying the fruits of Big Science found few *occasions* to defend it: if they were not too busy, they may well have been unfamiliar with the cultural resources that would have allowed them to offer such a reflective defense. The scientist was being morally and institutionally normalized—by default as much as by design.

The left wing of American academia, including what there was of a scientific left, produced the most vigorous criticisms of Cold War Big Science and its de-moralizing effects. But none was as eloquent as the criticism that emerged from the very heart of the State power: President Dwight D. Eisenhower's farewell address on 17 January 1961. The old soldier famously identified the corrupting effects on the American political system of the Cold War "military-industrial complex," but he also, less memorably, pointed out parallel sources for the corruption of science. Scientific research had become central to the military build-up and also much more dependent than it had ever been on the patronage and direction of the Federal government. The motive of intellectual curiosity had been substantially replaced by serving the State and securing the costly equipment that could be provided only by State funds:

> The free university, historically the fountainhead of free ideas and scientific discovery, has experienced a revolution in the conduct of research. Partly because of the huge costs involved, a government contract becomes virtually a substitute for intellectual curiosity. For every old blackboard there are now hundreds of new electronic computers. The prospect of domination of the nation's scholars by Federal employment, project allocations, and the power of money is ever present—-and is gravely to be regarded.

Nor was the scientist merely the innocent captive of State concerns, and Eisenhower acknowledged the growing political power of scientific expertise in encouraging what Jürgen Habermas later called "technocracy," the scientific prestructuring and coopting of the political process: "In holding scientific research and discovery in respect, as we should, we must also be alert to the equal and opposite danger that public policy could itself become the captive of a scientific-technological elite."[119]

Only a few months later, the physicist Alvin Weinberg, then director of the Oak Ridge National Laboratory—one of the biggest technoscientific installations in the country—responded directly to Eisenhower's sentiments by giving a name to the current pathological condition: "Is Big Science," he asked, "Ruining Science?" Weinberg thought that it was: "Whenever science is fed by too *much* money, it becomes fat and lazy." There was now "evidence of scientists spending money instead of thought." Money was easier to come by than lovely ideas, so elegance, imagination, and philosophical depth had given way to a brute force approach to problem-solving. The scientists were behaving like administrators, and many of them were, indeed, becoming administrators.[120] Within a few years, the left-activist physicist Ralph Lapp was vigorously agreeing: "Big science is not necessarily great science," and there were reasons to think that the "force-feeding of research may inhibit its sense of vitality."[121]

Eisenhower, Weinberg, and Lapp had only given rhetorical form to sentiments within American science that had been articulated from the end of the war and even before. The routinization of science had followed from its commercialization and organization. And, in the course of these changes, the character of the scientist had fundamentally altered. It was commonly said that science had simultaneously lost its intellectual and political integrity. Only a few years after the end of the war, the mathematician Norbert Wiener lamented "the degradation of the scientist as an independent worker and thinker to that of morally irresponsible stooge in a science-factory."[122] Addressing the Philosophy of Science Association at Columbia in 1949, he called on American scientists to resist becoming "the milk cows of power."[123] As chapter 6 will show, the transition from individualism to organization seemed to many scientific commentators the proximate cause of virtue lost. Wiener worried that, once the power of science was appreciated by the State, scientists would be subject to permanent mobilization: "At no time in the foreseeable future could we again do our research as free men." "From the bottom of my heart," Wiener wrote, "I pity the present generation of scientists, many of whom, whether they wish it or not, are doomed by the 'spirit of the age' to be intellectual lackeys and clock punchers."[124] And, if freedom was a necessary circumstance for both Truth and virtue, then science would have become the victim of its own success. Einstein was reflecting both on this circumstance and on the political interferences of the McCarthy era when—months after the Oppenheimer hearings—he told an American journalist: "If I were a young man again and had to decide how to make a living, I would not try to become a scientist or scholar or teacher. I would rather choose to be a plumber or a peddler, in the hope of finding that modest degree of independence still available under present circumstances."[125] But it was clear that Einstein reckoned the problem of de-moralization was endemic in the *very idea* of scientific research as a job: "Science is a wonderful thing if one does not have to earn one's living at it. One should earn one's living by work of which one is sure one is capable. Only when we do not have to be accountable to anybody can we find joy in scientific endeavour."[126]

In part, this was a generational matter, particularly so in physics. Even if most scientists either enthusiastically embraced Big Science or made their peace with it, there was a current of nostalgia, especially among the articulate elite. Physicists who had come of intellectual age in Rutherford's Cambridge, in Franck's and Born's Göttingen, or in Bohr's Copenhagen, who had experienced first the Nazification of German science and then the McCarthyite attacks, who had been mobilized for military work in the war and then experienced the vast scaling-up and bureaucratization of Big Science in the 1950s and 1960s, reckoned they had reason to lament the passing of a Golden Age of cosmopolitanism, autonomy, and integrity, of smaller-scale science conducted at a more leisurely pace.[127] The physicist Percy Bridgman, who had taught Oppenheimer at Harvard, described how much more difficult it was for the older generation of scientists to make the moral and practical adaptations to the change from small-scale artisanal to big organized science: "The older men, who had previously worked on their own problems in their own laboratories, put up with this as a patriotic necessity, to be tolerated only while they must, and to be escaped from as soon as decent." But the younger generation was different: they were used to "cooperative work in large teams" and unused to "individual initiative."[128]

Descriptions of the scientist's moral equivalence, then, figured not just in a generation's experience of loss-through-material-gain, but also in a specification of what science and the scientist *ought* to be. Interestingly, one of the youngest of the physicists at Los Alamos later became famous for his deflationary insistence that science required nothing special from its practitioners: "You ask me if an ordinary person could ever get to be able to imagine these things like I imagine them. Of course! I was an ordinary person who studied hard. There are no miracle people. It just happens they get interested in this thing and they learn all this stuff, but they're just people . . . So if you take an ordinary person who is willing to devote a great deal of time and work and thinking and mathematics, then he's become a scientist!" This did not, however, prevent Richard Feynman from becoming the focus of a late modern cult of personality and genius.[129] Again, while postwar physicists experienced more material gain, and expressed more sense of loss, than other sorts of scientific practitioners, laments about science de-moralized through bigness, bureaucracy, specialization, and routinization were not confined to physicists. The biologist Bentley Glass was concerned that the American scientific community as a whole was unprepared for the power thrust upon it, and that the distortions introduced into academia by preferential support for the natural sciences, projected scientists into positions of cultural responsibility they were unable or unwilling to accept. While making more money, and enjoying more political influence, than their colleagues in the humanities and social sciences, it was not clear to Glass that natural scientists as a whole had the moral maturity to respond as they ought to do: "The scientist passionately defends the freedom of science and fails to perceive that it and academic freedom are one. Academic scientists have been rather ordinary participants in the defense of academic freedom and the elevation of the standards of their profession . . . This growing and awakening giant, the academic scientist, has indeed much to learn as he moves toward leadership."[130] Like many other scientists of this period, Glass saw biology as still relatively unaffected by the political, intellectual, and moral problems of Bigness, but, nevertheless, there were biologists who came of intellectual age before the war who were sour about the nature of modern biology and the character of the modern biologist.

By the 1960s, several American life scientists were disturbed by the state of their science, causally relating what they saw as the intellectual decline of biology to professional success. At the Rockefeller Institute, the Austrian emigré developmental biologist Paul Weiss condemned the "irrelevance, triviality, redundancy, lack of perspective, an unbounded flair for proliferation" that were "just some of the symptoms of 'Big Science'" in biology. There was now a cult of "bulk ahead of brains, and routine exercises ahead of thought": instrumental possibilities and considerations of fundability shaped biological research, not questions of unrestricted scope. Biology had become a matter of answers driving questions, of equipment dictating research agendas, and, in general, of the tail wagging the dog. The art of the possible had taken over from genuine, free, and spontaneous intellectual inquiry. And, in Weiss's view, the proximate cause was the "crowd of mediocrity" that now made up the community of researchers, a crowd brought into being by the professionalization and routinization of the scientific career:

> Throughout the phase of history which we have come to survey [last three centuries], till very recently, to be a scientist was a calling, not a job. Scientists were men of science, not just men in science. They had come to science driven by an inner urge, curiosity, a quest for knowledge, and they knew, or learned, what it was all about. They were not drawn or lured into science in masses by fascinating gadgets, public acclaim, manpower needs of industries and governments, or job security.[131]

The biochemist Erwin Chargaff was another Austrian emigré who came to the States in the late 1920s, and, like Weiss—though more polemically and passionately—argued that the whole of modern science, including biology, had declined from a Golden to an Iron Age. The current generation of reductionist specialists was a race of intellectual and moral pygmies: "There [is] something wrong with ever-smaller people making ever-greater discoveries."[132] Chargaff constructed a dialogue between an "Old Chemist" (OC) and a "Young Molecular Biologist" (YMB) in which he made plain what he considered the constitutive link between good science and good people. The YMB urges the OC to reconcile himself to the success of molecular biology:

> YMB: You seem to have the romantically foolish idea that only a good man can be a good scientist.
>
> OC: It is always dangerous to use the argument *ad hominem*, and you should judge from yourself. But is it not a desperate situation when an old proverb must be reversed to read: Wherever the fish stinks there is its head?[133]

Chargaff's conception of the scientific calling was Romantic: "We are no longer used to passion in the natural sciences; it has been replaced by ambition. Our young geniuses are passionately ambitious instead of being passionately passionate; and it has become very difficult to distinguish between what is an ardent search for truth and what is a vigorous promotion campaign."[134] In biology, and elsewhere in science, the search for the Truth about Nature has been taken over by a search for results that can be reliably manufactured in the laboratory, and the moral outcome was embodied in the character of the scientific huckster: "It all started as a search for truth; but hundreds of thousands, at Hiroshima, at Nagasaki, paid with their lives for such lovable inquisitiveness. What an ivory, what a tower! . . . There is a real danger that our science may suffocate in its own excrements."[135] Where Max Gottlieb drew a crucial distinction between science and the stock market, Chargaff reckoned that such a distinction did not exist anymore: modern science "resembles much more a stock-market speculation than a search for the truth about nature."[136] This was a state of affairs directly ascribable to the transformation of science from a calling into a job: "The institutionalization of science as a mass occupation, which began during my lifetime, has brought with it the necessity of its continual growth . . . —not because there is so much more to discover, but because there are so many who want to be paid to do it."[137] Gunther Stent was one of the founders of the molecular biology that Chargaff despised, but he too found reasons to lament the loss of scope, wisdom, and integrity in his discipline. The very idea of individual scientists' having moral authority over their fields and followers had practically disappeared. By the 1990s, the presumption of morality had, in Stent's view, been taken over by the apparatus of institutional surveillance:

There are not many working scientists left who exert much spiritual power over their disciplines. However personally and professionally ethical prominent latter-day scientists may be, they have lost the halo of incorruptibility and unimpeachable integrity that many of their predecessors once projected. In fact, it seems pretentious nowadays for a senior scientist to act as the exemplar of an admirable person worth emulating, when the role of arbiter of integrity and good manners in scientific conduct has been taken over by impersonal organs such as the U.S. Public Health Service's Office of Research Integrity.[138]

In the early 1940s, Robert Merton had written confidently about "the virtual absence of fraud in the annals of science"[139] —after all, this structural, rather than motivational, honesty was what his "norms of science" were supposed to account for—but half a century later Stent's remark about the Office of Research Integrity indicates how close fraud and misconduct had come to the center of cultural consciousness about science. Well-publicized Congressional hearings on scientific fraud involving government-funded research were held in 1980 and were reported by *New York Times* journalists William Broad and Nicholas Wade. They were shocked by what they saw as widespread "betrayals of the truth," and in 1982 published a highly successful, but tendentious and carelessly argued, book on "fraud and deceit in the halls of science." Nevertheless, it is telling that the journalists set out their stall by first addressing what they took to be *still-current* views about scientists' virtue: "Scientists are not different from other people. In donning the white coat at the laboratory door, they do not step aside from the passions, ambitions, and failings that animate those in other walks of life."[140] Historical changes in the nature of the scientific life had produced both the presumption of moral equivalence and the necessity of external surveillance, since late modern scientists were not the kind of people who could be trusted to police themselves. That is to say, they were no longer to be treated purely as *professionals.* "Modern science is a career," and for Broad and Wade this counted as exposé: "Not only do careerist pressures exist in contemporary science, but the system rewards the appearance of success as well as genuine achievement." There was even a misdirected slap in Max Weber's face for seeing "science as a vocation. The individual scientist's devotion to the truth, in Weber's view, is what keeps science honest." Merton too came in for rough treatment, though Broad and Wade were not sharp enough to notice that Merton was the writer who denied both the fact and the necessity of individual moral constitutions.[141] Horace Freeland Judson's more recent exposure of scientific fraud laments that contemporary science is *no longer* a Weberian vocation and that it *no longer* conforms to Mertonian norms. Science was once a higher calling; now it is not; and the shift is indexed by what the author sees as the pervasiveness of fraud. The practitioner of Thomas Kuhn's puzzle-solving "normal science" is said to be morally normal too, and normal science is remarkably identified as a proximate cause of fraud: "It brought the scientific enterprise down with a thump, from a high Weberian calling to an everyday occupation like many others—however lit by occasional flashes of excitement."[142]

So, by the 1980s, the nature of Weber's and Merton's explanandum— the overall integrity of science, whether it was underpinned by individual virtue or by communal self-policing—was being influentially denied. Several years ago, historian of medicine Charles Rosenberg summed up changing orientations to writing the history of American science by drawing special attention to the scientist's moral equivalence as a matter of fact and to circumstances that made that equivalence undeniable to external commentators: "No man is a hero to his valet . . . , and few scientists are noble and disinterested seekers after truth in their ethnographer's account. This antiheroic perspective seems all the more plausible in a generation accustomed to charges of scientific fraud and to the spectacle of scientists rushing to patent their findings, incorporate themselves, and solicit lucrative consulting posts."[143] Scientists are driven by the same range of motives as businessmen, politicians, and churchmen—among whose ranks fraud and dishonesty are now expected—so "Why do we think," asked the evolutionary biologist Richard Lewontin, "that the devotees of Newton's Laws will be more saintly than those ruled by Cardinal Law?"[144] By the last decades of the twentieth century the commercialization of biomedical science and the general intersection of money and knowledge were being widely identified as a source of "ethical erosion."[145] It was now to be understood that individual scientists would do whatever they could get away with, and their collective institutions would close ranks and protect them rather than face their moral responsibilities. Practical consequences could and did flow from this change of sensibilities. If this is what you thought about scientists' virtue, then the only course of action was more and more vigilant external regulation.

Less apocalyptically, but more influentially, Daniel S. Greenberg, the news editor of *Science* magazine from the mid-1960s, and the most influential science journalist in America, was putting into circulation the character of "Dr. Grant Swinger," the scientist-on-the-make and on the plane to Washington; Prof. Morris Zapp in a lab coat; director of the Breakthrough

Institute and chairman of the board of the Center for the Absorption of Federal Funds (motto: "As Long As You're Up, Get Me a Grant"); recipient of the prestigious Ripov Award and the Segmentation Prize, annually given for the most publications from a single piece of research.[146] Greenberg was no Romantic, lamenting the decay of the philosopher-scientist stripped of virtue. He was not a scientist, and he valued his "outsider" status and his innocence of the technical language of science. So he simply took late modern science for what he thought it obviously was—a well-paid professional job like any other of the type, with its jobholders doing politics like any other interest group.[147] The scientist-on-the-make had now acquired iconic status. In his more straitlaced moments Greenberg analyzed the historicity of the situation and the consequences of postwar Big Science for democratic accountability:

> Science was once a calling; today it is still a calling for many, but for many others it is simply a living, and an especially comfortable one, for not only is it relatively affluent, but its traditions of freedom and independence provide a façade behind which government-subsidized liberties may be taken in the name of creativity.[148]

Greenberg had long been matter-of-factly convinced of the moral equivalence of the scientist, but he thought it was a sensibility that still went against the grain of the culture and that needed to be more widely distributed. (Indeed, a program director at the NSF reviewing Greenberg's first book in 1968 offered solid evidence that its distribution was still far from universal: "Most scientists will be unhappy indeed . . . with its apparent working premise that scientists are no more noble of purpose and performance than are ordinary men.")[149]

Certainly by the 1960s, a deflationary conception of scientific knowledge and of the character of the scientist had escaped from the scientific community to—some, but not all—general cultural commentators. The extent of that historical shift was a matter of degree. In the mid-1930s at the University of Chicago—an institution then strongly committed to the cohesion of a general culture—president Robert M. Hutchins was warning against the "trivializing" effects of ever-increasing scientific specialization and, at the same time, drawing a picture of what the new technical experts were like. "A university must be intelligible as well as intelligent," he wrote in 1936, and rampant specialization was at the cost of "general intelligibility." Scholars who abandoned that responsibility were demeaning their role and the institution that they inhabited: university professors are no longer what they were—"the charmingly eccentric old gentlemen" of tradition, "cloistered theoreticians who know nothing of what is going on in the world."[150] And that is why Hutchins wrote in 1963—with only superficial facetiousness: "My view, based on long and painful observation, is that professors are somewhat worse than other people, and that scientists are somewhat worse than other professors... There have been very few scientific frauds. This is because a scientist would be a fool to commit a scientific fraud when he can commit frauds every day on his wife, his associates, the president of his university, and the grocer... A scientist has a limited education. He labors on the topic of his dissertation, wins the Nobel Prize by the time he is thirty-five, and suddenly has nothing to do... He has no alternative but to spend the rest of his life making a nuisance of himself."[151]

TO THE PRESENT

American academic humanists and social scientists responded in different ways to postwar transformations in the scientific condition and career. As in science itself, the critics made the most noise and put their mark most strongly on High Culture. Criticism of contemporary science on the grounds of its intellectual and moral ordinariness had become a trademark of what academic resistance there was to militarism and commercialism. Eisenhower's military-industrial complex had been rightly rechristened the military-industrial-academic complex.[152] Some of the humanists and some of the social scientists—not sharing in the material rewards that their natural scientific colleagues were reaping from the Cold War State—were impelled to offer external commentary on the condition of scientific knowledge and knowers. They saw their relative cultural authority eroding as the natural scientists and engineers became a presence in the corridors of political and economic power. They once presumed that there were distinct values belonging to the life of inquiry, that they and the natural scientists had a common commitment to such values and to the academic institutions that uniquely housed them, that they and the scientists were positioned similarly with respect to money and power, and that the virtues associated with the life of inquiry flourished as much in physics as in philology. Now they were concerned that the natural scientists had broken free of that presumed intellectual commons, and by the 1960s some academic humanists and social scientists eloquently said so. Natural science, they announced, had been both normalized and de-moralized—drained of the virtues belonging to the intellectual life properly so-called. The "priests of nature" had become part-time Pharisees in the Temple of Intellect. And a new element was being introduced into reflective accounts of what had happened to the vocation of science—the *resentment* of nonscientific academic colleagues, even as the increasing imitation of what was taken as Scientific Method displayed the homage paid from the weak to the strong and even while equally vigorous criticisms of supposedly de-moralized science emerged from *within* the community of natural scientists.

The sociologist Lewis Feuer's *The Scientific Intellectual* (1963) noted that the scientific role no longer possessed any special virtues, nor were any such virtues attributed to scientists by the wider culture. Scientists are "becoming just one more of society's interest groups, lobbying for their greater share in the national income, and for the perquisites of power and prestige." The key moment was, of course, the Manhattan Project, but especially the presence of Oppenheimer, Fermi, Lawrence, and A. H. Compton on the Scientific Advisory Board that recommended targets for the atomic bomb. The new scientists had thrown over their historical legacy of moral authority: "They were no longer the guardians of the 'new philosophy,' the prophets of a new hope. They were technicians with the prejudices of ordinary men carried away with pride by their new technological accomplishment."[153] Lewis Coser belonged to the same generation of sociologists as Feuer. He was a German-Jewish emigré who had studied with Merton at Columbia, and his 1963 *Men of Ideas* worried about fundamental changes in what it was to be an "intellectual" and the particular forces affecting the modern scientific role.[154] The "modern intellectual" in general had come a long way from what he once was. The lineage had once connected the intellectual to both the transcendent and the moral domains. Intellectuals had been "the priestly upholders of sacred tradition . . . They question the truth of the moment in terms of higher and wider truth; they counter appeals to factuality by invoking the 'impractical ought.'" Given this description, it was hard for the late modern scientist to *be* an intellectual, for the notion of "higher and wider truth" had been substantially given over and the fallacy of moving from is to ought had been accepted. The mere "mental technicians and experts" of the present could count as no more than "distant cousins" to the historical intellectual.[155] Here, Coser endorsed postwar American scientists' own testimony about the nature and consequences of their condition. If intellectuals "live for rather than off ideas," then Big Science and the expansion and routinization of the scientific career were incompatible with the moral purpose and moral constitution of the intellectual role. The scientist had *once* been an intellectual but no longer was.[156] Science was now "one of the major industries of America." The shift from the artisanal to the organized production of scientific knowledge was assimilated to Weber's account of the routinization of charisma, but might have been equally linked to Eisenhower's formulation. Contemporary big government and industrial laboratories resemble "but little the small workshops that lonely discoverers and inventors habitually used in the last century": "The typical scientist today is a specialized research worker operating within a bureaucratic setting."[157]

For Coser, the university still remained a refuge and the natural home for science as an intellectual and moral pursuit. Just insofar as universities were *not* authoritarian, bureaucratized, and organized, they could and did provide a significantly different thought-environment from industry, "in which scientific work is determined by business interests." The contrast in ideals, Coser reckoned, was fundamental, even if he worried about academic bureaucratization looming on the horizon.[158] This change in institutional circumstance affected intellectual scope and quality, and, in the end, it eroded the moral authority that flowed from integrity and independence. The dependence of American scientists on the military had practically extinguished their ability and willingness to speak Truth to power. In the mass, scientists now were merely "expert technicians defending the differing positions of their employing agencies," and the "great majority" of such people mean to put bread on their families' tables rather than aspire to the intellectual's proper calling.[159] In 1959, C. P. Snow had insisted that scientists *ought* to be regarded as "intellectuals," admirable sorts of chaps with "the future in their bones," uniquely committed to the betterment of humankind, moral heroes (if any such were to be found in academia).[160] They had come to possess cultural authority; they ought to have much more of it; and, if they were not constitutionally virtuous in themselves, their contribution to the Good Life was unique. Now he had his answer.

Who Is the Industrial Scientist?

THE VIEW FROM THE TOWER

What is the attitude of the scientific man towards his vocation—that is, if he is
at all in quest of such a personal attitude? He maintains that he engages in "science
for science's sake" and not merely because others, by exploiting science, bring about
commercial or technical success and can better feed, dress, illuminate, and govern.

Max Weber, *Science as a Vocation*

THE RISE OF THE INDUSTRIAL SCIENTIST

Whatever they had to say about the nature of the late modern scientific
vocation, few American academic humanists, social scientists, or general
cultural commentators were focally engaged in giving it a close empirical
look. In the main, what they said about the scientific vocation was in
the course of some more general project—theorizing the modern order
of things, describing contemporary relations between knowledge and
power, giving an account of how these had changed in recent years, celebrating what the scientific life had come to, or, more often, lamenting
what it had become. Evaluations—celebration and criticism—were central to discussions of the scientific vocation and commentators mainly
traded in typifications sufficient to that evaluative task. As in the public
at large, it was Hiroshima and Nagasaki, more than anything else, that
drew attention to the scientific vocation, even if it was the changing nature of the American research university that most engaged the passions
and interests of academic commentators. Nevertheless, there were other
academic writers for whom attention to the scientific role and the virtues
associated with it was much more central, and these notably included
sociologists whose special subjects included science, organizations, occupations, and administration. And while, for these social scientists, the relations between science and State power signaled by the Manhattan Project
were also important, it was the much longer career of science in industry that was the focus of much of their work. How did these commentators conceive of the scientist enfolded in the institutions of commerce
and production? What stories did they tell about the role of the virtues in

industrial science? And what did their accounts reveal about more widely shared presumptions about scientific knowledge and the knower?

At the turn of the twentieth century, the figure of the American industrial scientist was a remarkably new thing. The material potential of industrial science excited; its institutional forms fascinated; but no one had a very secure understanding of what kind of people industrial scientists were or how their characteristics stood with respect to their academic colleagues. During the course of the century, industry became more and more common as an institutional home for scientific inquiry, while cultural commentators often struggled to make sense of the new phenomenon and to appreciate its significance for the authority of science and for the identity of the scientist. In part, industrial science was a form imported to the United States from late nineteenth-century developments in German chemical, pharmaceutical, and electrical industries (notably I. G. Farben, Höchst, BASF, Bayer, and Siemens). American observers of German developments were deeply impressed by what could be achieved through the recruitment of highly qualified academic scientists into corporate research and through the creative organization of their labors.[1] And in part American developments were responses to indigenous business and political conditions, where commercial competition, cost cutting, and anti-monopolistic political sentiment encouraged big, innovating industries to extend their commitments to in-house research beyond the routine analytic and testing functions that had been standard in much chemical and manufacturing industry.[2] It was thought that organized industrial research, including a small but very significant proportion of fundamental research, could result not just in cost cutting and improved production processes but also in the development of totally new products or even new industries.[3] And, following the work of Alfred D. Chandler on the history of business organization, some historians have argued that industrial research was just another element in drives towards vertical integration that characterized late nineteenth- and early twentieth-century American capitalism: "Just as United States Steel integrated backwards into the sources of its raw materials, so did firms like GE, AT&T and Kodak integrate backwards into another of their raw materials—knowledge."[4] By the 1920s, industrial science in America had emerged as a significant focus of cultural attention—one historian refers to a "national frenzy" for it—and something to be encouraged by government in furthering the national interest.[5]

The first natural homes of American industrial research were in the new and large-scale electrical firms following on Edison's work in the late nineteenth century—General Electric, AT&T, and Westinghouse—but by the early decades of the twentieth century huge investments were being made in organized research by the photographic pioneer Eastman Kodak; such large chemical companies as DuPont, American Cyanamid, General Chemical, and Dow; petroleum companies, including the various descendants of Standard Oil; and tire and rubber companies like Goodyear. In pharmaceuticals, the special concerns of the "ethical" drug industry to project a scientific image to physicians encouraged them in the early decades of the century to engage university-trained scientists, mainly to ensure that their manufacturing practices could withstand scrutiny, but not originally to use scientific research in new drug development.[6] In 1938, 57% of all industrial research workers were employed in the electrical, chemical, petroleum, and rubber industries, and half the total number of industrial research workers were employed by the forty-five largest companies—that proportion declining to a third by 1950.[7] Smaller companies with research requirements such as chemical analyses or process improvement might either make arrangements with contract testing firms like Arthur D. Little, Inc., of Cambridge, Massachusetts, or with independent research institutes, including the Armour Research Foundation (Chicago) and the Mellon Institute of Industrial Research (Pittsburgh). In addition, trade associations sometimes established their own cooperative research facilities or did deals with particular universities, such as the Tanners' Council research contract with the University of Cincinnati.[8] The close association between scientific research and small, nimble, entrepreneurial companies came much later, notably with the rise of the Silicon Valley electronics and biotech industries.[9]

The enlistment of science in the cause of commerce and production was not, of course, qualitatively new. The mobilization of skilled mathematicians, natural philosophers, medical men, and naturalists in furthering both productive and trade purposes goes back to Antiquity and the globalizing trade patterns of the early modern period crucially depended upon such skilled personnel.[10] Nevertheless, twentieth-century industrial science was quantitatively bigger and more significant than ever before and it sat astride major fault lines in the culture. It was celebrated, condemned, and recurrently treated either as a major achievement or a major problem. Who were the industrial scientists? What were their capacities, predicaments, and virtues? What was *science* such that it could or could not be organized, planned, and mobilized to achieve stated material outcomes? Was any such thing as organized science possible in principle, and, if it was possible, how did it work in practice? From its origins through the emergence of the late-twentieth-century "knowledge economy," the phenomenon of organized industrial research elicited an enormous volume of emotionally charged comment. Some American observers and participants celebrated organized industrial research as one of the greatest innovations of the modern age and among the most powerful forces in making a benign and bounteous modernity. It was a Very Good Thing. One of the founders and main propagandists for industrial research, Arthur D. Little, wrote in 1924 that "the laboratory has become a prime mover for the machinery of civilization . . . for research is the mother of industry."[11] In 1928, the director of the engineering division of the National Research Council announced that "research [is] now a universal tool of industry."[12] In 1946, Frank Jewett of AT&T declared that the coordinated research team was "the most powerful, effective and economical method of handling complex [scientific and technical] problems."[13] A scholar of technological invention, writing before the Second World War, endorsed the claim that the research laboratory was a major cause of broad civilizational advance,[14] and in 1958, one of the premier organs of corporate capitalism celebrated this still-new and noteworthy phenomenon, *Fortune* magazine blandly proclaiming that "the preeminent discovery of the twentieth century is the power of organized scientific research."[15] However the late modern order of things was to be conceived, the phenomenon of science in industry was seen as central.

Yet later sections of this chapter assess the views of twentieth-century American commentators who reckoned that science in industry was a crucial *problem* in the late modern order and that the very idea of industrial science might even be a *contradiction*. The goal of scientific inquiry was Truth; the goal of business was Profit. The natural agent of pure scientific inquiry was the free-acting individual; the natural agent of applied research and development was the organized team. The incompatibilities were treated as both important and evident. Organized research, whether in industry (where it had its natural home) or in government or university laboratories, was said to be a prostitution of the very idea of science and a visible index of how modernity was going disastrously wrong. How could a society and its institutions so badly misunderstand the nature of genuine inquiry as to attempt to organize it, to control it, and to direct it towards specific material ends? It was a cultural crime and a pragmatic catastrophe, for those who expected to enlist science in the proliferation of material goods were systematically killing the Goose That Lays the Golden Egg. As one industrial researcher, critical of "collectivist" and utilitarian tendencies, put it, if the public, government, and industry "can be induced to keep their hands off, and not wring the neck of the goose that lays the golden egg, the golden eggs will continue to be forthcoming. Penning up the goose and stuffing it may produce *pâté de foie gras*, but it will not increase the egg supply."[16] For these reasons, both the phenomenon of organized industrial research and the body of what people said about it are perspicuous sites for understanding changes, conflicts, and divergences in how sectors of American society understood the relationships between knowledge and the knower, between the good society and good knowledge, between virtuous people and valuable knowledge.

Much of this commentary embedded distinctions between some such notion as "pure science" and some such notion as "applied science," the former presumed to have its natural home in the university and the latter in commercial settings. The attractiveness of this scheme of things was pervasive and persistent; it was built into appreciations of what sort of person the scientist was and what sort of society scientists inhabited. Yet the clear divide between pure and applied, and between the institutions in which these supposedly different forms of inquiry were housed, was not accepted by all participants and commentators. It seemed to have made little sense to industrial scientists and those who managed their labors in the first part of the twentieth century. They might deny the legitimacy and pertinence of the distinctions between pure and applied science, or they might grant them but contest the notions that attached the pure and the applied to different institutional types or to different sorts of motive. So, among industrial scientists and managers who accepted the vocabulary distinguishing "pure" science (or, in alterative formulations, "basic," "fundamental," "pioneering," or "blue sky") and "applied" science, these categories were often said to be radically unstable, and research executives often insisted that such definitions and demarcations—important as they might be to accountants, compilers of government and corporate statistics, and academic critics and apologists—were difficult to make and might even be meaningless in practice. C. F. ("Boss") Kettering of the General Motors Research Laboratory, for example, testified both to the causal importance of pure research and to the difficulty of distinguishing between research modes: "I think it was Dr. [Harold] Urey who, when asked, 'What is the difference between pure science and industrial research?' made the statement that the difference was twenty years. I think that seems to be about right." There were some general problems with "knowing where to put the research," that is, knowing where in the corporate structure to locate it or knowing whether pure research was best done elsewhere, but, otherwise, the categorization of pure and applied was understood to be largely arbitrary.[17] James Fisk of Bell Labs said that "our fundamental belief is that there is no difference between good science and good science relevant to our business."[18] And a Minnesota Mining and Manufacturing (3M) research manager noted the inescapably conventional component of any such sorting scheme: "What is called applied research in solid state physics can often be described as basic research in electrical engineering. An application of mathematics to science is likely to be called basic research when it is done by a chemist or a physicist." A motivational distinction between pure and applied *could* possibly work, while such a distinction might not be particularly relevant to organizations interested more in outcomes over time than in states of mind.[19]

True, as time went on, many firms were obliged to institutionalize some sort of "pure-applied" distinction, for example, because of the practical need to place various activities in different parts of the organization and because of the expediencies of accounting for costs, capital investments, and profits. Industrial researchers and managers did have to make day-to-day judgments about timescale and about probabilities of outcome, and such judgments *might* be an important aspect of distinguishing the pure and the applied, but, within industry, such definitional exercises were rarely taken with great seriousness. The imperatives to embrace the pure-applied classification, to regard it as natural or absolute, and to distribute epistemic and moral virtues accordingly, were compelling elsewhere in American culture, but they were *not* notable features of thinking in and around the industrial research laboratory. At the same time, it is important to understand not just the burgeoning scale of industrial research through the course of the century but also the proportion of basic research (however it was defined) that was actually being done in corporate settings. The percentage of industrial research that was officially classified as "pure" or "basic" was indeed low: estimates from the 1940s to the early 1960s range from 4% to 8%.[20] And in that sense one could legitimately say that disinterested scientific inquiry—"science for science's sake"—had its natural home in academia while science with a clear commercial objective had its natural home in industry. However, and despite the relatively small proportion of the industrial research budget that was devoted to basic inquiries, as late as 1953 American universities were doing *less* basic research in dollar terms than was American industry.[21] And, as will be indicated later in this chapter, while the "idea" of scientific research might be identified with academia, the developing late modern reality was that more scientists did their work in other institutions.

This chapter describes the shape and scope of the industrialization of science in twentieth-century America and explores some of the effects of those changes on appreciations of who the scientist was and what characteristics were attached to the knowing subject. I concentrate here on responses to the industrialization of science emerging from the academy, and related strands of cultural criticism—responses that expressed various sorts of unease or alarm about the consequences of commercialization for the nature and vigor of science and for the character of the scientific researcher. I draw attention to the vocabulary of virtues used to express these criticisms and to the relations that commentators discerned between the virtues of scientists and their community, on the one hand, and the capacity of scientific inquiry to acquit its legitimate goals, on the other. Chapter 5 follows these sensibilities and tensions from academic commentary to the views and practices of industrial research managers, who confronted practical problems of organization and control on a day-to-day basis.

THE INDUSTRIALIZATION OF SCIENCE AND THE NATIONAL INTEREST

That the emergence of organized industrial science so intruded itself into American consciousness arises partly from the novelty of the phenomenon and partly from the challenge it represented to existing understandings about the nature of science and of the scientist. Because science had so substantially become the culture's official reality-defining practice, and because its practitioners enjoyed attendant authority, changes affecting the joint identity of science and the scientist were a focus for much commentary in the first part of the twentieth century about The Way We Now Live. Accounts of industrial science emerging from outside the corporate laboratory were almost always aimed at something different from straightforward *descriptions* of institutional and cultural realities: most often they were celebrations. denigrations, or expressions of various anxieties. But the resulting body of commentary nonetheless had a definite, if diffuse, connection to concrete institutional realities. Contemporaries had access to statistics that established the increasing scale and practical significance of industrial science in the American economy, as well as the changing relations between academia and industry as venues in which knowledge was produced. The referents of figures on industrial research are, however, often unclear and notoriously variable among sources. So, for example, figures about resources for industrial research may or may not include government subvention; the categories of "scientists" and "engineers" may or may not be confined to university graduates or to workers holding higher qualifications; the count of research installations may proceed from varying criteria and be variously enumerated. Accordingly, any statistics in this area have to be taken with more than the usually sized pinch of salt. Nevertheless, the explosion of industrial research from the first decades of the twentieth century was clear to American observers, and the gathering and distribution of pertinent statistics even before the Federal government got seriously into the business, testifies to industry's fascination with what it had wrought and external observers' appreciation that something consequential was happening.

In 1920, it was estimated that American industry was spending $20 million in about 300 research laboratories. Five years later, Bell Labs alone had a budget of $12 million, by far the largest R&D corporate facility in the U.S.; DuPont was laying out a bit less than $2 million; and GE about $1.4 million.[22] Industrial research exploded through the 1920s, and just before the stock-market crash of 1929 it was reckoned that about $130 million was then being spent in more than 1,000 laboratories.[23] In 1940, before World War II radically reconfigured industrial research, $234 million went to support work in industrial laboratories then said to number from 2,200 to 3,500, an expansion significantly assisted by the Revenue Act of 1936, which made corporate expenditures on research tax-deductible.[24] The name of the game changed with the Cold War expansion of government support of industrial research related to military concerns: in 1950, $2 billion, and, in 1954, $9.4 billion was spent on industrial R&D, most of this now coming from the Federal government, and creating the "military-industrial complex" that President Eisenhower's 1961 farewell address warned against.[25]

As a percentage of U.S. national income, industrial research expenditures increased spectacularly from 0.04% in 1920 to 0.87% in 1952. After World War II, major oil companies were committing about 1% of their gross income to research, and the most research-intensive industries (electrical machinery, scientific and control instruments, and chemicals) were gauging their expenditure on R&D at about 4% to about 6.5% of annual sales revenues.[26] The research workers followed the resources. Scientists were concentrated in large, innovating industries: in 1942, DuPont alone was employing 1,500 research chemists and research managers, and in 1958 Bell Labs maintained more than 10,000 employees, of whom one-third were professional scientists, and 15% of these were engaged in what the organization itself regarded as basic research.[27] American industry was sucking in research workers at an accelerating rate—there was a 60% increase between 1952 and 1959 alone. Industrial scientists, therefore, were important contributors to the general phenomenon of "the professional in industry" that, as I shall describe later, so occupied and distressed academic social scientists and even some business school faculty during the 1950s and 1960s.[28]

The Cold War generated grave anxieties about a shortage of qualified research workers—a shortage ascribed partly to the decline in birth rates during the Depression (though numbers gaining doctorates in the natural sciences continued to increase through the 1930s), and, to a lesser extent in the United States, to war deaths and a lost generation of technically skilled personnel.[29] In 1952, Peter Drucker noted in the *Harvard Business Review* that "professional manpower is the scarcest manpower in this country today—and is likely to continue to be very scarce for years to come. Korea and the defense program are actually minor factors. The real trouble is the tremendous increase in the demand for professional people both by business and by government, just when there is a very sharp dip in the number of men taking professional training as a result of the low birth rates of the 1930's."[30] The Cold War exacerbated worries about technical personnel shortages, but such anxieties were already being expressed in the late 1910s and during the boom decade of the 1920s. Technical people were being valued in unprecedented ways. "Never before," one commentator wrote in 1920, "has the outlook been so grave for the procurement and training of the required number of research men." Companies were "raiding the college faculties" to such an extent that academics warned industry against eating its own seed corn, and industrialists themselves were widely agreed that this sort of thing had to stop.[31] After the Second World War, the GI Bill of Rights—enabling more than a million ex-servicemen to attend university—was seen not just as an expression of national gratitude but as "a most significant means of increasing our supply of scientists."[32]

The "scientist-gap" preceded the "missile-gap" in American Cold War worries. "Scientific manpower," an AEC commissioner wrote, was a "war commodity" and a "major war asset" that needed to be "stockpiled."[33] With the outbreak of the Korean War, the director of Monsanto's atomic energy project wrote that the "national security and the strength of this country depend" not just on adding to the numbers of available industrial research workers but "upon the best possible utilization of the scientists and engineers that we have."[34] An editorial writer in a trade periodical expressed the shock that Sputnik gave to the American military-industrial research establishment: "With the advent of Sputnik, science has been thrust into the front as the force for world supremacy. Overnight an object whirling through space challenged the Free World to rally its scientific forces and embark on an extensive research and development program." America must urgently "develop a scientific reservoir," and the National Defense Education Act of the same year showed how swiftly the U.S. answered that call, endeavoring to expand the overall supply of scientists and engineers, but especially responding to the shortage of scientists able and willing to work in both industrial and government facilities.[35] In 1960, a research manager for Sun Oil was one of very many directly connecting a shortage of skilled scientific manpower to Cold War Doomsday scenarios: "It has become painfully clear that the communist world is relying upon its rapidly growing mastery of science and technology to outstrip, indeed to overcome, the free world in the struggle for survival of conflicting ideologies."[36] And a few years later, a chemical engineering executive stressed the importance of getting more technical people into industry, and getting more productivity out of them: "The American way of life if not the future of the entire world depends on it. If we do not find a successful answer to this question, the communist ideology will provide one. This will be catastrophic."[37] Worries about scientists' loyalties coexisted with portrayals of the American scientist as the savior of capitalism.

These sensibilities and practical requirements—especially, but not exclusively, in the Cold War period—were directly manifested in salary differentials between academic and industrial scientists, and commentators made much of such differentials. Thorstein Veblen's early-century celebration of "irresponsible science and scholarship" in the university, and his warnings about the administration of the academy by businessmen, matter-of-factly accepted professors' "somewhat meager" salaries, consequent high rate of academic celibacy, and, if married, small family size. If anything, the acceptance of these differentials was counted a badge of honor and a visible sign of scholarly rejection of commerce and its material rewards.[38] At Eastman Kodak in 1920, Kenneth Mees recognized that research was not then a well-remunerated career, but predicted that it would soon become so, obliging the research director to confront the issue of what motives were proper to research workers: "It may be thought that such [monetary] considerations will not produce first-class investigators, and undoubtedly a man who is attracted to research for money is unlikely to possess the necessary temperament for success." But that was a risk

Mees was nevertheless well prepared to take.[39] As late as the mid-1930s, the president of the University of Chicago, Robert M. Hutchins, wrote that the professor—including, of course, the academic scientist—"has entered the profession because he is interested in the pursuit of truth for its own sake. He has no vested interests which he is struggling to protect. His income is small. He knows it always will be, and he knew it when he decided to become a professor." It was that impartiality with respect to material things that secured his cultural authority.[40] Throughout the first part of the century there was a consistent strand of commentary insisting that it was not in the nature of the scientific man to care very much, if at all, about money or the material goods money could buy, and, therefore, that salary differentials were unlikely to lure scientists from the ivy to the smokestack. The *vita contemplativa*—in twentieth-century America as in medieval Europe—just needed fewer material resources than the *vita activa*, and the contemplative life was understood to attract those who were more internally than externally motivated.

In the 1920s, two executives of a small chemical company noted that "unreasonably high salaries are not demanded by men interested in scientific research." Such people, assumed to be motivated by the joys of the work itself, were unlikely to be tough negotiators on matters of salary, while they might care a lot about the conditions of work.[41] A 1942 article asking "Who Is the Research Man?" emphasized the salience of non-pecuniary motives: "The research man instinctively states his preferred mediums in the following order: ideas, things, and men, with the dollar a poor fourth; for research is primarily but not exclusively an interest in ideas, secondarily an interest in the embodiment of them in things for the use of man, and least a matter of money . . . This is the play instinct at work—the drive that makes one person, for the sheer fun at the time and without regard to consequences in money or credit, do something which another might consider work."[42] Just after Einstein's death, his friend Upton Sinclair retold a story he had heard from Einstein's second wife, Elsa. In 1933, Einstein was sought out by Cal Tech, which offered him an enormous salary, the exact amount of which he did not disclose, even to his wife. But Einstein reportedly rejected it, saying "That is too much; I will take $10,000 and no more."[43]

This kind of sensibility persisted—usually in attenuated forms—well into the post–World War II period of acute technical manpower shortages. Some commentators feared that basic science was being jeopardized by industry's lure of big salaries and big equipment budgets. A Chicago parasitologist retailed his colleagues' fears that "our scientific faculties will eventually consist of the overaged, the incompetent, and a few fanatics who prefer the academic atmosphere, no matter what the cost," while others continued to think that scientists with a genuine calling for pure research would not be much swayed by material concerns. For at least a few "fanatics or 'queer ducks'" willing to continue in university work, "there are compensations for the flesh pots of his life payable in the joy of teaching, in the advantage of close contact with scholars in other disciplines, and in real freedom and independence in intellectual pursuits. These benefits of academic life mitigate the lack of great material rewards."[44] For other postwar observers the increasingly uncommon figure of the "true scientist" was "only concerned with following his vocation." He was no ascetic but, for him, a middle-class allowance of 1950s consumer goods was good enough:

> He is not properly concerned with hours of work, wages, fame, or fortune. For him an adequate salary is one that provides decent living without frills or furbelows. No true scientist wants more, for possessions distract him from doing his beloved work. He is content with an Austin instead of a Packard; with a table model TV set instead of a console; with factory—rather than tailor-made suits . . . To boil it down, he is primarily interested in what he can do for science, not in what science can do for him. The breed, unfortunately, is dying out.[45]

Congressional hearings contemplating a National Science Foundation and an Atomic Energy Commission just after the war were sites where such sensibilities were consequentially expressed: a senator asked Karl Compton—physicist and president of MIT—"Do you think this is a correct statement that probably of all the professions in the world, the scientist is less interested in monetary gain—I am speaking of the pure scientist?" Compton agreed: "I don't know of any other group that has less interest in monetary gain."[46] In the early 1950s, the National Manpower Council, set up at the Columbia University Graduate School of Business to advise on the "partial mobilization" of scientists during the Korean War, said that "a spirit of public service or the attractions of an academic life hold many professionals in government or education, where they earn less than they could in industry."[47] So pay differentials between academia and industry for similarly qualified people were sustained by what was supposed to be the different moral makeups of those people, as well as by the resources available to different types of institution.

WORTHY OF HIS HIRE

But from early in the century there were other voices in this conversation, and some academic commentators struck by the contributions that organized scientists had made to both war and industry were unimpressed with the supposed honor of asceticism, arguing bluntly that "the research man deserves a living wage."[48] Those sentiments intensified with the end of the Second World War. An industrial scientist in 1950, writing a preface to a book about the nature of the scientific career, observed that the older generation of scientists would not understand the need for such a book, as "scientific research was, almost without exception, taken up 'for the love of it' and all other considerations were entirely secondary. They [the older generation] do not realise that to-day natural science is generally regarded, like medical science, as a career; at its highest, no doubt, a vocation, but nevertheless not to be divorced from all the other matters that a young man (or woman) takes into consideration when choosing his life's work." The rewards reasonably to be expected from any skilled, professional line of work might also be reasonably expected from scientific work.[49] The 1953 communication to *Science* magazine about the modest material wants of the "true scientist" quoted above elicited a volley of angry responses. One respondent rejected the picture of the "true scientist" as "a funny man in an ivory tower." Serious scientists cared as much as any other worker about proper rewards for their labor: "The professional scientist is not different from others in needing a satisfactory standard of living, in desiring rewards commensurate with his training and productivity, and in wanting to play a part in his own future and that of his family." Research today "is a major industry," and its workers are, of course, worthy of their hire.[50] Another letter writer argued that the "true scientist" uninterested in material rewards *should* become an extinct species. Americans generally had learned to respect occupations according to how they were materially rewarded, and it was, therefore, a disservice to the scientific profession for a research worker to settle for cheap cars: "The plumber who owns the new Packard and the salesman who owns the new Buick can only look with pity upon their learned neighbor, the professor, who can hardly afford to keep up his Austin."[51]

Just after leaving the Manhattan Project, the nuclear chemist Glenn Seaborg was offered a full professorship at the University of Chicago. He was struck both by the enormous salary—$10,000, nearly four times what he had been making before the war as an assistant professor at Berkeley—and by President Robert Hutchins's compliment accompanying the offer: "You deserve the good life."[52] Grappling with the post–World War II "shortage" of scientists, the President's Scientific Research Board staff viewed the current $10,000 limit on civil service salaries as leading to "a continuing exodus of capable scientists from Federal employment" to industry, and they proposed immediately to raise the maximum permitted pay to $15,000.[53] When more or less reliable statistics on these matters finally become available, they show that research workers in industry were indeed getting a wage that allowed them to live rather better than they would in a comparable academic position.[54] A U.S. Department of Labor survey reported in 1948 that the median salary for Ph.D. scientists working in higher education was $4,800; for those employed in government facilities, $6,280; and for those in private industry, $7,070. If you reached age sixty as an industrial scientist, your chances of making more than $10,000 a year were 62%, while in academia the highest proportion of professors in any age group making that kind of money was only 5%.[55] A few years later, a joint National Academy of Sciences—Department of Labor survey of Ph.D. scientists found median academic salaries ranging from $4,670 (for chemists) up to $5,700 (for engineers), compared to median industrial salaries from $6,250 (for biologists) to $8,000 (for engineers). So, it was noted, even the lowest-paid industrial group—then the biologists—was being paid more than the highest paid university scientists—the engineers. Moreover, when the median ages of scientists were taken into account, salary differentials between academia and industry were even greater. For workers under thirty years of age, median salary was not very different, but remunerative opportunities for those in senior positions were much larger in industry. Among Ph.D. scientists aged fifty to fifty-four, the median salary was almost $10,000 for industry versus $5,460 for universities.[56] The outcome implied the motive.

In industry, there was room at the top. *Fortune* magazine reported in 1948 that "top-grade scientists" could start in industry at an annual salary of $5,000, with the topmost rungs of scientist-managers commanding up to $50,000, while scientists started their careers in academia on as little as $2,000, increasing to just $7,000–8,000 for full professors, with only about a dozen top university scientists making as much as $14,000 a year.[57] While perceptions that industrial rates were far ahead of academic pay-scales were well-entrenched, it appears that the endemic problem of comparing like with like allowed for great statistical variability. When *Time* magazine talked about the image of the ill-paid scientific life as a hindrance to post-Sputnik recruitment, it emphasized that, with aggressive pursuit of industrial consultancy fees, the academic scientist could make up to $20,000 a year, so overlapping with salaries for the industrial scientist.[58] A few years earlier, *Fortune* conceded that the "rule" was that industrial salaries were higher than academic salaries, but, for top scientists, top institutions like Harvard, MIT, or Princeton could "match or exceed what industry is willing to pay."[59] And by the mid-1960s official government figures showed that differentials were not then so great: a median annual salary of $13,000 for industrial scientists, $12,100 for Federal government scientists, and $12,000 for academic scientists.[60] Nevertheless, and despite these caveats, *Fortune* worried that any overall significant salary differential between academia and industry might well cause America's strategically vital basic research enterprise to atrophy. As a postwar Berkeley physics graduate student said, giving one reason why he wished to go into industrial research: "[I] would like enough money so that I could eat at a restaurant without pricing the entrees & have a nice home, car, & plenty of life insurance . . . Most of these things can't be done on a University professor's salary."[61] Other industrial scientists—happy with their lot—nevertheless explained their institutional choice to an interviewer by noting that "I have four children and need an industrial salary"; "I decided to leave the teaching field because the pay was poor."[62] As early as 1951, a survey by the Engineering Manpower Commission found that some three hundred industrial organizations had 18,000 jobs on offer but were able to attract only 9,000 graduates to fill those openings.[63] "Creative scientists," wrote the president of Monsanto in 1955, "have not gone begging for at least fifteen years, and our best forecasts indicate the demand for their services will continue to grow."[64] It was a seller's market for qualified scientists and engineers in industry. They increasingly knew what they were worth and they were getting it.

The salaries of industrial research workers made up a very high proportion of corporate research costs, as high as 75% in some industrial laboratories.[65] Research managers and executives thought that the matter was ultimately settled by market forces—"the law of supply and demand"—but, in practice, all sorts of perturbing factors were acknowledged to affect research workers' rates of remuneration, notably including the idea that good pay (as well as job security) made for good morale, which, in turn, made for productive and motivated scientific employees.[66] Early in the century, the directors of an industrial chemical company thought that academic salaries were kept low because scientists valued the benefit of being buffered against the business cycle, shying away from industry because of economic risk. Even those research workers choosing to enter industry were said to be too soft-headed (and too averse to trade unionism) to demand high pay and organize to get it.[67] Kenneth Mees at Eastman Kodak was willing to admit that industrial salaries would probably always be restricted to reasonable levels "by the competition of the very badly paid scientific men of the universities," since at least certain sorts of academic positions might offer scientists social, cultural, and intellectual attractions that could not be compensated for by higher industrial pay.[68] However, I have found no research manager who explicitly agreed with Harvard Business School's Charles Orth in thinking that industrial salaries were higher across the board because scientists as a body had to be bought off from their natural academic inclinations.[69]

The experience of the General Electric Research Laboratory in these matters is instructive. During the 1890s, the physical chemist Willis R. Whitney had turned down an invitation to join Arthur D. Little's Cambridge consulting company at double his MIT salary, informing Little that he would prefer to be a professor "than be president."[70] A decade later, Whitney was still not notably keen to leave MIT for the industrial possibilities offered him at GE: "He doubted whether he would find enough interesting problems [at GE] to satisfy his active and eager mind."[71] But neither was he happy in academia, with his heavy teaching load, his limited research resources, and, indeed, his low pay. GE weaned him from MIT over several years by assuring him of their commitment to "scientific ideals," by allowing him initially to spend considerable time back at MIT, and by paying him the equivalent of a MIT full professor's salary while he was still on MIT's payroll. Soon, Whitney's worries about leaving the university dissipated, and he became GE's first director of research. He then set about poaching other academic scientists. In 1905, he won the experimentalist William Coolidge away from MIT. Coolidge rejected Whitney's first offer, but settled for improved terms that included doubling his MIT pay. Four years later, Irving Langmuir joined GE from academia, dissatisfied by inadequate university pay and other conditions. He intended to stay in industry for only a few years, but, as Leonard Reich writes, "finding everything he wanted at GE, he stayed for his entire career." This was a reasonable enough decision, since by 1916 "GE had the most fully equipped research laboratory in America, if not the world," and in 1932 Langmuir became the first scientist to win the Nobel Prize (in chemistry) for work done in industry.[72] In 1912 or 1913, the young James Bryant Conant recalled hearing Whitney lecture at Harvard, describing how he had "persuaded a young man"—probably Coolidge—"to leave an academic laboratory and join his group" at GE. Whitney, Conant recalled, "had offered the professor an opportunity to carry on in the GE laboratory the same line of research he had been pursuing," but Whitney was confident that the "erstwhile professor" would soon enough find the problems of industrial science tempting enough to abandon his existing research interests, and, as Conant wrote, "in a short time, [Whitney's] confidence was justified."[73]

After World War II, the Steelman Report's survey of American scientists asked them, inter alia, "Aside from money considerations, where do you think a person can get most satisfaction from a career in science—in the Federal Government, in an industrial laboratory, in a university, or somewhere else?" Even with the rather odd proviso that money was not to count, 31% said that they preferred to do science in industry. And when just industrial scientists were asked about their preferred work environment, again putting financial considerations to the side, 58% opted for industry. So, if it is to stand at all, the "buying off" hypothesis needs much qualification.[74] Industry recognized that it had to pay research workers the going rate; good research workers didn't grow on trees, and they might migrate to some other *company* if you didn't make them happy.[75] Still other research directors expressed conflicted opinions about the strength of the money motive among research workers. On the one hand, it was said that "incentives for the research worker do not hinge around incentives for him to do research. If he is any good, if he is worth his salt at all, he wants to do research in spite of anything else that you might do to him administratively. That is his fun." On the other hand, the same administrator insisted that "as is the case with any job, the financial incentive is important." Pension arrangements and bonuses, as well as salaries, had to be got right if you wanted to attract good scientists, keep them content, and retain their services.[76] And while, as already noted, the industrial going rate historically tended to be considerably higher than for academic positions, commentators were not agreed in explaining why this was so. Some scientists might indeed be happier on less pay in academia, but research managers in general seem to have found that money *was* an important consideration in securing the happiness of their scientific staff, and most companies that supported research had more money at their command than universities to make research workers content.

Whether or not scientists were attracted to industry only, or mainly, for the money, both the absolute numbers of research workers in industry and their proportion in the total population of employed scientists and engineers were expanding throughout the century, and the experience of working in industry was becoming an increasingly common feature of an American scientific career. In 1928, one estimate counted 30,000 qualified scientists and engineers working in industry; in 1940, there were 70,000, about half of whom possessed academic qualifications; in 1952, 250,000 (of whom 100,000 were then identified as "professional" scientists and engineers); and five years later the number was put at almost 750,000.[77] Certainly, by mid-century, the industrial, rather than the academic, scientist was closer to the institutional norm. Again, while the figures, and the criteria used to assemble them, vary significantly, both the trend and the general shape of the overall mid-century state of affairs are clear enough. In 1948, *Fortune* estimated that 42% of all scientists and research engineers were working in industry; in 1956 another Federal government survey counted 58% of all scientists and 88% of all engineers then in industrial employment. By contrast, 1955 figures had only 17% of all scientists and 2% of engineers then working in universities, their supposed "natural" home.[78]

Chemistry was probably the most "industrialized" scientific discipline. Historians have estimated that in the late nineteenth century 90% of American chemists were working in nonacademic settings, and in the late 1940s government statistics showed that 54% of all U.S. *Ph.D.* chemists then worked in industrial laboratories, 10% in government laboratories, and only 33% in academia.[79] In 1958, the *Harvard Business Review* massaged some Federal government statistics to derive "full-time equivalent" (FTE) research scientists then working in industry and academia, correcting for the proportion of their time that university scientists were obliged to devote to teaching and administration. In the mid-1950s, there were 8,262 faculty involved full-time in scientific research in American universities, and 23,192 engaged in research part-time, bringing the full-time equivalent researchers in academia to 16,534, while industry—where scientists and engineers were far more likely to spend their days actually doing research—could count 52,000 FTEs.[80] Image had become seriously disconnected from institutional reality. The normal site of scientific research by mid-century was not academia but industry.

THE VIEW FROM THE TOWER

This situation would be hard to guess if one relied solely upon dominant strands of academic social scientific commentary from the 1930s through the 1950s (and beyond) on the nature of the scientists, the scientific community, and scientific values.[81] Merton's story about how social structure and institutional control worked on the moral ordinariness of individual scientists was introduced in chapter 2. From the same setting of 1930s and

1940s academic social scientific and related cultural commentary there emerged a related picture of "conflict" between the moral economies of science and those of industry. Just as the scientist in industry was becoming an increasingly normal feature of the American social and cultural landscape, academic sociology described and theorized industrial science as problematic and possibly pathological. Sociologists drew a picture of an academic scientific community as a peculiarly Good Society, even while insisting that individual scientists were motivationally no better than anyone else. By contrast, they saw little if any virtue in industrial science. There was, moreover, a causal link between the deficit of industrial virtue, the distressed moral condition of the industrial research worker, and the limited possibility that industrial science could actually produce the goods that were so widely expected of it.

There was, of course, a sense in which scientists' natural home *was* the university, because, for all practical purposes, wherever it was that the mature scientific researcher eventually found an institutional home, every one of them passed through academia. That was where they acquired their skills and, according to Merton, whatever values were deemed appropriate to their role. As the twig was bent, so grew the tree: early socialization into such values was reckoned to be strong, consequential, and persistent. The values of academic science became, as Merton said, "internalized," where they formed the scientist's "conscience" or "super-ego."[82] Those scientists who secured university employment remained within the system of values into which they had been so powerfully socialized, and, for them, there was no tension between the "institutional ethos" of the scientific community and that of the organization in which they worked. They were one and the same. For the industrial scientist, however, the situation was different. The transition that scientists experienced when they entered industrial employment for the first time was, in the dominant social scientific account, fraught and emotionally traumatic, for on either side of this divide distinctively different and incompatible values resided. The result, it was said, was a population of industrial scientists who were unhappy, anxious, and maladjusted. Scientists took industrial employment because suitable academic research careers were in short supply or because they were just too low-paid to keep body and soul together for those not keen on an ascetic way of life. They found adaptation to incongruent industrial moral orders difficult, sometimes smoldering with resentment throughout their careers. Industrial scientists deeply disliked, if they did not actively rebel against, the violation of essential scientific values found in industry: secrecy, regimentation, hierarchy, constraint,

short-termism. The money—for the money was understood to be good—
never really made up for it. Scientists were being forced into a gray flannel lab coat, and it didn't fit. That was their central problem; it was a
problem for a society that, especially in situations of partial or total mobilization, required as many industrial and governmental scientists as could
be "stockpiled"; and it was a problem that industrial research managers
would have to deal with as best they could.[83]

This is a story about "conflict of interest," though here the texture and
vector of the putative conflict are rather different from what they are
understood to be in the present-day American research university and
in commentary on its relations with industry. In the earlier story, the
emotional pull of the unique scientific ethos is so strong that there are
grounds for worry that industry, or indeed government laboratories that
share some of industry's characteristics, can obtain a sufficient supply of
such people, or, when they are recruited, that they can be kept happy
and productive. It was vital for those unfamiliar with the authentic ways
of science to understand what sorts of people scientists were and what
kinds of demands organizations could not make of them if they expected
to enjoy the fruits of science. So even though Merton's early essays on
the norms of science were understandably preoccupied with such threats
to scientific integrity as those posed by the Nazi idea of "Jewish physics"
or the Soviet concept of "bourgeois genetics," his initial description of
the "institutional ethos" of science also remarked upon the tensions between scientific and commercial values. In science, Merton pointed out,
Die Gedanken sind frei, and the "rationale of the scientific ethic" whittles
down "property rights in science" to the "bare minimum" needed to secure recognition and esteem to the originator of a scientific idea. Science
is public knowledge or it is not science at all, and the very ideas of secrecy
and property rights in knowledge offend those who have internalized
scientific values: "The communism of the scientific ethos is incompatible
with the definition of technology as 'private property' in a capitalistic
society." And Merton here specifically alluded to late nineteenth-century
litigation between the Federal government and the Bell Telephone Company that established the inventor's "absolute property" in his invention and his right to withhold crucial knowledge of it from the public.
For mid-twentieth-century scientists, Merton wrote, this was a "conflict-
situation," and, while different scientists were responding to it in different
ways (by taking out patents, by becoming entrepreneurs, or, indeed, by
advocating socialism), nevertheless the friction arose from a deep conflict
of values between science and commerce. Conflict was always likely to

appear whenever science came into contact with institutions whose values differed from those needed for the pursuit of certified objective knowledge and which attempted to enforce "the centralization of institutional control."[84]

From these early statements about the norms of science, there emerged a *prediction* about what empirical research would eventually show, if, indeed, systematic empirical research was deemed necessary to confirming such a matter-of-course state of affairs: scientists socialized into this value system would suffer the "pain of psychological conflict" when presented with situations requiring or encouraging them to behave in ways that violated the scientific norms. To avoid or free themselves of this "pain," it was "to their interest" to conform to the ethos in which they had been socialized. Should the internalized "pure science sentiment" be put under pressure by "other institutional agencies" committed to the application of knowledge and concomitant organizational control, the result will be "the persistent repudiation by scientists of the application of utilitarian norms to their work."[85] So it could be deduced that, for social-functional reasons and for derivative psychological reasons, scientists would vigorously resist assimilation into the value system of commerce—such was the psychological grip of scientific values and such was the functional dependence of science on the embrace of these values. The scientist in industry would be in constant conflict with commercial values and corporate organizational structures. Scientists were too fiercely independent and mindful of their individual integrity, too skeptical, too hostile to authority structures, too loyal to science and too disloyal to local organizational values. Such persons would, quite naturally, pose a major problem for the smooth running of commercial organizations. The passage from the academy to industry was represented as a transition from a morally extraordinary to a morally ordinary community, from high to low. Given their socialization, scientists should, and (according to this story) did, rebel against that loss of virtue. That was just their nature.

THE ORGANIZATIONAL SCIENTIST
AS ACADEMIC RESEARCH TOPIC

The papers that Merton himself produced on this subject from the late 1930s to the late 1950s were not the upshot of systematic empirical work on contemporary science. Although his thesis work showed an impressive command of seventeenth-century English ideologies of science, and his references to views about science from a range of cultural environments testify to great breadth of reading, there is no evidence that Merton ever spent any time—and certainly not during this period—in natural scientific or engineering settings, or that he read systematically in the literature about the social forms and quotidian conduct of twentieth-century science.[86] But, inspired by his views on sharp and crucial value conflict between science and industry, a number of Merton's students, and those he influenced more indirectly, developed a large body of work that did have a substantial empirical base and that also gestured at the practical-policy relevance of the resulting findings. Value conflict was predicted on theory, and by the late 1950s and early 1960s sociologists steeped in that theoretical tradition, and sustained by a Cold War environment where there was an intense practical interest in problems of research organization and innovation, set out to assess institutional and psychological facts about the contemporary American scientist in industry. There might, indeed, be policy recommendations emerging from such studies, but the apparent primary impulse of these writers was to show the power of existing sociological frameworks for understanding organizational realities and to extend and develop such frameworks by bringing them to bear upon the contemporary predicament of the organized scientist. Governmental institutions were, as we have seen, deeply concerned about the character of the scientist and the conditions in which scientists' talents might be effectively mobilized, and the sociologists grasped the resulting research opportunities.

By and large, these sociologists started from the presumption that there existed *as a matter of fact* a fundamental conflict in the goals and values of scientists and businesspersons. Their main concern was to document the particular forms this conflict took. In *The Scientist in American Industry* (1960), the Rutgers University sociologist Simon Marcson wrote that the scientist's professional training "involved internalization of a set of norms and values" that gave pride of place to professional autonomy. "By education and professional training," Marcson explained, scientists' values are oriented to the community in which they were socialized: they are concerned with the sort of work "which will bring scientific recognition"; they respect "skill and achievement, independence of the individual in his work, and colleague relationships." Such goals and needs necessarily diminish the commitment that the scientist can bring to "making devices for missile systems, refrigerators, or computers. In short, the scientific community and the business community have different ideas about what is valuable and worth while."[87] Accordingly, companies are prepared for difficulties in recruiting talented Ph.D. scientists, and corporate efforts are made to reassure them that "some fundamental research is being conducted at the [industrial] laboratory," that "distinguished scientists who have made fundamental contributions are members of the scientific staff," and that the values into which they have been socialized will be respected and accommodated to some significant extent. That is to say, corporate officials were well advised to represent to future employees that their working environment would be as much like a university as possible, even though a small number of research managers expressed concern that this might, in some cases, be unethically misleading.[88] This was nevertheless a hard sell, and, while Marcson offered no evidence to support the claim, he wrote that "the recruit obviously is not convinced" by such assurances, "for when he has a wide latitude of choice, he chooses research and teaching in the university."[89] That was where scientific values were honored and that was where scientific virtues lived.

Lacking such latitude of choice, or selecting industry for idiosyncratic reasons, the newly recruited scientist faces a period of severe strain in adjusting to industrial conditions and expectations: "he is expected to be on time" (if not literally to punch a clock); he might like to work late, but when he finds his associates gone and the supply room closed, he learns the habit of leaving on time, working to clock rhythms rather than to the unpredictable temporal requirements of the research at hand. Should the industrial scientist, so to speak, "lose himself in his work," he is likely to find himself locked in, devoid of assistance, and criticized by his colleagues the next day for excessive displays of zeal. More importantly, he learns to modulate—if not entirely to abandon—his aspirations to undertake fundamental research and to have complete autonomy in choosing his research program—to which academia had accustomed him as a matter of course. "The divergence in expectations" and "the different norms held by scientists and by the [industrial] laboratory" express themselves in "strain and role conflict."[90] Making these adjustments is hard, if it is psychologically possible at all; it takes time; and a number of recruits are so strongly socialized into academic values that they just cannot survive in industrial organizations, ultimately abandoning corporate life. But, insofar as scientists *do* adapt, any such adjustment is evidence that they have been successfully, and with great difficulty, *resocialized*: since the scientist "is not prepared to be an employee, the industrial laboratory must undertake to make one out of him."[91] Under such conditions of difficult resocialization, scientists develop affective ties to the company that formally conflict with prior professional loyalties.[92] To some extent, Marcson noted, industry adjusts too. Under the pressure of skilled technical manpower scarcity, commercial organizations were beginning to relax the forms of control and mission-specificity that it is in their institutional nature to impose.[93] However, any such accommodation by industry is not to be taken as the adoption of new values, but as a tactical compromise within the distinct values to which industry is naturally committed. The scientist, it might be said, *believes in* the scientific ethos: that is his virtue. When industry becomes the scientist's employer, it *acts* in part *as if* it believes in some of the same values: that is its contingent prudence. In this way, the social scientist preserves the ideal-typical notion of role conflict between different institutional forms, while identifying those special contingent circumstances that lead to a modification in how values are actually made manifest in concrete working conditions.

Two years later, the Berkeley sociologist William Kornhauser (working with his junior colleague Warren O. Hagstrom) published a similar study of *Scientists in Industry*. Strains between science and secular organizations are endemic, appearing in the formulation of research goals, in the demarcation of research autonomy, in the implementation of controls and accountability, in the provision of incentives, in access to, and ability to influence, corporate policy: "Science needs autonomy to realise its purposes, but industry needs coordination to achieve its goals. There are, therefore, *inherent strains* between science and industry that find their most concrete expression in the industrial research organisation."[94] Scientists' academic socialization causes them to favor "deep-probing research and the advancement of knowledge. The business experience of industrial managers, on the other hand, initially disposed them to use professional specialists as mere technicians for routine operations."[95] It is "in the very nature of industrial research," Kornhauser wrote, "that high standards of intellectual excellence will continue to be threatened by organisational pressures for quick and easy solutions," with the resulting very real possibility that the creativity and integrity of science will suffer.[96] So industrial scientists who manage adaptation to their new institutional environment give up a certain amount of virtue—their happiness increasing in inverse proportion to their ability to contribute to the legitimate aims of science. Industrial scientists have to learn to submerge their professional interests in the greater good of the firm. That's a large part of what it means to work in industry, and it's a lesson that, because of the strength of academic socialization, is hard to learn. Moreover, industrial scientists have got to adapt, if they can, to commercial requirements for secrecy and the restricted flow of information, just those requirements that Merton identified as incompatible with scientific virtue: "The industrial scientist . . . compares such restrictions with the scientific community's norm of full and open communication and feels alienated from the organization."[97] The major difference in these respects between Kornhauser and Marcson lay in the greater extent that the former thought mutual adaptation of values and forms of authority had actually occurred in postwar America.[98] Nevertheless, Kornhauser agreed with Marcson, and with general sociological sentiment, that the *normal* relationship between scientific and corporate values was one of strain, that scientists acquired their values through effective academic socialization, and that adjustment to the generally hostile value system of industry was a difficult process of resocialization, sometimes succeeding, more often not.[99]

American academic contributions to the problem of the scientist in industry emerged in the 1950s from concerns central to the structural-functionalist sociological tradition. The founding in 1956 of the journal *Administrative Science Quarterly (ASQ)* testifies to the theoretical and disciplinary potential social scientists saw in problems associated with the organizational environment of the professional in general and the scientific research worker in particular. American academic ambition to build a fully general science of administration was not a wholly new thing in the 1950s.[100] Yet in the mid-1950s some academic social scientists were frustrated that this much-needed general science of administration had still not been produced. Coming out of the Graduate School of Business and Public Administration at Cornell, early numbers of *ASQ* criticized the current lack of any such fully general theory. What was wanted was a global and testable theoretical framework for how organizations of any kind functioned and were managed, ideally along the lines laid down by the work of Talcott Parsons.[101] But neither the editors nor the contributors to *ASQ* maintained without reservation that administration was fundamentally the same social phenomenon wherever it was found, and early numbers of the periodical were marked by a tension between the prize of a general theory of administration and the attractions of documenting particular cases.[102]

Among these cases, the management of research received a significant amount of early attention in *ASQ*, and the third number of volume 1 was wholly devoted to the administration of scientific and engineering research. The "editor's critique" of this number noted the special problems associated with the recent transformation of science from an individual to an organized phenomenon: "[Once] researchers operated autonomously, free to roam where their interests and rationality took them. When research projects become massive, the picture changes." From the perspective of an enterprise aiming at a general theory of administration, the point of interest was not that science was *special* but that it had so recently and importantly become so *similar* to other organized and administered practices. Once, science did not belong to the class of administered social phenomena; now, it did. Here the large-scale and disciplined organization of scientists during the war was iconic for academic writers in the 1950s: "That the Manhattan Project was successful called into question a number of beliefs concerning research which had been widely held by both scientists and laymen," specifically the belief that to organize science was to destroy its moral basis and its creative capacity.[103] One of the most influential of these contributions was Herbert Shepard's account of "Nine Dilemmas in Industrial Research," and among the key dilemmas was a crucial difference in patterns of social relations, of affective ties, of interests, and of loyalties that distinguished at least some professionals in industry from most of their corporate colleagues.[104] Here, Shepard took over a key notion developed some years earlier by Robert Merton in connection with patterns of community influence, and greatly elaborated in the second volume of *ASQ* by Merton's student Alvin Gouldner. The company works to ensure its employees' loyalty by satisfying their wants (salaries and benefits) and holding out the possibility of material sanctions for disloyalty. Employees thus satisfied are those whose affective ties and forms of association are "local": they identify with the company and are truly "good company men." Many scientists—not all—may, however, identify themselves with the values of their professional group. If they are organic chemists, it may be that what makes them tick is what is considered valuable behavior in the global community of organic chemists. Such employees are, in that sense, "cosmopolitans," and the paradox lies in the circumstance that, because of the unpredictability of scientific research, the disloyal cosmopolitan may ultimately prove of more material use to the company than the loyal local.[105] The application of the local-cosmopolitan distinction to scientists in industry was, therefore, an extension of a sorting resource developed within academic sociology, and, given the practical significance of scientific organization in the postwar period, it was natural that it be extended to that domain. But it is no less pertinent to note the special cultural and political resonance of this vocabulary in the immediate aftermath of the Oppenheimer security hearings of 1954, and of heightened Cold War concern about the loyalty of scientists in a National Security State.[106]

If the Merton-Marcson-Kornhauser genre was the most visible, coherent, and influential product of the engagement between American academic sociology and the problems presented by the scientist in industry, there was, however, a small number of academic studies less informed by structural-functionalist traditions and their presumptions of value conflict and inherent strain. Donald Pelz and Frank Andrews's *Scientists in Organizations* (1966) was the product of psychologists not notably driven by the sociologists' "value conflict" presuppositions, and there was little in the book about the problems of "strain" and "adjustment" that so centrally occupied the sociological literature of the 1950s and 1960s.[107] Anselm Strauss and Lee Rainwater's *The Professional Scientist* (1962) was a study of American chemists that came out of the Chicago sociological tradition, and, while it was concerned with questions about the professions and industrialization, its cautions against generalizing about the values and aspirations of chemists, let alone "scientists," set it apart from the Mertonian work. For Strauss and Rainwater, the central observation was the fragmentation and diversity of experiences, values, and career patterns.[108] By the early 1960s, a little American sociological criticism of the Merton-Marcson-Kornhauser genre began to appear, most vigorously from the awkward Merton student Norman Kaplan, while in Britain skeptical sentiments were expressed by Stephen Cotgrove and Steven Box (whose investigations did not support the view that all doctoral scientists embraced the values, or possessed the motives, that "socialization" was supposed to imbue in them), and, most acutely, in one of Barry Barnes's first publications (challenging core structural-functionalist assumptions about the nature of socialization).[109] Yet the American critics were effectively marginalized and British work in this area, and later developments in the "sociology of scientific knowledge," made scarcely any impact on American academic thinking for at least ten or fifteen years after their appearance, and by then the center of gravity of sociological studies of science had moved far away from the organizational and moral problems preoccupying writers in the 1950s and 1960s.

ORGANIZATION MAN AND THE LOSS OF VIRTUE

While academic views of the organized scientist clearly responded to sentiments pervasive in American thought and institutional practice, and while some of the pertinent work was a more direct response to governmental and corporate concerns, there is little evidence that this sort of writing traveled much beyond the bounds of academia. However, the same set of problems was central to works of cultural commentary and criticism that were very widely distributed in Cold War American culture. Indeed, preoccupations with organization and individuality, control and spontaneity, were right at the heart of American culture in the 1950s. Among the most representative and influential works of 1950s American cultural commentary was *The Organization Man* (1956) — not an exercise in academic sociology, but a semi-popular piece of social and cultural criticism by the journalist William H. Whyte, Jr. (1917–1999), who had been writing for *Fortune* magazine on business and the American scientific community since the late 1940s.[110] *The Organization Man* responded to the same cultural strains that in the 1950s produced such academic performances as David Riesman's *The Lonely Crowd* (1950) and C. Wright Mills's *White Collar* (1951), but which also gave rise to popular works of cultural and social commentary like Vance Packard's *The Hidden Persuaders* (1957) and Philip Wylie's *Generation of Vipers* (1955); Hollywood's *Executive Suite* (1954), *The Man in the Gray Flannel Suit* (1956), *Desk Set* (1957), *On the Waterfront* (1954), *The Caine Mutiny* (1954), and *Rebel without a Cause* (1955); the science fiction film genre that included *Invasion of the Body Snatchers* (1956); and Ayn Rand's cultish "novel" *Atlas Shrugged* (1957). For all of them, the central phenomenon-to-be-addressed and the social-problem-to-be-resolved was the struggle between authentic American individualism and the dark forces of conformity and collectivism. For anti-Communists, the threat of crushing collectivism came from the Soviet Union and Red China, its iconic form being the "brainwashing" that supposedly transformed the free-acting and spontaneous individual into an ideological robot. For those who stood up to McCarthyism, the source of anti-individualistic thought-control was the political witch-hunt that sought to preserve America from Communist conformity. For many social commentators, the menace to individual authenticity took the older social forms of small-town "Babbitry," boardroom cynicism, and the newer patterns of Levittown suburbia. For the Hollywood science-fiction genre, it was a takeover by hostile extraterrestrials. But for many concerned, like Whyte, to protect the creative scientific spirit that resided in the unique, autonomous, and free-acting individual, the threats were those bureaucracies and industrial organizations that, paradoxically, hoped to benefit from the creativity of their scientific research workers.

At the time he wrote *The Organization Man*, Whyte had no institutional ties with academic social science — though he was later a professor at the Hunter College of the City University of New York and became influential in academic urban geography — but he was conversant with much classic and contemporary sociological literature. There are references in *The Organization Man* to the work of Max Weber, Gunnar Myrdal, Elton Mayo, Lloyd Warner, Robert and Helen Lynd, F. J. Roethlisberger, and Reinhard Bendix—but none at all to Merton's work in the sociology of science or to that of his student Bernard Barber, whose *Science and Social Order* had appeared four years previously. Nevertheless, the centerpiece of Whyte's book is a set of three chapters on "The Organization Scientist" identifying the pressures brought on scientists in American industry to conform to industrial values, work conditions, and structures of authority—pressures that were well on their way to eroding the nation's capacity for technological and commercial innovation at just the historical juncture when those capacities were most needed. For Whyte, as for Merton, the crux of the matter was a conflict of values between science and those institutions called upon to support science, and in particular a failure on the part of sustaining institutions to comprehend the unique values that alone would encourage scientific geese to lay their utilitarian Golden Eggs. Organization was stifling scientific genius and only recognition of the integrity of the free-acting individual could halt the drift to mediocrity. Utilitarian concerns were voiced in the vocabulary of virtue.

VIEWS FROM THE BUSINESS AND ENGINEERING SCHOOLS

The problem of the scientist in industry, and, more generally, the professional in organizations, was one that fundamentally shaped both the sociology of science and the sociology of organizations in the Cold War decades. Treatment of research management and of the place of the scientist in industrial and governmental laboratories centrally occupied the two major American anthologies of the sociology of science to a degree that present-day academics may find remarkable, given the virtual disappearance of these subjects from contemporary foci of interest. As Norman Kaplan acknowledged in his 1965 anthology *Science and Society*, subjects including "the internal organization of the research laboratory, the administration of laboratories, the optimum size of research units, problems of teamwork and individual effort as well as organizational productivity, optimum organizational atmospheres or climates for research" have "received more attention than [any] others."[111] However, in the immediate postwar period, members of sociology departments were not the only academics commenting on the practical problems supposedly associated with the scientist in industry. So too did more practically oriented academics and quasi-academics. Many Cold War researchers wrote in a "managerial" idiom, accepting the responsibility of what might be called "social-science-in-action" to identify and assess problems-on-the-ground. How did you effectively recruit scientists into those industrial and governmental organizations whose flourishing was so vital to national security? Once recruited, how did you maintain and maximize their morale, creativity, and productivity? "Full utilization of our most highly developed scientific talent," a military personnel bureaucrat wrote, "is more than an idealistic dream—it is a national necessity."[112] Certainly, before the establishment of the National Science Foundation in 1950, and for years thereafter, much of the funding for this sort of work came from governmental agencies that had a pressing practical interest in the organization of scientists in nonacademic settings. This was work well supported by the military, especially the Human Resources Division of the Office of Naval Research (ONR), but also the Air Force Personnel and Training Research Center; the Army Materiel Command, Research Office, and Corps of Engineers; the Pentagon's Advanced Research Projects Administration; such nonmilitary Federal agencies as the Public Health Service; and industrial sources like the Industrial Research Institute and the Standard Oil Development Corporation.[113]

Much of the research in this managerial idiom during the late 1940s to the early 1960s was conducted in the relatively new university schools of management, business, and industrial relations, as well as in engineering schools, which were developing strong interests in what later came to be called "human relations."[114] Among the most prolific of these researchers was David Bendel Hertz, professor of industrial engineering at Columbia, who had previously been director of engineering at the plastics division of the Celanese Corporation, and who later became manager of operations research at the Popular Merchandise Company in New Jersey. Hertz's work on creativity in industrial research was supported by the ONR as well as the American Chemical Society and the trade publication *Industrial Laboratories*.[115] Among other investigators in this area whose work spanned the divide between academic social science, industry, and the quasi-academic world of business schools were Herbert A. Shepard (who took a Ph.D. in industrial economics from MIT in 1950, and who remained at MIT on the industrial relations faculty until 1957, when he became a research associate with Esso Standard Oil Company, leaving in 1960 to become professor of behavioral science at the Case Institute of Technology in Cleveland); L. E. Danielson (Bureau of Industrial Relations, University of Michigan); and Lowell H. Hattery (professor of public administration at American University in Washington, D.C., and before that with the Office of Scientific Personnel, National Research Council).

At the Harvard Business School, Robert N. Anthony's monograph on *Management Controls in Industrial Research Organizations* (1952) was sponsored by the ONR; the private business consortium, the Industrial Research Institute; and funds from the school's corporate sponsors. Its intended readers were industrial research managers, and its case-study methods—famously associated with the Harvard Business School— promised managers a sound empirical and comparative basis for inferring best practice in research management.[116] From the same academic stable, Ralph Hower and Charles Orth had sensed in the late 1940s that the practical problems associated with managing scientists in industry "might well become the subject of systematic study within the framework of the Harvard Graduate School of Business Administration," persuading the school's director of research to support such work on a large scale. Hower and Orth explicitly contrasted their work with more purely academic investigations in this area, which had been "aimed at measuring attitudes, testing hypotheses and concepts, developing theories of behavior, and generally adding to our store of basic knowledge about organizations." Their own approach was, they said, "wholly clinical," seeking to provide a usable account of the relevant managerial skills. While disavowing the hubristic idea that social scientific study could "offer any set of rules or formulas to guide industrial research administrators," they nevertheless claimed that through systematic and rigorously conducted case studies, and warranted inference from those studies, "the task of administering R&D personnel can be made easier and the results usually improved."[117] In this ambition, there was some acknowledged continuity with such classic early investigators of the conditions of industrial productivity as F. W. Taylor, Elton Mayo, and F. J. Roethlisberger, but it was now generally thought by business school writers that research workers were a breed apart, requiring quite different frameworks than productive workers for understanding their motivation and management.[118]

With the possible exception of the operations research genre, prac- tically oriented social scientists had few theoretical axes to grind: they sent out survey forms soliciting scientists' and engineers' responses to questions about their motivations and satisfaction;[119] they looked for empirically observable indices of communication within organizations; they produced easy-to-remember six- or seven-part lists and tables of the mental and personality traits that marked out the creative research worker; they developed simple and vivid graphic representations of the internal structures and external relations of the corporate research laboratory; they offered short checklists of helpful hints on how to assess and monitor the health of research organizations. A typical product of this sort of work was the bureaucratic report, submitted to the sponsoring organization—relatively robust, but not notably abstract, account on which practical policy might be based. There was no evident intention here to address issues in the theory of social structure or action, or to test existing social scientific models. Yet, at the same time, the overall tone of this literature makes it plain that some strands of academic social science did possess practically relevant expertise. If it was not notably theoretical, then at least it was an expertise proceeding from social scientists' ability to design and carry out rigorous and objective large-scale surveys, to process the resulting data with statistical rigor, to test explanatory hypotheses, and thus to replace anecdote with science. This "managerial" social science literature accepted completely that the administration of research functions was fraught with a large number of day-to-day problems that needed to be addressed in order to secure greater productivity and creativity. There were, indeed, "strains and stresses" in research organizations, and these were of great practical significance.[120] But, with few exceptions, the problems identified by writers in this idiom were taken to be mundane and concrete features of organizational life: methods of effective recruitment, salary scales, the costing of research and its benefits, lines of communication within the research organization, the transformation of researchers into managers and the special awkwardnesses associated with the role of research administrator.

ORGANIZATION AND THE VIRTUES

Organized, and, specifically, industrialized, science became normalized during the course of the twentieth century, but external cultural commentaries responded to that changing state of affairs in very different ways. Some adopted a naturalistic idiom, accepting as a matter of course the transformation of individualistic, ascetic, inconsequential, and priestly Truth-seekers into organized, well-remunerated, instrumentally oriented, and morally ordinary research workers, while other commentaries either treated the industrialization of science as a marginal phenomenon or viewed it as a worrying pathology whose consequences might be disastrous, for science and for the culture as a whole. Different idioms manifested different attitudes to the relationship between technoscientific knowledge and the virtues. The "managerial idiom," practically responding to the stimulus of governmental and industrial concerns about personnel shortages and national security during the Cold War decades, was broadly neutral about the effect of organizational life on scientific knowledge and the moral character of the scientist. The organization of science and the accelerating change in habitat from the university to the corporate or government laboratory were simply taken as facts of late modern life. Such changes threw up a range of practical problems of integration, motivation, and internal organization that had to be confronted and resolved, but, with rare exceptions, commentary from the business and engineering schools showed little anxiety about the new order. Both the Federal government and industry wanted more, more productive, more innovative, and more content scientists and engineers, and academic social scientific assistance was readily available to seek out means by which these goals might be more effectively realized. The flavor of this work was instrumental, pragmatic, and amoral. The scientist in industry was a research worker, an expensive and valuable resource, perhaps in need of a degree of special handling, but neither constitutionally, motivationally, nor morally different from anyone else. Questions about the virtues and their constitutional distribution just did not arise.

But for other academic commentators the industrial organization of science was a phenomenon that *could not*, by the very natures of both science and industry, be treated as normal, since it threatened both the possibility of objective knowledge and the virtues of knowing subjects. For Merton and his followers, the scientist-socialized-into-virtue was a condition for the production of certified objective knowledge and any interference with the expressions of those virtues, such as those demanded by industry or the secretive government laboratory, not only *would* be resisted by the genuine scientist but *must* be so resisted if science itself was to thrive. Scientists are internally motivated; dedicated, even called, to their work; they are selfless; resistant to convention and authority; intentionally blind to social convention and prejudice; unconcerned for fame and material reward; open. Their virtues are a pastiche of the heroic, chivalric, Stoic, and Christian. Put such people into the moral environment of corporate capitalism, and the resulting tensions are not merely mundane and contingent but ideologically essential. For William Whyte, the pertinent opposition was between the "social ethic" projected by both industry and American civil society and the integrity of individual thought. Science, properly speaking, was individualistic in its very nature. Attempts to transform the genuine scientist into the Organization Man were misguided and self-defeating. Just as American industry did not understand the nature of scientific inquiry and the conditions in which it could be acquitted, so it did not appreciate scientists' moral constitution. They were exceptional individuals, and attempts to treat them as "average Americans" were at once practical, epistemological, and moral errors.

I have called each of these strands of commentary "external" because they emerged from various sites outside of the industrial research laboratory itself. They each arose from different cultural segments; they had different constituencies; and they responded differently to currents in the wider culture. I now turn to accounts of industrial scientists that emerged from the sites within which they worked during the first six or seven decades of the twentieth century. How did *participants* talk about industrial scientists and their virtues? Did they see an essential tension between virtuous science and amoral industry? Did they see virtue residing in the active or the contemplative life, in civic engagement or in solitude? What, if any, distinctive virtues were industrial scientists reckoned to possess and what role did these virtues play in the making of knowledge and things?

* 5 *

Who Is the Industrial Scientist?

THE VIEW FROM THE MANAGERS

Nowadays in circles of youth there is a widespread notion that science has become
a problem in calculation, fabricated in laboratories or statistical filing systems just as
"in a factory," a calculation involving only the cool intellect and not one's "heart and soul"
... Such comments lack all clarity about what goes on in a factory.

Max Weber, *Science as a Vocation*

INDUSTRIALIZATION AND THE SCIENTIFIC VIRTUES

By the middle of the twentieth century several ways of conceiving the scientific vocation and its virtues coexisted in American culture. One insisted, against acknowledged presumptions to the contrary, that the individual scientist was morally and motivationally little different from anyone else. The vocation had either shifted from a calling to a job, or, it was said, had *always* been a job like any other, and it was only historical legend that presumed otherwise. Individuals were not drawn to science because they were morally better than anyone else, nor did the life of science make them better. In a de-magified world, nothing about the object of scientific inquiry, and nothing about the means by which scientific knowledge was discovered, was accounted morally uplifting. The moral ordinariness of the scientist was a projection of democratic sensibilities, suited to both the quantitative expansion of the scientific role and to the changing institutional circumstances in which scientists increasingly found themselves—handmaids to the creation of wealth and the enhancement of power. It was a sensibility expressed as much by commentators on science as by scientists themselves, though one must leave open the question of how exactly far it was diffused in the culture and among the different institutions of twentieth-century American society.

Another sensibility emerged strongly from the ranks of academic social scientists and from other cultural commentators on science and its place in society. This sensibility accepted the moral ordinariness of the individual scientist, but insisted on the unique virtues attached to the *communal life* of genuine science. Moreover, against the background of profound changes in the institutional geography of the scientific life, this was a sensibility that uniquely associated the scientific virtues with one sort of institution and saw them pathologically lacking in others. Specifically, it celebrated universities as the natural home of the scientific virtues and condemned the life of organized industrial and State science as leading to both moral danger and epistemic error. Neither of these sensibilities systematically addressed the texture of social interactions between individuals doing science, but both seemingly shared a disposition—already evident in Weber's work—to see modernity, in part, as the replacement of familiarity with faceless institutions and impersonal rules. Notions of Method encouraged a view of scientific inquiry as regulated by impersonal criteria, while the bureaucratization and industrialization of much scientific work also contributed to an emerging picture of science in which the characteristics of familiar people played little, if any, role.

External commentators might celebrate this state of affairs and the directions in which the scientific life was moving; more commonly, as we have now seen, changes occurring during the twentieth century were a source of anxiety, most especially during and after the Second World War. The mass recruitment of scientists into industry, and their mobilization in government laboratories, proceeded from the powers' wish to have the goose's Golden Eggs. However, the enforcement on science of supposedly alien modes of organization, planning, and discipline was thought to sap the virtue that allowed genuine scientific knowledge to be produced. At the same time, these changes provoked a cultural crisis, because the link between communal virtue and the authority of society's most reliable and powerful knowledge was widely thought in danger of being broken. By making scientists into morally ordinary figures, and by changing their preferred institutional setting from the university to the corporation or the government organization, you were at the same time threatening the cultural authority of science. You could, as some academic social scientists did, substantially ignore such massive changes in institutional setting, or you could, as other cultural critics did, acknowledge them and express alarm about their consequences. As one observer wrote in the 1960s, "The awe that scientists now inspire, and the patronage they command, have inevitably changed the nature of their calling. They have become richer and more caught up in worldly affairs."[1] Nevertheless, bodies of external commentary were widely agreed about the unequal distribution of communal virtue between academia and industry. This was the case in the middle of the twentieth century, and it is a sensibility that persists, in even stronger form, in the early twenty-first century.

By comparison to external commentary, we have so far heard little about what industrial scientists and those who managed their inquiries thought about their work, the environment in which it was done, their own personal characteristics and those of their colleagues. In fact, they did have quite a lot to say about these things in the first two-thirds of the century, even though their views have been little noticed by historians of science, technology, and business, or by sociologists of organizations and professions. One can only speculate about the relative neglect of these views, though it is likely that a number of considerations are involved. First, in contrast to such towering figures as Einstein and Oppenheimer, the mass of scientists working in industry were rarely reflective or articulate, or, if they did seek to describe the world in which they lived, they almost never did so systematically or eloquently. Whatever views they had about their institutional practices were scarcely ever offered in major public forums, or, when they were so offered, taken seriously by the mainstream of American intellectual culture. Second, as we shall see, what commentary emerged from these quarters scarcely engaged at all with the views of academic social scientists described in the preceding chapter, and, when it is juxtaposed with those views, more often than not it conflicted with them.

This internal commentary can be found in a variety of places. From the early 1950s, industrial consortia and managers established journals of their own in which to trade experiences about problems encountered in the new, and fast-changing, world of commercial research. *Industrial Laboratories* was started in 1949, as a trade monthly designed to inform research managers about new equipment, processes, and trends in their field, and in 1960 it was renamed *R/D: Research Development*. *Research Management* commenced in 1957, sponsored by an industrial consortium, the Industrial Research Institute (IRI), and continuing in 1988 as *Research Technology Management*. These two periodicals were the major postwar general vehicles in which research managers, and allied staff, addressed their practical problems. Somewhat more specialized journals in the area included the *IRE [Institute of Radio Engineers] Transactions on Engineering Management*, intermittently publishing practical commentary on problems of research management in issues from 1955 to the early 1960s, and several periodicals of older vintage, such as *Chemical and Engineering News*, the *Journal of Industrial and Engineering Chemistry*, the *Journal of Chemical Education*, *Electrical World*, and *Mechanical Engineering*. The *Technology Review* (published by MIT), *Personnel* (the periodical of the American Management Association), the *Scientific Monthly*, and *Science* (organs of the American

Association for the Advancement of Science) also occasionally provided outlets for reflections on the management of industrial research. Conferences and publications on research management convened and sponsored by such large firms as Standard Oil of New Jersey, Standard Oil of California, and the IRI are further sources for such stories; and, from the 1920s through the 1960s, a small number of the more reflective research executives published books on the organization of industrial research facilities and the administration of science.

The majority of this commentary comes from research managers rather than those whose inquiries they managed. It is not difficult to understand why this should be the case: it was the managers' job to speak for the laboratory, and it was the managers who had the means and the occasions to do so. There is a small amount of testimony from bench scientists referenced below, but there are no significant and systematic ethnographic studies of working industrial researchers. Nevertheless, there are reasons why managers' views are pertinent to present purposes in their own right, and, while it would be wrong simply to equate their views and experiences with those who worked for them, it would be equally wrong to presume that the sensibilities of laboratory personnel and managers were seriously disconnected. Few executives of big corporations had been production workers, but practically all industrial research managers had themselves once been bench scientists, and some maintained an experimental presence in the laboratories they administered. Moreover, it was a vital part of the managerial role to speak up for the researchers under their supervision, partly because they might think it right to do so and partly because protecting the integrity of laboratory researchers was a way of securing their own authority within the firm. No doubt there were tensions between research managers and research scientists—and this chapter and the following one will discuss some of them—but there were sharper conflicts between research managers and such officers of corporate headquarters as accountants and executives in charge of strategic planning.

Then, there is the tone, style, and evident purpose of research managerial observations. In marked, if unsurprising, contrast to the external academic commentary on industrial research, writing by research executives displayed little, if any, interest in making arguments of general sociological interest, in scoring theoretical points, or in using passages of research management as case-studies for purposes other than coming to some more or less robust findings about recurrent practical problems in and about the industrial laboratory and proffering some more or less plausible practical solutions to those problems. It is writing that deals, for example, with how things tend to go in large corporations versus smaller; in chemical companies versus electrical companies; in product-oriented laboratories versus discipline-oriented facilities; and so forth. A research director at RCA in the early 1950s, for example, rejected the pertinence of general theories of administration—such as those being proffered by academics publishing in the *Administrative Science Quarterly*—in terms typical of practical managers, insisting that the requirements and problems of research management "will vary among different units of industry" and will even depend upon the "differing outlooks towards research" taken by "individual research administrators."[2] And, as an agricultural research administrator succinctly put it, "It is a condition and not a theory that confronts us."[3] This material is not "academic" in tone or appearance: there are rarely, if ever, footnotes or literature references. For writing in this genre that is contemporary with, or subsequent to, the work of Merton and his followers, it is as if such work never existed—and the evident purpose is not apologetic, defensive, and only rarely celebratory, but rather in the spirit of trading pertinent "war stories" among congenial colleagues. For the most part, the historian is overhearing an internal conversation among research managers, not an edgy *justification* of institutional practices addressed to some possibly skeptical, or ideologically hostile, external audience. The question here is almost always practical management, not sociological generalization: how can one make industrial scientists more productive, more creative, happier, more likely to stay with the firm? what forms of organization work best for the industrial research laboratory, or for particular types of laboratory and in different industries and in firms of certain sizes? how does one attract the most able research workers and how can one recognize the signs of ability in potential recruits?

That is to say, while the managerial literature cannot possibly be confused with contemporary academic sociological accounts of organized and industrial science, it is ostensibly *about* the same social world, and, therefore, one might expect that the realities the two genres describe would be broadly similar. In fact, they are not. If the world of organized industrial science described, and condemned, in "the view from the tower" was amoral, drained of virtue, homogenized, closed, and constrained by bureaucratic rule, the realities recounted by those in charge of its quotidian administration resembled that picture in few respects. Specifically, while the academic sociologists predicted pervasive and serious role conflict attending the passage of university-socialized scientists into the amoral realm of industry, those research administrators who might have been expected to confront such problems as a substantial predicament in personnel management show practically no sign that such issues of value conflict even existed. It is not the case, as I shall show, that there were *no* distinctions made between academic and industrial conditions for producing knowledge: a range of distinctions *were* made, but the identification of institutional differences was widely understood to be problematic, the amorality said to be characteristic of industrial research was systematically rejected by what seems to be the great majority of research managers, and the geography of virtue presumed or described in much academic commentary looks very different when viewed from within the corporate world. Why should that be? Why should the same social realities sustain such different accounts of it? These matters are central to conceptions of knowledge, knowing subjects, and the cultural authority of knowledge in late modernity, and they should be addressed in concrete, rather than abstract, terms.

RESEARCH UNCERTAINTY AND ITS CORPORATE CONSEQUENCES

One of the most mundane, yet characteristic, features of any research properly so called is *uncertainty*—uncertainty in its outcomes and uncertainty in the procedures employed to secure outcomes. If one defines research as an inquiry into the relatively unknown, then neither the exact shape of the eventual results, nor the methods which will be successful in securing those results, nor the time and resources required for success, nor the likelihood of success, nor, finally, the consequences of findings can be exactly specified in advance of undertaking the research. In the early days of the Eastman Kodak Research Laboratory, its director Kenneth Mees wrote that the efficiency of research work is necessarily "very low ... since it is very rarely possible to arrange any research so that it will directly proceed to the end required." Most research fails to achieve its goal, so industrial research is "justified only by the great value of the [few] successful attempts," not by its average rate of success.[4] Of course, as Thomas Kuhn and others have argued, much "normal" scientific research is in the nature of puzzle-solving, proceeding with strong expectations about the general form of the outcome sought.[5] And, while there are important exceptions to any such generalization, many relevant scientists expressed their view that "applied" research—commonly said to be more characteristic of industry—was more likely to be of the "normal" sort than the "basic" research whose natural habitat was supposed to be academia. Accordingly, uncertainty would be a much less radical feature of inquiry conducted in industry than inquiry conducted in universities. From the point of view of some occupants of the "tower," including both scientists and nonscientific commentators, industrial applied science, engineering, and development was intellectually undemanding, if not trivial, and was marked in no significant way by the radical uncertainties attending academic basic research.

The Federal government's post–World War II Steelman Report volume on the *Administration of Research* drew concrete policy implications from these presumptions: "Applied and developmental studies can be planned and programmed in advance, at least to a very great extent. It is difficult, if not impossible, to plan or program the processes by which fundamental advances in scientific thinking occur. These are rare phenomena. They depend upon the swift reaches of the gifted mind . . . It is impossible to predict when or how or where they may occur. In one sense, they cannot be bought or sold, they cannot be ordered in advance in the market place. What may come from them, or be built from them, cannot (as in the case of nuclear fission) well be foretold . . . By definition, fundamental research is a venture into the unknown. Unanticipated results are the normal expectation."[6] It followed that the working environments and freedoms of action of the pure and applied scientist had to be designed with this distinction in mind. An astrophysicist-turned-operations-researcher endorsed these sentiments in the early 1950s: it was vital, he said, "to distinguish research from *mere* application. In some problems all the research . . . has been done, and it is *merely* necessary to apply known formulae to specific situations. This might be called engineering, . . . and should be distinguished from research."[7] Many people thought that the two could and should be managed differently: it was wrong to attempt to "schedule" basic research, but "in the less creative fields of applied research, or engineering development, or production research, the *when* factor can frequently be dealt with."[8] So, if one equated what happened in the industrial laboratory with applied research so conceived, there was supposed to be little, if any, uncertainty associated with it. Significant consequences for the imputed virtues of practitioners flowed from views about the nature of different sorts of inquiry.

There are, however, substantial problems with such views. First, although the great majority of industrial research was avowedly "applied," not involving the intentional search for fundamentally new knowledge or for abstract scientific principles, some of it was as "basic" as much science occurring in academia, and was recognized to be so by eminent academic scientists. As early as 1923, Carl Barus, a physicist at Brown University, remarked on the quantity of scientific publications now originating from industry: "These papers are by no means wholly of an applied or utilitarian character. There is an abundance of pure science, of abstract discussion, obviously encouraged by the business administrations in question."[9] In the early part of the twentieth century, resources for basic research in American universities were so limited that the absolute quantity of such research being done in such corporate laboratories as General Electric, Westinghouse, AT&T (later Bell Labs), Eastman Kodak, and DuPont was at least comparable to that produced by the universities. In any case, industrial laboratories were in the business of *both* applied and basic research, so an "apples with apples" comparison of institutional environments is apposite. Second, uncertainty is an irreducible property of *any* real-world inquiry: knowledge of the relevant features of the context is always imperfect; predictions of the future are always likely to be confounded by unforeseen developments. It is sensible to say that the quality and degree of uncertainties differ in different sorts of inquiry, but it is not sensible to say that uncertainty can *ever* be eliminated.

Research, of any sort, is always action-under-uncertainty of a high order compared to many aspects of everyday life. Finding out whether the plasma in a fusion reactor can be successfully contained, whether you can make a battery for laptop computers that lasts twelve hours, or whether the cost of wind-generated electricity can be brought down to that of natural-gas-produced electricity are all inquiries marked by substantially more uncertainty than deciding where to locate a new fast-food franchise, what is the quickest way to drive to work today, or even how to select students likely to be successful at Harvard. And, from the point of view of the commercial organization, many aspects of corporate life are more predictable, more routinized, and therefore more powerfully accountable to central control than the research function. For example, the conditions for producing and marketing the next batch of Ektachrome film were more certainly known by Eastman Kodak than whether any such thing as Ektachrome film could be conceived and brought into being. Those differences in uncertainty have important implications for the relationship between the corporate research laboratory and other corporate structures, and also for the conditions in which industrial research—pure or applied—can be *managed*. Whatever the quality of uncertainty, those entrusted with doing industrial research might be thrown into tension with corporate segments whose understanding of, and tolerance for, uncertainty were at odds with those of research personnel. For these reasons, even uncertainty that was arguably far less radical than that involved in many sorts of fundamental inquiry might become morally crucial to the life of the industrial scientist. After World War II, this is how the official history of the Office of Scientific Research and Development (OSRD) accounted for the initial conflict between industrial modes of organizing research and the tendencies of the armed services: in 1939, "the services had not learned yet, as American industry had, that it is fatal to place a research organization under the production department." Seen from the point of view of military neophytes in the art of scientific organization, American industry was considered to have long experience, and toleration, of research uncertainty.[10] In 1945, Congressional testimony by Vannevar Bush, the director of the OSRD, underscored this point: "Basically, research and procurement are incompatible . . . Research . . . is the exploration of the unknown. It is speculative, uncertain. It cannot be standardized. It succeeds, moreover, in virtually direct proportion to its freedom from performance controls, production pressures and traditional approaches."[11]

So uncertainty is crucial to understanding the place of the virtues in the conduct of industrial research, and the relative visibility of uncertainty to internal and external commentators needs careful interpretation. How did uncertainty figure in the culture of twentieth-century industrial science? How did research managers think about uncertainty? How was uncertainty implicated in the relations between research and the rest of corporate life? Finally, how did appreciations of uncertainty relate to the moral life of the industrial laboratory and to the attributed moral character of individual industrial scientists? Uncertainty in the research function meant that corporate headquarters were committing resources for outcomes whose contribution to the corporate bottom line and to shareholder value could not be precisely, or, at times, even approximately, known at the time the funds were obligated. That is the fundamental fact about industrial research and almost everything else of present interest in these connections follows from it.

ACCOUNTING FOR RESEARCH

Some industrial research undoubtedly was marked by less uncertainty of outcome and procedure than others: for instance, routine testing, assaying, and many sorts of product and production improvement, even if some corporate scientists were reluctant to give all such activities the name of research. Testing and assay functions that were especially common in chemical companies had to march to the same rhythm as production, and outsourced work of this sort—as, for example, that performed by Arthur D. Little, Inc., from 1886—was arranged through precisely timed contracts or might even be done overnight, though Little himself became an eloquent advocate for in-house industrial research, arguing that American manufacturers "will find their balance sheets deeply dyed with red" unless they embraced the "flood of new knowledge pouring in from the laboratories."[12] For testing and assay work, the imposition of strong control procedures and of relatively strict time disciplines and resource allocations were hard to resist—and some of those in charge showed few signs that they wanted to resist them. But at the other extreme there were many types of industrial research activity which escaped such controls, some to a remarkable extent. When the first director of the Eastman Kodak Research Laboratory, the English expatriate C. E. Kenneth Mees (1882–1960), was solicited in 1912 by George Eastman to found a new photographic research laboratory in Rochester, N.Y., he told his boss that commercializable results were not to be expected for ten years (figure 1). Eastman magnanimously replied that he could wait and wrote a check.[13]

In general, Mees wrote, the industrial research laboratory of the relatively pure sort that he planned demanded a different timescale from the routine assaying or development types: it will be "for many years unremunerative" and will "for a considerable time after its foundation . . . obtain no results at all which can be applied by the manufacturer."[14] And when Mees came to reflect on the organization of industrial research in general, he made it clear that Eastman Kodak was not to be regarded as a special case. Mees got ten years to show payoff, but for any industrial research operation, four to five was widely considered to be a minimum. This kind of research was a long-term commitment and it was not to be expected that it should demonstrate that it was paying its way on the same timescales used to evaluate other sorts of corporate functions.[15]

Early in his administrative career, Mees wrote that "those with the most experience of research work are all agreed that it is almost impossible to say whether a given investigation will prove remunerative or not," and a fortiori when it might do so.[16] Despite massive changes in many aspects of industrial research in the decades following the founding of Eastman's research laboratory, Mees's views remained unchanged into the 1950s:

Research work, and usually most development, cannot be scheduled, since the scientist sets his own pace according to his enthusiasm and interest at the moment. Perhaps some types of development work can be scheduled closely, but even this is questionable. In actual practice, the individual

FIGURE 1. C. E. Kenneth Mees (1882-1960), founder and first director of the Eastman Kodak Research Laboratory from 1912, and, later, vice-president in charge of research and development at the company until his retirement in 1955. Mees was probably the most articulate and influential of all early American industrial research directors. The son of a Northamptonshire Methodist minister, he took a D.Sc. in chemistry at University College, London, under Sir William Ramsay, and then worked for about six years at the Croydon photographic company Wratten and Wainwright before accepting George Eastman's invitation. He was one of the few research directors to write extensively and systematically about both his own experience and about how industrial research in general ought to be managed, and his views were very influential both in the U.S. and in Britain. (Reproduced by permission of the American Institute of Physics, Emilio Segrè Visual Archives.)

can be assigned a problem or problems on which he is expected to report regularly and is allowed to spend the remainder of his time on work of his own choosing as long as it is in the field of the laboratory's interests.[17]

In 1919, when Charles "Boss" Kettering was recruited by Alfred Sloan to head the General Motors Research Laboratory, he laid down strict conditions before accepting. He told Sloan that "I would never be held accountable for the money I spent . . . You can't keep books on research, because you don't know when you are going to get anything out of it or what it is going to be worth when you get it."[18] Kettering broadly agreed with Mees on timescales: "In our business, research is something that is concerned with things as much as ten years or more in advance." Anything shorter than that ought to be called by its proper name: "simple experimental engineering."[19] Just after World War II, a consortium of oil company executives counseled against expecting returns from industrial research in less than five to seven years; and one survey of industrial research managers in the late 1960s found expected average "payback" times from R&D of about four years—with big firms having a more generous time horizon than smaller ones—though another survey by the management consultancy firm Booz, Allen & Hamilton noted with alarm that less than a quarter of American large companies had *any* formal method for evaluating research, still less calculating a payback time.[20] RAND Corporation economists in the late 1950s drew importantly upon such sentiments in arguing *against* the military's desire to impose rigorous cost-benefit and temporal control regimes on research and development.[21] It was, in their view, industrial research managers who had the largest stock of relevant experience in such things and who functioned in environments of the greatest cost discipline, yet "despite talk of close controls, budgetary and otherwise, much industrial research is conducted under very loose control. In general, the greater the technical advance represented by the object of research, the looser are the controls. There appears to be recognition in practice of the great uncertainty attached to major inventions and research findings."[22]

Industrial research managers commonly acknowledged that it was the nature of genuine research for the unexpected occasionally to turn up, and that these unexpected outcomes might be commercially consequential, sometimes enormously so. Inquiries with no clearly foreseeable outcomes could be worth far more than those with outcomes whose value might be calculated. Research worthy of the name was always to a degree unpredictable, and expressions of that sort of sentiment were absolutely standard among industrial research managers. At the beginning of DuPont's research efforts, Charles L. Reese, in charge of the company's Chemical Department, sent the director of the Experimental Station a copy of Frederick Taylor's just-published *Principles of Scientific Management* (1911), and, as JoAnne Yates tells the story, Reese queried him "about whether the principles presented could be applied to any Experimental Station work. His reply was that the principles applied only to routine work, not to experimental work. Although certain aspects of the station's work could be and were standardized, many others could not. Thus, there were relatively few directives conveying procedures or rules from the station director to the division heads." DuPont executives understood from the outset that "research work was by nature less standardizable than production work."[23] In 1950, Mees and his associate John Leermakers at Eastman Kodak insisted that "it is really not possible to foresee the results of true research work," and in 1958 a GE administrator, Malcolm Hebb, writing about "Free Inquiry in Industrial Research" defined research as "systematic inquiry into the unknown," drawing from that dry definition the morally important conclusion that the "detailed course of a scientific inquiry" is always subject to unpredictable vicissitudes, and, consequently, that "a certain amount of freedom on the part of the investigator" is implied by the very idea of research.[24] It was very widely understood that the firm would get only marginal benefits from research whose outcomes *were* wholly predictable. Moreover, it was appreciated that some—not all—research workers were the sorts of *people* who did their best work when allowed substantial autonomy, and chapter 6 will discuss how industrial research managers actually thought about the possibilities of and limits to the planning of scientific research.

Some research directors did, indeed, reflect on what were seen as fundamental differences between research in universities or in free-standing research institutes, on the one hand, and research in industrial settings, on the other.[25] While academic research could be, and through the middle of the twentieth century generally was, relatively cheap, research problems there and in nonprofit institutes could be "chosen with complete disregard of the dollar sign"—not in the sense that academic resources were abundant but in the sense that research products did not have to *pay* and were not normally expected to do so. Although there were some notable exceptions, American universities did not routinely take ownership of the intellectual property produced by its staff and then manage it to produce revenue until about the 1960s and 1970s, and attention to a revenue stream that might be generated from such intellectual property was not widespread until after the Federal Bayh-Dole Act of 1980.[26] But in the typical industrial research laboratory, as Norman Shepard, the director of chemical research for American Cyanamid, wrote in the late 1940s, "it is quite a different matter—a 'horse of another color.'" There are limits and constraints that flow from the necessity of justifying research through its contribution to the corporate bottom-line. Some research workers, and even the research director, might "chafe under this restriction, but it is a necessary one." And if that restriction made them unhappy, "then those men and that director should return to a university or institutional laboratory, where there is complete academic freedom."[27] The directors of an industrial chemistry company insisted that "in commercial undertakings everything must pay for itself in one way or another," offering schemes by which the payoffs of all sorts of industrial research—from routine assays to fundamental inquiries—might theoretically be calculated.[28] And just after the Second World War the counsel for Standard Oil of California was very sharp with those who thought industrial research of even the most pure sort could possibly have noncommercial motives: "The satisfaction of scientific curiosity is not the fundamental basis upon which modern industrial research is founded. Considerable sums are expended on basic research, from which there is no immediate prospect of profit, but by and large industry's justification for expenditures must satisfy its stockholders, its customers, the scientists who do the work and, in the long run, the public."[29]

Indeed, the necessity that industrial research be seen to pay its way was institutionally axiomatic, since it was inconceivable that an internal argument could be made for its corporate support on any other grounds: "A board of directors," the head of research at the ammunition manufacturer, Western Cartridge Company, blandly noted, "does not approve a research budget on the basis of the pleasure that some ultimate consumer is going to experience, but on the profit which is more immediately to be received. If the research effort does not increase the financial returns to the company, it will not be regarded as successful." Management must have "an objective basis for evaluating research," in just the same way that it evaluates expenditures in the production or sales departments. Research expenses in an industrial laboratory had to be approved in the usual way, and due diligence had to be exercised in controlling costs.[30] Even such a vigorous exponent of laissez-faire laboratory non-organization as Kenneth Mees made no bones about the importance of cost discipline: "A satisfactory cost accounting system is extremely valuable in a laboratory, both for the control of current expenditure and for the preparation of the budget," and Mees's book reproduced the Eastman Kodak Research Laboratory's Assignment Cards and Daily Report forms that allowed the accounting office to translate time and materials into objective research costs (figure 2).[31]

Yet at the same time research managers and associated executives knew that the contributions to corporate profits of only a small fraction of industrial research could be rigorously evaluated and that the commercial value of much industrial research that was being done, and that should be done, could not be evaluated at all. When Mees wrote in 1916 that it was "impossible" to tell his masters whether or not a specific piece of

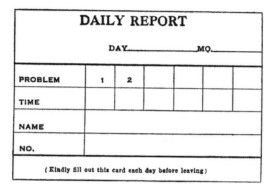

FIGURE 2. Specimen research assignment and daily report forms used in the early years of the Eastman Kodak Research Laboratory. While celebrating many aspects of research spontaneity and disorganization, Mees wanted it understood that the director of a properly managed research laboratory had to know who was doing what, when, and how much it was costing. Good bookkeeping was not, in Mees's view, incompatible with significant research freedom. (From Mees, *The Organization of Industrial Scientific Research* [1920], p. 128.)

work would pay off, he noted that the only "general conclusion" that could safely be drawn was that "the deeper a given investigation goes towards the fundamentals of the problem the more likelihood there is that the results will be of value."[32] Robert E. Wilson, research director at Standard Oil of Indiana in the late 1940s, and otherwise among the more hardheaded spokesmen for industrial research, articulated a common theme when he bridled at efforts at excessively precise budgeting of research projects and, most particularly, of fundamental research: "It is an awful accounting headache all the way along the line. Why spend so much money to find out exactly what this, that, and the other problem costs when you can not accurately appraise its value in any case?"[33] And in 1960 the president of the Shell Development Company, whose research laboratories then employed about 1,000 professional scientists, engineers, and mathematicians, wrote that while "efficiency in most [industrial] operations can be measured, efficiency in research cannot yet be measured."[34] In practice, some accounting convention was settled on to fix the research laboratory's budget and to fit it into the customary cost-benefit environment of the company, as it necessarily had to be.

For all that, many managers freely conceded that scientific inquiry was "a gamble," or at least that its dividends were extremely hard to measure.[35] Accordingly, for the full value of research to be realized, cost disciplines had to be loose. At Eastman Kodak, Mees stipulated that "research is a gamble. It cannot be conducted according to the rules of efficiency engineering. Research must be lavish of ideas, money and time."[36] "Boss" Kettering of General Motors called one strand of legitimate industrial research the "Monte Carlo" type: "We take a chance on spending so much for research in the hope that the boys may stumble onto something and we may make a little money out of it. That is just shooting craps with progress."[37] And at Standard Oil of Indiana, Robert Wilson said that "my own philosophy of handling research has been to give a good research director a stack of chips and tell him to get into the game and do his best. It is a gambling proposition."[38] Urging ever-closer ties between academic "fundamental" research and industrial concerns, that great British Marxist advocate of scientific planning J. D. Bernal spoke in the same idiom as many capitalist American research managers when he insisted on the inherent uncertainty of research decisions:

> If work were done on a product about which we knew in advance, then it could not be called research; true research had to work into the unknown all the time. Since, therefore, we did not know what we wanted to find no one could blame research workers for not finding the right answers, nor praise them for finding them. In running a research establishment one was really running a gambling concern and taking un-calculable risks for unassessable rewards . . . Although we might have a very good system of accountancy in determining our expenses, the expenses and the rewards were totally unrelated.[39]

But for all that frank talk, you could not expect executives to avow on all public occasions that they were actually just gambling with corporate funds. That's not the kind of thing that would play well with shareholders at the annual general meeting. Industry was neither a casino nor a scientific sandpit in which highly qualified scientists were well paid to amuse themselves. So in 1948 the industrial chemist D. H. Killeffer noted that each time some new research program was undertaken, "the researcher must establish in some manner that the probable value to be realized will be greater than the probable costs," but, while executives yearned for such a "sure indicator," he was aware of no formula securely to establish the future commercial values of research.[40] In the 1910s, Willis Whitney mused that improvements in electric lamp manufacture that might be plausibly attributed to research had saved consumers and producers $240 million a year between 1901 and 1911—all for an annual GE research budget of $100,000—but he did not pretend to know exactly how to quantify the specific benefits of industrial research.[41] As Mees and Leermakers unromantically put it in 1950, "In the absence of a generally accepted system [of evaluation], the compromise adopted by many organizations is one in which the [research] director makes his best guess as to the cost of a program or project and the operating and sales departments make their guesses as to the probable savings in cost or profits to be derived if the investigation is successful. From these two guesses, a decision is reached."[42] The uncertainty inherent in costing research and reckoning its benefits was one reason why Willis Whitney at GE judiciously chose to keep a certain amount of production work in the laboratory to generate income. He had one of the laboratory's top officials put together an annual report of "articles made regularly by this Company which had their origins in the Research Laboratory," estimating the revenue from their sales, and thus, as Reich writes, justifying the regular losses which nevertheless appeared on the laboratory's annual balance sheet.[43]

Of course, once you had agreed on some conventions for calculating the benefits of past research, you might use these conventions as locally accepted warrants for the inherently unknowable future, and so justify whatever laboratory disciplines or freedoms seemed prudent. At Western Cartridge Company in the 1940s the practice was to treat 3% of the gross sales of a new product for three years "as a fair measure of the value of the research" done by the company's research laboratory, and for improvements in products to attribute to research 3% of gross sales of the improved product for just one year. The practice was freely conceded to be "arbitrary," whatever justification it possessed deriving from "the recognition by many companies that [3%] is a desirable average ratio of research expenditures to sales," even though the same writer noted that in 1940 the average research expenditure of companies maintaining a research function was in fact just 0.6% of sales. Another argument held that a proper analogy for research was the company's insurance policies—research for the company's insurance against a technological future—and, therefore, that the research budget should be fixed at about the same level as the total insurance premiums for the year.[44] Through the middle part of the century, at least, research managers were of two minds whether research expenditures should be charged to expenses or capitalized, and, while accounting practices apparently differed widely, it was said that there was a general preference to have research treated as a current expense: "there are very real difficulties in establishing a sound basis for capitalizing research expenditures which may some years later prove to have an application with a return many times that of the original expenditure or may be completely without return." The more fundamental the research, the greater the uncertainty about return on capital, so research uncertainty impelled managers to have expenditures written off as they occurred and to find other ways to show productive yield.[45]

It was also widely recognized that research that was subject to severe accounting disciplines sacrificed its identity as research and, therefore, its potential corporate value. If the research was of the testing and assaying sort, you might easily make such calculations, and, for all sorts of industrial research, you could determine the test tubes, the reagents, and the personnel you needed tomorrow, next week, even, in some cases, next year. But as hardheaded a manager as the Western Cartridge man admitted that the time frame in which industrial research could show a commercial impact overran the corporate norm: only a small percentage of such work could be "completed, accepted by the factory staff, and put in to operation in less than one year. Probably two to three years is the average time elapsing between the starting of a [research] project and the appraisal of the returns by the auditing division."[46] When managers were forced to guess about the time frame in which accounting for research costs *should* be made, estimates ranged from as low as the two or three years cited by the ammunition factory up to five to seven years, and very few were as fortunate as Mees in freeing themselves from the accountant's rule for ten years. But even the tough-minded study of industrial research commissioned in 1946 by a consortium of oil companies insisted that research was necessarily a long-term commitment, for at least five years in many cases, and must be, for entirely practical reasons, buffered from the vicissitudes of the business cycle: "Research cannot be turned on and off. To preserve any semblance of continuity, it must be maintained as a going concern; the successful termination of many projects may be several years away. The organization can be readily expanded, but not so readily contracted without harmful results."[47] As one oil company research director concisely put it, management had to supply "patient money."[48] The early Depression years were indeed very hard for research personnel at such science-intensive companies as GE and AT&T, where almost half of laboratory personnel were let go, but such was the sensitivity to corporate dependence on research that the number of American industrial research laboratories actually continued to increase through the 1930s.[49] And it was during the Depression that Westinghouse, under the prompting of newly appointed research director Edward U. Condon, moved to establish the in-house Westinghouse Research Fellowships to support young physicists free "to do what they wanted to do rather than what the laboratory wanted them to do."[50] So there were some wholly practical considerations bearing upon the freedom of action accorded industrial researchers. And this freedom of action was consequentially linked to views about what motivated industrial scientists, how they ought to be treated, and what their personal characteristics were.

UNCERTAINTY, INTEGRITY, AND INTELLECTUAL OPENNESS

At the opposite pole from corporate acknowledgment of the uncertainties attending research was the repeated insistence that industrial inquiries had to be relevant to the company's specific goals—that they should produce practical results and pay their way. The resulting tension was endemic. I have noted that many research managers and related executives were quite happy to draw a contrast between an academic environment—in which scientists supposedly could do just as they liked (provided they could secure the resources to do so and that their work was deemed pertinent to their disciplinary and departmental communities)—and industry—where the scope of research was legitimately constrained by organizational purposes. Industrial research managers were frank about the fact that the scientists in their employ were free only within limits. Recognition of those legitimate constraints was consistent from the very origins of American industrial scientific research. In 1919, the physicist Frank Jewett of science-friendly AT&T remarked that "the performance of industrial laboratories must be money-making . . . For this reason they cannot assemble a staff of investigators to each of whom is given a perfectly free hand."[51] And a few years later, his colleague John J. Carty insisted that "unless the work promises practical results it cannot and should not be continued." Corporate scientists must ask themselves, and be required by their superiors to answer, the fundamental question "does this kind of scientific research pay?"[52] There were many industrial research directors who disapproved of the idea of basic research in their laboratories, and, of course, plenty of industrial laboratories didn't do any.

Research administrators at Owens-Illinois Glass Company noted that there was no point in directing fundamental research if commercial benefit was not reasonably to be expected of it. True, you could not *know* that benefit with certainty, but common sense and pertinent precedent offered more or less plausible expectations. Research executives were firm in their position: no research in industry could be supported *for its own sake*, nor did industrial research workers enjoy the freedom to pick just any kind of problem they pleased; glass companies could not be expected to support employees' fundamental research in, say, oceanography.[53] Creative people were, of course, always likely to be tempted into "fascinating but irrelevant side alleys," and it was the supervisor's task to get them back on organizational track, but not before checking that the byways really were devoid of commercial potential.[54] At the same time, there was no reason for research workers to imagine a *necessary* conflict of agendas where none really existed. James Fisk of Bell Labs stressed that all commercial organizations had to ask themselves "what business are you in?" and to adjust their research programs accordingly, but "this concept, the conscious setting of objective, does no violence to basic research. It simply implies that the research man will understand the purpose of the organization, know what is technically feasible in the business, and have a criterion of relevance for his work. Because research is relevant to the main aims of the overall organization, it need be no less basic and no less a contribution to science, as experience has frequently shown."[55] "Fisk insisted that there was no substantial difference between good science and commercially relevant science," and by the late 1950s Bell Labs had the Nobel Prizes to support their claim.[56]

While some external commentators made much of the supposedly coerced reorientation of industrial scientists from the knowledge-driven concerns of the academy to the corporation's commercial goals, quite a lot of internal evidence makes this problematic. Some companies enabled, or even encouraged, research workers to maintain and expand their professional ties, allowing them time to attend academic conferences and to visit university laboratories, sometimes for extended leaves. At Bell Labs, Fisk viewed research workers' attendance at scientific meetings and symposia, and "occasionally a 'sabbatical' term" in academia, as "a kind of 'preventive maintenance,'" keeping research scientists' skills in top working order.[57] A Sylvania manager wholly agreed: "The output of organizations whose research workers have full opportunity to grow professionally and avail themselves of it, will reach the market first and usually be superior. There is no substitute for imagination and resourcefulness, which are the basis of creative thinking. These flourish only when the investigator has ample opportunity for developing."[58] And an oil company research manager in the late 1940s fully recognized the desirability of research workers keeping up their professional ties, even while he expressed measured skepticism about their special enthusiasm for attending conferences in salubrious settings.[59]

Related concerns informed industrial management policy towards secrecy and publication. Again, it was a commonplace among academic social scientific commentators around midcentury that one of the basic divides between universities and industry was academia's norm of absolute openness and commercial commitment to great secrecy. (The same contrast is more familiar from commentary on science and the military.) But, once again, industrial laboratory realities are very far from supporting such a strong distinction. Of course, no research manager commenting on what is now called intellectual property thought that scientists could possibly be allowed freely to publish corporate secrets. Their work was understood to belong to the company, and it was for the company to decide what could or could not be published in the open scientific literature. Scientists joining DuPont, for example, were eventually obliged to sign nonnegotiable agreements that all "inventions, improvements, or useful processes" made while in the company's employ were the "sole and exclusive property" of DuPont, and they agreed not to "disclose or divulge confidential information or trade secrets."[60] Just a year after Churchill's "Iron Curtain" speech in Fulton, Missouri, a meeting of scientists concerned with patent policy used the phrase-of-the-moment to express their anxiety that "many industrialists" were drawing "an iron curtain around their research laboratories."[61] However, in practice, many research managers vigorously endorsed the commercial prudence of a quite free publishing policy and argued for the barest minimum of secrecy. The aphorism "when you lock the laboratory door, you lock out more than you lock in" comes not from Robert Merton or from Michael Polanyi but—in the same temporal and cultural context—from "Boss" Kettering at General Motors.[62] The free flow of technical information, or, at least, the freest flow compatible with broad corporate interests, was widely acknowledged in these circles to be a net benefit to all parties.[63] Patents were commonly seen as a form of open publication that nevertheless protected commercial interests. A properly managed patent system ensured communication, since its absence would necessitate the "iron curtain" of secrecy that scientists were supposed to fear: "With our patent system the industrial scientist may enjoy all the freedom to give and take which is possessed by his peers in the universities." The constraints of securing intellectual property to the company were deemed "trifling."[64]

Some research managers thought they had to advertise their laboratories as workplaces attractive to the most talented research workers, many of whom were found in academia, and there was no better way to do this than to encourage their scientists to participate in professional society meetings and to publish in the same journals as their academic disciplinary colleagues. From the very beginning of the Eastman Kodak Research Laboratory, Mees prided himself on an open publication policy. His obituarist noted Mees's feeling that such a policy "would enlarge the knowledge of the basis of the subject, and even though it might help his competitors in business the resulting advance in the field as a whole could not fail to benefit his Company and science." At the time of Mees's death in 1960, reports in the open literature from the Eastman facility numbered over 2,000.[65] These policies, and their rationales, were widely imitated. In 1948, the director of research at Sylvania wrote that "the reputation of the organization whose research men do publish their findings will be enhanced according to the caliber of the work done"; an oil company research executive agreed that encouraging research workers to publish their results—subject, of course, to "adequate patent protection"—brings "substantial indirect value to a company, which gets a reputation for being willing to have its men present papers and for being progressive. It is not just a matter of humoring the man, but can usually be justified from the company viewpoint"; and a research manager at GE pointed out (without qualification) the likelihood and advantages of reciprocity: "Any laboratory that must do good basic scientific research must also encourage open publication of research results because this policy insures access to the work of other laboratories."[66]

Even at DuPont—a company that some academic scientists reckoned to have become unduly secretive—official policy was massaged by research directors who recognized that a liberal publication policy was simply necessary for attracting first-rate chemists and maintaining their morale. So in the late 1920s two leading DuPont research managers recruited scientists by assuring them that "the work . . . shall be published almost without restriction," and, as Hounshell and Smith document, that open policy with respect to DuPont's *fundamental* research was effectively realized.[67] In addition, publication was understood as a vehicle for getting disciplinary communities to take up your problems, whether in their relatively pure or relatively applied forms. The more people working away in your area, the better for you. (And that is one, sometimes unappreciated, reason why Edward Teller at Livermore in the 1950s urged the freest possible dissemination of information concerning thermonuclear weapons.) An RCA research director writing about recruitment problems noted that publication *was* important to young scientists, but so were patent and incentive awards, as each of these was a sign that the company recognized individual contributions.[68] Moreover, in fast-moving fields, it was appreciated that you won not by locking up all possible intellectual property but by keeping one step ahead of the competition, and, as one moves closer to the present in many knowledge-intensive high-tech and biotech industries, such sensibilities become increasingly common.[69] A high degree of secrecy in such fields was of little concrete value. Even in the 1940s, a Sylvania research manager said that whatever necessity there might be to embargo publication was usually only "temporary": commercial commitment to publication could and should remain both as a principle and as a substantial reality.[70]

Such a widely accepted principle could and did coexist with the practical possibility that publication might, at any moment, be prohibited or delayed for commercial reasons—and industrial research workers understood this very well. The delay need not be very long, and ideally should not be. At Bell Labs, Fisk reckoned that industrial scientists did appreciate "the need for prompt patent applications, and a well-organized laboratory can get such work done expeditiously. Hence, patent questions need cause little delay in publication. But in any event secrecy is unattractive in industrial basic research—and is very seldom necessary. The communication of knowledge is a responsibility of scientists, a most important mechanism of scientific advance."[71] The relevant director of research at Bell Labs at the time of the invention of the transistor by William Shockley, John Bardeen, and Walter Brattain in 1948 (figure 3) celebrated the "traditional" company policy of "promptly publishing the work after a decent interval had ensured for verification and for filing initial patent applications."

What later came to be called the advantage of technological "first entry" gave additional justifications for the free publication of industrial research:

> We hold that a laboratory that draws knowledge and trained people from
> the world of science is thereby under some obligation to return value in

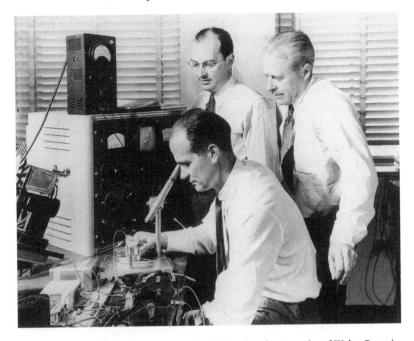

FIGURE 3. William Shockley (1910-1989), John Bardeen (1908-1991), and Walter Brattain (1902-1987) at the Murray Hill, New Jersey research campus of AT&T's Bell Labs. In 1956, they won the Nobel Prize in Physics for the invention of the transistor nine years earlier. The team was both effective and dysfunctional. Shockley, a spectacularly difficult man, fell out with Bardeen and Brattain over credit for the invention: Bardeen went back to academia; Brattain refused to work with Shockley again; and Shockley eventually left Bell Labs, famously going on in 1956 to found the Shockley Semiconductor Laboratory as a division of Beckman Instruments. A further falling out between Shockley and his researchers led to the departure in 1957 of the "Traitorous Eight" employees (including Gordon Moore and Robert Noyce) to found Fairchild Semiconductor, which was, in Silicon Valley legend, the origin of the South Bay high-tech industrial complex. (Reproduced by permission of the American Institute of Physics, Emilio Segrè Visual Archives and Bell Labs.)

kind. By prompt publication of its scientific advances a laboratory makes this return to the common fund. I think that prompt scientific publication is a sound policy for any research enterprise to follow as much as it can . . . This policy, furthermore, is not based wholly on a sense of moral duty but contains a modicum of self-interest. Science and technology advance through competition and the interplay of bright minds. No one laboratory has more than a fraction of the talent that can operate usefully on a new scientific idea. Exposing the idea to other minds will lead to more advance. And who is in a better position than the originator himself to recognize and profit from further advances?[72]

Some potential recruits to industry may have bridled at such secrecy as there was in industrial laboratories, varying as this did from company to company and from one type of research to another. But many other able scientists clearly did not. Reich writes that even with substantial "restrictions on communication and publication, Whitney had little trouble finding researchers willing to work at GE," with even such a prestigious and well-equipped institution as MIT being a rich hunting ground for recruits.[73]

Finally, and most tellingly from the point of view of academic theories of socialization, a number of industrial research managers were worried that their newly recruited academic scientists became *too quickly and too totally accepting of the values and research agendas of what they took to be corporate, as opposed to academic, culture.* At Eastman Kodak, Mees judged it very important that personal credit for research be given to individuals and that publication be under their names. "The publication of the scientific results obtained in a research laboratory is quite essential in order to maintain the interest of the laboratory staff in the progress of pure science . . . When the men come to a laboratory from the university they are generally very interested in the progress of pure science," Mees wrote, "but they rapidly become absorbed in the special problems presented to them, and without definite direction on the part of those responsible for the direction of the laboratory there is great danger that they will not keep in touch with the work that is being done in their own and allied fields. Their interest can be stimulated by journal meetings and scientific conferences, but the greatest stimulation is afforded by the publication in the usual scientific journals of the scientific results which they themselves obtain."[74] Research men, Mees thought, "*naturally* want to engage in work which will result in direct and visible financial gain, and hesitate to carry on fundamental work for which no commercial application can be seen."[75] It was the *company's* job to show them the continuing importance of disinterested inquiry. A British industrial research manager agreed: just because of the accumulating temporal distance from the pure research experience of university training, "there is almost invariably a tendency [for any research worker] to move in the direction from fundamental towards applied work."[76] The practical problem pointed to here was not the strong and persistent socialization into academic values presumed by academic sociologists but its opposite—the matter-of-fact willingness of research workers trained in universities to *abandon* such putatively distinct values. And it was that spontaneous abandonment that concerned research managers like Mees—not for moral or ideological, but for wholly practical, reasons.[77]

FLEXIBILITY AND FREEDOM

Giving industrial scientists significant amounts of company time to do their own research is not the late twentieth-century invention of the Silicon Valley high-tech and biotech "knowledge economy." At Eastman Kodak, Mees encouraged such autonomous research in the 1910s, and a survey of industrial practices in 1950 found that an allowance of 10% to 20%—that is, a half-day to a day a week, with company resources to match—was then common, even in American "smokestack" industries, though in some industrial laboratories studied by sociologists the take-up of that allocation of free time was low.[78] Mees recognized that the research men working under him might have a range of scientific interests and that, "provided that it does not interfere too greatly with the general work of the laboratory, it is most desirable that a man should be allowed to follow such a hobby to some extent." Nor was this just a matter of morale-boosting indulgence, "since in this way many of the most valuable discoveries will be made."[79] The president of Dow Chemicals said that he had "learned that if a research laboratory is to produce results, the men must be allowed the freedom to be a bit crazy," and the American Cyanamid research director quoting this remark approved up to 20% paid company time for research personnel "to work out their own ideas."[80] In 1953, *Fortune* magazine chided the general run of chemical companies for not allowing their scientists as much free time as Bell Labs and GE did, but noted that "at least two of the big chemical companies give their first-string researchers 25 per cent of their time to go fishing."[81]

For very highly prized scientific employees with strong track records of achievement, that free time could amount to considerably more. The terms that won William Coolidge for GE from MIT in 1905 included one-third—in other versions one-half—of his time to continue an existing personal research project.[82] One academic with close ties to industrial re-search managers summarized current sentiment on this subject in 1950 by noting that one way of ensuring "high morale" for the creative industrial scientist was allowing him "to devote up to half of his time on a project which he considers 'fun' and distinguished from those which he considers to be 'chores.'" The company should just "write off" the costs of scientific "fun" as a necessary expense.[83] When DuPont tried to recruit the organic chemist Louis Fieser from Bryn Mawr in 1927, Fieser was struck by the research freedom on offer: "I never expected to go into industrial work but the thing which makes a decision so difficult in this case is that I don't have to sell my soul at all; they even said I could bring my quinones along and continue my present work."[84] And a 1952 survey reported that, for some individuals, in some industrial organizations, "no plans are made as to how they shall spend their time."[85] This industrial policy was well known among academic scientists. In 1952, the Harvard physical chemist E. Bright Wilson, Jr., noted a widespread sentiment "which holds that every applied research laboratory should set aside perhaps 20 to 30 per cent of its resources for long-range fundamental work in the field with which it is concerned. The choice of this work should be left largely to the more experienced research workers themselves." It was a philosophy of research management that, Wilson said, "has justified itself in many organizations."[86] While David Hounshell is surely right to identify this sort of research latitude as a recruiting and retention tactic for some "academic elitists" who saw industrial research "as a poorer career option than that offered by a university or a private basic research institute," he does not claim that all research workers felt that way, and there is abundant evidence of *commercial* justifications for considerable amounts of industrial research freedom.[87] "The purpose of this freedom" in industrial laboratories is not, Wilson wrote, "philanthropy but a hard-headed realization" that a significant degree of researcher-directed basic inquiry paid off.[88] Recognizing the integrity of industrial research workers and ensuring degrees of free action for them was not, therefore, incompatible with corporate goals, and the extent to which such free action was effectively realized flowed from contingent circumstances, including the nature and stage of the research agenda, the resources available to underwrite dispersed lines of research, and the wishes of researchers themselves.

In the 1940s, the chemical engineer F. Russell Bichowsky wrote about the differences in temporal frame between industrial and academic research. In academic settings, resources for research might be in short supply, but time was rarely a limiting factor, while in industrial research the situation was sometimes the reverse: in industry "research is, by nature, fugitive. Problems seldom last more than a year."[89] Yet it was clear that corporate researchers must be given far greater freedom to organize their labor than factory-floor workers. And that difference in time frame was one reason why it was sometimes thought that the research laboratory ought to be physically separated from production facilities. Insofar as industrial scientists saw themselves as workers "apart" from other categories of corporate employees, they might choose publicly to display that distinction by, for example, "starting work an hour or so later than the production and office staff."[90] But if research workers' relative temporal freedom became visible to timed labor, friction between sectors of the workforce could arise that might cause practical problems for management: "To increase the number of new ideas, research personnel must be given much more freedom to come and go, to gossip, smoke and go to conventions, than would be customary in any other part of a closely managed company. They become a privileged or elite corps and this is one of the very best reasons for a separate research organization, since otherwise these 'privileges' are apt to cause difficulties."[91] A 1948 survey on the physical siting of industrial research facilities stressed the considerable psychological benefits that came from separating the laboratory from "the factory environment." As one respondent said, "I think one should remember . . . that research work is a psychological matter. Many research workers are impelled by pride to a considerable degree. From this angle it is good to have them off by themselves, where research is the business of the day, in a place where nothing is more important than research" (figure 4).[92]

Several industrial research directors noted that research workers generally tended to resent the very idea of clocking-in or other forms of work-discipline, particularly if these disciplines were seen to be the same as those imposed on production workers. "Punching of time-clocks, pettiness relative to time off, . . . criticism for apparently doing nothing but looking out of the window" rightly "incenses research men," and increases the possibility that the company's most valuable assets will walk out the door.[93] A student of the personnel policies of industrial research laboratories in the early 1950s commented on how hard it was to evaluate the output of research workers on an individual basis. It was, for example, almost impossible to say for such people "what constitutes work and what constitutes idling. Research is a creative learning process in which stereotypes of efficiency and productivity do not apply. The researcher must have time to reflect, to review his progress to date, to check his method of approach, and even to relax and get away from his research for a time. There is no necessary correlation between physical activity and productivity in research."[94] One cannot *look and see* whether researchers are working efficiently or, indeed, working at all. Hence, one cannot monitor their work through visible signs or indicators. The research person staring out the window might turn out to be working very hard and productively. A National Research Council administrator arguing for abundant free time for "a truly creative research worker" insisted that such time must allow not only for personal projects but time when the researcher could "just sit and think": "He may not look as if he is doing anything. Of course, it is difficult for an administrator to determine whether he is just sitting, or sitting and thinking." It had to be a matter of trust in the scientist's

FIGURE 4. The Westinghouse Research Laboratories in Forest Hills, Pennsylvania, ca. 1937. The facility had been built in 1916, several miles distant from the company's East Pittsburgh production plant. To the left is the company's "atom smasher," a Van de Graaff–type accelerator, which had just been constructed and which signified leadership in industrial nuclear physics. The physicist E. U. Condon had been hired as associate director of the laboratory, given broad authority to build up a large-scale program in fundamental research. (Reproduced by permission of the American Institute of Physics, Emilio Segrè Visual Archives and Westinghouse.)

dedication and integrity.[95] A British commentator with thirty years of industrial research experience, much quoted by American research managers, wrote that "a man engaged on research is at his best when he feels completely free and 'at home' in the laboratory . . . A research worker will work better and not worse because a cigarette, a cup of tea, a sandwich or a chat, can be indulged in during working hours. A man used to his pipe will do far better research if he need not remove it from his mouth because of a regulation." The professional's self-discipline makes externally imposed organizational discipline not only unnecessary but destructive.[96] Research, after all, is not just a staff function but salaried rather than wage labor, and unless—which is highly unusual—the executives and general management submitted *themselves* to such clock discipline, research personnel could not effectively be subjected to them either. If temporal

controls were desirable at all for research workers, they might best be implemented through the normal research practice of being obliged to keep
a diary or notebook treated as essentially corporate property, "having
official status" (as the chief engineer for the Hoover Company put it).[97]

And even this control was considered to be more a token of diligent
time-use than a coercive means to ensure it through external surveillance:
"With regard to the use of their time, research workers, in general, consider that they are on their honor."[98] At the GE Research Laboratory,
Willis Whitney initially tried to enforce diligent and disciplined notebook keeping by research staff, but by 1920 he acknowledged the limits
of a strict policy in these matters: careful record-keeping had "always
been desired—used to be expected—but it has not been attained and
therefore is not required ... Some of the best men are the poorest record
keepers."[99] The problem of what later in the century came to be called
research "milestones" was often identified as a potential source of friction
between research workers and managers, although one that might be
harmoniously resolved in a wide range of ways. In the late 1950s, research
managers at the National Cash Register Company observed that research
people often regarded "the urging of management for target dates as 'just
not scientific,' ... 'you can't do research that way.'" But when management's leadership was "sincere and impartial," and when management
took pains carefully to explain organizational goals, the technical staff
were content to go along. After all, "a sense of urgency is not a deterrent
to creativity when the problem is well defined."[100] It was a question here
of legitimate personal leadership, not of rule-books.

UNCERTAINTY AND THE ORDERING OF RESEARCH

The presumption of role conflict described in the preceding chapter now
appears to lack much evidential support. Yet there is no reason to deny
that the management of research workers in industry was intensively
problematic or that research managers were well aware of such problems. From early in the century, industrial research executives were fully
cognizant of a wide range of problems in their domain, and their commentary was overwhelmingly geared to the identification and practical
resolution of those problems. There is scarcely any area connected with
the organization of industrial research that was not recognized in managerial commentary as a real or potential problem. From the turn of the
century to the post–World War II period, those responsible for the establishment, organization, and daily management of industrial research

were reflectively aware that their enterprise was both highly consequential and radically novel; that there were few, if any, patterns available to be taken off the shelf and successfully imitated; that such successful examples of organized industrial research as there were might be of limited utility beyond their specific scientific, technological, commercial, or corporate settings; and, importantly, that the very idea of research undertaken in a corporate environment was a hybrid entity, not quite belonging to the moral economies of commercial production or of a university or free-standing research institute. Industrial research in the United States from early in the twentieth century was therefore characterized by a high degree of *normative uncertainty*. When Kenneth Mees retired in 1955 as head of Eastman Kodak's Research Laboratory, he reflected back on the state of affairs obtaining when he took up his position in 1912: "I knew nothing about running [an industrial] research laboratory, of course; nobody did." Mees visited Willis Whitney at GE's Research Laboratory in Schenectady, N.Y., one of the early twentieth-century industrial laboratories most celebrated for its support of fundamental research, but he came away impressed more with lessons about what *not* to do than with concrete ideas about how things *should* be structured.[101] "Whitney's laboratory," Mees said, "was not so much organized as inspired."[102] Participants constantly discussed how the thing should be done, and, even as industrial research expanded so spectacularly in the aftermath of World War II, no stable template emerged for any important aspect of its management. Almost everything about it was seen to be problematic. While practically minded participants rarely reached out to academic social science to address their problems, they actively looked for relevant experience and tried to see whether there existed some middle-range principles of organizational structure, some modestly generalizable techniques of managing and motivating research workers, or some relatively robust rules of thumb, heuristics, or maxims in any area that might be successfully transferred from one sort of industrial research laboratory to another. This search accounts for such reflectiveness as one finds in managers' commentary.

The "view from the tower" importantly predicted serious difficulties of adjustment for academically trained scientists entering the very different normative world of industry. Specifically, much external commentary saw such conflict arising from the transition between a virtuous academic world and an amoral industrial world. It would be very wrong to suggest that industry experienced no substantial problems in adjusting at least a portion of their recruits to the corporate work-world. However, with vanishingly few exceptions, unhappy industrial scientists, made unhappy by the strength of their socialization into unique academic values, *just do not exist* in the commentary of research managers and allied executives.[103] Industry could be as much a "natural" home for inquiry as academia. The virtues associated with inquiry could flourish as much in corporate as in university settings. Indeed, some of the rare reflections on the transition process offered by practicing research workers themselves drew attention to disorientation in leaving the *more* "regimented," and *more* conservative, world of academic thesis research, in which you had just one supervisor to satisfy, for an industrial laboratory in which the lines of control were not nearly so clear. True, orientation to commercial outcomes is not something that the newly minted natural science Ph.D. was accustomed to in the period much before the 1980s, but those who chose industry might come to embrace those goals (if they had not already done so), highly valuing them and setting aside what they might see as the inconsequentiality of academic research. Moreover, irregardless of orientation to commercial outcomes, many scientists might find that industrial research was just *interesting* to them and that much of it was highly thought of in the universities. "In general," as one newly recruited corporate electrical engineer said, "industrial research is well-respected in the academic world."[104] If industry happened to be "where the action was" in a given line of inquiry, then that was where science might have a natural institutional home. Whatever virtues were recognized in industry might then be the virtues appropriate to doing that science.

Sometimes it was thought that there were problems getting new industrial research workers accustomed to a more team-oriented style of work than was typical in academia.[105] Others saw "strains and stresses" arising from the contrast between academic specialization and the necessity for the organized research worker to submerge *disciplinary* criteria to interdisciplinary *project* goals:

> The intense specialization necessary to the performance of a rigorously defined task often produces a viewpoint and background that is a barrier to understanding the importance of considerations outside the realm of the specialist . . . Emphasis on mathematics, rigorous methods of inquiry, attention to physical phenomena, freedom to follow where reason leads, and devotion to detail are characteristic of research. These requirements sometimes give rise to problems when the scientist must work in an environment where his attitudes and objectives must be weighed against other objectives that are also important.[106]

If that were the case, then such tensions as existed were those between strongly disciplinary and interdisciplinary inquiries, and not strictly those between "science" and "industry."

So the commentary of industrial research managers is about little else than problems and their possible resolutions—problems of recruiting, assimilating, remunerating, retaining, motivating, organizing, and directing the labors of research workers. It was widely recognized that research workers moving from university to industry might go through processes of adaptation, but, for the most part, such adaptation was seen in terms of getting people and their families settled in (schools and Little League baseball for the kids, canasta partners for the wives, churches for the family, etc.); getting research workers familiarized with organizational culture, routines, and expectations.[107] New recruits to industrial laboratories sometimes reported an initial sense of "feeling lost," but much of that feeling derived from a series of mundane uncertainties: should they extend their thesis research in the new corporate setting? if they were attached to a preexisting research group, what was expected of them now? how were they meant to balance technical and commercial goals on a day-to-day basis? who should they talk to in order to sort out institutionally proper from improper conduct in general?[108] Members of industrial research facilities were not necessarily seen as "one big happy family"—though such attitudes *were* sometimes expressed—and serious tensions were sometimes acknowledged to exist between companies' research functions and such of their other arms as accounting and production.[109] In managerial commentary, the industrial research laboratory was full of tensions, just as its place in overall corporate culture continues to be problematic—at least for those early twenty-first-century companies where the distinction between the research function and corporate goals still makes sense. But to recognize such tensions and conflicts was much the same sort of thing as it was to recognize endemic tensions and conflicts between, say, firms' production and marketing divisions or, on the production floor, between supervisors and skilled workers.[110] Just as these sorts of tensions and conflicts were acknowledged and dwelt upon in internal business commentary, so research managers were obsessed by the organizational problems of the industrial research laboratory. It is just that the persistent and consequential problem of value socialization so precisely and persistently identified in the academic literature did not exist in corporate managers' views of their own laboratories. And neither did the moral geography posited in the "view from the tower."

TALKING ABOUT INDUSTRIAL SCIENCE:
NATURALISM AND LEGEND

If Cold War academic commentary on the organizational and moral regimes of industrial research seems so problematic, how can one generalize about the corporate laboratory while capturing the experience of those who directed it? Managers lacked interest in abstract and global accounts since few of them saw the industrial research laboratory as a homogeneous natural kind. While all such facilities were parts of larger corporate wholes, while all the corporations to which they belonged were profit-seeking entities, and while this circumstance bound all industrial laboratories formally to a corporate profit-motive, both the nature of such bonds and a range of other circumstances meant that the experiences of managers and scientists working in different laboratories varied enormously. At one extreme, the research function might differ relatively little from the production function: what went under the name of research might, from another point of view, look something like troubleshooting and process improvement. At the other pole, what transpired in the industrial research laboratory might not only be comparable to academic fundamental inquiry but could, in important ways, be *more* free and spontaneous than research carried out in *any* contemporary American university, especially if one remembers the ability of such large industrial firms as GE, Bell Labs, Eastman Kodak, and DuPont to offer a huge supply of material and human resources, freedom from teaching and routine administration, a stimulating environment of research-committed colleagues, and the capacity to explore research agendas beyond the bounds of constraining academic disciplines. That is to say, one can come up with a range of conclusions about the respective work environments and moral economies of academia and industry, depending upon one's choice of exemplars. The great majority of external commentaries contrasting "the nature of the university" and its values with industry are evidently modeled on elite universities, and not on provincial teaching institutions. American institutions of higher education, certainly at the beginning of the century, and arguably to the present day, were not globally regarded as natural homes for research. Most were under-resourced; most had a primary commitment to teaching;[111] many experienced cultural, political, and religious pressures which seriously compromised any notion that universities, *as such*, were communities of free, open, and suitably resourced inquirers.[112] With due respect, the University of Southern Mississippi is not MIT, and, as we have seen, GE was well able to attract distinguished researchers away from even MIT because, among other reasons, the company offered more effective freedom of action.[113]

Towards the end of the nineteenth century, a New York University geologist worried about the pressures for immediate practical results experienced by government scientists, but then warned that "the conditions are even more unfavorable in most of our colleges and none too favorable in our greater universities." All these institutions wanted "magnetic teachers" rather than creative researchers, and none of them had the resources increasingly required for a serious research program.[114] The dean of Brown University's graduate school, writing in 1923, even foresaw a time when industry would completely take over the research function from academia—including the pure research function. Surprised to find that so much pure science was being supported in American industrial research laboratories, he worried what effect this would ultimately have on a professorate inevitably and unalterably committed to instruction, turning "a faculty of high aims and specifically equipped scholarship into a body of schoolmasters." The day will soon come when industry "will have absorbed and assimilated l'élan vital, the soul of a university." And then industry will become the natural site of research: "The university will be the humble expository mechanism of the intellectual accomplishments of commercial enterprise. In brief, there will be a complete inversion of the method by which the world's knowledge has deepened in the past."[115] Nor should one think that such sentiments disappeared after the first decades of the twentieth century; they were expressed even after the conclusion of the Second World War, and after the establishment of the National Science Foundation, when vast governmental resources had begun to flow into academic science. Writing in 1954, an eminent cardiologist advised professors to reconcile themselves to their lot and to concede the prosecution of pure science to industry:

> The university teacher who carries on research as a side line to his teaching and utilizes untrained or partly trained fellows has as his most important function the training of his assistants in the methods of research. He is pointed not at the production of new *facts* so much as at the production of new *researchers*. The university professor must recognize that his first function is education . . . Big business has in recent years attacked the problems of pure science with the organizational precision that American business knows so well.[116]

Free action in research was and remains a matter of material resources and time as well as of rhetoric and ideals, so it seems reasonable to insist on an "apples with apples" comparison, and to draw much evidence about the nature of industrial science and the industrial scientist, as I have done here, from those large firms that dominated the world of industrial fundamental research in the early to middle part of the century.

Talk among research managers about the role of the personal virtues was rarely theorized and even more rarely surrounded by the halo of ideology. Rather, it largely flowed from the mundane circumstance of intellectual, material, financial, and temporal *uncertainties*. The greater the acknowledged uncertainty of the research process, the greater the scope of autonomy, and the greater the reliance upon the integrity and vocational dedication of the individual researcher. This substantive relationship between uncertainty and integrity made itself manifest at several levels. Research managers themselves often assumed the role of assuring corporate executives that treating the research laboratory, in crucial respects, differently from other arms of the company was sound business practice. If the laboratory were to be held accountable for its expenditures and outputs in the same way as other parts of the firm, it would not perform its profit-boosting functions better than if it were granted considerable autonomy: it might not be able to perform such functions at all. Research was a *different* bit of the company and had to be treated accordingly. As a search for the unknown, it could not, by its very nature, report what that search was going to cost and what the dollar value of its benefit would be. One job of research managers was, so to speak, to keep the accountants off the backs of their research workers.[117] Even if this minimalist conception of the research director's task was sometimes disputed by others in the profession, it nevertheless pointed to the possession of quite special personal characteristics.[118] Such people had to be unusually persuasive, ideally building up over time a relationship of trust with corporate headquarters, and chapter 6 treats issues of personal leadership in some detail. Research directors also had actively to manage the moral regimes of their laboratories. If they insisted that the uncertainty of the research process implied that skilled research workers' judgments had to be trusted, they had also to ensure that trustworthy, as well as capable, personnel were recruited, that morale was high, that personal problems were dealt with, that the right persons were selected for the right projects, and, crucially, that the channels of communication that so many directors deemed essential to the creative process be monitored and maintained.[119] When Westinghouse made a major commitment to fundamental research in 1937, its executives recognized that they had no alternative but to find a trustworthy research director and then to trust him: "On a fundamental research program of the type we are now considering, Management cannot offer very much in the way of concrete assistance in directing the work because of the lack of scientific knowledge and the difficulty of evaluating accomplishments. We must depend largely on the vision, judgment, and inspiration of the man selected to lead this activity."[120]

For all that, it would be a mistake to conclude that the moral regimes that I have identified in industrial research laboratories were produced simply and straightforwardly as an attempt to *imitate* academia, taken as the paradigmatic site of scientific virtue. Such imitation was the conclusion to which such academic sociologists as Marcson and Kornhauser were drawn in their treatment of "strains and stresses" in the industrial research laboratory of the 1960s. In a period of labor scarcity and Cold War "partial mobilization," scientists socialized into academic "norms" demanded that their industrial employers oblige them by providing a university-like working environment, as much as it was possible to arrange such a thing within a corporate framework. It was unnatural for industry to offer such concessions to the virtues, but—it was said—circumstances obliged them to do so. Yet these conclusions derive from a dubious set of assumptions, many of them perceptively identified in some now-neglected work by the sociologist Norman Kaplan (1923–1975), himself a Merton student. Kaplan thought that the Merton-Marcson-Kornhauser genre was substantially wrong in its assumptions about (1) the nature of science and scientists, (2) the nature of the research environment, and (3) the nature of organizations.[121]

Kaplan was almost alone among academic social scientists during the Cold War in commending a variegated view of science, scientists, and the institutions in which science was housed. He challenged the widespread assumption among academic commentators "that all or most scientists would like to remain in the university for the rest of their scientific careers." There was, Kaplan judged, no empirical warrant for such an assumption, nor for the pervasive view that "those who take positions in government or industrial laboratories must be doing so with great reluctance," nor, again, for the belief that scientists coming to industry or to government laboratories do so "expecting to find a university environment, or at least a reasonable facsimile of it."[122] It is undeniably true, he said, that all scientists are *familiar* with a university environment, just because they all pass through one. But it was abundantly clear that "at least a fair proportion" of scientists entered graduate school with the clear intention of working in industry, and with a pretty good idea of what they might be getting themselves in for. Recruits were therefore sometimes puzzled by industry's efforts to promise them an environment is like that of the university "plus a little more money." And "all of this is happening to a recruit who wanted to come to industry in the first place, did not expect to find the university there originally, and often is not especially concerned with how much like a university industry is." Some research workers wanted to work autonomously and to maintain ties with the academic disciplines in which they were trained; others did not. Some were vitally concerned with publication and their disciplinary reputations; others identified strongly with corporate goals, wanted to see an idea through to a product, and found both intellectual and emotional satisfaction in such work. Nor did much social scientific writing about university-industry relations pay great attention to changes rapidly occurring in the nature of American *universities*. The idea that academia offered absolute freedom and autonomy was made a nonsense by the necessity of raising funds, the willingness of journals to publish only work of a certain kind, and the constraints of discipline-based departments. Autonomy is never absolute.[123]

Kaplan's naturalism and particularism stand out among the academic commentators on science and its institutional forms. He wanted to *describe* the moral orders in which twentieth-century science was being done and he reckoned that they had been insufficiently well described. His social scientific colleagues clearly also aimed at description, but their descriptive impulses were richly mixed with impulses to celebrate and accuse. Kaplan was *right*—his descriptions *were*, for the period, unusually detailed and sensitive—but one still needs to appreciate why it was so hard to write naturalistically about the moral regimes in which late modern science was being done. It was difficult because of the peculiar relationship that obtained between knowledge and virtue, between the social forms in which valued knowledge was produced and the social forms that were accounted virtuous, between the moral character of the knower and the cultural standing of what was known.

The Scientist and the Civic Virtues

THE MORAL LIFE OF ORGANIZED SCIENCE

Charismatic domination transforms all values and breaks all traditional
and rational norms: "It has been written . . . , but *I* say unto you . . . "

Max Weber, *Economy and Society*

"BIG SCIENCE": ITS ORGANIZATIONAL FORMS AND ITS MORAL CONSTITUTION

In some strands of twentieth-century American cultural commentary, the
phenomenon of organized research was frankly celebrated. The organization and effective control of inquiry were pointed to as momentous,
modernity-defining and modernity-making inventions. Organization
speeded up discovery and the production of valued goods consequent
upon discovery: more knowledge, more profit, more power, more allevi-
ation of suffering, hunger, and want. But in other sensibilities the organization of inquiry counted as a violation of the very idea of science. Organization sapped the power of science to yield objective knowledge; it stifled
innovation; it reduced the scientist to the level of the hireling; it subjected
the researcher to the dominance of the organization rather than to the authority of Truth, definitively transforming science from a vocation to a job.
Just as organization eroded the scientist's virtue, so it distorted the capacity of inquiry to produce its proper object. The immorality and the imprac-
ticality of organization were two vocabularies for condemning the same
thing. As earlier chapters have indicated, the contrasting values placed
upon scientific organization proceeded from differing conceptions of the
identity of knowledge, of the moral status of knowing subject (both individual and collective), and of the objects of knowledge. This chapter offers
a focused account of how both external commentators and participants
thought about aspects of scientific organization, and, in particular, of how
virtue was seen to be threatened, lost, relocated, or enhanced within the
configurations of organization.

There are two poles to scientific organization: the mundane experiences of individuals forged into collective actors and of those whose wills create, maintain, and modify the forms of collectivity—so to speak, those who are organized and those who organize, the led and their leaders. Accordingly, this chapter considers both scientific teamwork and scientific leadership. Leadership and the collective structures that are led may be in tension, but they also define each other. What is thought about organized inquiry is in dialogue with what is thought about social and intellectual relations within the organization and, therefore, about those who invent and enact those relations. The chapter starts with twentieth-century commentary on the phenomena of scientific teamwork and the related issue of scientific planning, and then considers the question of the individuals whose leadership molds collective action. The material here spans the period from the early twentieth-century origins of American industrial research to about the middle of the Cold War period. Aspects of organized science in later periods, and up to the present, are considered in chapters 7 and 8.

When Alvin Weinberg, director of the Oak Ridge National Laboratory and, under Eisenhower, a member of the President's Science Advisory Committee (figure 5), reflected on the phenomenon of Big Science in 1961—indeed when he gave it its name—he identified four related features that constituted the novelty of scientific bigness: big funding; big instrumentation; big industry and, especially, big government as its patron; and, lastly, big organizational forms in which science was conducted.[1] Six months before Weinberg's essay in *Science* magazine, Eisenhower's farewell address had picked out, and worried about, each of these features of the New Scientific and Technological Order. In the "technological revolution," Eisenhower said, "research has become central; it also becomes more formalized, complex, and costly. A steadily increasing share is conducted for, by, or at the direction of, the Federal government. Today, the solitary inventor, tinkering in his shop, has been overshadowed by task forces of scientists in laboratories and testing fields."[2] The president was concerned about the effect of these changes on democracy, but the physicist worried about their effect on science. The free-acting individual was being replaced by the government-funded organized research team. There were real possibilities here for the corruption of science, even for the destruction of the very idea of science.

The sentiments voiced by Eisenhower and Weinberg pointed to something supposed to be quite new under the Sun: nothing the like in science had ever been seen before.[3] The contract system binding academic

FIGURE 5. Meeting of the President's Science Advisory Committee, 19 December 1960. This picture was taken a month before Eisenhower delivered his farewell address, warning of the dangers of the military-industrial complex. Alvin Weinberg is standing behind the president (seated, center), just to his right. John Bardeen, formerly of Bell Labs and then at the University of Illinois, is standing third from the right. James B. Fisk, then president of Bell Labs is seated on the left. (Reproduced by permission of the American Institute of Physics, Emilio Segrè Visual Archives and the Dwight D. Eisenhower Presidential Library.)

research to the Federal government, industry sponsorship of science, the scale and expense of scientific instruments, the specialization of scientific knowledge and the division of scientific labor, and, above all, the planning and organization of scientific research as collective work that flowed from these other conditions—they were all new: the offspring of the Manhattan Project, the MIT Radiation Laboratory, the Johns Hopkins Applied Physics Laboratory, the postwar weapons laboratories, and the teamwork of big particle physics, which went on to spawn the 138-scientist authorship of the paper announcing the discovery of the W and Z particles, and the 500-scientist team planned for a single experiment on the aborted Superconducting Supercollider.[4] The novelty of organized Big Science was insisted upon as much by its celebrants as by its critics. Just after the War, "Boss" Kettering—simultaneously director of research at General Motors and president of the American Association for the Advancement of Science—approvingly announced that one major, and probably permanent, lesson of wartime mobilization was the value of teamwork: "We learned how to cooperate."[5] But the Berkeley zoologist Richard Goldschmidt linked the introduction of "organized teamwork in research" not just to the war but to large-scale twentieth-century changes in economic life, whether formally socialist or formally capitalist: "The trend everywhere is away from individualism and uncontrolled economy toward collectivism." Scientific teamwork was just a species of the genus collectivism, a pathological form in society and in the social configurations that produced knowledge.[6]

Such were the scale and the dramatic outcome of World War II organization that the significant experiences of scientific mobilization in the First World War were put in the shade. But in the immediate post–World War I period, military mobilization and the early triumphs of organized industrial research were already giving occasion for frank celebration of these new modes of scientific production.[7] They were efficient and they proceeded from a just conception of how researchers ought to deal with one another. At the Westinghouse Research Laboratory, director P. G. Nutting observed in 1918 that "cooperation and team work" in industrial research facilities are now "carried to an extreme heretofore unknown," and he celebrated "the interchange of ideas" and the "excellent spirit of cooperation and comradeship": "Such effective team work in scientific research is new, but the results indicate that it has come to stay."[8] The chairman of the War Emergency Board of Plant Pathologists announced that "cooperation among scientists for the solution of problems must come" and that scientific individualism was not only unproductive but "feudal" and "autocratic."[9] The chairman of the National Research Council, a psychologist at the University of Chicago, similarly criticized the outmoded "fetish among scientists that we must rely upon individual inspiration and initiative, and that the individual worker must be safeguarded in every possible way from the corroding influence of administrative organization."[10] Elihu Root of General Electric declared in 1919 that "scientific men are only recently realizing the effective power of a great number of scientific men may be increased by organization just as the effective power of a great number of laborers may be increased by military discipline . . . The prizes of industrial and commercial leadership will fall to the nation which organizes its scientific forces most effectively."[11]

In the mid-1930s, a student of technical invention wrote passionately against traditional individualistic assumptions and celebrated the fact that "the twin *Zeitgeist*" of the "modern age" was constituted by "Science and Organization." The equation between the unique, free-acting individual and the authenticity of knowledge not only *should* be broken, but late modern developments—properly understood—were vividly showing that there *was* no such link. Individualism was just not a good description of how proper knowledge was made.[12]

Even when, after the Second World War, the far larger scale of scientific organization gave rise to more intensive reflection, observers drawing attention to the phenomenon of Big Science teamwork were by no means exclusively on the side of the angst-ridden opposition. In 1946, Frank Jewett of AT&T announced that the coordinated team was "the most powerful, effective and economical method of handling complex [scientific and technical] problems," and, in 1953, Earl Stevenson, the president of the industrial consulting firm Arthur D. Little, Inc., wrote that "the greatest revolution in the long history of mankind" had occurred within his own lifetime, and that this revolution was "largely due to the development of a system for utilizing the results of science—a system of teamwork among scientists, engineers, and manufacturers."[13] But the critics were probably more vocal in the wider culture and among academic scientists.

In 1925, the most influential portrayal of the American scientist before Watson's *Double Helix*—Sinclair Lewis's *Arrowsmith*—concluded with Martin Arrowsmith's joyful escape from the organized scientific production-line of the McGurk (i.e., Rockefeller) Institute in New York to the solitude of the New England woods, where, with his buddy Terry Wickett, Arrowsmith would "experience a great release of his creative energies."[14] In 1959, Merle Tuve, with vast wartime experience of organized research directing proximity fuze work for the Office of Scientific Research and Development, bitterly complained about the current tendency to lump together for accounting purposes "the intensely personal activity of individual professional workers in search of scientific knowledge"—what is properly called basic research—with the sort of routine fact accumulation, measuring, and development undertaken by "organized groups of technicians." He was scathing about the fashionable idea "that teams and big instruments create new areas of knowledge": more often than not they are making work for scientists, not making scientific knowledge properly so called.[15] By the 1960s, Alvin Weinberg was seeing clear causal connections between intellectual specialization, the collectivization of research, and the erosion of creativity. Big Scientific teamwork is pathological:

Growth and fragmentation impair the efficiency of science by forcing science to become a team activity, because a single knowledgeable mind is in many ways a more efficient instrument than is a collection of minds that possess an equal total of knowledge. The act of scientific creation, no less than any intellectual creation, is largely an individual act . . . I simply cannot imagine the theory of relativity, or Dirac's equation, coming out of the teams that nowadays are so characteristic of Big Science.[16]

The Manhattan Project physicist Robert Wilson believed that "most of us" have a knee-jerk reaction to the very words "team research": "We have a suspicion that team research is superficial, uncreative, and dull; that it is overorganized and overfinanced." By contrast, the individual researcher was "doing creative, poetic, and enduring work — true intellectuals they, not bureaucrats enslaved by a computer. Team research, the cliché tells us, is bad; individual research is good."[17]

One now begins to see a family resemblance in the identification, and the criticism, of knowledge responsive to social interests and, particularly, of socially organized knowledge: Truth is more solitary than social; a voice crying in the wilderness; in relation to the individual genius, society is, as Swift suggested, but a confederacy of dunces. Much of the rhetoric of Eisenhower's farewell address closely parallels *Signs of the Times* by the nineteenth-century Scottish Romantic essayist, Thomas Carlyle: "No individual now hopes to accomplish the poorest enterprise single-handed and without mechanical aids; he must make interest with some existing corporation, and till his field with their oxen."[18] The issue here is the relationship recognized to obtain between the value of knowledge and the social condition. Knowledge is the product of genius; genius is irredeemably individual; attempts to organize the production of knowledge worthy of the name is a recipe for disaster; a camel is a horse designed by a committee; and mediocrity is the necessary consequence of collectivity.[19] The Nobel Prize committee makes a significant gesture in that direction by restricting the award of all its prizes — barring the peace prize, which has gone to organizations — to no more than three persons. In a sacred idiom, scientific discovery is divine inspiration; in a secular idiom, it is spontaneous and serendipitous. If one follows Michael Polanyi and his associates in the Society for Freedom in Science, writing against J. D. Bernal and his Marxist friends, science cannot be organized for the same reason that it cannot be planned: it is not a product of formal rational method but an emergent property of the individual human mind — personal, not social or rationally formalizable, knowledge. Polanyi's colleague, the Oxford zoologist

John Baker, liked in this connection to commend Einstein's remark about himself: "I am a horse for single harness, not cut out for tandem or team-work," noting that "some degree of withdrawal from social life is recorded again and again in the biographies of great scientists."[20] Writing in the context of the Oppenheimer hearings and the imposition of loyalty tests, Vannevar Bush—with his intimate experience of large-scale organized science—stipulated that scientists "are an individualistic lot; otherwise they would be of little value as scientists."[21]

Detlev Bronk, physiologist and president of both the Rockefeller Institute and the National Academy of Sciences, warned against the postwar fetish of administered science and equated the individual with the radically innovative:

> No one directed Newton to discover the laws of gravitation. No one organized Faraday's discoveries in electricity for the benefit of the modern electric age . . . No one instructed Niels Bohr to pave the way for the production of atomic energy. Many scientific discoveries will elude direction and organization as surely as would the creation of great music or poetry, or sculpture or art. Much of scientific research is exploration of the unknown and I, for one, do not believe that it is possible to direct the course of an explorer through unexplored territory.[22]

And a Johns Hopkins psychologist likewise condemned the current tendency to "forget that in the past great discoveries have with few exceptions been made by individual workers, often working in great isolation; that some of the most important discoveries have been made without any plan of research—largely by accident." "Team research" is all very well for articulating and refining existing ideas, but "it rarely produces new ideas."[23] Organizing genuine scientists is like herding cats, and that's just as it should be. The sentiment even emerged from the heart of American postwar high-tech business. Just weeks before Eisenhower made his skeptical remarks about organized science, William O. Baker, vice-president for research at Bell Labs—one of the fabled successes of American scientific organization—told the American Association for the Advancement of Science that "the ideas of science come one at a time from one person and one mind at a time. Sometimes two or three can aid each other. But scientific discovery cannot be collectivized, and it does not flourish in collectivized settings."[24] Similarly, Ralph Cordiner, the CEO of GE in the late 1950s and early 1960s, wrote that

> if you can name for me one great discovery or decision that was made by a committee, I will find you the one man in that committee who had the

lonely insight—while he was shaving or on his way to work, or maybe while the rest of the committee was chattering away—the lonely insight that solved the problem and was the basis for the decision.[25]

Scientists—so the classical trope goes—lack the social virtues: the history of science, it is commonly said, establishes that beyond doubt, and scientists should wear their individualism as a badge of epistemic honor, as it is a corollary of their antiauthoritarianism. You cannot organize scientific production into teams, for teamwork makes no sense without external command and control. It may work for automobile manufacture, and it may even work for applied science and engineering, but not for science properly so called and certainly not for basic inquiry. In the immediate aftermath of the First World War, American scientists traced the spirit and forms of scientific organization to Germany, contrasting these to the intense individualism of science in Britain. Organization might be well adapted to the development of novel ideas, but the true "original genius" worked alone: "Such a thinker is far enough separated from others to allow nature to work out her own impression on him . . . Work of highest value can be done only as an expression of the most marked individuality." In that setting, individuality or organization in science were not just matters of historical circumstance; they might also be stipulations about racial essence.[26] After the Second World War, many of those academic scientists who had for the first time experienced massively organized large-scale research insisted on a distinction between genuine science, on the one hand, and applied research or development, on the other. Merle Tuve, reflecting on and celebrating the teamwork of the wartime Applied Physics Laboratory, wrote that "it was engineering development, not basic research. 'Team work' is always development, research is an individual activity."[27]

The sentiment was echoed by academic scientists who had not been involved in war work and who now saw a threat to pure science in post–World War II large-scale government sponsorship of university research. A University of Chicago parasitologist warned of the danger to basic science that flowed from misplaced notions of what could be successfully organized. Applied science and development work were predicated on knowing what you were looking for or were trying to produce, so directed applied research groups might work, but basic science was "largely dependent on lucky guesses (inspiration if you like) and often just plain fumbling": this kind of thing "could neither be organized nor directed."[28] By the 1960s, cynical or embittered scientists whose careers had spanned the transition from small-scale artisanal to big organized science were pronouncing anathemas on the new heretical order.[29] The biochemist Erwin Chargaff, for example, wrote that "the institutionalization of science as a mass occupation, which began during my lifetime, has brought with it the necessity of its continual growth . . . —not because there is so much more to discover, but because there are so many who want to be paid to do it." "I see only one salvation," Chargaff concluded: "the return to what I should call 'little science.'"[30] And in 1965 the physicist Ralph Lapp, who had worked on the Manhattan Project, sounded an alarm about a technocratic "new priesthood," writing that "big science is not necessarily great science . . . [The f]orce feeding of research may inhibit its sense of vitality." Bureaucracy "may ensnarl scientists and turn them into paper-pushers and sterile administrators." Were Galileo and Einstein alive today they might never get themselves funded by a Federal grant: too individualistic.[31] So a large body of emotionally charged twentieth-century American commentary identified the capacity to produce genuine scientific knowledge with the virtues of the free-acting individual. The authentic scientific community was not an organization; it was a spontaneous assemblage of free actors, and attempts to transform the making of scientific knowledge into a formally organized endeavor could succeed only by sacrificing Truth, Progress, and, ultimately, Power. With these generic considerations in mind, I want to retrieve some pervasive and detailed features of how teamwork in twentieth-century Big Science was characterized. First, I describe how it was widely talked about by academic social scientists and cultural critics in mid-century America; then I describe how teamwork was perceived from within the research laboratories of big industrial corporations.

ORGANIZATION MAN AND HIS VICES

While Robert Merton's sociological writings about science had nothing to say about scientific teamwork as such, the tension between individualism and the forces of social structure lie right at the heart of his enterprise. Merton reckoned that the behavior of the scientific community cannot be sufficiently accounted for by aggregating the innate characteristics of its individual members. Chapter 2 has already drawn attention to Merton's insistence that there was nothing special about the innate temperaments or personalities of the individuals recruited into the role nor that their motives, as practicing scientists, were in any way "distinctive."[32] Nevertheless, Merton thought that scientists' communal behavior was unique, and so the gap between moral and motivational ordinariness, on the one hand,

and communal exceptionalism, on the other, was bridged by a special kind of *socialization*. Once one appreciated the force of this socialization, one understood why individualism could not account for patterns of communal conduct. But here lay the essential tension of this sociological story: the forces of social structure, as articulated and enforced by the scientific collectivity, produced society's most individualistic actors—skeptical, antiauthoritarian, respecting the relevance of no group identifiers (race, religion, gender, age, or status). In this way, Merton's sociology, far from giving sole explanatory role to the collective nature of scientific activity, actually found a new idiom for emphasizing its individualistic character. So Mertonian practitioners in the early 1950s found themselves making the sociological case in markedly psychological terms: "The canons of validity for scientific knowledge," Bernard Barber wrote, "are individualistic: they are vested not in any formal organization but in the individual consciences and judgments of scientists who are, for this function, only informally organized."[33]

By the mid-1960s, American academic sociologists broadly in Merton's tradition were engaging with the phenomena of Big Science teamwork that had already received so much attention from postwar scientists, administrators, and politicians. On the whole, they didn't like it, seeing it overwhelmingly as a problem. At Berkeley, Warren Hagstrom began his study of "Traditional versus Modern Forms of Scientific Teamwork" by announcing that "basic science is an individualistic enterprise."[34] "Individual independence [as a norm] in making decisions about research programs probably contributes to the efficiency of scientific research, even in the short-run": "The ethos of science can be described as individualistic." And this individualism "makes teamwork of all kinds more difficult."[35] Teamwork in science, even in pure science, where it was considered a highly unnatural form, Hagstrom recognized, was nothing new: observational astronomy, for example, had been a complexly organized activity for centuries; Bacon called out for organized inquiry in *The New Atlantis*; and in the 1790s Lavoisier said that "most of the work still to be done in science and the useful arts is precisely that which needs the collaboration and co-operation of many scientists."[36] But what, in Hagstrom's view, was worryingly new were the forms now taken by organized teamwork and the consequences these forms had for scientific integrity. The twentieth century, he said, had witnessed a shift away from traditional hierarchical professor-student teams and ephemeral teams of freely interacting, equal-status professionals, the first of which ended with the graduation of the student and the second with the acknowledged completion of the project.

What we are seeing now in scientific teamwork was exactly analogous to changes occurring in modern corporate capitalism, where "free partnership and apprenticeship" have been supplanted by "a more complex form of organization." In science, free "collaboration and the professor-student association" are being undermined by similarly complex organizational forms, the division of intellectual and practical labor, and — echoing both Weberian and Marxian sentiments — "the separation of the worker from the tools of production, and greater centralization of authority." In characteristically modern manifestations of scientific teamwork, both the structures of coordination and control and the division of labor conspire to turn previously free-acting scientists into what Hagstrom called "professional technicians," permanently alienated from the products, indeed from the authorship, of their intellectual labor. They give up their autonomy, working at others' bidding; they solve problems "for money and not for recognition in a scientific community. In other words, the professional technician, like most workers in modern society, is capable of alienating himself from his work." Ultimately, the creation of this cadre of professional technicians has the capacity disastrously to reconfigure the normative structure of science: "The technician is alienated from his work products; he cannot be expected to be strongly committed to the norms and goals of science since he is not paid to have commitments which get in the way of others."[37] It is, he wrote, a "nightmare" scenario in pure science, realized to its greatest extent in the experimental sciences, and especially in experimental physics, but, fortunately, not *yet* the norm in all areas of academic science.[38] However, in industrial research, even in its minority pure forms, the situation is much worse, where command and control requirements have already effectively resulted in renormalizing the scientist, undermining the scientific virtues, and eroding the quality of scientific knowledge.[39]

These sorts of concerns about scientific teamwork were not confined to academic social science. The special, and specially dismal, place of science in the institutional setting of American industry was what most unsettled the former *Fortune* journalist William H. Whyte, criticizing scientific teamwork in his 1956 *The Organization Man*.[40] Chapter 4 sketched the cultural context in which Whyte's work appeared and its relationships with other contemporary sentiments about scientists' individual and collective character. As Whyte saw it, the individualism of the Protestant Ethic was being systematically subverted by what he called the "Social Ethic," the deadening hand of corporate collectivism. So far as industry, government, and now even academic tendencies were concerned it was all part of the pathological "Fight against Genius": "Occasionally, the in-dividual greats of the scientific past are saluted, but it is with a subtle twist that manages to make them seem team researchers before their time." Whyte reckoned that Americans in the 1950s had "a widespread conviction that science has evolved to a point where the lone man engaged in funda-mental inquiry is anachronistic . . . Look, we are told, how the atom bomb was brought into being by the teamwork of huge corporations of scien-tists and technicians." But they were being misled: only occasionally does someone "mention in passing that what an eccentric old man with a head of white hair did back in his study forty years ago had something to do with it." Whyte's target was an ignorant, and ultimately self-destructive, un-willingness on the part of industrial managers to recognize authentic sci-entific values and to accommodate those values in the industrial research laboratory. His fundamental argument was that "between the managerial outlook and the scientific there is a basic conflict in goals." The current "orgy of self-congratulation over American technical progress" attributed it overwhelmingly to "the increasing collectivization of research." This was, however, a mistake that was storing up enormous future trouble for national welfare and security. If America organized the individual researcher out of existence, it would put a definitive end to scientific creativity and, of course, the valued goods flowing from that creativity.[41]

In the tensions surrounding the idea of Organization Man, and es-pecially in those associated with the scientist in industry, what is being played out are contrasting notions of knowledge, of an innovating society, and of the virtues of the creative person. American industrial management was working remorselessly to "mold the scientist to its own image; indeed, it saw the accomplishment of this metamorphosis as the main task in the management of research." If it succeeded, Whyte wrote, it would be com-mitting suicide, for "every study"—none was here cited—has demon-strated that the "dominant characteristic of the outstanding scientist" is "a fierce independence" that will not tolerate control, interference, or col-lectivization: "In the outstanding scientist . . . we have almost the direct antithesis of the company-oriented man." The company must understand, as it currently refuses to do, that the primary loyalty of the first-rate sci-entist "must always be to his work": "For him, organization can only be a vehicle." That is why American industry was attracting only the scientific "mediocre" and "second-rate." In all of America, Whyte announced, there were only two corporate research groups—General Electric and Bell Labs—that allowed their scientists the time and resources for "'idle curios-ity'": virtually all the rest drew no lessons from the success of these two, and were hell-bent on transforming the spontaneously creative scientist into a predictable and routinized Organization Man. They had contempt for the value of disinterested inquiry, thinking that project-oriented applied research would provide an endless source of creative ideas, or that fundamental breakthroughs would take care of themselves on the cheap. Industry failed to recognize "the virtue of purposelessness"; it actively discouraged the free play of curiosity; it hindered, and sometimes forbade, the publication of their scientists' results; it budgeted their scientists' time so rigidly as to make impossible the pursuit of individual lines of inquiry.[42] A booklet from Socony-Vacuum Oil Company, setting out its corporate policy, stood proxy for this obtuse misunderstanding of the conditions for scientific creativity: there was "No Room for Virtuosos," it announced; no place for the individualist. And a company film produced by Monsanto to entice young people into an industrial research career underlined the sentiment: "The film takes us to Monsanto's laboratories. We see three young men in white coats talking to one another. The voice on the sound track rings out: 'No geniuses here; just a bunch of average Americans working together.'"[43] Whyte saw American industry efficiently transforming the scientist from a virtuous, free-acting individual to a de-moralized, externally controlled team-player. Yet, Whyte maintained, in so doing it was making a fundamental mistake about the nature of scientific inquiry and the conditions in which it could produce either Truth or Power. As a like-minded critic of organized, planned research put it in *Harper's Magazine* just two years before the publication of *The Organization Man*, the American obsession with organization was already resulting in "fewer really new discoveries": "[We] may be killing the goose that lays the golden eggs."[44] In the Cold War and McCarthyite context, the defense of scientific individualism was a powerful way of reminding American society how much its security and welfare depended upon some of its least sociable and least conforming members.

Those professional cultural critics, academic social scientists and humanists who wrote about the matter at all in early to mid-twentieth-century America were overwhelmingly agreed that the idea of scientific teamwork was at least problematic and possibly subversive of the very ideas of genuine scientific inquiry and of its essential virtues. But the enlargement and differentiation of the social forms in which science was being done accelerated after World War II. Almost everyone observing the phenomenon considered it both novel and culturally significant, something that both reflected and embodied changes in the cultural structure and institutional ties of science, and something that was having remarkable effects on the practice of science and on the moral identity of the scientist. The organized scientist was accounted a new sort of person in the world, an embodied manifestation of a new form of modernity. The rise of the scientific team represented a cultural crisis for science itself, a crisis for the identity and authority of late modernity's most valued form of knowledge. If you took the defining virtues of the authentic scientist to be those of the heroic individualist, then scientific Organization Men appeared stripped of their appropriate virtues.

"FOOTNOTE-TO-FOOTNOTE VERSUS FACE-TO-FACE": INTERDISCIPLINARITY AND THE SOCIAL VIRTUES

The cultural critics and academic social scientists did not, however, have it all their own way on this subject. While they were criticizing and lamenting, the collectivization and organization of science were proceeding with little reference to their views. Those responsible for managing team research and for recruiting its members recognized the team as posing a set of practical problems, and, while they rarely theorized those problems in the approved academic manner, they were well aware that the social forms of scientific teamwork were relatively new and that the outcomes of teamwork, for both knowledge and knowledge workers, were unpredictable. If theoretical discourse on organized science was rare among practitioners, recognition of its uncertainties was not. Reflections on the practical problems of team science among practitioners and those closely allied to them took several forms. What do such reflections mean for twentieth-century conceptions of the nature of scientific knowledge, scientific practice, and the virtues of the scientist? Practical commentary on scientific teamwork emerged from the same range of institutional sources discussed in the preceding chapter: from industrial research managers and executives in charge of research and from the growing ranks of quasi-academic specialists in such subjects as "industrial relations" and "industrial economics" populating the business and engineering schools of American universities from the postwar period and receiving important research funding from the Federal government and industrial consortia. I concentrate here on the views of industrial research managers, since they were more intimately involved with the day-to-day realities of team research, with specially long experience of it as it emerged from very early in the twentieth century in such large innovating industrial firms as General Electric, AT&T, DuPont, Eastman Kodak, and Westinghouse. This is where both the practical forms of team research and reflections on its conditions developed most strongly, long before academic scientists had major experience of it in the Manhattan Project, the Rad Lab, or Johns Hopkins's Applied Physics Laboratory proximity fuze project.

So what did industrial research managers think about teamwork in the creation of knowledge? What practical understandings of the character of the scientist and of the scientific process were inscribed in their working notions of teamwork? Unsurprisingly, one does not find here any blanket condemnation of collectively organized inquiry. The possibility and the value of organization were, to a very large extent, *presumptions* of capitalist corporate culture about its highly qualified research workers as they were of any other class of corporate workers. This is not to say, however, that research managers took scientific teamwork for granted: many of them recognized special tensions bearing upon the organized research worker, and many were unsure of how organized industrial research was done. They reflected on the phenomenon, they sought out relevant corporate experience, and even, very occasionally, called on the assistance of social scientists to help them understand how to do the thing better, or at least to justify preferred policies already in place. *Why* the scientific team? For the industrial research manager, one crucial response was that corporate research naturally centered not on the competencies and career interests of any one worker, or on any one group of specialized workers, but on the *project*, which typically called on the skills of research workers from a variety of scientific disciplines, willing to communicate, and placed in organizational structures where communication beyond that obtaining within a specialized discipline was facilitated and encouraged.[45] So research on photographic emulsions for Eastman Kodak—and this even before the development stage brought scientific researchers and production engineers into contact—necessarily enlisted the labors of colloid chemists, physical chemists, organic chemists, and physicists specializing in geometrical and physical optics (figure 6).[46]

Bell Lab's invention and development of the transistor mobilized physicists, chemists, and engineers—organized in subdisciplinary groups, but with coordination and control being effected both by personnel interchange, by periodic internal meetings in which each group was kept informed of other groups' progress, and by a multidisciplinary coordinating committee.[47] Thomas Midgley of the Ethyl Corporation spoke for many research executives when he observed that "the 'team' is better adapted to the solution of problems," whether pure or applied, "involving two or more fields of science."[48] After the war, scientists and research managers with huge experience of this aspect of teamwork could point

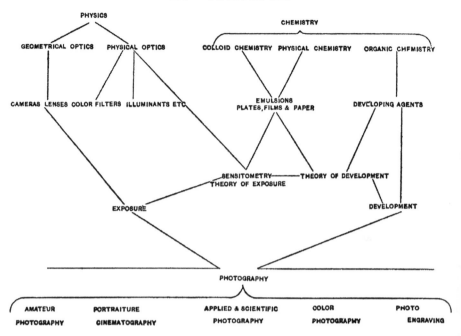

FIGURE 6. Kenneth Mees's depiction of the relationship between scientific specialties and the research problems of a photographic research laboratory, such as the one at East-man Kodak that Mees directed. Mees categorized his laboratory as "convergent," one in which a range of scientific disciplines and subdisciplines were organized to bring their expertise to bear upon a specific production or development problem that might not itself belong to any one discipline. Note that this is not an organization chart but a rendition of how various forms of technical expertise relate to one another and to the overall goal of the research effort. (From Mees, *The Organization of Industrial Scientific Research* [1920], p. 84.)

to the development of radar and, especially, of the atomic bomb, as the most spectacular and public evidence of teamwork's necessity and success. Edward Teller famously insisted in 1955 that such technoscientific projects as the design of the hydrogen bomb are inevitably "the work of many people," not just because they were large but because they were *multidisciplinary*.[49] The president of Arthur D. Little, who was also chief of the Chemical Engineering Division of the National Defense Research Committee during the war, wrote that "the development of the atomic bomb is probably the outstanding example of what can be done when creative effort is organized and focused upon a single objective—when engineers and scientists work closely together as a team to bridge the gap between scientific knowledge and engineering requirements."[50] And a director of research with a chemical company argued that "co-operation has become more and more necessary as industrial research has become more and more a teamwork job . . . All who have read the reports of the history of the development of the atomic bomb have no doubt been greatly impressed with the brand of teamwork which was brought into play. Here were a considerable number of independent groups, specialized in different fields, and each doing a very special task, but all carefully co-ordinated and directed toward the same common end."[51] Interdisciplinary aspects of Manhattan Project organization were widely pointed to after the war as a pattern for industrial, and even for some academic, research, even as many of those organization patterns owed much to the experience of industry.[52]

Just as natural as project organization was the *ephemeral* nature of industrial projects. Teams had to be continually constituted and reconstituted, as promising projects emerged and others proved disappointing. Hence, the very idea of the project team implicated a notion of organizational *flexibility*.[53] So an industrial chemist wrote in 1942 that "research is, by nature, fugitive. Problems seldom last more than a year," and, accordingly, that research men had to be flexible, easily transferrable between projects and laboratory divisions.[54] The development of the transistor at Bell Labs, it was said, showed the restricted value of specialist disciplinary groups because there must also be "flexibility for dealing with unusual situations as they arise."[55] In 1953, an aeronautics research director wrote that the research worker "does not have a free choice in the paths he follows but must shift his lines of emphasis as demanded by the over-all objective . . . This redirection of scientific and engineering effort can only result from the concept of many people working as a team."[56] Physical chemists at Eastman Kodak could find themselves working on emulsions one day and switched to high-vacuum distillation techniques the next. Academic physical chemists might stick to their disciplinary last, or even their chosen problem area, throughout their careers, but industrial scientists, whether working on pure or applied problems, had to be more intellectually and socially adaptable, as they could not know with any certainty on what, or with whom, they would be working from year to year.

And here, indeed, industrial research managers from early in the century pointed to differences between academic research, with the work routines and personality types it supposedly fostered, and those that made sense in corporate settings. It was not that the training offered by universities was *theoretically* individualistic—despite all the principled talk of academic autonomy, the master-student and master-technician relationships were understood to be very significantly hierarchical. Rather, it was appreciated that students-in-training reported *only* to their supervisor, and that professors reported to no one *locally* who could effectively tell them what they could or could not do. The *academic* team in anything like the industrial sense before World War II was accounted a relative rarity, and that is what was meant when industrial managers referred to the university experience as individualistic. So in 1953 the president of Arthur D. Little acknowledged that "the great edifice of scientific knowledge" had historically been constructed through "unorganized and highly individualistic effort," and an industrial consultant writing a few years later said that the traditions of academic science conflicted with those of even people-friendly new organizational theory: "The traditions of scientific organization prescribe formal, impersonal relations but give little direct guidance for close, collaborative relations. They are more footnote-to-footnote than face-to-face rules . . . A relatively low value is placed upon collaboration in much of scientific education: the student is taught to do independent work. From this he learns the virtue of not being dependent, but at the same time it contributes little to his skill in relations of close interdependence."[57] Interdisciplinary research was natural in industry, and some commentators thought it would become so in academia, but universities still offered poor preparation for what it involved. In 1952, Peter Drucker wrote about a difficulty for organized work arising from the scientist's "deeply ingrained working habits—so deeply ingrained as to have become almost a part of [his] personality. He has been trained—and rightly so—to work on his own." He *can*, for all that, work in a team, just on the condition that he is left alone in his specialized domain, not subject to supervision from those lacking his specialized knowledge: "In his own field he is apt to insist on having complete control of the entire job."[58] Other commentators were even more forthright in describing the inherent individualism of the academic scientist. *Fortune* magazine in 1953 wrote that "the scientist, particularly the most gifted, is, by almost any definition, a maverick. His endowments, drives, interests, political opinions, and even religious beliefs are not, in most cases, those of the majority of society."[59] And a pair of academic human biologists writing in *Science* in 1944 noted how university mores supposedly reinforced the scientist's constitutional individualism: "Successful execution of cooperative research requires modification of competitive work habits which have been fostered by the hyper-individualistic philosophy of life expressed in the traditions of university research . . . Thus the young scientist . . . is poorly prepared to participate in the activities of a committee or a research team . . . Cooperative work is a social art and has to be practiced with patience."[60] The industrial scientist was therefore, in Aristotelian terms, *zoon politikon*, a social animal, and the virtues such persons required were those social virtues that fit them for technical work *with others*.

MOBILIZING THE VIRTUES, MAKING KNOWLEDGE

Industrial research managers were well aware that individual scientists varied enormously in their technical skills, their cognitive and practical habits, and their temperaments. They knew that teams could not be thrown together arbitrarily without regard to such individual considerations, and that those research workers lacking well-developed social skills were not necessarily useless to the corporation. Early in the century, several research managers did indeed express a wariness about both the idea and the necessity of genius. In 1916, Kenneth Mees at Eastman Kodak, one of the most articulate defenders of freedom of research in *industrial* laboratories, insisted that "it is necessary first to dismiss from the mind completely the idea that any appreciable number of research laboratories can be staffed by geniuses." It was not, however, that such towering original talents were undesirable, but that they were too scarce and too expensive. Accordingly, "all we have a right to assume is that we can obtain at a fair rate of recompense, well trained, average men having a taste for research and a certain ability for investigation." The task of the research manager was to combine a few "geniuses" and many "average" research workers, and to get the best that could be obtained from all of them, and this notably involved creative forms of social organization.[61] A research director for Arthur D. Little, Inc., similarly insisted that effective organization could mitigate the need for individual genius: "Organized research does not depend upon individual genius; it is a group activity, as distinct a business activity as selling merchandise; it is as capable of organization and direction; so-called business methods are equally productive in its administration. Supermen are not required." Accordingly, effective group action depended upon searching out and mobilizing the appropriate social virtues.[62] In the 1920s, the permanent secretary of the National Research Council, trying to overcome resistance to organization, and celebrating the achievements of team science in industry, invited his scientific audience to admit that "most of us are not geniuses"; we are sound, talented men "with more or less gregarious instincts, socially minded, and willing to play and work together." Research executives at a 1920s New York chemical company argued that their top researchers "should be 'gentlemen,' not in the snobbish sense, but in the broad meaning of the term, involving the qualities of fairness and consideration for others."[63]

Genius *might* very well be worth the trouble, but it had to be the real thing, and research managers were in general skeptical of the popular mythology that held genius necessarily unaccountable to the social virtues. "Sometimes," a GE research manager wrote, "it is possible to isolate a single individual of exceptional qualifications and treat him as a prima donna, but only a genius is of much value as a prima donna. As a rule, no laboratory can afford to hire men who lack the generous spirit of cooperation."[64] Yet, especially in the research laboratories of larger American companies, the willingness to accommodate whatever counted as genius, and its accompanying eccentric personality, seems to have became more pronounced over time. In the 1930s, even a director of a major industrial research laboratory—albeit one as dedicated to basic research as Bell Labs—had no problems in announcing that "research is in its very essence individualistic."[65] And in the very same year that *The Organization Man* appeared, *Fortune* magazine ran a piece about contemporary industrial research that gave the lie-oblique to its former assistant managing editor's key contention: its title was "Industrial Research: Geniuses Now Welcome."[66] But research managers tended to avoid the abstract lumping tendencies characteristic of academic social science, and in the post–World War II period this sort of sentiment was routine: "Not all scientific workers are of the same caliber," not even all Ph.D.s; some have specific strengths that others lack; some have temperaments and motivations that differ from others'; some are stubborn; others are not. "The scientific virtues are present in different proportions in different people, and some virtues when carried to an excess may be detrimental."[67] It takes all kinds to make up an effective research team—all kinds of intellectual skills and all kinds of temperaments and capacities for social interaction.[68] By the 1950s, it was common to hear industrial research directors both recognizing the organizational accommodations that might freely be made under certain conditions to asocial eccentricity *and* to insist that several appropriately organized men of modest abilities, working together, might constitute "a very good substitute for a genius."[69]

The importance attached to the social virtues in the life of the industrial scientist is evident in research managers' commentary on *recruitment*. At least as early as the mid-1920s, managers were asserting the importance of "personality" in effective organized research, and making lists of the requisite virtues (see figure 7), of which the last was typically that appropriate to teamwork: "imagination, initiative, resourcefulness, energy, persistence, judgment, honesty, accuracy, dependability, loyalty and cooperativeness."[70] By the post–World War II period, a fairly standard list of the twelve "personality traits" that industry looked for in its recruits had evolved, and these included "the ability to cooperate," "to work on a team," "to get along with the fellows," as well as "courage" and "integrity."[71] The director of an innovative industrial chemical and biophysical research laboratory wrote about the absolute necessity of integrity and honesty: dedication to the internal rewards of inquiry and not just to its external compensations, the willingness to deal with one's colleagues frankly and openly, to give criticism forthrightly and sensitively, and to receive it without affront.[72]

A manager of an engineering laboratory defined the requisite virtue of "scientific integrity" as "the ability to consider another man's work as favorably as you would your own or another group's needs as favorably as you would those of your own group."[73] And a research director at the Ethyl Corporation said that the characteristics he wanted in young researchers were "'honesty' first, then, 'cooperation' which includes willingness to submerge personal desires in joint accomplishment."[74] In 1946, a GE physicist writing in *Science* described the virtues looked for in hiring and retaining industrial research personnel: honesty is "the basis of true friendship and teamwork, and hence is an essential requirement in a cooperating group." And "Generosity, an old-fashioned virtue, is also necessary in the modern research laboratory. It is essential for cooperation, without which a laboratory to-day is primitive."[75] The managers were aware that these virtues were traditional, even as they were mobilized in creating some of the most novel social configurations of the late modern order. They were traits considered to belong both to the masculine and the social virtues. Some academics training chemists for industrial work accounted the virtues of "learning to live together" so valuable that they recommended that excessively "introverted" or "extroverted" students be encouraged to join Alpha Chi Sigma, the professional chemical fraternity.[76]

The question of gender is central to practical questions about the makeup and functioning of the scientific team and, at the same time, deeply problematic in its operation. Statistics for the proportion of women among industrial scientists are patchy, but it is a fair assumption that, for the period from the beginnings of American industrial research to the immediate post–World War II period, the figure was near or below the 8% of all employed scientists cited by the 1966 National Register.[77] In 1951, out of 3,000 American physicists with Ph.D.s, just 66 (or about 2%) were women.[78] This is just to say that the industrial research laboratory in the first two-thirds of the century, like almost all American science, was very much a man's world. Though absence of evidence is not in itself evidence of

SAMPLE OF LETTER TO REFERENCE

John Smith and Company
65 Blank Street
New York, New York

Confidential

Gentlemen:

....has applied to us for a position in our Research and Product Engineering Laboratories. We would prefer calling on you in person for information pertaining to this applicant's personal history and background, and regret that we are not able to do so. However, your frank and candid answers to the questions on the attached sheet would be greatly appreciated and naturally would be kept in strict confidence.

For your convenience in replying, please use the enclosed envelope. Should you have information you do not care to put in writing, you may telephone us collect.

Thank you very much for your co-operation.

Yours very truly,

RESEARCH AND PRODUCT ENGINEERING LABORATORIES

By:...

CONFIDENTIAL

Applicant's Name.. .

Date...

REPORT FROM APPLICANT'S REFERENCE

From:..Address...

1. How long and how well have you known applicant?...

   ....

2. Do you believe that applicant is:

   Honest...Sober...........................Dependable........................

3. Is there anything which would tend to reflect unfavorably on applicant's character or reputation?..

4. Any question of loyalty to the United States?..

5. If a former employer please fill in the following:

   A. Employed by you beginning...............................Ending..............................

   B. Nature of duties...

   C. Reason for leaving your employ...

   D. Would you re-employ?..

6. Do you have any information regarding applicant's family background, along the lines indicated above, which you believe would be of interest to us?................

   ....

7. General Comments:

FIGURE 7. A specimen letter of reference for an industrial research worker, as presented by the research director of a carpet factory in 1948. Note both the traces of Cold War anxieties (as in questions about the applicant's "loyalty to the United States") and also the many questions about character and personality: can the applicant "get along" with

PERSONALITY

Consider manner-
isms, appearance,
general impression   Undesirable ( )   Average ( )   Good   ( )   Exceptional ( )

ABILITY TO LEARN

Consider quickness to
learn and retain new
methods, ideas and
directions, capacity
to think   Slow   ( )   Average ( )   With Ease ( )   Exceptional ( )

QUANTITY OF WORK

Consider amount of work accomplished and speed of doing it	Low Output	( )	Average Output ( )	High Output	( )	Very High Output	( )

QUALITY OF WORK

Consider accuracy and thoroughness	Careless	( )	Passable ( )	Good Quality	( )	Highest Quality	( )

DEPENDABILITY

Consider reliability, willingness, consistent industry, and honesty	Unreliable	( )	Usually Reliable ( )	Reliable	( )	Absolutely Dependable ( )

JUDGMENT AND COMMON SENSE

Consider ability to see things to do, resourcefulness, aggressiveness	Needs Constant Supervision ( )	Routine Worker ( )	Resourceful	( )	Original	( )

CO-OPERATIVENESS

Consider ability to get along with people in various capacities, willingness, loyalty	Requires Prompting ( )	Indifferent ( )	Co-operative	( )	Helpful	( )

Check the type of position for which you would recommend the candidate in our Laboratories.

Synthetic Organic.....( )   Routine Lab. Work.....( )   Pilot Plant........( )

Analytical..........( )   Administrative.........( )   ....( )

Physical.............( )   Engineering...........( )   None............( )

What are the candidate's eccentricities or oddities?

Signed..

Title or Occupation..

others? is he "honest"? Other portions of the form (not reproduced here) inquire about the applicant's "physique" ("pleasing and vigorous" versus "weak and sickly"), his "family background," and his "physiognomy" ("attractive" to "repulsive"). (From Furnas, ed., *Research in Industry* [1948], pp. 233–236.)

FIGURE 8. Katharine Burr Blodgett (1898–1979), a physical chemist employed by the GE Research Laboratory from 1920, and the first female scientist to work for the company. Blodgett (center) is demonstrating surface chemistry technology on visitors' day. She graduated from Bryn Mawr in 1917, and then took a master's degree in chemistry from the University of Chicago and a Ph.D. in physics from Cambridge University in 1926 (the first woman to be awarded such a degree from Cambridge). At GE, Blodgett frequently worked with Irving Langmuir, with whom her father, George Blodgett, had also been involved as head of GE's patent department. She was an inventor of a range of processes for applying monomolecular coatings on metal and glass, known as the Langmuir-Blodgett film, and retired from GE in 1963. She was a "starred" scientist (picked out as a notable researcher) in the 1944 edition of *American Men [sic] of Science*. (Reproduced by permission of the American Institute of Physics, Emilio Segrè Visual Archives, *Physics Today* Collection.)

absence, I could find the names of only a small handful of female industrial scientists before about 1960 (one of whom, Katharine Burr Blodgett, appears in figure 8); I have no record of the views of any one of them; and I do not have any knowledge of a single female research director.[79]

As was common usage until recent decades, the masculine pronoun referring to the industrial scientist was usually meant to refer to people in

general. Few participants or commentators bothered with the "or she" formulation, suggesting that the scientist—industrial or academic—might
be other than male. Yet there was more at issue in treating the characteristics of team researchers. Just because notions of cooperation carried
the baggage of an overwhelmingly masculine body of industrial scientists and engineers, the social virtues were arguably projected onto, so to
speak, "people like us," people with whom one would like to work, who
presented few awkwardnesses for the organization in the smooth performance of its day-to-day functions. And, almost needless to say, that
criterion bore upon all sorts of groups underrepresented in the industrial
research laboratory—women, of course, but also African-Americans, His-
panics, Asians, and Jews. We do know that, with the Cold War "shortages,"
the relative absence of women from the scientific workforce in general became a matter of serious practical concern.[80] One research director, writing just after the end of World War II, wholly approved recruiting many
more female industrial scientists, but the section of his essay discussing the
matter was titled "The Woman Worker as a Special Problem"—the main
"problem" being their tendency to abandon their careers to marry and
have children, thus losing the company's investment in their training and
necessitating new hiring, and a secondary difficulty being the necessity
of recruiting a female personnel representative to deal with delicately
unspecified "disciplinary and social problems."[81]

Industrial science, and science in general, gradually opened up to women,
the pace accelerating from the 1960s and 1970s. But as the industrial laboratory was a man's world, operational sensibilities towards the virtues
required to succeed in it were informed by men's relations other with men.
And, almost needless to say, wherever the attribution of virtues and the
texture of sociability were molded to the contours of ethnicity, religion,
and class, then it might be sensible to talk more particularly about the
relations between industrial scientists who were, before the post–World
War II decades, not just male but overwhelmingly white, Anglo-Saxon,
Christian, and middle-class. Yet stereotypes of women's capacities and
characteristics that flourished in the early and middle parts of the twentieth century do not globally militate against their place in the scientific
team. After all, the same set of stereotypes that cast males as combative
and aggressive portrayed women as sociable and supportive. And if the
social virtues were indeed deemed essential to the work of the scientific
team, then, it might be thought, women were *better* positioned than men
to benefit from a stress on the ability to cooperate. So there is a patron-
izing triteness, but perhaps more than mere triteness, in a GE research

director's comment about women in the laboratory: "All, by their helpful participation in the laboratory's work and their promotion of its social affairs, added to the congeniality of the atmosphere."[82]

WHAT IS A TEAM?

Academic critics of scientific teamwork seemed to have assimilated their understanding of the team to military command and control models, and that is one reason why so many of them feared and condemned it. But, apart from identifying the team with some general notion of coordinated collective inquiry, industrial research managers were far less certain what kind of social form this was, if it even had a fixed identity, or if it was strictly necessary for conducting industrial research. And so throughout the early and middle part of the century they debated whether the proper analogy was the army platoon (though few agreed that such strong forms of control were appropriate), the team of horses (roughly similar to Durkheim's mechanical solidarity), the rowing crew, the football or rugby team (much favored for its specialized elements striving towards a common goal), or the symphony orchestra or opera production.[83] At Bell Labs, James Fisk was insistent that "[science] is an institution of men, a structural community of experts. In the face of the universally accepted precept that scientific (and almost all other scholarly) achievement is the result of individual effort exerted by particular, gifted minds, it is striking that a community of effort constructed as an institution of men is the basis for the most spectacular progress we have made in the application of scientific knowledge for human use." But he was equally adamant that such a "structural community" need not be thought of as a "team" at all: "I am not talking about teams, as diligent and efficient as they have been in the application of knowledge in many vital areas. . . . [including nuclear projects]. Rather I am saying that the best application of science and the synthesis of new knowledge turn out to be by a community of gifted people working intimately but independently, with each free to follow his own mind." By respecifying the "team" in this way, Fisk hoped to shed any connotation of command and control, while preserving the sense of an organically integrated community of inquiry.[84]

Some commentators commended organization while acknowledging that the forms of team research were sui generis, and could not be straightforwardly modeled on *any* existing social pattern: it "must involve something substantially different from organization in enterprises of other kinds, for example, war, industry, sport, and exploration."[85] Widely quoted administrators like Kenneth Mees at Kodak were adamant that organization charts (figure 9) purportedly describing the structure and authority patterns of a team were meaningless fictions and that anyone who took them seriously was likely to erode the possibilities for genuinely creative work: "As soon as you get a group together to do research, you find that you have to have an organization. You don't have to plan or organize the research itself, but you must have an organization among the men. This is usually expressed in a chart, and, frankly, I don't think that those charts mean anything, though recently I found them very valuable to meet the requirements of the War Labor Board and the Treasury Department."[86] Fisk grasped a similar nettle, while expressing his reservations in a more politic way: "It is sometimes said that formal organization and research are not compatible. I do not agree with this. Orderliness in relations among people, smoothness in function, and a free flow [of] ideas within a conceptual framework are helped, not hindered, by a charting of function and responsibility. But the organization chart must not become master, and organizational position must not be confused with scientific stature."[87] Again, the crucial point about research deemed worthy of the name— pure or applied—was that its outcome and trajectory were uncertain, and you simply could not determine in advance or with any stability what organizational forms were appropriate. There were, indeed, types of inquiry valuable in corporate settings where any sort of team approach was viewed as unsuitable.[88] Despite all the rhetoric identifying the university as radical and corporate settings as conservative, so far as organizational forms for conducting inquiry are concerned, the opposite is far closer to the mark, all the more so as one approaches the present. It was industry, rather than the universities, where the most radical experiments were undertaken in the institutional forms of inquiry and where institutional uncertainties were greatest.

The organization of American industrial research was never monolithic and never attained a fixed pattern that might become a template for just any form of industrial inquiry. It is hard to generalize about how the thing was done, just because it was done in so many ways. From the 1910s and 1920s, the most reflective commentator on the concrete forms of organization was Eastman Kodak's Kenneth Mees, and his writings on the subject were widely cited throughout much of the century. First, Mees insisted that organization be made to fit both the forms of inquiry and the personnel available. There was no one-size-fits-all formula for such

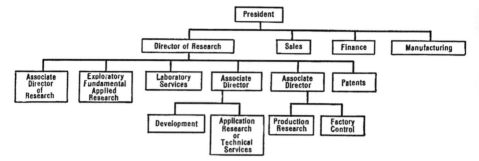

FIGURE 9. A specimen scheme showing the typical place of the research function in the overall organizational structure of a small company in 1948. Note that the director of research reports directly to the corporate president; the division called "exploratory fundamental applied research" reports straight to the director and is independent of "development," "production research," and "application research." This chart is one of a range of schemes deemed appropriate for different types of company, prepared by a research director for a domestic and industrial heating and cooling equipment manufacturer. (From Furnas, ed., *Research in Industry* [1948], p. 84.)

organizations. Yet one basic choice, Mees noted, was between what he called a "department" versus a "cell system." In the former, there were to be sections organized mainly around disciplines, each headed by one individual, known for his specialized disciplinary knowledge and skills. Accordingly, there would be a hierarchy *within* such sections and between the laboratory director and the section chiefs. In the cell system, there would be "a number of investigators of approximately equal standing, each of them responsible only to the director, and each of them engaged upon some specific research." There was no way of saying that one was superior to the other; each had its virtues and vices. The department system might deliver a high degree of intra-section coordination and communication, while running some risk of stifling the initiative of younger researchers. The cell system tended to be good for "men of original initiative and of the self-reliant type," but it was inclined to "exaggerate the vices of such men. They tend to become secretive, to refuse co-operation, to be even resentful if their work is inquired into; while if a man who has developed a line for work for himself in a cell leaves the laboratory it may be very difficult for anybody else to take up the work." In practice, Mees concluded, one probably wanted to take elements of both systems, and to combine them as occasion warranted.[89]

In addition, you had to decide whether your research laboratory was of the "convergent" or "divergent" type and organize it appropriately. The convergent laboratory was one in which all inquiries are connected with one common subject, while the divergent laboratory embraced many kinds of inquiry, with no obvious connecting bond of interest. University laboratories, Mees observed, were almost of necessity divergent because their primary function was said to be training. They they could not afford to be too specialized, while industrial laboratories tended to the convergent type, since they "must necessarily be prepared to deal with any problems presented by the works," focusing their energies on such problems—at least for a time—to the exclusion of others. Mees offered his own photographic laboratory as an example of the convergent type, while advertising the substantial freedom afforded its scientists within that framework.[90] There was, indeed, a temptation in a convergent laboratory for scientists trained in a specific discipline to stray from their corporate focus: "The men working in such a laboratory will often be intensely interested in the whole of the science with which they are associated, and will have original ideas of considerable value which they will naturally wish to test experimentally but which will have no direct relation to the general work of the laboratory. As a rule, it is necessary to resist this tendency," Mees cautioned. However, it would also be a mistake too strictly to constrain scientists' natural inclinations, and, within the boundaries laid down by the fact that the laboratory belonged to, say, a photographic company, there was, and ought to be, plenty of scope to pursue leads as the researchers saw fit.[91] Organizational theory—such as it was—took the form of rules of thumb rather than laws and principles. Industrial research laboratories had to remain organizationally flexible. Unlike academic departments, they could not commit to pursuing any given line of inquiry as long as its members liked. Company interests might change; researchers might be shifted to more promising lines of research and away from those that the director decided were not panning out. So whatever organizational forms were in place at any given moment could be reconfigured tomorrow, and it was essential that individual scientists understood this. Teams, that is to say, were, to a degree, ephemeral, and their value was understood to flow importantly from their flexibility.

The core defense of the scientific team in industry was just that the whole should be, and often was, greater than the sum of its individual parts. Sometimes this epistemic organicism was simply assumed as a commonsensical matter; less frequently, it was the object of managerial or even academic reflection. A director of a cancer research laboratory, surveying the growth and conditions of team research just after World War II, observed both "how much more effective in research a team can be than the sum total of the efforts of the same people working separately as rugged individualists," and how uncertain was the knowledge on the basis of which you could organize a research team.[92] An industrial bench scientist interviewed in 1953 emphasized the relationship between team organization and what he saw as industrial leadership in science: "When you work for industry, you are at the forefront of research. Industrial research is done by a team possessing various skills. The team member can accomplish more than the individual scientist in the university."[93] The director of a medium-sized technical company in New England noted the dominant presumption of individual creativity, but went on to argue a contrary case: "We usually think of creative effort as the activity of an individual. Actually, joint efforts involving two, three, or more of the right people can sometimes be more productive than the efforts of the same number of creative people working separately." Individual members of a research team "seem to act as catalysts for each other's creativity."[94] Research managers at Eastman Kodak wrote repeatedly that "men who are only average when dealt with singly may become extremely able by the mental stimulus provided by association with other men working on similar problems," and they experimented ceaselessly with the semipermanent but significantly fluid social forms that might allow such synergies to happen.[95] At Westinghouse, a research director acknowledged that it was the managers' job to set research goals, but immediately pointed out the superiority of collective judgment: "If we believed ourselves to be omniscient, we would of course set these goals ourselves. Personally, I am extremely conscious of my own limitations, and therefore strive to create an environment within the laboratory such that the wisdom which resides collectively within the research personnel is used not only to prosecute research projects, but also to choose the directions these projects should take."[96] As early as the 1920s, research managers were writing about the epistemic vices of both individualism and a single-personality-dominated research organization: it was prone to parochialism, narrowness of view, egocentrism, and dogmatism—all of which might be corrected by more collaborative forms of work.[97] (In more recent years, the aphorism of choice for such sentiments comes from the search-engine company, Google: "Nobody is as smart as everybody.")[98] These sorts of organicist sentiments—and, indeed, a range of practical policies informed by them—were absolutely standard among American industrial research managers from quite early in the century.

In a more recognizably reflective idiom, one of the founders of experimental social psychology, Sir Frederic Bartlett, described in the 1930s

what he called the "social constructiveness" of technical knowledge—
possibly the first academic to use such a term—and his model was the
teamwork he had experienced in research on anti-aircraft technology
during the Great War.[99] No existing technology, Bartlett argued,

> can be said to be the work of any single man, but of a number of men . . .
> Not only is no complete instrument the result of the foresight of any one
> person working alone, but it is not simply the aggregation of the contri-
> butions of a number of different men, all belonging to the same army unit,
> or to related units. A, perhaps, proposed this; B that; C the other thing;
> and E, very likely proposing no specific detail himself, worked all the de-
> tails derived from the various sources into a practical form, so that the A, B
> and C details are not any longer exactly as A, B and C thought of them.[100]

As a practical, not programmatic, matter, industrial research managers
throughout the twentieth century were conducting ongoing experiments
in the social construction of knowledge; and as a practical, not as a theoretical, matter, they took a social epistemology for granted.

THE SCIENTIFIC TEAM AND THE GOOD SOCIETY

As with views on the socialization of scientists treated in chapter 4, early
to mid-twentieth-century American culture contained diverging sensibilities to scientific organization and the virtues. Here too there was a
remarkable gap between strands of commentary on scientific teamwork
produced by academic social scientists and cultural critics, on the one
hand, and the practical reflections of industrial and government research
managers, on the other. It is tempting to say that actual experience with
the realities of organized research must be a major part of any explanation
of these differences—academic humanists and social scientists had, and
have, little experience of organized inquiry compared to their natural scientific and engineering colleagues—but there is probably something more
fundamental at issue. Ostensibly different *descriptions* of the knowledge-
producing team inscribe different *evaluations* of the good society. The
constitutive relationship between valued forms of social organization (or
the lack of organization) and valued intellectual products has been much
written about in the sociological history of science recently, and this is a
late modern manifestation of that phenomenon. In interpreting American views on the links between the good society, good knowledge, and the
virtuous knower in these periods, the cultural fault lines induced by reac-
tions to Fascism, Communism, corporate capitalism, the Cold War, and

McCarthyism are unavoidable considerations.[101] Writing at the height of these tensions, the president of Arthur D. Little, Inc., ingeniously erased any notion of tension between individuality and collectivity in innovative technical processes. The easy creative merger of the individual and the group was both the American way and the guarantor of a prosperous and secure national future:

> Organized creative technology falls easily within the framework of our American political and social concepts and reflects our way of life in giving full play to the genius of the individual, in giving him freedom to exchange ideas with his fellows, and opportunity without artificial restraints for joining with others in creative work. European observers today note what that astute traveler Alexis de Tocqueville remarked about American democracy in the mid-nineteenth century: that the home training of the American child, and his subsequent schooling, engender in him an attitude toward community action that knows no counterpart elsewhere in the world. The American has a strong sense of individuality, but at the same time an urge to work with others in the solution of a common problem. The concept of civic responsibility has nurtured in the typical American an ability to act responsibly and creatively in a group—and without government decree as the motivating force.[102]

A few years after the end of the Second World War, a leading member of the GE Research Laboratory noted early difficulties that Willis Whitney experienced in integrating foreign-trained scientists into the culture of industrial research: "They were imbued with the German idea of individualistic, secretive research and did not take kindly to the policy of teamwork." Voluntarily chosen teamwork was reckoned to be deep in the American character and close to the heart of national virtue.[103]

Differing American views of the scientific team were also situated in a broad sweep of cultural history. The organization of science in the twentieth century, especially, but not exclusively, in its industrial settings, formed a sharp challenge to appreciations of the identity of the scientific knower that had developed since Antiquity, and chapters 2 and 3 traced some of that pre-history. If the idea of genius, and individualist and disengaged forms of scientific inquiry, formed a more or less coherent evaluative repertoire from Antiquity through the early modern period, the organization of scientific labor could seem to some threatening, unnatural, and immoral. Nevertheless, what was happening to science from the late nineteenth century onwards was a massive change in social realities that re-situated the truth-speaker from the solitary to the social condition.

The statistics presented in chapter 4 firmly establish that change: I have already noted that, from quite early in the twentieth century, the typical American scientist, at both bachelor's and advanced degree levels, worked in nonacademic organized settings, most commonly in industry. Those realities might be only patchily recognized by many cultural commentators, and, when realized, they might be deplored, but they were nevertheless effecting changes in appreciations of who the scientist was: "The scientist today," one commentator wrote in 1965, "is typically an 'organization man.'"[104] For many scientists themselves, organization was not a point of theoretical dispute; it was a substantial fact about their conditions of existence, and the only matters of controversy concerned the exact nature of organization that they experienced.

SCIENTIFIC PLANNING AND THE INTEGRITY OF THE SCIENTIST

For those who feared organization and those who celebrated it, organization implicated some notion of *planning*—the control and direction of scientific inquiry. Why else would one organize a group of scientists, if not deliberately to coordinate their labors to some specific end? In that sense, the organized scientist was bound to be executing some sort of *plan*. Whether science was the kind of activity that could or could not be planned embedded views on what kind of persons scientists were, and, especially, views on the nature of their vocation. Were they autonomously dedicated to the pursuit of knowledge, and were they the best judges of how to achieve knowledge, or did they require external discipline, direction, and assessment similar to that experienced by other sorts of workers? In this way, debates over the planning of science were, of course, highly political, but at the same time they inevitably enfolded notions of the scientist's virtue. Some of the most voluble critics of planning, including those opposing planning imposed by both socialist and capitalist governments and by corporate capitalism, rejected planning as antithetical to the very idea of science. So, for example, Michael Polanyi influentially wrote that "any attempt at guiding scientific research towards a purpose other than its own is an attempt to deflect it from the advancement of science." He acknowledged the realities of wartime scientific organization but not the legitimacy of any enduring lessons flowing from that experience: "Emergencies may arise in which all scientists willingly apply their gifts to tasks of public interest. It is conceivable that we may come to abhor the progress of science and stop all scientific research, or at least whole branches of it, as the Soviets stopped research in genetics for twenty-five years. You can kill or mutilate the advance of science, you cannot shape it. For it can advance only by essentially unpredictable steps, pursuing problems of its own, and the practical benefits of these advances will be incidental and hence doubly unpredictable." The uncertainties of genuine scientific research, as opposed to the routine applications of fundamental science, meant that inquiry just had to be spontaneous and free. The mobilization of science during the war was an aberration, and neither from that experience nor from the half-century long record of industrial organization should *any* positive lessons be drawn: "In saying this, I have *not* forgotten, but merely set aside, the vast amount of scientific work currently conducted in industrial and governmental laboratories." Industrial science and equivalent forms of organized government science just weren't science, properly so called.[105]

By now, it should come as no great surprise that some American industrial research managers actually agreed with sentiments like Polanyi's. Kenneth Mees became famous in management circles for his celebration of research *disorganization*. The epigraph to his influential 1920 book on *The Organization of Industrial Scientific Research* boldly stated that "there is danger in an organization chart—danger that it be mistaken for an organization," the source of which was not a scientist fed up with corporate bureaucracy but one of the founders of American technical management consultancy, Arthur D. Little.[106] Disorganization, and the research autonomy consequent on recognition that planning in these areas was naturally constrained, was justified as commercially functional. Mees was quite hardheaded enough to insist that "the primary business of an industrial research laboratory is to aid the other departments of the industry," and that its central responsibility was to contribute to the corporate bottom line,[107] but he rejected the distinction—central to the views of both the British anti-planning Society for Freedom in Science and to the practical orientations of the post–World War II Steelman Report—that only "pure" science was incompatible with planning.

Commenting on early writings by Polanyi, Mees said that "I take issue with [the] description of applied science as a field in which freedom of science might conceivably be undesirable. I have been engaged in applied science for forty years, and in that period I have come very definitely to the conclusion that the prosecution of applied science in its most efficient form is identical with that of pure science. I don't think for a moment that it is desirable that applied science should be directed *except in times of emergency*."[108] Mees's own much-quoted aphorisms include: "When I am asked how to plan, my answer is 'Don't,'" and "No director who is any good ever really directs any research. What he does is to protect the research men from the people who want to direct them and who don't know anything about it."[109] As his obituarist put it, "Mees recognized the significance of the individual and brought into his sphere as many as he could, and protected them from outside interference."[110] And the only remark by Mees that finds its way into the books of quotations is an eloquent condemnation of research control:

> The best person to decide what research work shall be done is the man who is doing the research. The next best is the head of the department. After that you leave the field of best persons and meet increasingly worse groups. The first of these is the research director, who is probably wrong more than half the time. Then comes a committee, which is wrong most of the time. Finally there is the committee of company vice presidents, which is wrong all the time.[111]

Not all research managers were willing to credit the literal truth of Mees's self-denying ordinance. "You obviously plan," one of his colleagues said. He had looked through the published reports of the Eastman Kodak Laboratory and concluded from their topical continuity that "it is obvious that they are planned"—to which Mees had a ready response: "As a matter of fact, our scientific work *is* continuous, not because it is planned but because our scientific men continue to work along the same lines." That's just what they are interested in doing, and, to that extent, the scientist's commitment to inquiry was in no necessary conflict with corporate goals.[112]

Writing in the aftermath of the Steelman Report, Mees and Leermakers were worried that policy makers were drawing the wrong conclusions from the history of industrial research and that they were ignoring *limits* on the planning of science. "Only those things can be planned which can be controlled, and plans are carried out only as long as control is effective," they wrote. Production may be planned, and the last stages of development may be planned, but very little of what is called science—fundamental *or* applied—is subject to effective planning. Scientific knowledge "proceeds from the free operation of the minds of scientists."[113] It was right that a company should gauge the success of its research installation by its contribution to the development of new products, but it was quite wrong to think that ideas for such novelties could emerge through command and control apparatus. "The initiative shown by the laboratory staff members in proposing and undertaking new work naturally depends upon the amount of freedom they are given to follow up their ideas. If the work of the scientists is rigidly restricted as to problems and methods of attack, their productiveness in new ideas will necessarily be low ... New products are not found by attention to them on the part of the management of the laboratory. They arise spontaneously from the mass of ideas and suggestions coming from the research workers themselves."[114]

This sort of sensitivity to the pathological consequences of attempts rigorously to control the conduct of research was, according to the president of Bell Labs in the 1940s and 1950s, something that "all successful industrial research directors know ... and have learnt by experience": the "one thing a director of research must never do is to direct research, nor can he permit direction of research by any supervising board."[115] A Minneapolis-Honeywell research director who insisted on the bottom-line criterion for judging research results concluded his piece in the trade journal *Industrial Laboratories* by conceding that, while "the amount of freedom or control of research projects is probably the most difficult question in its administration, ... I tend toward the principle expressed by Thomas Jefferson for government: 'The least government is the best government.'"[116] And in the early 1950s, the president of Arthur D. Little, Inc., approvingly quoted the sentiment that made pure research, even in its industrial manifestations, antithetical to organization: "Scientific research in its most elemental form is a very private occupation which eludes all attempts to bring it under the control of conventional management. The rule for the organization of pure research is, 'Don't try to organize it.'"[117]

The politicization and the moral charge of debates over planning tended to mask what planned science actually looked like, so to speak, at ground level. Some research managers made a distinction between the possibility and even necessity of planning what they called the "function" of research over a number of years—that is, organizing commitments to it and its place in corporate activities—and planning the "act" or "conduct" of research, in which considerable freedom of action was deemed simply necessary.[118] The Steelman Report echoed that sentiment: it was important to "distinguish between 'planning for research' and the 'planning of research.' It is well accepted that consistently productive research must be 'planned for,' in the sense that competent men must be assembled, facilities provided, and equipment installed for their use. The actual 'planning of research,' the scheduling of the detailed operations carried on, is quite another matter" (figure 10).[119] One might freely concede a sense in which research was "largely a matter of inspirational guidance," and, for that reason, something that could not be planned. Yet one could at the same

FORM ENG. 126
PRINTED 1-42-25P

REQUEST FOR ASSIGNMENT OF { JOB ☐   PROJECT ☐   N⁰ 11232

ASSIGNED TO:- ___K. R. Moore___ JOB NO. _____ PROJ. NO. ___E-8691-7___

ESTIMATED COST___$1600 - 500 Hours___ NOTE BOOK NO. ___4718___ PAGE NO. ___1___

CLASS NO. ___1-92___ CHARGE TO ___5209 - III___ WORK NO. ___IV-K___

TITLE:- ___Test Apparatus - High Frequency Armature Tester (Pulse Type)___
___Development and Tests of___

ASSIGNMENT: ___Please develop final calculations for a pulse type high___
___frequency armature tester as outlined in preliminary calculations___
___of project E-8410-7. Construct such laboratory samples as may___
___be needed to prove theory and a final unit for standard laboratory___
___use.___
___Sufficient number of tests are to be made to prove the capability___
___of this unit in determination of weak insulation as well as___
___production errors in construction.___

AUTHORIZATION: ___Executive Action Project E-8410-7___

REQUESTED BY _____ DATE _2-7-46_ APPROVED BY _A. D. George_ DATE_2-10-46_

FIGURE 10. Mundane planning in a post–World War II industrial research laboratory. While ideologically charged debates raged over whether or not planning was compatible with the idea of science, many aspects of day-to-day research not only could be planned but had to be planned for the inquiry to proceed. This is a specimen project request form, submitted by an industrial research section head and approved by the relevant research director. It says what the inquiry is about, what its purpose is, who's going to do it, what part of the organization it belongs to, what it's supposed to cost, and how long it's supposed to take. Bureaucratic codes are assigned to allow the request to be tracked and filed. Note, however, the language of uncertainty and approximation even in this bureaucratic document—"Construct such laboratory samples as may be needed"; the number of tests needed to establish the technology's capabilities are specified only as a "sufficient number." (From Furnas, ed., *Research in Industry* [1948], p. 164.)

time insist that "planning" was properly and uncontroversially used to designate "many of the preparedness steps preceding research . . . as well as many of the assistance programs," and that these actions were vital to a successful research program. What kind of equipment and personnel did you need? What sort of time frame did you envisage?[120]

In such contexts, the emotions and ideological responses still generated by debates over the planning of science often pass over consequential distinctions between its concrete forms. David Noble's survey of early American industrial research correctly cites occasional corporate rhetoric pointing to the desirability of tightly managing the creative process, while paying scarcely any attention to the large body of managerial commentary that frankly acknowledged *limits* to any such control.[121] In practical terms, widespread industrial recognitions of the limits to planning translated into a significant tolerance for the spontaneous and unpredictable emergence of novel research agendas among industrial research workers who enjoyed considerable freedom to formulate their courses of work. Early in his career at Eastman Kodak, Kenneth Mees observed that "the choice of investigations must necessarily be made largely at random and will be influenced to a great extent by the ideas of the scientific workers themselves; if any worker has a desire to take up any particular line of work, provided that it is associated with the general trend of work in the laboratory, it is usually wise to let him do so."[122] In the late 1920s, Mees noted that he had to warn a scientist who developed an interest in high-vacuum pumps and gauges that this was work in no way compatible with the concerns of a photographic company, and one lesson drawn from this story was that scientists in Eastman Kodak's research laboratory could *not* do just anything they wanted. But Mees then repeatedly went on to relate that, when he saw what a splendid technology was being developed, he secured the resources to spin off a distinct commercial laboratory, one that ultimately became a highly profitable vitamin-producing joint venture with General Mills. And, as elsewhere, the general lesson that Mees wanted research managers to learn was that autonomy and trusting scientists' judgments were, after all, good business practices. There was no alternative to a significant degree of autonomy, and attempts to be unremittingly controlling were ultimately self-defeating.[123] It wasn't any desire to celebrate academic values or to make academically oriented scientists happy that was the crucial issue for Mees. Rather, it was the capacity of industrial research to pay big commercial dividends that argued for a high degree of autonomy, trust, and spontaneity.

Robert K. Merton's recently published book on the career of the word *serendipity* shows that early to mid-twentieth-century industrial research managers were among the most important users of the term and did much to put it into general circulation. At the GE Research Laboratory, Willis Whitney (one of many alleged "discoverers" of Horace Walpole's eighteenth-century term) and Irving Langmuir were at pains to stress how much scientific discovery was, and must inevitably be, spontaneous—something whose outcomes could never be foreseen by the planners. Langmuir repeatedly noted that the inherent uncertainty of research was the most powerful, and legitimate, argument against planning: "The

American public likes to think of supermen...They are supposed to know how to plan. Now the examples I want to give you are all examples of unplanned research where things happened in a way that nobody could have planned and arranged and as a result we got results which were most satisfactory." Planning is possible when you know the definite cause-effect relations, but this is not the case in genuine scientific research: "Then planning does not get us very far. All we can do is, like serendipity, put ourselves in a favorable position to profit by unexpected circumstances."[124] By the late 1940s and early 1950s, the serendipitous nature of scientific discovery and its implications for both the planning of science and the free action of the scientist were being widely advertised by leaders of American industrial research, including executives at Arthur D. Little, Inc., Standard Oil, and Merck.[125] (Even in the early twenty-first century, the much ballyhooed, and hugely popular, "Six Sigma" managerial regime of quality control and process management developed by Motorola in the 1980s has generated a counterreaction from corporations whose identity is centered on research and innovation: at Raytheon, for example, a notable Six Sigma expert acknowledges that the "'define, measure, analyze, improve, control' mind-set doesn't entirely gel with the fuzzy front-end of invention. When an idea starts germinating," he concedes, "you don't want to overanalyze it" or to submit it to rigid efficiency disciplines.[126] And another commentator agreed that the attempt of such managerial regimes as Six Sigma to "replace subjectivity with objectivity and intuition with data" is "detrimental to exploratory research and design, which depend on subjectivity and intuition.")[127]

WHO IS THE RESEARCH MANAGER?

What kind of person was capable of leading an organization like the industrial research laboratory, planning what could only problematically be planned, organizing what could only problematically be organized, establishing the legitimacy of a facility whose place in corporate life was so ill-defined? What did such a person's leadership consist of, how could it be implemented, and with what consequences? It is in these connections that the normative uncertainty of the industrial laboratory is so pertinent. At the origins of industrial research, no one knew what a corporate research laboratory was, and so plans for its establishment and routine conduct could not be found ready-made and suitable for application to the case at hand. Chapter 5 noted that when Kenneth Mees was handed the job of instituting Eastman Kodak's Research Laboratory, he "knew nothing about running a research laboratory, of course; nobody did." So he went to visit Willis Whitney at GE, the one major figure who might be reckoned to know something about the business, and from Whitney Mees learned, as he later said, much about creative *disorganization* and the importance of a light hand on the tiller.[128] From one end of the century to the other, the nature of the industrial research laboratory and, therefore, the nature of its appropriate leadership remained uncertain. The variables were too many and too complex, including the size of the firm, the identity of the personnel, the type of inquiry, the stage of work on the spectrum from fundamental research to development, and, of course, the personality and vision of the individual research director.[129]

Mees repeatedly insisted that the proper job of the research director was *not* to direct, and, although his sentiments were widely cited throughout the century, several industrial research directors called him on this claim, pressing the case that the role was, after all, a vital one. Early on in his career at Eastman Kodak, Mees himself had insisted on a highly individualistic interpretation of the history of technological industry. The origins of most innovative industries were, he wrote, "dependent upon some one man, who frequently became the owner of the firm which exploited his discoveries."[130] And so, much later, when Norman Shepard of American Cyanamid wanted to say that Mees was being slightly disingenuous in insisting that the research director did very little, he reminded Mees of what he himself had implied: "[The research director] is the most important factor in the success of a laboratory. A research laboratory is usually the shadow of a man and that man is the chief of research." The research director—Shepard himself didn't like the word and preferred "chief," "manager," or, following DuPont practice, "adviser"—was a vital force because he gave the laboratory its institutional identity, setting its "mood" and "tone."[131] Technical knowledge was important, but it was not enough. The research manager had to "stimulate, inspire, encourage, and lead his men. He must know them personally; be able to call them by name and give freely of his time in personal conference, both in his office and in the laboratory. This is a very difficult assignment in a large laboratory, but can be accomplished by making the rounds in the laboratory, chatting with the workers, and showing interest in their experimental work and results." He had to, in more current business parlance, "walk the corridors," showing himself to his researchers and making face-to-face contact with them. He had to show each worker "the implications of his problem; how it ties into the whole research program; what it will mean to the company in improved quality of product, a new product, or in economies, if solved . . . A word of praise here, a word of encouragement there, means everything to a research man." The research manager not only spoke *to* the scientists under him, but he spoke *for* them to headquarters, protecting workers from "the harsh and often undeserved censure that comes down from top management—especially nontechnical management—when the laboratory is apparently slow in solving a problem" or making a new discovery before some competitor. His task was to inspire his subordinates but also to make sure that his superiors appreciated the inherent uncertainties of research.[132]

So Shepard insisted, against Mees, that the director *did* direct scientific work, but reckoned that Mees actually understood that. Moreover, the two managers shared a sensibility about what was involved in effective direction. No research workers were really effective if they were told what to do in any very forceful or direct way: "The initiative shown by the laboratory staff members in proposing and undertaking new work naturally depends upon the amount of freedom they are given to follow up their ideas. If the work of the scientists is rigidly restricted as to problems and methods of attack, their productiveness in new ideas will necessarily be low."[133] The laboratory is ultimately dependent upon ideas spontaneously thought up by those working most closely on a problem, and, for that reason, direction that was too aggressive was ultimately destructive: "In a laboratory operating efficiently, there is automatically an ample supply of ideas arising in the operation of the laboratory, and a director should be no more concerned as to the supply of new products than a well man should be concerned about the state of his heart. Any attention to the matter is a sign that something is wrong. New products are not found by attention to them on the part of the management of the laboratory . . . If there is a lack of ideas in a laboratory, it is generally to be traced to a lack of encouragement of this development of ideas by the scientific workers themselves. This condition may occur in departments run by a dominating personality who discourages the ideas of others even though he may do so unconsciously."[134] Running a laboratory, in Mees's view, was largely a matter of ensuring effective communication between relevant individuals and groups, and, to this end, he adapted a GE and Nobel system of weekly "conferences" (figure 11)—lasting between an hour and ninety minutes—in which technical matters arising were discussed and ideas canvassed: "Instead of the work being settled in a personal conference between the director and the worker, it is thrashed out at this weekly discussion with the assistance of all those who are working along allied lines."

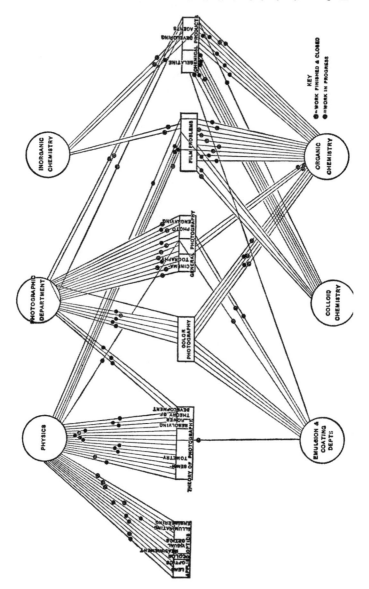

FIGURE 11. Lines and forms of technical communication in an industrial research laboratory. Here Eastman Kodak's Kenneth Mees sketches the departments of his research laboratory (shown as circles) and the weekly morning "conferences" (shown as rectangles) in which their work is discussed by all relevant members of the laboratory, no matter what department they belong to. Each "conference" has its day of the week. "In this way," Mees wrote, "the conferences really direct the work of the laboratory." (From Mees, *The Organization of Industrial Scientific Research* [1920], p. 126.)

The conferences ranged in size between four and ten people, that is, slightly smaller than the military squad, and ensuring dense and effective face-to-face communication. It was the director's responsibility continuously to monitor intra-laboratory communication, to ensure that blockages and breakdowns in communication were addressed, and to make certain that personal conflicts that might obstruct communication were resolved.[135] The director made the groups *happen* and monitored their healthy functioning, and that was the nature of his seemingly effortless superiority—paradoxically to organize self-organizing groups:

> If you want to catalyze the research of the groups, all you have to do is to get them together and talk about what you want. For instance, if you want better boots and shoes, you just tell them so. Then they go away and presently they come back, after they have thought about boots and shoes, and they tell you what they have been thinking. They certainly haven't all been thinking the same thing. It won't be long before your research laboratory has organized itself, and then your job consists of keeping the men talking to each other.[136]

The job of the director was to be, so to speak, a laboratory Leviathan, guaranteeing the equal intellectual entitlements of all those underneath him. At Westinghouse, a research director acknowledged that it was, indeed, managers' job to set research goals, but immediately pointed out the superiority of collective judgment: "If we believed ourselves to be omniscient, we would of course set these goals ourselves. Personally, I am extremely conscious of my own limitations, and therefore strive to create an environment within the laboratory such that the wisdom which resides collectively within the research personnel is used not only to prosecute research projects, but also to choose the directions these projects should take."[137]

The research director's person was, therefore, an embodied solution to the normative uncertainties of the institution over which he presided. The solution might not always be a satisfactory one—there were, of course, good and bad directors, and much of the rhetoric about effective leadership that has just been reviewed may be partly self-serving and may correspond incompletely with quotidian realities—but there was *no* effective way in which the goals and practices of the industrial research laboratory could be set, maintained, or modified without these being enunciated or embodied by the research director. In that sense, and while it is necessary to take into account a whole range of structural realities and contingent circumstances, it is quite right to say that the research laboratory *was* the "shadow" of an individual. Neither at the beginning of industrial research nor at present is there a rule-book that is a sure guide for how to do the thing. Accordingly, it is wholly proper to see the management of many industrial research facilities as a matter of "charismatic authority." If, as Weber argued, the characteristic gesture of charismatic authority is to say "It has been written . . . , but *I* say unto you," then those in charge of almost all industrial research laboratories but the most routine have said "It is *not* written, and I say unto you."[138] Institutional life came *personified*, and some of the most characteristic institutions of both late modern science and late modern corporate capitalism depended greatly upon the moral authority of individuals and the social relations of familiarity. It is a dependence that grew even stronger as technoscientific institutions developed through the later twentieth century and into the present.

The Scientific Entrepreneur

MONEY, MOTIVES,
AND THE PLACE OF VIRTUE

*The antagonism between charisma and everyday life arises
also in the capitalist economy, with the difference that charisma
does not confront the household but the enterprise.*

Max Weber, *Economy and Society*

WHO IS THE SCIENTIFIC ENTREPRENEUR?

By about the 1970s, it had become common to think that to be a scientist
was to do a job much like any other professional job and that scientists
were morally and motivationally pretty much the same as anybody else
with their backgrounds and in their station of society. It was no longer
news in American culture that scientists were laborers worthy of their hire,
that scientific knowledge might be produced in an organizational mode,
that scientists might be accounted valuable—even uniquely valuable—
workers whose products might contribute not just to the growth of
knowledge for its own sake but to the generation of wealth and the projection of State power. It was understood that you could make money
doing science, even academic science. The monetary rewards were still not
widely thought to be vast—or even on a par with those received by such
other professionals as physicians and lawyers—but still the profession of
science now might plausibly be thought of as a route to a comfortably
bourgeois style of living. Some of these sensibilities were the upshot of
the industrialization of science from the early years of the twentieth century; others can be largely attributed to the institutional, political, and
economic changes ushered in by the militarization of academic science
during the Second World War and institutionalized during the Cold War.
Some sensibilities were contested, and remain contested to the present;
others were, and are, widely accepted as matters of course.

It is against this background that the rise into American cultural awareness of the figure of the *entrepreneurial scientist* is so noteworthy. Broadly

construed, the role of the scientific entrepreneur is not wholly novel—for example, it makes some sense to describe the seventeenth-century experimentalist Robert Hooke as an entrepreneur—but it was only from about the 1970s that entrepreneurial activity began substantially to shape American appreciations of what kind of person the scientist was and what the scientific life might lead to. For present purposes, the scientific entrepreneur is defined as one who is both a qualified scientist and, like all commercial entrepreneurs, a risk taker.[1] And the specific sort of person with which this chapter is concerned takes risks to commercialize knowledge that they themselves or others have produced. They have one foot in the making of knowledge and the other in the making of artifacts, services, and, ultimately, money. They may or may not be aware of any "conflict" between these aspects of their identity, but they embody drives and activities that, during the course of the twentieth century, were widely held to be in tension, and sometimes in opposition. This chapter is not concerned with the situation of researchers in large industrial research laboratories— where the firm may be regarded as entrepreneurial but where salaried researchers assumed little personal risk and commonly expected few of the rewards of successful entrepreneurial activity by the company. The entrepreneurial scientists who became distinctive characters of the period from the 1970s were individuals who sought, *by their own efforts*, or those of a small number of coworkers, to turn knowledge into profitable goods or services. Exemplars are the academic scientists who produce potential intellectual property (IP), who—with or without the assistance of their university—secure rights to that IP, and who then take a substantial role in transforming the relevant knowledge into profitable goods or services. This might mean leaving the academy to help found a business, or it might mean—less lucratively, but more securely—remaining within the university and seeking to produce additional commercializable IP.[2]

The emergence of the figure of the scientific entrepreneur from the 1970s was spectacular, but it built on foundations laid down earlier in the century and even, as in the case of some agricultural research, in the late nineteenth century.[3] The Berkeley chemist Frederick Cottrell invented an electrostatic precipitator in 1907, for which he eventually secured a patent, only to assign it to the nonprofit Research Corporation, which he established to generate a revenue stream for the support of academic research. In the 1920s, Harry Steenbock, professor of biochemistry at the University of Wisconsin, personally patented a method for producing vitamin D using ultraviolet light, then worked with several companies to commercialize the technology and, in 1925, used the proceeds to help found the Wisconsin Alumni Research Foundation (WARF), a nonprofit research-support organization legally independent of the university.[4] At Stanford in the late 1930s, a professor in the physics department, William Hansen, and two unpaid department researchers, Russell and Sigurd Varian, invented a microwave tube with evident commercial potential in aviation guidance. Unlike Berkeley or Wisconsin, the Stanford administration established a policy of taking ownership of all IP produced by its professors. It then entered into an agreement with the Sperry Gyroscope Company to commercialize the technology, some of the revenue from which went to the department and the individual inventors. After the war, the Varian brothers struck out on their own, founding the company Varian Associates, which enjoyed close links with Stanford scientists, several of whom sat on the company's board of directors and held stock in it.[5]

This sort of thing was not uncommon in the first half of the twentieth century.[6] Universities comfortable with the idea of patenting professorial discoveries argued that the public interest would best be served by academic ownership of IP, protecting society from "pirates" who might appropriate inventions wrongfully and develop them irresponsibly. Yet, despite these examples, from the beginning of the century to about the 1960s and 1970s, the generality of academic scientists and administrators remained unfamiliar with the academic production and management of commercializable knowledge. Many of them—but by no means all—saw a series of problems associated with institutional involvement in such things. Even those academics and academic administrators advocating university patenting thought, nevertheless, that universities should keep the management of patents and licenses at arm's length.[7] Some of these recognized difficulties were wholly practical: many administrators did not believe that their scientists and engineers were at all likely to produce commercializable knowledge and so saw little reason to institute formal management procedures; some foresaw concrete political and financial problems associated with ties between academia and for-profit corporations, for example, resulting tensions in their relations with state legislatures and the commercial concerns that sponsored some academic research; and some—but, again, not all—professors and university administrators even saw such activities as morally wrong, in conflict with a fundamental idea of the university's identity. There *was* some ideological discomfort with the very idea of commercial concerns within university science—more among the scientists than the administrators—but statements that look like ideological defenses of Ivory Tower academic science can often bear more pragmatic interpretations.

So Frederick Cottrell decided against assigning his precipitator patent to the University of California because of his concern that this would raise the "possibility of growing commercialism and competition between institutions and an accompanying tendency for secrecy in scientific work."[8] This is one of the reasons why—the examples of Stanford and MIT notwithstanding—American universities through the middle of the century sought to keep the commercialization of IP at an administrative distance, and why some academic scientists were uncomfortable with the idea of commercialization, save as a way to generate revenues for academic research in a period when such funding was scarce. In the mid-1930s, as Grischa Metlay shows, only two American universities had formal patent policies, and by 1942 that number had grown only to twenty.[9] Although academic scientists clearly differed in their adjacency to potential commercial activities—engineers and chemists were traditionally in a rather different position from invertebrate zoologists and paleontologists— university science was not, on the whole, thought to have much to do with wealth creation. Academic scientists did not, by and large, see themselves, nor were they seen by the culture, to be in the business of business. Sheer unfamiliarity with the process of transforming knowledge into profit goes a long way to explaining the attitudes of pre-1970s American academic scientists, though some role has to be ascribed to a sense that the commercialization of academic science was morally wrong—an assault of the idea of the university as a site of intellectual virtue.

But by the late 1960s and 1970s, and certainly by the last decades of the twentieth century, what had once been seen as anomalous now was well on its way to becoming a normal feature of the American university scene. Some commentators and participants actively celebrated academic entrepreneurship. Clark Kerr, writing *The Uses of the University* in 1964 as president of the University of California system, probably did not—in this precise connection—have commercializing scientific entrepreneurship specifically in mind when he described the modern research university as a "series of individual faculty entrepreneurs held together by a common grievance over parking."[10] Academics were individual entrepreneurs in Kerr's sense mainly because their commitment to what might have once been called the university's intellectual commons was fragmenting, and their major concern was now directed towards making out in their own special fields. But by the dawn of the present century academic entrepreneurship took on special meanings that pinpointed inquiries with significant commercial consequences. In the late 1950s, the University of California instituted what was bureaucratically called "Regulation no. 4," which recognized the increasing importance of university-industry relations but which insisted that the university must be distinguished from industry through academia's commitment to basic rather than applied research. Individual professors' ties to industry had to be consistent with the university's teaching and fundamental research missions, and, while "appropriate public service" was mentioned among these missions, assisting industry in realizing its goals was not then counted as such public service, and freedom to publish research results openly was deemed a criterion of suitability for industry ties. As late as 1982, the University of California reaffirmed those sentiments.[11] But circumstances affecting the research university and its employees were changing through the 1970s and 1980s.

THE POLITICAL ECONOMY OF SCIENTIFIC ENTREPRENEURSHIP

During the 1970s, American industry and its political allies became increasingly alarmed by what was seen as a crisis of international economic and technological competitiveness. The U.S. was seen to be surrendering technology leadership to Japan and the east Asian "tiger" economies, and proposed solutions to this crisis included enhancing the supply of skilled personnel and encouraging technological innovation. With increased government investment in research universities in the post–World War II period, that meant new attention to these institutions as potential sources of the innovations that industrial laboratories were not themselves making at the rate required to retain national economic and military dominance. In 1968, Congress authorized the National Science Foundation to sponsor *applied* research, and universities competed for the new sources of funding. There was money to be had in serving industry much more vigorously than had once been the case, and American research universities in the 1970s were feeling the financial pressures of Vietnam-era cutbacks in Federal and state support. Industry might fill that gap.[12] Industrial commitments to university research contracts and grants increased from about a quarter of billion dollars in 1980 to almost two and a half billion in 2000, amounting to about 13% of government commitments.[13] Academic rhetoric and institutional realities both began to change: research universities increasingly offered themselves as handmaids to industry; helping industry was identified as public service; and the sorts of professorial commercial ties that were once a source of administrative unease were now enthusiastically encouraged. Moreover, in many areas of scientific research valued by academics—of which biotechnology, electronic engineering, and computer science are prime instances—basic advances might no longer be the prerogative of universities. A reconfigured industrial sector was, for these and other fields, where the action was and where the resources were located.

By the 1990s and the early years of this century, these shifts in circumstances and sensibilities were unmistakable. In 2001, presiding over the University of California, San Diego (UCSD)—one of the most entrepreneurially minded of all present-day American research universities—chancellor Robert C. Dynes applauded the university's role in spurring the growth of a high-tech region—"We are the proud parent and grandparent of 150 or more spin-off companies" founded by UCSD faculty and graduates—a boast that preceded his bland announcement that academic inquiry had *no* intrinsic value: "As scholars, we should not seek knowledge for its own sake."[14] At about the same time, the dean of the university's Engineering School—soon to depart to become a venture capitalist—declared that, besides teaching and research, the modern university now had a new "key mission"—"to ensure the effective transfer of research results and discoveries to the sectors of our society, usually the private sector, that can translate such discoveries into products and services for the benefit of society as a whole." Technology transfer to profit-making companies, that is to say, should be understood as having exactly the same status in the life of a university as teaching and the search for Truth, though the dean seems not to have noticed, or cared, that this left almost all academics who were not scientists and engineers—and, indeed, scientists and engineers of a certain sort—failing in one of their "key missions."[15]

Such sentiments were expressed more baldly at this California institution than was customary at more traditional research universities, but they nevertheless articulated a conception of the role of academic scientists and the institutions in which they worked that had been developing for several decades. Arrangements for university ownership of professorially produced IP had become more standardized, as had the apportioning of whatever royalties flowed from commercial licensing. Universities dreamed of growing rich through the licensing of blockbuster technologies produced by their science and engineering professors, and the professors themselves might hope to share modestly, but significantly, in the resulting revenues. Technology transfer offices had become normal institutional features of the U.S. research university; centers for encouraging scientific entrepreneurship multiplied; university-industry partnerships of various kinds came into being, including the proliferation of large-scale corporate sponsored research and the establishment of distinct organizations within the university to house such research. During the Reagan presidency, Congress signaled its active approval of the commercialization of Federally funded academic research by passing the Bayh-Dole Act (Public Law 96-517, the Patent and Trademark Act Amendments of 1980), which strongly encouraged universities to take ownership of any commercializable products of such research, to seek out companies to license the IP, and so to transform knowledge into profitable goods and services. Some universities intervened more directly, by taking equity shares in spin-off companies and by using some of their own resources to provide capital to them.[16] A few universities stood out for their entrepreneurial aggressiveness. In the late 1980s, Boston University famously made a series of major investments—totaling about $85 million—in the Norwegian biotech company Seragen with which some of the university's scientists had ties.[17]

Many sorts of academic scientists were confronted for the first time with an emerging array of possibilities for making very large sums of money for themselves. Industrial consultancies were a continuation of long-established patterns, as was sharing in licensing revenue. But leaving academia to found, or to work and take an equity share in, small start-up high-tech and biotech companies was a substantially new thing. In standard genealogies, the biotech industry was created in 1976 "over a couple of beers" at a San Francisco bar in a conversation between twenty-nine-year-old venture capitalist Robert A. Swanson (who had just joined the fabled Palo Alto firm of Kleiner Perkins Caufield & Byers) and biochemistry and biophysics professor at the University of California, San Francisco, Herbert W. Boyer. Working with Stanford genetics professor Stanley N. Cohen, Boyer had helped develop some elegant recombinant DNA technologies, patent rights to which were, as was then normal, assigned to the university, from the licensing of which Stanford and the University of California derived about $200 million until the patents expired in 1997.[18] Swanson saw vast commercial potential in licensing, developing, and exploiting these technologies for commercial drug discovery, and, while Cohen was unwilling to leave academia, Boyer and Swanson developed a business plan, put in $1,000 of their own money, secured $100,000 in seed capital from Kleiner Perkins, and incorporated a company, known as Genentech, which is still one of the world's leading biotech companies. In 1980, Genentech went public, opening at $35 a share and tripling on its first day of trading. In 1982, the FDA approved genetically engineered human insulin, developed by Genentech in collaboration with the giant pharmaceutical company Eli Lilly. Important science got done at Genentech, published in the open scientific literature; in 1989 Boyer and Cohen won the Nobel Prize for Medicine and in 1996 the $500,000 Lemelson-MIT Prize for scientific entrepreneurs (the "Oscar for Inventors"); patients were helped; Boyer became very rich and Cohen quite comfortably well-off. And this sort of spinning-out of biomedical science from universities to start-up companies established a new set of possibilities for scientists: new ways of making large sums of money; new institutional forms for doing science; new goals and ways of acquitting them.[19]

Ringing less radical changes on existing patterns of work in high-tech engineering disciplines, similar developments laid the basis for Silicon Valley's and Route 128's computer and electronics industries. Scientific and engineering professors could not only become entrepreneurs but entrepreneurship of this sort was being increasingly applauded and encouraged. Scientists growing rich by leaving academia to start up a company were culture heros to many politicians and fellow academics. They grew the economy; enhanced national economic competitiveness; expanded employment opportunities; increased the efficiency of work; helped to cure dread diseases. *These* were the people whose activities made the university politically strong; philologists and medievalists were considered to be going along for a ride essentially paid for by the commercially minded entrepreneurs. After all, a university education was increasingly justified in America through its effects on graduates' future earnings, so why shouldn't professors show what their own knowledge was really worth? Making large amounts of money through science might simply be approved of, but, at the same time, the attainment of riches could be seen as the natural, and praiseworthy, accompaniment of acquitting such virtuous goals as increasing economic productivity and curing cancer. If it motivated more scientists to do more of such things, then all the better, and, anyway, scientists might wish to grow very wealthy, just like anybody else. Why shouldn't they? From a certain point of view, in the life of academic science, and in the changing identity of the academic scientist, commerce and virtue were in no necessary conflict.

At the same time, however, other commentators, and, indeed, participants, took different views of the same developments. Academics with commercial "conflicts of commitment," it was said, might neglect undergraduate teaching and the best interests of graduate students in favor of those activities promising greatest monetary reward for themselves. They would gravitate towards applied and away from basic research, thereby endangering both intellectual and commercial futures. As a result, entrepreneurial activities would undermine the university's traditional "key missions." Just as Cottrell worried much earlier, from the 1970s concerns were also expressed that the intrusion into the university of commercial considerations and commercial ties would lead to a wall of secrecy where once there had been an unchallenged commitment to openness. Others feared for the objectivity of science. Scientists would, it was thought, produce not Truth but the results wanted by their sponsoring commercial concerns. Commercial sponsorship or subvention might be the condition for certain research programs being carried out at all, and so academics with conflicts of interest had the motive to produce biased knowledge. The consequence of commercial influence would be a loss of objectivity, and so, for some, the very idea of science was at stake in discussions over the nature, conditions, and outcomes of entrepreneurship.[20] Accordingly, late twentieth-century and contemporary American debates over the virtues and vices of scientific entrepreneurs implicate understandings of what kind of persons they were, of the nature and quality of scientific knowledge, and of the kinds of institution—academic and commercial—in which entrepreneurs worked. In an age of entrepreneurship, where did virtue reside?

HAVING FUN, MAKING MONEY

Between the ascetic natural philosopher of the early modern period and the commercially minded scientific entrepreneur of Silicon Valley, Route 128, and San Diego's Biotech Beach were many intermediate stages, and previous chapters in the book described some of them. But, for the period from the 1970s, special reference has to be made to cultural currents that have not yet been clearly identified. Chapter 3 briefly alluded to Cold War concerns that the recruitment of large numbers of scientists was being made difficult by a widespread image of science as "a gray, austere calling"; rather, it should be appreciated—as *Time* magazine lectured in 1957—that the scientist, properly understood, was having an enormous amount of "fun."[21] Fun and funds are not the same thing, but they are related. To say that scientists are motivated by money is ceteris paribus to say that they are motivated by the goods and services that money can buy, and by their desire to enjoy those goods and services. The root sentiment, then, is hedonism, rather than asceticism, and, as chapter 3 noted, by the early 1960s, such sociologists as Lewis Feuer were contending, contra Weber and Merton, that the motive force in the rise of modern science was hedonistic.[22]

In the 1960s, that claim still had the capacity to startle, and James D. Watson's *The Double Helix* was treated, when it appeared in 1968, as very startling indeed. Without doubt, this was the single most vivid and influential account of what it was to be a scientist that was produced in the second half of the century, and so it probably remains in the early twenty-first century. The picture that Watson offered of both real and ideal scientists stressed their highly competitive nature: instead of an insouciant disregard for fame, the scientist was in a fierce race for breakthroughs of Nobel Prize quality, and much, if not all, was understood to be fair in winning the race. For some commentators, the picture Watson offered of competitive modern science was as deeply unpleasant as it was inaccurate: the *Science and Engineering* reviewer found it "unbelievably mean in spirit, filled with the distorted and cruel perceptions of childish insecurity . . . a world of scorn and derision." For others, it was startling just because it told the truth. *Life* magazine's reviewer briefly noted that "the story should kill the myth" of scientific impersonality or the moral perfection of scientists: "These young scientists covet, lust, err, hunger, play and talk about it loud, well and long," and the *Chicago Sunday Sun-Times* noted that "what every scientist knows, but few will admit, is that the requirement for great success is great ambition. Moreover, the ambition is for personal triumph over other men, not merely over nature."[23] It was a race that was also understood to be enormous fun. Figuring out the structure of DNA was fun, but, like anybody else, the scientist could have fun outside of work. It came as a shock to some commentators that *The Double Helix* displayed the (male) scientist to be as interested in "popsies" as any other red-blooded twenty-three-year-old American man in the 1950s.[24] Watson wanted to be seen as being like other men, only more so. He has become iconic for genius, but also for ambition and moral ordinariness. (And so it is fitting that the commodified form of the icon can be bought for $21.95: it is a James D. Watson bobblehead doll; figure 12.)

But, for all the startle value of Watson's confession that he cared for celebrity and sex, there is nothing in *The Double Helix* to indicate a concern for money, beyond the modest fellowship emoluments necessary to keep body and soul together. In later life, Watson professed himself as interested as any other academic in the precise size of his salary: after winning the Nobel Prize in 1962, Watson was expecting "a larger than ordinary" increase in his $15,000 Harvard salary, and when dean Franklin L. Ford gave him no raise at all, he was majorly miffed ("Instantly, I went ballistic"), a wound that, on the evidence of his most recent autobiography, continues to fester almost half a century after the event. Nor, despite

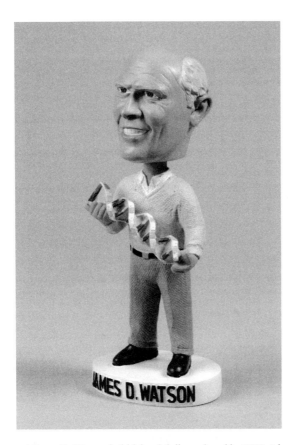

FIGURE 12. A James D. Watson bobblehead doll, produced by DNA Adventures Inc., a division of Von Enterprises, and available for purchase at the Cold Spring Harbor Laboratory bookstore and through several Web sites. "It may not be art," writes a trade journalist, "but it could be the perfect holiday gift for the DNA lover who has everything: the James D. Watson bobblehead doll. The eight-inch-high doll depicts the famous scientist wearing his trademark yellow sweater and holding the DNA double helix—the structure he helped determine 50 years ago. The bobblehead is the brainchild of Rachel von Rauschloeb, who runs the DNA microarray facility at Cold Spring Harbor Laboratory in New York. She came up with the idea while trying to find ways to focus attention on genetics research. It was a joke, at first. 'But then, I thought, "Hey why not?"' says von Rauschloeb. 'Dr. Watson didn't know what a bobblehead is, but it didn't take long to convince him to do it.' ... More than 1,000 were sold in the first week": http://www.genomenewsnetwork.org/articles/12_03/bobblehead.shtml [accessed 14 August 2007].

Watson's subsequent career as an academic entrepreneur, is there any trace in *The Double Helix* of an awareness that the new genetics that he was helping bring into being might have commercial consequences.[25]

Much the same could be said of a man who became Watson's rival as America's most celebrated scientist, the Caltech physicist Richard Feynman. Winning the 1965 Nobel Prize for Physics, Feynman cultivated a public reputation as a charismatic scientific performance artist (figure 13). Projecting his public image through a series of television programs, and producing several volumes of anecdotes about both his science and himself, Feynman wanted it clearly understood that science was "fun" and that he had achieved what he had through having fun—or, as he put it, "the pleasure of finding things out." In Feynman's account, the work for which he won the Nobel Prize was done to satisfy the play instinct: "I'm going to *play* with physics, whenever I want to, without worrying about any importance whatsoever."[26] But Feynman pointed out that "importance" might well follow satisfaction of the play instinct; indeed, it was more likely to do so than if the scientist actively sought some higher or more practical purpose. Feynman was having fun outside of, as well as in, science. Advertising himself as an integrated character, the childlike wonder and hedonism that were at the root of his scientific imagination expressed themselves in the whole of his life. And so his much-publicized bongo playing, his toying with art, and, above all, his sexual adventurousness served to show just how much fun a scientist could have. If scientists were once conceived as "gray," now they came in psychedelic color. A Swedish correspondent may have gotten too close to the mark when he evidently suggested that the drumming was part of an act to show that the scientist could have as much fun as anyone else. Feynman erupted: "Theoretical physics is a human endeavor, one of the higher developments of human beings—and this perpetual desire to prove that people who do it are human by showing that they do other things a few other humans do (like playing bongo drums or juggling) is insulting to me. I am human enough to tell you to go to hell."[27] And yet, like Watson, there is almost nothing in Feynman's many stories about himself that indicates any concern for the commercial consequences of his scientific work or for more money than attached to his, undoubtedly ample, academic salary. If Feynman is to be accounted an entrepreneur, it would have to be in the traditional academic sense of someone highly skilled in transforming intellectual ability into reputation.[28]

Watson and Feynman, for all their virtuosity in science and in self-promotion, and for all their emphases on the hedonism of the scientific

FIGURE 13. The great theoretical physicist Richard Feynman (1918–1988) — "the best mind since Einstein" — juggling at the beach in Malibu, ca. 1950. Feynman seized every opportunity to advertise the playfulness of doing physics his way. A much-reproduced iconic image is of Feynman playing the bongo drums, and a biography is titled *The Beat of a Different Drum*. Bongo playing was a practice he started as a very young man while working on the atomic bomb at Los Alamos, but he was also fond of juggling, magic, and practical jokes. The physicist Freeman Dyson once said that Feynman was "half-genius, half-buffoon," but eventually changed that to "all-genius, all-buffoon." (Reproduced by permission of the Caltech Institute Archives and the Melanie Jackson Agency on behalf of the Estate of Richard P. Feynman.)

life, nevertheless did not cross the line into identifying commercially consequential entrepreneurship as central to that life. (Indeed, when Watson was heading the Human Genome Project in the early 1990s, he exploded in anger at the idea that gene sequences might be patented, saying it was "sheer lunacy.")[29] However, the public culture soon became well aware of the figure of the commercializing entrepreneurial scientist, and not just in the abstract, but as particular and celebrated individuals. While the computer scientists, software artists, and electronic engineers of Silicon Valley showed early on what fun could be had, and what vast sums of money could be made, by those with the appropriate, academically acquired scientific and technological skills,[30] it was developments within the biomedical sciences that most powerfully projected into general consciousness the new entrepreneurial opportunities opened up to those whose major, or even sole, prior institutional home had been academia. And that heightened the sense of surprise: with the new post–World War II awareness of the fundamental changes affecting physics, many commentators who fretted about the effects on science of access to power and wealth both expected and hoped that the biological sciences would remain forever unaffected by such things, calm and quiet disciplinary spaces where traditional scientific virtues might continue to flourish.[31]

The spectacular growth of the biotech industry gave that sensibility such a serious jolt because it touched science at just the point that had seemed immune to the entrepreneurial way of life. A journalistic account of these changes, as they appeared in the late 1970s and early 1980s, captures the startling nature of the changes affecting life scientists: "For biology, fear and secrecy were largely new. Until the middle of the 1970s, the scientists in molecular biology seemed little different than monks toiling in scholarly poverty and obscurity. Openness was one of the chief virtues practiced. They discussed results openly. They sent one another not only ideas, but also samples of their work in tissue cultures and extracted genes. Graduate students who wanted to work in the field were expected to live on salaries at the poverty line . . . As a compensation [compared to lawyers, doctors, and businessmen], the biologists had their intellectual purity. There was a special honor in the poverty of the dedicated researcher, and a suggestion that money could not tempt a talented biologist from the rigors of his work. Industrial laboratories were full of secrets and unimagination." All this changed from the mid-1970s. Biologists, or at least certain sorts of biologists, "found themselves in demand. Their science has not only begun to master the principal mechanism of life, but the mechanism turned out to be exploitable. It might clone dollars as easily as genes . . . The best academic researchers could now name their price."[32] A Harvard molecular biologist went so far as to offer a cynical historical account of the idea of academic purity: in the beginning of entrepreneurial biology, scientists who consulted for, or worked with, gene companies were criticized; the idea of academic "purity is something which developed out of necessity. Since there was no money, a sense of sainthood was required in the situation. Now it's not required."[33]

Two public figures eventually gave special force to the developing picture of the entrepreneurial biomedical scientist: Craig Venter and Kary Mullis.[34] Venter achieved renown through the 1992 founding of The Institute for Genomic Research (TIGR)—a nonprofit genomics research organization—and, in 1998, the NASDAQ-listed Celera Genomics, of which he was president until his controversial removal in 2002.[35] In June 2000, Venter "won" the race (or, more accurately, shared in a politically brokered tie) with the publicly funded NIH-led effort to sequence the human genome, becoming the cover boy for entrepreneurial biomedical science, its most visible embodiment. Venter was featured both as one of *BusinessWeek*'s "Great Innovators" of the past seventy-five years and as one of its "Top 25 Managers." He was *Time* magazine's "Person of the Year" for 2000, earning a full-length *New Yorker* "profile," and, ultimately, becoming the subject of an adulatory biography and one of the dramatis personae in several other accounts of the genome project (figure 14).[36]

The standard way of introducing Venter to the public culture was to stress how unconventional, ornery, and bloody-minded he was; the *New Yorker* profile starts out by quoting an unnamed senior scientist in the publicly funded genome project—"Craig Venter is an asshole"—but also how visionary and how *right* he was, with respect to both his preferred "shotgun" sequencing methods and his conception of how science might flourish within the bowels of capitalism.[37] Accounts of Venter's life, notably including those he offered of himself, insist on the unconventionality of his route to a scientific career—more an accident than a vocation. Growing up in a working-class neighborhood south of San Francisco, he surfed more than he studied. Bored with school, and not particularly good at the rote memorization he identified with schooling, Venter muddled partway through a California community college, becoming a surf-bum—"I was a surfer in high school, I was a surfer in Vietnam, and I'm still a surfer"—and enlisting in the Navy (under threat of being drafted), where he served as a medical corpsman in Vietnam and was a serious discipline problem.[38] The Vietnam experience taught him, in *Time*'s account, both about "the fragility of human life and the colossal ineptitude of big bureaucracies." In his own words, "If you suffered fools, you died. I dealt with thousands of people dying because of stupid government policies."[39] Front-line experience with injury and death gave Venter a vocation—not for scientific research but, he said, for medicine, and whatever science could enhance

FIGURE 14. J. Craig Venter, the leader of one of the two groups (the other was the NIH effort led by Francis Collins) racing to sequence the human genome. When, in the issue of 13 December 2004, *Business Week* celebrated Venter as one of America's "Great Innovators," they elected to pose him wearing a white lab coat on his right side and a businessman's suit-jacket on his left, the ideal confluence of scientific knowledge and the cash nexus. "Who could ever have imagined," *Business Week* asked, "that a surfer working as a night clerk at Sears, Roebuck & Co. would eventually become the driving force behind the race to read the genetic code of humanity? That's the unlikely story of J. Craig Venter, a brash biologist who engineered a major leap in scientific knowledge—and earned millions—by masterminding efforts to probe the DNA of everything from microbes to man." (*Business Week*, 13 December 2004, p. 18.)

the power of medical care. Venter was of the same mind as *Arrowsmith*'s Max Gottlieb: biomedical research was more powerful than medical practice. "A doctor can save maybe a few hundred lives in a lifetime," Venter told his younger brother, "A researcher can save the whole world." He took first a bachelor's degree in biochemistry, then a Ph.D. in physiology and pharmacology at the University of California, San Diego, and, following that, spent almost ten years teaching at the State University of New York and the Roswell Park Cancer Institute, both in Buffalo, before joining the NIH in 1984, where his career in genomics took off, and which he left in 1992 to become a scientific entrepreneur.[40]

Venter's position about the patenting and commercial exploitation of gene sequencing knowledge was, and is, more nuanced than is sometimes represented by his academic critics. Attacked by the enemies of commercializing biomedicine, Venter defends himself by pointing to what he's done that is wholly compatible with altruistic virtues. A story on Venter in *Wired* noted that he had "promised that he would give away the basic human code for free. Celera would make money by selling access to gobs of additional genomic information and the powerful bioinformatics software tools needed to interpret it. His critics claimed that he was trying to have it both ways, taking credit for providing the world with the code to human life and reaping profits for his shareholders at the same time. Venter cheerfully agreed." "My greatest success," he said, "is that I managed to get hated by both worlds."[41] Within Celera Genomics, Venter was apparently often at odds with the lawyers who wanted the company to take a more aggressively commercial approach. But just as Venter was a hero in entrepreneurial circles, so he was feared and disliked by those who thought it morally wrong and scientifically destructive to patent gene sequences or to take private ownership of the entire human genome, even though it was the NIH—while Venter was working for it and under the commercializing leadership of Bernadine Healy—which first pushed the idea of patenting gene fragments.[42] "[Celera Genomics'] fundamental business model," Venter said in 2000, "is like Bloomberg's. We're selling information about the vast universe of molecular medicine," and this was the sort of remark that, on the face of it, might identify Venter as a straightforward commercializing entrepreneur and that appalled the defenders of a virtuous scientific commons.[43]

Like Watson, Venter didn't care whose feelings he bruised; didn't want any part of a gentlemanly conception of science; thought that science was in general suffering from a deficit of individualism and boldness; didn't hide his lights under a basket or disguise his immense amour propre; couldn't abide what he saw as the hypocrisy of denying that there *was* a race to be first. Unlike Watson, however, Venter freely acknowledged the cash nexus as central to the scientific enterprise, both as a motive to effective research and as a condition for acquitting its objectives. Even after the puncture of the stock market bubble in spring 2000, the drying up of venture capital financing for the next several years, and his controversial ousting from the presidency of Celera Genomics, Venter remained an entrepreneurial icon: "king of the startup biotech sector," as *Wired* put it in August 2004.[44] He is currently president of the nonprofit J. Craig Venter Institute for basic genomics research and in 2005 cofounded another commercial firm, Synthetic Genomics, which aims to use genetically altered microbes as alternative fuel sources. From 2003 to 2006, Venter was on a partly government-funded round-the-world tour on his racing yacht *Sorcerer II*, combining science and pleasure—the science bits dedicated to the collection of marine microorganisms and the sequencing of their genomes, with the resulting information made freely available on the Internet. "This time around," *Wired* wrote, "he's doing everything he can to convince the world that he has no commercial motive: Here, take it all, I ask for nothing in return."[45] The flavor of the enterprise is indicated by the expedition's Web site, which introduces the project with a quotation from Khalil Gibran: "In one drop of water are found all the secrets of the oceans."[46]

Kary Mullis resembles Venter in certain respects: a fondness for surfing, an enormous ego combined with an abrasive personality, an intolerance of what he sees as the herd instinct of the modern scientific community, a deep involvement in commercial biotech, and a status as one of the public icons of entrepreneurial science. In 1993, Mullis won the Nobel Prize for Chemistry for his role in the discovery of the one of biotech's basic tools, the polymerase chain reaction (PCR). This is a technique for taking even small amounts of relatively impure DNA and getting it to make multiple copies of itself, so that specific genes can be produced in quantities suitable for investigative or forensic purposes. Mullis also wants it understood that he was having an enormous amount of fun, outside of science and inside it too: one of the chapters of his self-promoting semi-autobiography is titled "A Lab Is Just Another Place to Play"; the book's cover shows Mullis emerging Poseidon-like from the Pacific, his wet-suit dripping, his bare chest and taut belly proclaiming vigorous middle-aged health, his massive surf board thrusting manfully forward from his hip, oozing testosterone from every pore (figure 15).[47]

Both in this book and in interviews, Mullis advertises his super-charged sexuality and boasts of his sexual success rate.[48] Mullis took his Ph.D. (in biochemistry) at Berkeley in 1972, but, apart from one year lecturing there and postdoctoral positions at the University of Kansas and the University

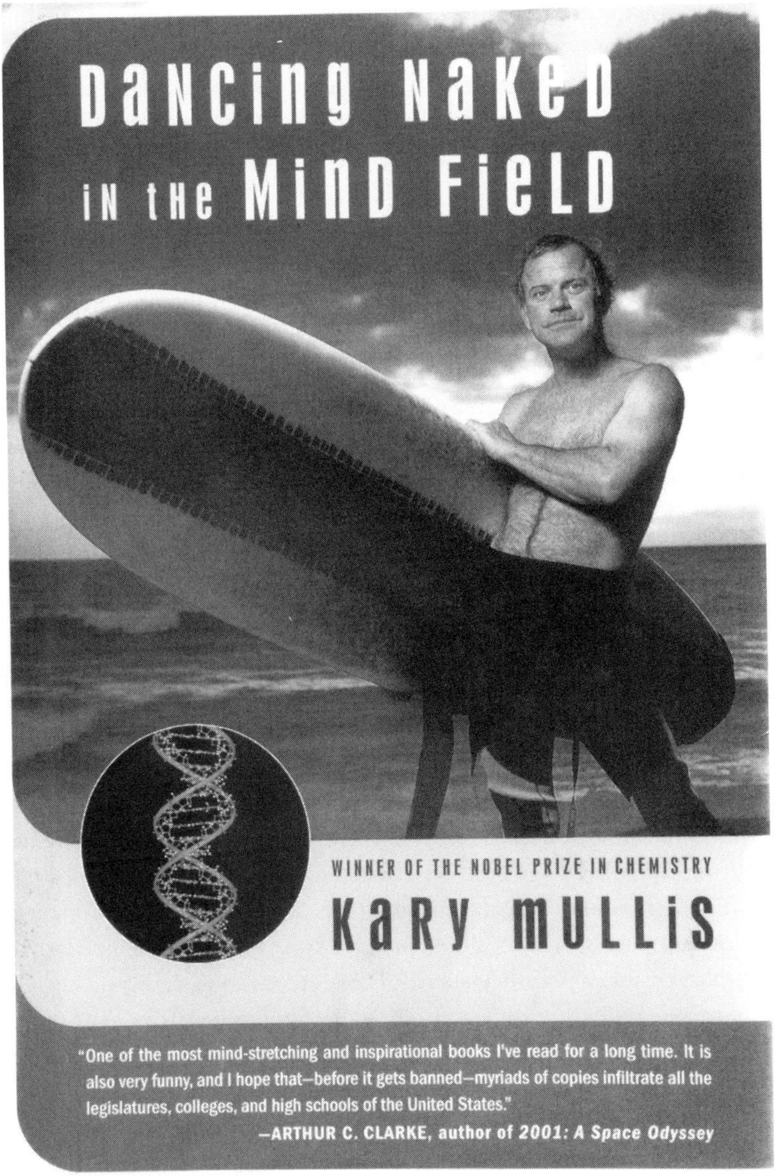

DANCING NAKED IN THE MIND FIELD

WINNER OF THE NOBEL PRIZE IN CHEMISTRY

KARY MULLIS

"One of the most mind-stretching and inspirational books I've read for a long time. It is also very funny, and I hope that—before it gets banned—myriads of copies infiltrate all the legislatures, colleges, and high schools of the United States."
—ARTHUR C. CLARKE, author of *2001: A Space Odyssey*

FIGURE 15. The cover of Kary Mullis's autobiographical book *Dancing Naked in the Mind Field* (1998). On the day he won the 1993 Nobel Prize for Chemistry, Mullis went surfing. The camera crews tried to follow him down the southern California coast, "asking everyone who came out of the water whether he was Kary Mullis." "As it turned out," he writes, "none of the other Nobel laureates that year were serious about surfing, and 'Surfer Wins Nobel Prize' made headlines."

of California, San Francisco, he has spent the whole of his productive scientific life either in industry or consulting for industry. Mullis ascribes his choice of biochemistry over astrophysics partly to its astrological compatibility (he is a believer), partly to his sense that anything related to human biology would continue to be funded, and partly to his thought that biochemistry would make for better chat-up lines than astrophysics. He makes no reference whatever to altruistic motives or the notion of calling.[49] In 1979, he joined the staff of the Cetus Corporation in Emeryville, California, where he was employed when he discovered PCR, leaving there in 1986 for a San Diego biotech company. Latterly, he has served as a consultant for many other biotech companies and has forged a career as a writer, provocative public lecturer, and professional scientific gadfly—opposing the HIV theory of AIDS, maintaining the innocence of O. J. Simpson (recruited to serve as an expert witness on DNA fingerprinting for Simpson's "dream team"), and expressing skepticism about the reality of global warming and other manifestations of what he takes to be scientific fashion.

When Mullis had the idea for PCR, he was instantly aware of the celebrity that awaited him—"I would be famous. I would get the Nobel Prize"—as well as its commercial potential, even though he underestimated the enormous sums of money that would eventually be made from it. Mullis wanted the money that his discovery was worth, and he was not shy in saying what he wanted and how he felt when he did not get what was owing him. Obliged by the terms of his employment to assign the patent rights to Cetus, Mullis got a $10,000 bonus, while Cetus eventually cleared $300 million when the patent rights to PCR were sold to Hoffmann-LaRoche—possibly the most ever paid for a patent. Mullis repeatedly asserts that he was ripped off financially by the greedy confederacy of dunces who were his colleagues at Cetus. The wound continues to fester: "Screw Cetus," Mullis writes. He regrets not listening to a friend in biotech who urged Mullis to contest Cetus's rights to the PCR patent, arguing that the invention was not made on company time or as an upshot of company-assigned duties.[50] So, while much about Mullis's public character has to be ascribed to his diligently cultivated idiosyncrasies, for many lay observers he came to represent both the virtues and vices of entrepreneurial science. Venter and Mullis are public figures, taking positions, and having positions ascribed to them, on the public stage. That's part of what it means to be an iconic figure of entrepreneurial science. It is important to talk about such actors-on-a-public-stage, for they constitute a focus for much abstract talk about scientific virtue and vice. But, just because these figures are iconic, they are not always the best access points for other sorts of questions, for example, what it is like to live and work in the world of entrepreneurial science.

THE LIFE-WORLD OF SCIENTIFIC ENTREPRENEURS

Late modern entrepreneurial science is sometimes celebrated and sometimes condemned. Often it is represented either as all virtue (helping the economy, making science relevant to human needs, curing disease) or all vice (corrupting the university, distorting the integrity and objectivity of science, setting profits over human well-being). But never does it escape the vocabularies of moral value, and rarely is it *described* in much detail, especially with respect to the experiences of those who live within its opportunities and constraints. Who are entrepreneurial scientists? What do they think about the life they lead and the choices they confront? How do they manage the institutional and personal possibilities presented to them? What motivational languages do they use to make sense of their own lives and those of their associates? How, if at all, do they engage with the celebratory and accusatory rhetoric surrounding their work? I want here to retrieve some of the moral and practical texture of late modern American entrepreneurial science, by letting participants speak, to a large extent, for themselves. Some of these scientists have spoken in their own right, by writing books and articles about what they do; a few have been written about by other scholars with aims roughly similar to my own; others I have personally talked to and formally interviewed.

Among contemporary American entrepreneurial scientists there is no single coherent story either about what that life is like or where its virtue resides. This needs to be insisted upon: heterogeneity is not the same thing as ambivalence. Different kinds of scientists—in different institutions, in different disciplines, with different intellectual and social backgrounds—experience the entrepreneurial world in different ways. Neither "science" nor "industry" nor "the university" are any longer—if they ever were—homogeneous natural kinds. The American public research university is not the same institution as the private university; among private universities there is a world of difference between, say, Harvard, MIT, Chicago, and Stanford; small start-up companies differ in crucial respects from more mature corporations; biotech companies face quite different circumstances from software companies, and scientists and engineers within them live and work differently. Biology is a different sort of practice from physics; within biology, the evolutionary biologist confronts the world of entrepreneurial science differently from molecular biologists or genomicists. And within genomics or molecular biology research, the idiosyncratic agendas, personalities, and life histories of laboratory directors make a world of difference to those who work with and for them. As much as critics and celebrants presume the stability and reliability of their references about "entrepreneurial science" and its practitioners, the sheer heterogeneity of the phenomena and how they are experienced give the lie to much of this talk.

One source of historical and contemporary heterogeneity must be addressed at the outset: neither "industry" nor "the university"—as sites of virtue, vice, and the routines of scientific inquiry—are now what they once were, and much of what has been written about the imputed characteristics of these institutions even in the first part of the twentieth century should (as earlier chapters have shown) be viewed with some skepticism. A contemporary cliché has it that universities have become more like commercial corporations while certain companies have become more like universities. What those who use this cliché probably mean is something like this: "The *idea* of the university has become increasingly incompatible with the reality of universities, while the *idea* of the commercial company is now often at odds with the reality of many companies." The American research university has become frankly corporate in its institutional structure, its scale, its financial routines, and in many of its ways of recognizing merit. At the same time, much of what counts as "industry" has become more like the "idea of a university" and even more like academic realities, partly as a result of shifts in the theory and practical management of skilled personnel and, more importantly, through the rise of the "knowledge economy": the recognition that economic growth depends crucially on science-driven technological innovation and that firms' competitive advantages flow from the intellectual capital they can command. Added to that is the increasing importance in advanced economies of profit-making companies whose products are more accurately defined as new intellectual goods than as material artifacts.

Chapters 5 and 6 described aspects of how the research function was managed in those large American companies that supported research laboratories in the first part of the twentieth century, but, by the 1970s and 1980s, the leading sectors of the economy were more and more populated by companies in which the distinction between "the research function" and corporate goals was hard or impossible to make. Accordingly, questions about the corporate management (or nonmanagement) of highly educated personnel whose job it was to discover new knowledge became central to important sectors of late modern capitalism. So tendencies to recognize and value the relatively high degrees of autonomy characteristic of many industrial research laboratories in the first half of the twentieth century were enhanced in the high-tech and biotech corporate environments of the late part of the century. In seeming paradox, American research universities were increasingly drawn to management thinking and practices that emphasized greater individual accountability, hierarchical supervision, and institutional planning, and these practices were rhetorically associated with being more "businesslike." As early as the 1960s, and considerably before Genentech-type phenomena presented his university with new realities, California's Clark Kerr was one of the first to draw attention to institutional convergence: "The university and segments of industry are becoming more alike. As the university becomes tied into the world of work, the professor—at least in the natural and some of the social sciences—takes on the characteristics of an entrepreneur. Industry, with its scientists and technicians, learns an uncomfortable bit about academic freedom and the handling of intellectual personnel. The two worlds are merging physically and psychologically."[51] The "convergence" of academic and corporate environments for inquiry should, however, be subject to all sorts of pertinent qualifications, but there is a great deal of truth in the claim—and, in the decades following his pronouncement, more truth than Kerr realized—as an account of 1980s biotech makes plain: "Like the intermingling of foreign cultures, academia and commerce adopted a few of the characteristics of each other. The companies had their academic scientists, their campus environments, and the academic tendency to wander off in interesting research directions. Academia, on the other hand, became more responsive to commercial stimuli." Once, academic life scientists responded almost entirely to peer pressure, but by the mid-1980s "a more complex interchange with biotechnology, including actual commercial ties, now supplemented" peer review. Commercial biotech was widely sensed to be "where the action was" and academics became increasingly willing to do work of interest to the industry.[52] Contemporary research universities and high-tech companies differ enormously—among themselves and over the course of their careers—and these differences are often the relevant considerations in shaping individuals' decisions about where and how to do their work. So what is needed in these connections is a move from institutional abstractions to the concrete realities of individuals' lives and choices. What does the embrace or rejection of entrepreneurial science look like, and feel like, to individuals confronted with these sorts of choices?

For many individuals whose work might have commercial consequences, either personal entrepreneurship or taking a position in industry
may not present itself as a matter of focal awareness. There are contemporary scientific practices in which entrepreneurship or other commercial
activities are facts of life, the conditions in which scientific inquiry of
certain types is routinely done. If, for example, you want to take drug development beyond a "proof of concept" stage, then either entrepreneurship or working in biotech or "pharma" is what you do. It is just hard or
impossible to do this sort of work within the confines and constraints of
a university: if what you want to do is to discover drugs and take them
through clinical trials, academia is not a feasible place to do it.[53] Similarly,
for many types of software and Web development or genomics research,
commercial settings are indeed "where the action is," whether one helps
start up a company of one's own or joins an existing high-tech or biotech
company.[54] And, again, the changes of recent decades are probably best
viewed more as matters of degree than of kind. In 1953, one industrial
scientist straightforwardly observed that "when you work for industry,
you are at the forefront of research."[55] And, from one end of the twentieth
century to the other, in many sorts of engineering the choice to conduct
inquiry in an academic setting is what needs special explanation. For
other individuals, engaged in other sorts of science or engineering, the
heterogeneity of contemporary academic and industrial settings means
that choices can, or must, be made, though whether such choices implicate abstract "ideas" of the university or of industry is a contingent matter.
Individuals differ enormously in the extent to which they think about the
institutional conditions of scientific inquiry as abstract matters. Some do;
many do not. And, while external cultural commentators evidently see
the industry-academia divide sitting astride a major institutional, intellectual, and moral fault line, it would be massively inaccurate to imagine
that pertinent practitioners necessarily do so.

ACADEMIA VERSUS INDUSTRY AS SITES OF VIRTUE

Late modern American scientists—more in some specialties than others—
tend to look at the geography of institutional virtue in particularistic
and fine-grained terms. But, just for that reason, it is good to start by
confronting evidence that *some* researchers continue to talk in a morally
charged way about "the idea of the university," to contrast it with "the
idea of industry," and to take or justify career decisions on the basis
of such apparently "traditional" contrasts. To that end, I talked to, and

carefully listened to, a number of my scientific and entrepreneurial colleagues as they tried to make sense of the world they live in, its characteristics, its moral textures. I formally interviewed over twenty-five scientists and engineers at various stages of their careers, and I more informally talked to dozens of others. They were chosen because each had either contemplated, or was then contemplating, a move to or from an entrepreneurial setting. That is, they were confronting institutional types, purposes of inquiry, and the social forms of inquiry not as theoretical matters but as concrete decisions in their own careers. This is where the chicken of "the idea of scientific vocation" comes home to roost. I tried to find out how these men and women thought about the institutions in which they did, or might do, their work, and how they experienced the routines, rhythms, rewards, virtues, and vices of scientific life. Accordingly, much of the rest of this chapter derives from what they told me as I asked them about their work, their careers, their motives.[56]

Take, for example, Professor Sean O'Reilly, a natural products chemist in his mid-sixties working at an oceanographic institute of a major Southern California research university.[57] O'Reilly's research involves the development of culturing techniques for marine microorganisms, and, in 1998, he was a scientific cofounder of a drug discovery company that licensed from the university IP he had created, on whose scientific advisory board he sits, and for which he consults on a regular basis. The company aims to discover potential drugs—including drugs for cancer—through "high-throughput screening" of substances isolated from the cultured microorganisms. Nevertheless, despite O'Reilly's intimate knowledge of entrepreneurial science, and despite the active interest he takes in what he sometimes refers to as "his" company's affairs,[58] he draws a sharp distinction between doing science in industry and in academia. He never had any interest in leaving the university, even though inducements had apparently been offered him to do so. His attitudes towards industry were strongly felt and they were based on personal experience. After his Ph.D. and postdoctoral research, O'Reilly was employed in the 1960s for several years in the laboratory of a large petrochemical company in the San Francisco Bay area, working on "improving oil products through basic research": "It was a very, very academic industry ... It was soft, fuzzy research. It was academically challenging. It was publishing." But O'Reilly nevertheless reacted strongly against the experience, and it shaped his subsequent attitudes: "Well, I tell you, I had no prior bias toward industry or academia when I accepted the [industrial] job ... What I knew was that I wanted to do research. I was a research-trained individual and

I wanted to do research and I saw the opportunity at the university; I saw the opportunity in industry; they were equal. I went there [industry] with a completely unbiased view," and he could just as easily have chosen academia at that early stage of his career. "I was really totally unaware of any comparisons between the two." It soon became apparent to him that the company was "inflexible" in its attitude towards research: "Sure, they wanted to do research, but they had guidelines that I thought were unreasonable and there were things like 'If you speak your mind, you'll never get promoted.' And they weren't so obvious to everyone, but, as I was there, I started to see them myself. 'My God, we're in a place here where they essentially bought us.' They bought us wholesale. And if something is wrong here, we can't say it's wrong. We have to shut up and do our job, and it would be things like 'is Barnacle Oil polluting the bay in San Francisco?' And if they came to that company and we knew that Barnacle Oil was polluting that bay, we'd be [out of] there in a microsecond, right? . . . No free speech. No democratic recourse. No way to deal with that company."

O'Reilly saw no problem in generalizing about conditions in institutional types known simply as "academia" and "industry," insisting that the differences between doing science in academia and in industry were, and remain, profound, and mentoring his numerous graduate students about those fundamental differences. Based on his own experience in a large company, and on his familiarity with contemporary biotech, he had concluded that there were three key differences between industry from academia. First, industrial research workers had to know how to compromise and had to accept the reality of compromising about research agendas: "People who are inflexible and very strong minded do not do well in industry." Second, they had to know how to "interact with dissimilar people, . . . realizing that you are not the boss, so you have to walk into industry with a very positive team-oriented capacity": "There's nothing more disruptive than someone who doesn't play a team game." Finally, industrial research was not egalitarian: "You're not in a democratic society . . . You do not have recourse. You do not vote. You do not question higher authority. If you do, you may not survive because you are susceptible to what I would call the vagaries of a more dictatorial kind of environment . . . People who really know how to play their cards and play their cards well, who will be very positive, never go against other people, tend to do reasonably well in industry. Rogue people who don't take orders, who have a lot of other less than perfect attributes [laughs] don't do well in industry. I realized [that at the petrochemical company], well,

I didn't look good in a suit." The research he was doing at the company was interesting enough; he was even willing to say he was "having a good time" doing it. "But something wasn't there to make me feel as though I was doing something valuable, receiving a reward, a career reward for my achievement, and I realized that other issues, other criteria, were being used to select the people who were going to move up. I said, 'I can't possess those criteria.' Literally, it was just looking good in a suit, keeping your mouth shut beyond anything else, literally having this company pay you to be an absolutely loyal, unquestioning individual."

So far as O'Reilly was concerned, the industrial scientist was, so to speak, an "organization man" in the 1960s, and substantially remained so in early twenty-first-century entrepreneurial biotech. He granted that such small start-up companies were sometimes congenial to "rogue," non-suit-wearing scientists at the beginning of their careers: scientists "are not really rejected in terms of how they look in suits." Some scientists could function in this sort of atmosphere, but O'Reilly did not want to, and, ultimately, he ascribed his aversion to a combination of a "personal ethic" and a view about the conditions for doing good science: "One thing you can say, I think, is that industry does demand a degree of loyalty to the company. How that loyalty may or may not compromise a person's personal ethic is unclear." But elsewhere he expressed a view that was far more decisive. "Industry runs on dollars and not on science"; when industry decides that the science being done in its facilities does not have clearly foreseeable commercial potential, it's the science that goes. For O'Reilly, it was something of the *atmosphere* of work that was decisively important, an atmosphere related in diffuse ways to the judgment of "good science": freedom to choose problems, of course, but also freedom from intrusive displays of hierarchy and of formality. The "suit" symbolizes much of this. There is not much suit wearing among the bench scientists of the small drug discovery company with which O'Reilly is associated, but he intermittently makes himself a nuisance—according to some of the company's executives—by trying to encourage a more informal atmosphere than they wanted. Even though he manages a large group of graduate students and postdocs and this management absorbs a lot of organizational energy, in O'Reilly's view scientific virtue necessarily resides in the university, where research can be free and spontaneous, and where you can wear whatever you like. It is a vision that both Robert Merton and William Whyte would have instantly recognized.

O'Reilly is nearing the end of his career, and, like many biomedical scientists, he has lived through a revolution in the commercial possibilities open to life-science researchers. So it is plausible that the sharp distinction he draws between the virtues of academic and the vices of industrial science might trace back to the scientific culture in which he was originally trained, as well as to the evident unhappiness of his brief early experience in industry. Nevertheless, broadly similar sensibilities are occasionally expressed by younger researchers, including those working in disciplines that historically have had much closer ties to industry. When I talked to Nikolai Metzger in 2002, he was a twenty-eight-year-old computer science superstar, soon to recognized as one of the "100 top young innovators" by MIT's *Technology Review*, with notable expertise in computer vision, pattern recognition, and digital signal processing.[59] Then an assistant professor of computer science at the same university as O'Reilly, Metzger had cofounded a company while still an undergraduate at Caltech and before taking his doctorate in electrical engineering from Berkeley. The Silicon Valley biometrics company Metzger helped found made computer security devices based on fingerprint recognition. It is now more than ten years old; it flourishes; and Metzger remains on its board of technical advisors. But Metzger himself eventually had enough of industry, and, while he is quite possibly sitting on enough shares and stock options to forgo remunerated employment, he began curtailing his involvement in 1998—"just not coming in as much . . . and then I announced that I wanted to get an academic position somewhere." In 2001, he did become a junior academic, at a fraction of the remuneration he could have obtained from industry. At the university, Metzger is heavily involved with a multi-million-dollar project to develop software for the automated monitoring of laboratory animals—the so-called "Smart Vivarium" project—which interests him greatly on intellectual grounds but which is obviously of great potential commercial interest—in medical research and also in emergency responses to biological or chemical terrorism.

For Metzger, the sentiments identifying the university as a site of scientific virtue are both diffuse and visceral. Some of his feelings about the kind of life he wanted trace back to his family; some to his undergraduate days. Metzger grew up in Sacramento, the son of parents who had not gone to university but who were deeply committed to all four children obtaining degrees. For reasons he cannot specify or explain, Metzger always seems to have had a vocation for academic science, and it wounded him deeply to hear professors disparaged: "I have this somehow intrinsic respect for professors. I wanted to be one for a long time . . . So I really liked academia: that academic lifestyle and the culture, learning, and so forth." Metzger recognized that this was a sweeping generalization, but when it all works as he reckoned it should, then "those people seem like the greatest people in the world." At the time Metzger graduated from Caltech in the mid-1990s, it was not—in his view—an entrepreneurial environment—explicitly *not* Stanford or Berkeley. At the graduation ceremony, Gordon Moore (of Intel) was speaking, and he "talked about us [Metzger and his cofounder] as examples of entrepreneurialism. And basically we were the only ones." That was then, but, as Metzger insisted, "it's just completely changed [now]." During his time, "Caltech was all about pure science. The nerd culture there made us very proud of that, suspicious of patents, [we] didn't like the professor that did a lot of consulting . . . because it always seemed that their mind was somewhere else. Like, we really admired the professors that seemed to lead very simple lives but were really into it. You know, it's really sort naïve [and] old-fashioned. Caltech is its own universe." And this conflict between entrepreneurial and commercial orientations, on the one hand, and responsibilities to students and to disinterested inquiry, on the other, was a recurring theme for Metzger and several of his colleagues.

Involvement in the biometrics company he helped found made Metzger "miserable": "It just sucked the life out of me." Part of the misery was "the materialism of the Bay Area" at the height of the dot-com boom of the 1990s: the real estate fetish; the condescension towards academic scientists and engineers who were not playing the entrepreneurial game; entrepreneurial professors' neglect of their students; the necessity of chatting up venture capitalists. "This was a world that I was not good at. I didn't know how to schmooze in that world at all." It was a world he found "just suffocating . . . just totally out of whack." "What really made me want to run away as fast as I could to academia is what the typical day was like, what I did in the typical week." In the start-up phase, "everything was exciting because everything was new. There was still research going on, and it felt good to be viewed as this whiz kid . . . It felt like we were in the middle of the universe." "At the beginning there's the fun R&D that gets things going. But then we need to get a shrink-wrapped product on the shelves." And soon you work not to produce what you know is possible, or best, or good, but what you are told the market requires: "These are things that are not intuitive to the human spirit to do." Ultimately, you get surrounded by people "who *hate* what they are doing." He had to learn to deal with people for whom money was everything and science only a means to a monetary end. This was not freedom, but a form of "slavery." Metzger's taking up a professorial position was an expression of his sense that industry and academia were fundamentally different kinds of places, that virtue lived and flourished in universities, and that it expressed itself significantly in a commitment to *teaching*, to passing on the vocational torch. He was not naïve, nor, given his background, was it likely he would be. He understood that professors might have entrepreneurial drives, that they might wish to start up companies, but they should "get it out of their systems early in a clean way"—as he himself did. If they don't, then "it kind of pops out awkwardly in odd ways" and interferes with their proper academic responsibilities. The students must "feel like they are number one," or the professor is not doing his job. Although Metzger liked to talk about his commitment to "basic science," he was a member of a department in the engineering school, and the research he did there continued to be of interest to companies. Accordingly, he was well aware that, whether or not he ever again took any active entrepreneurial role, his freedom of action was bound up with his sources of funding. His work was not, in relation to, say, robotics, very expensive, and his preferred source of financial support was the National Science Foundation: "The more you can get funding to do basic research from agencies like NSF, the more autonomy you get." But if you start to get funding "that's closer and closer to industry that is more tied to a specific deliverable, [then] you might as well not be a professor . . . I came here because I wanted to direct my own research group and if my funding starts to get tied [to specific practical ends], then what am I doing here? . . . That's the idealistic picture I painted in my mind before coming here." For Metzger, academia meant autonomy, and autonomy, together with a commitment to one's students, was *the* cardinal scientific virtue.

Metzger had a departmental colleague, Lee Marvyn, who felt so strongly about entrepreneurial science and the entrepreneurial university that in 2001 he sent a letter to his dean resigning his position in protest and placing the letter on his personal Web site.[60] Marvyn is a computer scientist in his mid-forties, working mainly in the fields of computational biology, genetic algorithms, machine learning, and free-text information retrieval. What prompted Marvyn's resignation was his feeling that the boundaries between commercial and academic imperatives were being blurred, with disastrous consequences for the university's research and teaching functions. Academia was losing its virtue and he was compelled to register a personal protest:

> With respect to research, I believe our attention has become confused about the relative roles of the INTELLECTUAL PURSUIT OF QUESTIONS worthy of research and the FUNDING necessary to pursue them. Most research costs something, and funded research plays a pivotal role in the

support of graduate students. But it is not an end unto itself. During my time here I have seen this confusion deepen and expand, to the point that activities appropriate within a university and those typical of a commercial setting are almost interchangeable. Involvement with commercial enterprise has gone from anomaly, to common-place, to a badge of honor. It is no wonder "conflict of interest" has become a confused, artificial charade. Worse, our research agenda is being skewed towards questions that can be connected to "thrusts" of short-term economic consequence. University research must retain its focus on difficult, long-term research questions of foundational consequence; innovations that will make someone money will happen on their own.[61]

Like Metzger, Marvyn believes that what makes universities—at least in their ideal form—completely distinct from industry is their proper commitment to teaching, but he reckoned that the encouragement of university-industry links, and the related stress placed on raising research funds, had pushed teaching commitments into a distant second place. The university advertised that its "world-class research" would translate into superior and dedicated teaching, but students were being systematically short-changed.

Having taken his Ph.D. in 1986, Marvyn's sole job since then had been at his current university, but, during a break in his graduate education, he had worked full-time for a small database design company where he "made a little nest egg" before going back to finish up his doctorate, as he had always intended to do. Between taking up his academic position and the time I talked with him, Marvyn worked on various industrial research contracts (Apple Computers, Encyclopaedia Britannica) on his free time, but never did contract research in his academic laboratory, which continued to get its major funding from the NSF. Before and while working at the company, Marvyn "had a lot of positive imagery about creativity that happens in the commercial workplace," and it was a creative environment that he was looking for, whether it turned out to be in academia or in industry. But he was committed to what he called an "aesthetic" of inquiry, in which the research agenda would follow from some notion of internally generated, inherently interesting problems, not responding to short-term commercial considerations. It is an aesthetic that "has to do with an accumulation of knowledge as opposed to any practical outcome whatsoever," making it inappropriate to ask of such work "what this is good for." The inquiry's virtues were internal, not external. Realistically, even in universities, Marvyn thought that it was a good idea "to have a portfolio of deliverables available; you have to have [projects] that are safe, money-in-the-bank sort of results, the utility of which are clear to everybody," and, at the other pole, risky projects whose payoffs are hard for anyone to visualize. Industry might have once, and in a few specific places, supported such goals—he acknowledged that there were industrial laboratories "where they had as much autonomy and as much freedom as I do"—but such industrial places were atypical and increasingly rare. "Industry, as a whole, is about shrink-wrapped product, so to speak," and universities were obliged to take up the burden of pure research: "I absolutely value my ability to decide where my research agenda goes myself." He was deeply skeptical of much talk about the willingness of industry to support genuinely fundamental research programs on any sustained, long-term basis.

One aspect of university-industry ties to which Marvyn strenuously objected was their tendency to restrict graduate students' freedom of action. Taking industrial money binds the principal investigator (PI) to specific goals and timetables, and, while it offers important resources for the support of graduate students, "I'm not comfortable with that level of expectation and [linkage] between dollars and students. So in general I try to avoid it." The proper educational purposes of a university, in Marvyn's view, made many forms of industry-university cooperation deeply problematic and morally suspect. For his own part, Marvyn did not "sense any real immediate attempt to have me knuckle down and, you know, work on homeland security. It's herding cats, right?" But he was clear that his academic superiors would be much happier if his work was of more immediate practical significance, and he chafed at that: "You know, they say, 'You know, it would be great if you guys thought about printers a little bit more,' or something like that." The field of computer science in which he worked responded illegitimately, in his opinion, to the joint pressures of industry and the political powers.

Marvyn's decision to resign was therefore informed by his conviction that the university was making itself a less and less virtuous place: "The level of conversation has been pitched significantly toward pragmatics of an industrially defined notion of progress." The department was being "pitched" to local industry, under pressure from the entrepreneurial dean, in terms of its short-term deliverables, and its recruitment strategies were being shaped by those same considerations: "I found that offensive ... It didn't use to be that way." Marvyn was not convinced that university responsiveness to industrial concerns, and the blurring of the boundaries between academic and industrial research, were even achieving the goals advertised. There was, he thought, confusion about "what universities are good at and bad at," a confusion exacerbated by the Bayh-Dole Act and subsequent proliferation of university technology transfer offices: "It is [an] inefficient use of this state-funded resource, . . . to have us doing what, really, other enterprises—commercial enterprises—do better than we do." So Marvyn had a very strongly marked ideological, as well as practical, sense of what universities should be about as environments for inquiry. University research had the capacity to contribute to large-scale and long-term economic growth, just on the condition that it was *not* too strongly influenced by commercial goals: "On the research side of things . . . the university is great at doing something. Guys, you shouldn't kill the goose that is trying to lay these eggs." On the other hand, he was able to recognize that pressures bore on different sorts of researchers in different ways, and that, were he (as a senior and well-respected academic) now go to work in industry, "it would be [only] marginally different than the activity I do now . . . I wouldn't feel very different. I—you know—would have more money at the end of the day. There'd be—there would absolutely be more meetings. There would be more sales. I know that." But the real reason he was fighting to preserve his sense of a distinctive academic space and its attendant virtues was that "I bemoan the loss of a place that fostered what I see as a really valuable activity—something that used to be, I believe, more embraced . . . in nontechnological situations . . . That notion of what the university is, is what drew me to it. It's a beautiful vision . . . a romantic ideal." And Marvyn felt called to give witness to the blurring of that vision: "The university has been sold for golden strings. The university is about technology transfer. I think this is roughly the Bayh-Dole argument. 'We want our universities to accomplish practical things. We love our universities for that.'"

His is, nevertheless, a moral vision that Marvyn *knows* is not shared by the majority of his own academic colleagues. Most of them thought—in his own formulation—that he was tilting at windmills, that he was "bitching about being successful," that he was rocking the boat, that the situations he was complaining about were "not that bad," even that he was "crazy," and that he was "making this stuff up." Marvyn thought it possible that there was a generational distinction in sensitivities to such things. You had to be around maybe fifteen years or more in his field to see the changes taking place; most younger academics took the commercialization of the university very much for granted as just the "way things now are," "how the game is played." Indeed, the ideal, and to some extent real, distinction between institutional virtues and vices gestured at by O'Reilly, Metzger, and Marvyn are probably no longer the norm for scientists and engineers at major American research universities. Those sentiments are one way of conceiving scientific virtue in an entrepreneurial age, but not the only way. The geography of late modern technoscientific virtue is just not that simple.

IVY AND IVORY: THE MORAL HETEROGENEITY OF LATE MODERN TECHNOSCIENCE

For many scientists, the modern research university—or, at least the sectors of it with which they are familiar—is a deeply imperfect site in which to do the science they want—science of integrity, science done in a way, and for a purpose, that strikes them as *right*. For such scientists, there are other sites in which to live the Scientific Good Life, and the sensibilities expressed by O'Reilly, Metzger, and Marvyn make little sense. So it is not uncommon to find academic scientists complaining bitterly about the support structures and processes that shape their conditions of inquiry. James Hawicke, a professor of psychiatry who had tried unsuccessfully to leave the university to start up a genomics company, noted that "I certainly spend a *lot* of my time writing grants and even more time *worrying* about grants. And it's *really* stressful. The grant-writing process is so tedious." Moreover, its outcome is so uncertain that "people frequently experience it as almost capricious . . . the luck of the draw . . . and that is *very* frustrating and stressful, and [the] number-one complaint of pretty much *all* faculty, everybody hates it . . . It's awful and painful and stressful."[62] Where were research autonomy and integrity if these were the conditions of possibility? Even O'Reilly, who saw the university overwhelmingly as the site of scientific virtue, was frustrated by the grant system that controlled what he could and could not do: autonomy meant little without resources, and, if resources happened to be more abundant in industry than in academia, then there was a legitimate sense in which autonomy followed those resources. He reckoned that his own graduate students saw this aspect of the professorial role as profoundly repellent: "They tend to develop the impression that being a faculty member is painful and they don't want to do it." Students see the faculty member "working hard to support graduate students and seeing grants be rejected, seeing people essentially being belittled because they didn't dot some i's and cross some t's, they get a [low] score and a negative letter, and then they're embarrassed and their students have to go look for some work." The price of integrity to keep O'Reilly's group of about twenty-five

researchers going is about $1.8 million a year, and it's a continuing strug-
gle to secure those resources. Like Hawicke, O'Reilly also saw his research
freedom limited by the instrumental resources routinely available to him
from the grant system: "In one sense, [industrial scientists] are more facilitated to come up with that interesting stuff than we are in academia
because they have the resources, they have the support. If they've got to
have a new machine that costs a half a million bucks, they're going to get
it. Contrary to over here where we are, we say 'Gee, if I only had a $500,000
machine. Well, I guess we could write a grant and we're going to hear in
nine months if it's approved and then we're going to get the money in four-
teen months and then we're going to finally end up with the machine, in
two years, we might have one.' So that's a limit." The principle of academic
research freedom was limited by the practical realities of getting research
supported.

Diane Springfield, a junior scientist then working as associate director
for genetics for a (failing) Bay Area biotech company, fully acknowledged
that research agendas in industry were largely set by the company, but
recognized that autonomy in academic research was *also* significantly
constrained: "You work on what you can get funded and so you are
constrained by that . . . You may have a great idea but if you don't have
the money to pursue it you can't." "In my company," she said, "we have
a ridiculous number of [DNA] sequencing machines," and so she and her
colleagues could do things that would either take much longer to do or
be impossible to do in academia.[63] André Schumann, a young scientist at
a start-up bioengineering company, and another of MIT's "100 top young
innovators" for 2003, saw industry as clearly better resourced for doing the
research he wanted to do than academia: "They can just do science that
nobody else can do. That's one thing that was really amazing to me when
I came from academia to industry. There are certain types of research
that industry is far ahead of academia, and they probably always will be
in certain areas, just because of the fact that, you know, the problems
that they can address, they have the resources to do it, you know. They
can scale things up."[64] At Celera Diagnostics one of the chief scientists
decided against an academic career because of what he had seen during
his doctorate training, where "the only thing you really get to do is bring
in grant money. That is your goal in life, to bring in grant money, and
the pressures are apparent. I questioned whether or not that was what I
wanted for my life. Even at twenty-two years old, I had seen investigators
who lost grants and the effect it had on their lives."[65] A journalist writing
of the San Diego academic biologist-turned-entrepreneur Ivor Royston

said his decision to leave the university in 1990 "wasn't the occasional collegial back-biting and political sparring that ruined academic life for him—it was the bureaucratic plodding."[66]

The Nobel Prize–winning biologist Arthur Kornberg, who in 1980 helped start up a biotech company called the DNAX Institute of Molecular and Cellular Biology, was well aware of academic condescension towards industry that marked the 1970s and 1980s, but reckoned that it was both misplaced and much exaggerated:

> Able scientists are interested in industry. Some are discouraged by the atmosphere often encountered in university departments: the emphasis on entrepreneurial skills of grantsmanship, the inevitable clashes with university bureaucracy, the obligation to serve on committees, the burden of heavy teaching loads, and the pressure to choose a safe, fashionable research program that will produce publications for the next grant application and academic promotion. In the face of these problems, one might see an industrial setting as offering several advantages: excellent resources, research objectives in interesting areas of science, fewer distractions, and a team spirit united for achievement.[67]

Mark Jones's superb account of the birth of the San Diego biotech industry in the late 1970s and early 1980s notes that at least one prominent scientist recruited to the then-unfamiliar start-up environment had concluded that the academic rules of the game, notably including the "publish or perish" system, now constituted a substantial obstacle to scientific progress. They did more to discourage than to facilitate open scientific communication, and, because they stressed quantity over quality, they put a brake on genuine creativity and innovation. In the scientist's own words,

> Most [scientists] out there will admit to this. In order to ensure their futures, they are forced to publish things that are based on less than complete information, less than complete experience. They are forced to do a lot of experiments fast, get a lot of data, write it up in as many ways as possible, and put out a volume of publications. Or else they aren't going to get their next grant.

He thought that quite a lot of the scientists he knew were starting to believe "that academic science wasn't what we thought it would be. You didn't have the freedom to do the kind of research that you wanted, and it was getting harder to get grants, and certainly there was a lot of administrative shit to put up with."[68]

Both Paul Rabinow and Martin Kenney estimate that 30–40% of academics' time is spent, one way or another, in the grant application process and that the drag, the uncertainty, and the limitations of the system for funding academic science are substantial inducements for some biomedical scientists to leave the university for commercial entrepreneurial organizations.[69] As the next chapter will indicate, securing private finance for start-up companies can often be more uncertain than the NIH or NSF grant system—as O'Reilly asked, "Is the company going to be there two years from now?"—but the resources may be larger and they may be the only kinds of financing available for the sort of science that researchers want to do. It has already been noted that if you want to do drug development beyond "proof of concept," then academia is not the place to do it. Universities can go some way towards getting drug development off the ground, but they do not have the resources to take it more than a little way to realization—certainly not when the time from concept to market is about ten years and when the resources needed per drug are pushing $1 billion.[70] Moreover, the experience of autonomy with respect to academic structures may look very different to senior and junior scientists, to those who have established a track record and an accommodation with their funding agencies and to those who have not. Very eminent academic scientists I talked to expressed little frustration with the grant system; more junior ones, or those with comparatively modest reputations, were more likely to find the process constraining and irritating. Facing the possibility of failure in winning a grant, and failure in securing tenure, the autonomy and integrity of academic science may not be obvious to researchers at the start of their careers.

Heather Yellowlees, a young scientist who had taken her doctorate with Hawicke and who was mulling over her career options, felt massively insecure about almost every aspect of her scientific life. She did not feel that she had freedom of choice in her dissertation project (which was "a huge gamble" pressed onto her by her thesis committee); she did not feel she had adequate mentoring and support in doing the work; and, partly because of these feelings of insecurity, constraint, and isolation, she "never wanted to be faculty . . . I wanted to go into industry. I thought that's where I needed to be to do the kind of research that I wanted to do. You know, finding cures." Yellowlees had serious concerns about whether the biotech companies she had looked over were genuinely interested in her sort of research, but she was biding her time and looking for an ideal fit between her research concerns and a company's agenda. She had no desire to follow in Hawicke's footsteps. She had seen Hawicke lose funding, and so lose people from his research group, and she disliked that sort of insecurity: "I don't ever want to be responsible for somebody else losing their job. I don't want to deal with that or worry about, like, where my money's coming from or am I going to keep my lab, you know."[71] Finally, a number of scientists working in universities and nonprofit research institutes complained bitterly about what they saw as the conservatism of the NIH and NSF grant process, expressing their view that eccentric, bold, or just unfashionable research proposals had little chance of getting funded in the current scheme of things. If what you wanted to do fit with the preferences and priorities of funding agencies, then you might well experience the process as autonomous; if not, you might describe the world of academic science as seriously constrained. And if you equated scientific virtue with devil-may-care curiosity, then the virtue attached to academia was deeply compromised.[72]

Arthur Kornberg mentioned "heavy teaching loads" as one "burden" that might drive academic scientists into industry, and chapter 5 indicated that such considerations go back to the origins of industrial research as a career option. For some scientists, teaching is, indeed, regarded as a burden; for others, it may be a joy—one of the unique pleasures of working in a university; for still others, it can be a bit of both, or something in between. Nor is "teaching" a homogeneous practice, with a unique and stable set of demands, structures, and rewards. Heather Yellowlees would not even consider an academic career, in large part because she "never ever wanted to teach. That was never my thing. I wanted to do research." Her mentor James Hawicke didn't mind teaching, though, as is common for medical academics, the dividing lines between teaching, research, and the administration of his research group were much less clear than they were for, say, academic historians or sociologists. Hawicke wanted me to understand what "the deal" was for people in his position: "The deal is you come in the school of medicine, it's usually clinical responsibilities, and the rest of the campus it's teaching responsibilities. And if you want to be successful at research you raise money and sort of buy your way out of those responsibilities." In his type of medicine, about half of his time was absorbed in taking care of patients and teaching. "And the rest is yours to do research, but throw in administrative meetings, etc., etc. If you can get more research money, you can support more of your own salary and less of your salary has to come from clinical income . . . Not that anyone gets away without teaching completely. Nor would we want to, why would we be here otherwise? But you reduce the amount of it that you have to do."[73] Alfred Byster, a senior graduate student in electrical engineering, recorded an exchange between two of his professors about the role of teaching in university science: "Like I heard my advisor make a comment and he had a friend who was thinking of going into university to be a professor [so he could do] whatever he wanted, and my advisor kind of laughed and said 'Yeah, maybe I have a slight bit more academic freedom, but you don't understand, I have to teach classes, but I probably would get more research done in a company than I would directing ten students or teaching classes.'"[74] The same range of considerations—and others— determined the young biophysicist Jon Moore to reject two decent academic job offers in favor of joining the start-up Cambridge biotech company Vertex. As Barry Werth relates, "looking at the landscape of academic research, Moore recoiled. The more he thought about being an assistant professor and what it entailed—uncertain funding, chronic job insecurity, the need to sacrifice research in order to teach, dependence on better-known collaborators—the more appealing Vertex looked." Moore was particularly attracted by the opportunity that the company offered him, "practically impossible for someone at his level outside of industry, to plunge directly into a hot area." Industry was less hierarchical, more rich in scientific opportunity, and, in those senses, more free."[75] If you see teaching as a constraint and obligation, then industry promises relief from such duties, as well as other inducements.

Still other academic and industrial scientists insisted vehemently that there was *no* material difference between the sort of graduate teaching and postdoctoral mentoring that went on in universities and the less structured training regimens that characterized many high-tech and biotech companies.[76] Many scientists working in industry made repeated and matter-of-fact references to their roles as givers or receivers of advanced training, and, indeed, of their occasional role in co-supervising doctoral students. Entrepreneurial activities and a strong commitment to teaching need not stand in conflict. Eidur Gudjonssen, a spectacularly entrepreneurial professor of bioengineering, was much admired by most of his students for adapting research projects to their particular interests, and for taking account of whether they wished academic or industrial careers. But several of his students were valuable resources for him in starting up his newest company; one of them (mentioned below) became the CTO of that company; and others—in dubious compliance with university regulations, but with their professor's evidently enthusiastic consent—implied that they crossed the road from the university laboratory to the company's premises in order to do some of their research. There is much talk among critics of the commercializing university and of entrepreneurial professors about the exploitation of graduate students, set to work on their professors' corporate projects and given commercially routine, rather than intellectually challenging, projects to do. Henry Lyons, a senior electrical engineering professor, was seriously worried about the influence of commercial ties on graduate student research. In places where industrial influences are strong, graduate students would be handed routine projects that made no scientific, but much commercial, sense: "There's a lot of grubby work associated with [going] very far down the commercial road. And who's going to do the grubby work? The answer is graduate students. They're not going to get the same education, so that's my main problem with it . . . You don't become a creative [scientist] by working on a project that somebody else gave you," and his professorial responsibility was to ensure that his students *did* become creative people, that they shared his sense of a research vocation.[77] The advanced engineering graduate student Byster found industrial research "just not that interesting. Sitting in front of a computer and doing very detail-oriented work . . . Whereas it's the exact opposite at university." Byster *knew* that some graduate students in his area were being used to further their professors' commercial interests: "They were getting their Ph.D.s and essentially just working cheaply for their advisor's company." But he was also capable of seeing advantages in what others might call an exploitative relationship: "If things go well they get to join the company and they're going to get rich." He had nothing against that, nor did he think it was plausible that anyone else really did.

Doubtless, some of the criticism of commercializing academic science is wholly justified, and alarm expressed by some faculty at proliferating tendencies in these directions has already been noted. However, senior graduate students—both in engineering and in commercially relevant natural science—commonly told me how much they *valued* the opportunity to get hands-on entrepreneurial and commercial experience. This is what they *wanted to do*, and they freely chose academic supervisors who could introduce them to the entrepreneurial world. Michael O'Mair, a young Irish graduate student in electrical engineering, could think of few disadvantages to having graduate students work in their professor's company, or do corporate-sponsored research, provided that it was "sensitively" done.[78] Lyons took it for granted that his students, and especially his *undergraduate* students, were aiming for industry, and, while he strongly disapproved of academics founding companies, and even more strongly of universities encouraging technology transfer, he insisted that his own active industrial consulting work was abundantly justified by its contribution to showing students the practical relevance of what he taught.[79] What his engineering colleague Lee Marvyn reckoned was culpable neglect of the university's teaching responsibilities, Lyons celebrated as a valuable contribution to acquitting his moral and instrumental obligations to students. And among local students—both graduate and undergraduate—there was plenty of sentiment to support both.

It has been common to construe scientific virtue as the unconstrained and disinterested search for new knowledge, and this is a view that has been embraced over many years by academic commentators from inside as well as outside the scientific community. In this sensibility, a virtuous scientific motive just is inquiry "for its own sake," and virtuous sites are those that encourage the realization of that motive. But scientific practitioners, in fact, may have a range of desires, and chapter 5 documented the fact that the majority of researchers, throughout the past century, have worked in institutions whose formal purpose was the production of goods and services and in which theoretical knowledge may or may not have been recognized as a relevant resource or objective. Entrepreneurial science, therefore, may appeal to scientists who want to *make* something, to see ideas become embodied in products, and, perhaps, in products contributing to the social good. Why, apart from a Manichean conception of the relations between academia and industry, should that be surprising? O'Reilly, who was "never tempted" to leave academia for the small drug discovery company he cofounded, nevertheless is keen on its success and remains actively involved in it as a consultant. It is very important to him to "see the fruits of [my] long research make a difference in the community, make a difference in terms of employment . . . , a new set of things being developed." A drug coming out of his research, a drug on the market, helping cancer patients, and, of course, making money as a mark of its therapeutic success—that was important to him, and that would, he said, "validate" his life's work. His active role in the company reflected his belief that the university's technology transfer office did not attract "the best and the brightest," and so he would have to become some sort of entrepreneur himself if he really wanted his life's work ever to yield a drug. Joan Rhodes, a bioengineering Ph.D., who had worked for two Southern California biotech companies and had never held an academic job, commented on one of the non-monetary satisfactions that industry offered and that academia rarely, if ever, could. What many scientists coming out of academia liked, and that they didn't get in university employment, was the opportunity to "see a project through," that is, from an idea to its implementation in something that helped real patients. She said this without a tone of high-minded altruism, just as an item in a job-satisfaction list.[80] Similarly, Jon Marx, a distinguished cardiologist with positions both at a research university's medical school and at a Federal government hospital, remained in academia, while taking an extended sabbatical to start up two companies, explaining his industrial interludes because he wanted desperately "to get [my biomedical devices] into patients," which he thought he could do only by acting as an entrepreneur. Why, he wondered, was it thought virtuous to be useless?[81]

André Schumann, the young bioengineer with a strong entrepreneurial bent, explained the sort of interest he had in research in terms of the sort of person he was: "Any research project that goes on, I will have a very difficult time being successful at it if I can't answer right at the beginning—at the outset—what's the value of it? What's it good for? . . . What will it help?" And the Cetus and Celera Diagnostics scientists interviewed by Paul Rabinow commonly found it appealing that their discoveries should *amount* to something in the practical world of health care, feeling, in Rabinow's words, "some degree of discomfort that ideas in university settings rarely led to health-oriented results."[82] Amgen CEO Kevin Shearer said that he joined the company in 1992 because biotechnology seemed the answer to "big questions about the meaning of life." But he didn't mean questions of purely theoretical interest: "At the end of your career," he told a journalist, "when you ask 'What did you do that counts?' you can say 'I saved lives.'"[83] Nikolai Metzger, who escaped his Silicon Valley company because its quotidian routines made him so unhappy, nevertheless mentioned the delight he experienced when he was walking around town after a conference in Corfu, "and I see this rickety software store. And there's my company's product with Greek writing on it sitting packaged on the shelf . . . And I thought, I just came halfway across the world and looked in a window, and something that was just a tiny idea in . . . my head back in '93 is now for sale in Greece. And I thought, that is *amazing*." He knew this was not a feeling he could have experienced had he never been an entrepreneur.

The notion that some scientists, qua scientists, might wish to assist in cultivating the material fruits of their conceptual and manipulative labors, is nothing new. It probably goes back to the origins of industrial research, or even farther. In the early 1950s, a scientist working for a consumer products company made plain why he preferred industry: "Many universities don't do research except of the ivory tower type. I prefer to see the results of my research. It gives me a great thrill every time I see a package of 'Dot' on the grocery shelf because I mixed the first batch of it."[84] Max Weber thought that the very idea of a scientific vocation, and of scientific virtue, could not encompass commercial goals and entrepreneurial means, but there were scientists in 1918 who disagreed and still more who do so today. Many scientific entrepreneurs reject any notion that the transformation of knowledge into material products or marketable services is any less intellectually demanding, or that it requires any lesser degree of intelligence, than so-called pure science. They do not see acquitting entrepreneurial ends—the production of a shrink-wrapped object—as a straightforward deduction from pure science. Rather, they see achieving entrepreneurial goals as requiring a quite special, even superior, sort of mental ability. The entrepreneurial engineer, Professor Gudjonssen, who had started up three companies in his time, wanted it understood that getting companies established demanded all the intelligence going: "How do you make a business operate is a big challenge. How to make a product? How do you get customers? How do you build a business model? It's a challenge," an *intellectual* challenge.[85] The problems may be diffusely framed—how to raise finance, recruit and motivate people, organize the corporate environment, locate markets and identify competitors—but, because of that, they can plausibly be seen as *more* intellectually demanding than the well-framed problems of academic science. Entrepreneurs may see themselves as having a broad vision of the world, contrasted to the narrowness and inwardness of their purely academic colleagues. They know how to do things about which their colleagues are clueless. It's a matter of experience, of course, but it may also be seen as a form of constitutional intelligence. The first thing that Gudjonssen wanted me to understand, in order to appreciate the special kind of person he was, was that he came from a *family* of entrepreneurs and businesspeople: "So maybe one thing I would like to say . . . from my background is my dad ran a company. My grandfather ran a grocery store and it stayed in the family." He found risk taking natural, and he believed that the intellectual abilities enabling him to understand business, and how it worked, were bred in the bone. The ascription of entrepreneurial abilities and instincts to family background is a recurrent theme. It is a way of naturalizing what seems, to some, but not all, still in need of special explanation and special justification.

THE TEXTURE OF THE ENTREPRENEURIAL GOOD LIFE

The texture of quotidian life in entrepreneurial science, like that obtaining in more traditional academic research, is also too heterogeneous to summarize concisely or to be captured by any one coherent narrative. Idealizations of what that life is like abound—streams of idealizations produced by academic commentators, journalists, and sometimes by participants with particular points to make—though these idealizations usually function more as accusations and celebrations than as serious attempts at description. Nevertheless, the phenomenon, and its associated institutional patterns and routines, are sufficiently novel and noteworthy that some participants are keen to volunteer their thoughts about what it is like and how it differs from traditional patterns. They want to talk about it, to make sense of it, even, on special occasions, to justify it to themselves and others. So what sorts of things do entrepreneurial scientists, and those considering taking up that life, have to say?

Some of them address topics that have been traditionally seen to *distinguish* academic and commercially oriented science. They talk about secrecy and about autonomy, and the relative differences in these as they appear in academia and industry. Much of this talk can resemble the academic social scientific and humanistic criticisms of industrial science described in chapter 4: academic science is open and free; commercial science is secretive and constrained. Even in the mid-1990s, a celebrated and otherwise sensitive account of the world of biotech was drawn to a strongly dualist account of institutional norms: "Scientists in industry and scientists in academia tend to be brutally dismissive of each other," Barry Werth writes—though little in his excellent book *The Billion-Dollar Molecule* actually supports any such global imputation. "Academic researchers thrive on publication [while] industrial scientists'... success most often depends on keeping their best work secret."[86] Some of the scientists and engineers I talked to, including both industrial scientists and university researchers with significant industrial experience, spontaneously produced versions of such distinctions and, accordingly, of the relative virtues of the two forms of life. The computer scientists Metzger and Marvyn volunteered that sort of thing, and the academic electrical engineer Lyons bridled at corporate attempts to direct his research agenda—even at a university institute whose sole source of funding was industry. Lyons told me an iconic story of the sinister possibilities inherent in taking industrial funds. One corporate executive "called me up and said 'I want you to work on this specific project.' I looked at it and I didn't think it was really interesting, the research, and I said, 'No, I don't want to work on it.' And he said, 'You don't understand, I'm *telling* you,' and I said, '*You* don't understand, I am not working on it.'" Lyons, and his sense of virtuous scientific autonomy, triumphed: "[So] what was he going to do? OK?

So I didn't work on it, and they never pulled out of the [industrially sponsored academic] center. But you have to dig in your feet every now and then." The academic cardiologist Jon Marx, with extensive entrepreneurial experience, explained that "I'm not really interested in leaving the university...I like the autonomy...In an academic setting what I liked was that I could get this wild idea in my head and say, geez, you know, I have a lab here and I want to try this out. And I didn't have to go and ask anybody permission to spend time and money doing this pile of experiments," while, at the same, conceding that *his* command of NIH and other government funding for his work was far more secure than that of many of his more junior, or less well-known, colleagues.

At the same time, autonomy as a deep fault line between academia, in general, and industry, in general, was systematically denied by other researchers, and we have already heard from academic scientists who felt strongly *constrained* by what they saw as the capriciousness of the funding system, and from others who looked with envy at the superior resources, and relative abundance of time, available for their sort of work in the commercial sector. Chapter 5 noted the "day a week" free time—with corporate resources—commonly granted to researchers working at a number of early twentieth-century American industrial laboratories, and in some present-day companies that freedom of action has been expanded. At the search-engine company, Google, these are currently known as "20% time projects" and they are highly valued as sources of corporate innovation: "They allow rank-and-file engineers to spend 20 percent of their time on projects of their own choosing, on the theory that hundreds of brilliant Google employees pursuing their individual interests will come up with products and services that [the company's three top executives] could never dream up on their own."[87] At Genentech, CEO Art Levinson similarly draws attention to the pragmatic basis for a high degree of researcher autonomy: "People who like a hierarchical environment don't do well here...My job is to get the best scientists in and let them do their thing."[88] Not all present-day high-tech or biotech companies are alike in these respects, and one is not obliged to credit every story about freedom of inquiry in Genentech or Google, but, then, neither do all university environments match the ideals of autonomy widely ascribed to them. It is not uncommon for junior scientists—graduate students, postdocs, and academic scientists in the early stages of their career—to complain about the high degree of *control* they experience in universities. The young entrepreneurial bioengineer André Schumann escaped to Gudjonssen's laboratory for his Ph.D. work because he initially found himself in a highly controlling academic environment: "I found myself in a lab where . . . you know, it was kind of like, 'OK. Here are the projects you can choose from. Pick one. And this is what it will be. And you'll do it.' And, you know, 'I need you to write this paper; I need you to get this out.'" Whenever, in Schumann's telling, he tried to assert some version of limited research freedom, he kept getting "brushed off," and that frustrated him immensely. It wasn't what he wanted or what he expected research to be like. The experience of being "given a problem" by one's advisor is common in late modern academic science and, while it is not at all unknown in other parts of the university, it is one of the more consequential differences between practice in the natural sciences and in the humanities and social sciences.

Now consider secrecy as an institutional fault line, on either side of which scientific virtue resides. Again, it was not difficult to find both academic and industrial scientists articulating some version of the "Mertonian" story about the essential and functional openness of genuine science and the corrupting effects of any degree of scientific secrecy, about the university as the natural habitat of scientific openness and about industry as institutionally opposed to such openness. But the complications in any such neat story described in chapter 5 persist into the present, and contemporary entrepreneurial scientists also have a vocabulary for talking about secrecy and openness that is far more complex, fine-grained, and particularistic. Dennis Carlo, an early employee of the La Jolla monoclonal antibody company Hybritech, was blunt in his attitudes towards industrial secrecy, combining mercenary and medical motives: "Don't give me anything about *sharing information* . . . That's bullshit! We're not in the business of sharing information. We're in the business of making money and of curing disease." But, talking in the early 1980s, Carlo reckoned that the same sort of attitude now characterized *academic* bioscience. It was just the way the world was: "Some of the university people have tightened up. Everybody wants to be a businessman. Everybody thinks they are going to make a million dollars. The people in academia are trying to start companies . . . So those kinds of guys are tightening up and they aren't saying too much."[89] But entrepreneurial views of secrecy were typically more nuanced. In his role as CTO of a start-up bioengineering company André Schumann understood very well that "if we're going to write a paper, I've got to make sure that what's getting out into that paper is not confidential, is not proprietary." Against that consideration, he also acknowledged real benefits from a free-as-possible publication policy: "I think if you do the best scientific work that's out there—technically you do the best stuff, that's what's going to get you the best deals; that's what's going to build the business. Certainly, it needs marketing and all that, but if the science is solid, and it's well respected, then that goes a long, long way here, and publishing papers is a fantastic way to do it, you know." As in the past, companies that publish freely can, for that reason, attract top-class scientists who *want* to publish, but there is no basis for an a priori judgment that scientists, by their very nature, *do* want to publish, that this is their sole, or even major, desire. For some, the most effective proprietary realization of an innovative product may be more important than the disciplinary reputation that comes from publishing.

O'Mair, the Irish doctoral student in electrical engineering who had already helped start up a successful chip design company in his native country, thought the idea that academia was open and nonhierarchical compared to industry was "funny." He had heard people say that sort of thing, but he could cite as many instances as I was willing to hear of authoritarian and secretive *university* labs.[90] Similarly, Ken Loche, a small drug discovery company CSO, was well familiar with academic secrecy, which he found exactly "parallel" with industrial secrecy: "You will [get university] people who are saying 'I'm sorry. I can't talk about that until I get my grant.' . . . Or 'until I get my publication.' . . . Academia is not absolutely free exposure. Again, it's understandable for parallel reasons, [academics want] to protect that vast research that [they've invested in] or they want to make sure they get the grant based on this information or whatever . . . You want to be able to protect it. But it's the same thing, the originality of your discovery has to be protected," whether it's in a university or in industry.[91] Whether the goal is financial capital or symbolic capital makes little difference. Academic scientists can and do have "proprietary" concerns: many of them like the reputation that comes with publishing but take for granted the reasonableness, and the moral innocuousness, of "keeping something back," something that ensures that published knowledge is *incompletely* public.

For many scientists thinking about what sort of life they wish to lead, and where they wish to lead it, decisions do not present themselves in a binary mode: academia versus industry. The matter is often more fine-textured and contingent. Not all academic laboratories, even in the same line of work, are morally, temperamentally, or technically equivalent, and young academic scientists and engineer routinely evaluate the experiences of working in one PI's laboratory versus another. Much the same pertains to scientists' sense of what kind of industry, or what kind of company within a given industry, they might wish to work for. Diane Springfield, who had chosen to work in a genomics company over a variety of other options, including offers from academia, said that she had never thought of her decision about where to work in binary terms: "Really, what I kind of went with was not really a conscious decision of industry versus academic. It was more out of all the places I interviewed, which did I find the most interesting." She accepted her current company's offer because "it was a growing company, they were hiring a lot of people, kind of a fun environment, very young, a lot of people just like me, you know, . . . and, at the time, had a lot of money . . . large projects to get involved in." Schumann, the bioengineer, thought there might be something in the idea that people who liked a structured environment opted for industry while individualists liked academia, but he saw little sense in making that the basis for a rigid institutional divide: it's "the type of lab that somebody goes into or the type of company that someone goes into . . . What dictates the mood or feel of the lab, I think, is the [person] who runs the lab." Some industrial research directors, and some professors, are authoritarian; others are permissive; some industrial projects are very well framed, but some are not, and academic research projects span the range from highly focused to open-ended.

While preferences might have to do with the personality of the research director, or with the inherent interest of the projects, a vocal group of entrepreneurially minded researchers made a strong evaluative distinction between early-stage companies and companies that had grown to a certain size. Kary Mullis complained bitterly about the bureaucratization, and consequent loss of both autonomy and common sense, attending Cetus's expansion from start-up to relatively successful small biotech company.[92] Natural products chemist O'Reilly was one of many scientists I talked to who commented on the change in atmosphere attending corporate growth. If they are lucky or good at what they do, small biotech companies become bigger, and then they quickly come to resemble Big Pharma, with their hierarchical forms and their stifling of free expression. At the outset, the attitude was "'We don't give a shit if you look good in a suit because nobody even can afford a suit here.' You know? [laughs] That goes down and what comes up is 'Who is the most creative? Who can make our company more valuable? Who has got those skills?' But as that company grows, it finds itself [like a big traditional company] because then, all of a sudden, the public affairs becomes important. The image becomes important. They change a little more toward hard industry from soft industry." The change is from the congenial, egalitarian environment of "a few-people-in-a-garage kind of mentality, and we're working eighteen hours a day, to a CEO, all this hierarchy, a director of human resources." In his own continuing involvement with the start-up drug discovery company he cofounded, O'Reilly did all he could to try to retain the face-to-face character, the camaraderie, and the informality of its earliest days: softball games, trips to the beaches in Baja California—just the sorts of things he was used to doing in team-building exercises in his academic unit. "They [the company] don't think of that. I took my whole group down to San Felipe on a bus and bought them a whole weekend and that group was just on fire for the next two months, excited about their job and stuff. These guys [now] don't get any days off. They give them a twenty-buck picnic. It's not team building. Sometimes it's lost. The concept of how you build people to be committed to working together, you have to personalize people and sometimes industries just think that salary is enough. It's not. A salary is a part of creating something that will work. But it is not in any way the major part. It's the team. It's the commitment. It's the personal respect for individuals who work together, and how do you get that? You get that by personalizing people."

Jon Marx, experienced in starting up two successful biotech companies, noted that as soon as his first company "got to be forty or fifty employees, I could feel a different push within the company, [people] pulling and pushing against each other, and complaining against each other." But with his second company that didn't happen until there were seventy people: "And then there were some group conflicts." Awkward aspects of *Gesellschaft* replaced harmonious *Gemeinschaft* at different points, but the transition was seen as inevitable. Bioengineer Gudjonssen was probably the most purely entrepreneurial in his comments on the differences between the start-up and the more mature company: "My interest in building a company, what I find most interesting, is to be able to figure out how it should work, the structure of it, what kind of people you need in, what the product should look like, and I've also noticed when these companies become twenty to thirty people, I get disinterested because then there is more and more *management*." He didn't *want* to be a manager; he wanted to start companies on their way: "All the constraints of the trials and tribulations that came with *managing* or running a company on a daily basis, I did not like. Very regimented schedule, lots of personnel issues. Lots of legal issues, you know, from sexual harassment to knowing the evacuation routes if there is a fire. The things associated with *managing* a business on a daily basis, I did not find attractive. So I, as a founder of a company or a scientist, realized that once a company grows beyond fifteen to twenty people, you should get out of there because it

258 * CHAPTER SEVEN

becomes a real *company* all of a sudden." Managing "real" companies, for Gudjonssen, represented no intellectual challenge at all, no fun, no good.

Some of the texture of quotidian scientific life is given by the relative durability of research projects and the social forms in which projects are pursued. And many—though not all—of the researchers who spoke to me made much of the institutional differences in such things, what they valued about them, and how they affected their views of where the scientific good life might be led. So, for example, bioengineer Joan Rhodes, who worked for a small Southern California biotech company, reckoned that the fundamental problem some academics had in leaving behind academic work rhythms was getting used to flexibility. You might come with special expertise in, say, the physiology of the central nervous system, but if the company needed you to work on pulmonary physiology, then you had to adapt, getting on or getting out. That, indeed, was her own story. This might mean sacrificing the accumulation of intellectual capital (as in publication and reputation in a specialized area), and you had to recognize the possibility—if not the inevitability—that these changes might burn your bridges back to the academy and endanger disciplinary reputation. In the universities you could, if you wanted, spend your whole life accumulating specialized expertise, rising to the top of the reputational heap, while the need for adaptability made this hard in industry. Here, as in the patterns discussed in the preceding chapter, the "project," not the "discipline," was the paramount consideration, and, while researchers were, of course, trained in specialized academic disciplines, industry regarded the scientists they hired as embodied skill sets whose value was largely their ability to contribute to the success of projects organizationally decided upon. The flexibility often said to be characteristic of industry might be seen as a loss of autonomy—in which case, Rhodes said, you probably weren't temperamentally suited to industry. Alternatively, many of the scientists and engineers I talked to said they *liked* aspects of such organizational flexibility, that it prevented you from "getting stale," "narrow," or "too specialized," even that flexible patterns of work were more genuinely *intellectual*. Max Weber saw virtuous vocation as coextensive with rigorous specialization, but, again, there is no reason why this should necessarily describe the psychological states of researchers at either end of the twentieth century.

The notion of flexibility was often closely associated with that of teamwork, and the contrast between the imputed individualism of academic science and the collectivism and collaborative nature of industrial science described in chapter 6 persists into the present. Joan Rhodes was one of many researchers who matter-of-factly distinguished industrial science from academia on the basis of the former's team orientation. When you're in the university, and you're the PI, you are "God in your realm," she said (using a common formulation). In industry, you work in groups towards goals set by the company, though there are, of course, important degrees of autonomy. Here, a pertinent consideration might be the differing ways of *experiencing* control: PIs in a university department do not experience control, or the necessity of coordination, locally and on a day-to-day basis. Their "masters" are typically geographically removed—members of the discipline's core-set, gatekeepers at journals and funding agencies—and so PIs may well experience, or talk about, their work as autonomous and uncontrolled, despite the fact—already pointed out—that their research agendas are constrained by these distant peers and superiors. By contrast, when one works for a company, agents of control are local, and dealing with them is often a daily occurrence. Accordingly, industrial researchers may well experience, or talk about, their work as controlled and coordinated when there are substantial domains of free action.

Nevertheless, teamwork—whatever moral and scientific values were placed upon it—was a substantial feature of how many researchers described the life of industrial science. In many respects, these sensibilities were traditional. Some scientists, like O'Reilly, had little good to say about the corporate life of science, and much—not all—of his commentary on even start-up biotech companies could have come directly out of Whyte's *Organization Man*. Jon Marx declined to leave his senior academic appointments, explaining that "in a corporate world, there's a lot larger team and there's a lot of different players in it . . . And you have to work together with all of those different [players]. In a university setting, you have a much smaller team and . . . you can be the captain." He wanted very much to be a "captain," but he also wanted the companies he cofounded to be successful, so he stressed how important it was to bring on board "people who had a team spirit, people who knew how to work on a team. We didn't need any headstrong individuals. You know, people who have a great compulsion to run things their way." He himself *preferred* what he experienced as academic individualism, but he recognized the virtues of coordination and collaboration in the corporate setting. Industrial success has its requirements that have to be satisfied if you're going to get a drug or device onto the market and help patients, as Marx said he so much wanted to do. But that didn't mean he had to like it, or that he thought the team style of research had inherent virtues. Others had few or no such reservations about corporate teamwork and many criticisms of academic individualism. Michael O'Mair was definite in his view that the academic style was more individualistic. For his own part, however, he liked the team nature of industrial style. He expressed little or no concern about "credit"—whose name was on patents or papers. Reputation in his line of research, as in much software work, was propagated through informal channels rather than through formal publication or formally assigned ownership. Everyone knew who was a "wizard," and he knew that everyone who counted knew that he was one. The structural submergence of individual authorship carried few costs in O'Mair's world, while acquiring the reputation of an impossible person to work with had very significant costs. Henry Lyons had worked at Bell Labs before taking up his university appointment, and he distinctly remembered the virtues of its collaborative research structures. "I was in a team of really, really bright people," he recalled. "They'd come in and ask me questions. I would go to them and ask them questions. Somehow, at a university, professors don't do that as much and I think maybe it's ego and the notion of asking somebody else a question is in some sense saying 'I don't know the answer to this question.'" Academics could lose face that way. Despite the idealization of universities as questioning, critical places, Lyons said "I'm not sure at a university people like to say that." Maybe, he speculated, that's because of the teaching duties that he so greatly valued: "You get up in the classroom and there is no team there; it's you. So in the classroom you get to develop the notion that you are there to answer questions and not to ask them . . . There is much more team effort at a company. As a matter of fact, the first thing you do at a company when you have a good idea is to bounce it off a whole bunch of other people." So, if sharing and openness among colleagues counted as scientific virtue, then Lyons was saying that you could sometimes find that more easily in corporate than in academic settings.

As O'Reilly put it, "There's nothing more disruptive [in industry] than someone who doesn't play a team game. So . . . your value in a company is constantly being weighed from the point of view of what a pain in the ass you are versus what a positive contributor you are. If you are more of a pain in the ass than the amount of contributions, you know, [makes a throat-slitting gesture]." Still others refused to see any clear distinction in institutional mores. Lyons was prepared to generalize about academic individualism and commercial cooperation, but he was one of many academic scientists and engineers I talked to who insisted that their own laboratory was a collaborative space and that its scientific virtues importantly flowed from its teamwork. Lyons typically had about six or seven Ph.D. students in his laboratory at any one time: "I encourage them to interact. I meet with them individually but...once a week we have a team meeting and that's where people ask questions and I try to make it a little bit more like in industry. I think the students like it." And O'Reilly could see the positive aspects of teamwork when the thing was arranged *properly*, as it was in his laboratory: "I am actually a pretty good team player...But when I look, what I see are a bunch of fiefdoms in the university with almost no team players and when you go to the industry, you'll see a whole team...I think being a team player is critical, and in fact my research group requires a lot of team playing and I teach team playing." O'Reilly's research group at the university contained about twenty-five people, and that was a common number in his area: "Half of my job is managing the psychology of people who are working in teams who have some serious shortcomings and limitations in how they can be and what they can do, and that's a big portion of what I do." Having started out by making an evaluative contrast between individualistic academia and collaborative industry, O'Reilly wound up talking about many of the ways in which the organization of work is similar. There was much passion, but little stability, in his talk about the institutional virtues.

ENTREPRENEURIAL MOTIVES

What motivates the entrepreneurial scientist? What drives do entrepreneurs recognize and talk about as ways of explaining—to themselves and to others—who they are and what the virtues and vices of their lives are? Much talk about motives has already cropped up in connection with other imputed attributes of the entrepreneurial life, for example, freedom of action, openness and secrecy, interest in material consequences, and the quotidian rhythms and routines of scientific inquiry in different sorts of settings. Occasionally, participants foreground the question of motives. The entrepreneurial biomedical scientist Jon Marx was one of several researchers who parsed commercial and academic science through clearly opposed motives. If you're a real entrepreneur—and, despite what Marx had achieved in founding two biotech companies, he didn't think he was "brave enough" to cut ties to the academy—"you have to be motivated to make a lot of money. If you're not, you're not going to be successful...You have to truly be motivated to make a lot of money." The contrast with academic science was rendered as absolute: in the university, while your income rightly allowed you to live comfortably—Marx had a new Mercedes and was extremely proud of his collection of Robert Parker–approved fine wines—the motive was intellectual discovery. "I hold the university up on a pedestal. It's some ideal of free thinking; it's not directed toward making money or developing products . . . Company goals seem very different to me." This was a common way of talking about the geography of motives and, while it mapped easily onto a body of external criticism of entrepreneurial science, one could hear it occasionally articulated by scientists and engineers with considerable involvement in industrial science: Nikolai Metzger talked in similar terms about his displeasure with his time in his biometrics company. Yet this motivational geography was radically unstable in participants' sense-making vocabularies, and, just a few minutes after Marx insisted on the purity of the entrepreneur's mercenary motivations, he fluently embarked on a much more textured and nuanced account. You did not, Marx announced, start up a company such as his just to make money; you wanted to help patients by getting a biomedical device onto the marketplace as efficiently and quickly as possible: "There's no conflict between wanting to do that and wanting to make money. But to be a real entrepreneur you have to be motivated by both. If you're only motivated to make money, you're not going to do it because you're going to jump to a new idea that looks like it's going to make you more money. Or you'll give up on the [original] idea too easily. You have to be driven by the *idea*. And the word that, the word that's been used, a couple of people that I know that are incredible entrepreneurs, is *driven*. You've got to be *driven*."

Marx took leave from the university to work with his companies, and his return to academia may well be responsible for this motivational pastiche. But there is also significant overlap between Marx's vocabulary of motives and those articulated from within the bowels of high-tech and biotech business. Entrepreneurial scientists accept as a matter of course that capitalism is now, so to speak, "the only game in town," and that, should one wish to do certain sorts of science, and should one wish to "make the world a better place" through certain sorts of scientific and technological innovation, then making money is a necessary concomitant of acquitting those desires. Cetus scientists told Paul Rabinow that it was wholly possible to find projects that had commercial potential and which, at the same time, had "real fundamental passionate interest to you," and, in so saying, they were echoing sentiments that circulated in industrial research laboratories from early in the twentieth century.[93] Arthur Kornberg's account of DNAX, the biotech company he helped found, stressed the opportunities that venture-capital-fueled commercial research opened up for basic inquiry. He put a traditionally "academic" interpretation on his recruits' motives: "The financial inducements of equity in the company and a higher salary surely mattered, but they were mentioned only in passing." When the company was taken over by Schering-Plough in 1982, there was general skepticism about the continuation of the company's original free and open structures, but Kornberg became satisfied that there were no necessary conflicts of interest: "For their considerable investment in DNAX, Schering-Plough management wanted something tangible in discoveries for product development beyond the enhancement of their image as a biotechnology-oriented pharmaceutical company. They recognized that they needed to attract and retain a world-class scientific staff, as well as to sustain the allegiance of the DNAX founders and scientific advisors. To achieve this, [the new CEO] nurtured an academic atmosphere that encouraged prompt publications; the sharing of reagents, cell lines, and techniques (outside DNAX as well as inside); unstinted physical resources; and generous perks and compensation."[94] The passage of time, and, to a large extent, the realities of corporate success and failure, now make such sentiments seem a bit quaint, while other accounts of the realities of the scientific life in biotech—even in the golden 1970s and '80s—are at odds with Kornberg's.[95] But there is no reason to deny their general plausibility, not merely as justificatory resources but as sense-making resources, and enough has been said about contemporary cynicism concerning the realities of academic science to hesitate to dismiss them.

The desire for a "free space" in which to conduct the inquiries that one wants to conduct, that one might even feel oneself driven to conduct, is probably the major item in scientists' motivational lexicon. However, the institutions in which such free spaces may present themselves map only problematically onto the divide between academia and industry. Scientists may say they want to extend their intellectual curiosity, to have a challenging problem and the resources to address it, to acquire reputations as skilled and intelligent researchers, to help people by bringing a product or service into the marketplace, to have fun, to work with an interesting team or to have autonomy, to make money and support a comfortable lifestyle, to live in a place with good schools and good weather, to work in an organization that is well run and well led, to serve as a mentor, to be in the company of other bright and driven researchers. Hybritech's Howard Birndorf was widely credited with "genius" just for the early hire of the protein chemist Gary David, "a brilliant young scientist whose presence at the company served as tantalizing bait for several other key scientists who were lured away from academia," some of whom confided that their reasons for joining this initially implausible company was "Gary

David's integrity and scientific brilliance." It was organizational common sense that a talented scientist might just want to be where other talented scientists were.[96] The immunologist Joanne Martinis—employee number twenty-four at Hybritech—moved to San Diego from an academic environment in Philadelphia "on a lark": "I never expected the stock to be worth anything . . . I really viewed it [making monoclonal antibodies in quantity] as a challenge and a lark."[97] It is a well-known saying in West Coast high-tech and biotech circles that scientists and engineers "have more loyalty to their technology than to their companies," and that they'll work at whatever firm or academic unit allows them the greatest latitude in developing these technologies.[98] These, and many other desires, serve as sense-making resources that researchers use to account for career decisions, and, if necessary, to justify them by explaining what legitimate and recognized concerns move them. But what these motivational items and their use do *not* achieve is a stable distinction between institutional types and their characteristic virtues and vices. The map of institutional virtue in late modern science has to be drawn with a finer tool.

NORMATIVE UNCERTAINTY AND THE
ENTREPRENEURIAL LIFE

That particularism must flow, at least in part, from the very high degree of organizational heterogeneity and flux that now characterizes regions of the research university and, especially, of innovating industry. As Rabinow has written, the last two decades have seen "a general reshaping of the sites of the production of knowledge." New forms of knowledge are emerging together with new knowledge-making forms.[99] The conditions of much scientific and technological experimentation are now, in themselves, social experiments, attempts to explore what novel configurations of people, space, knowledge, material resources, and external support can best bring about wanted intellectual and technological futures. They are, *inter alia*, experiments in what motivates people and in how a range of motives—some of which have traditionally stood in conflict—might be satisfied together. More than ever, we appreciate that solutions to the problem of knowledge are solutions to the problem of order. But now the repertoires of choice for what knowledge is and what orders best advance it are of bewildering complexity. Universities have changed, most especially in the institutional configurations in which interdisciplinary scientific, medical, and engineering research gets done, but industry, nonprofit research institutes, and the forms of association between all of these have changed even faster. And new forms and sites continue to emerge—various sorts of contract research organizations, "virtual companies," globalized research groups, and the like. Indeed, one of the causes of the undeniably poor understanding that academic humanists and social scientists still have of their natural scientific and engineering colleagues, and of their continuing attraction to a simplistic language of binary opposition, must have something to do with the intense organizational conservatism of the former compared with the vertiginous organizational innovation and heterogeneity experienced by the latter. Present-day historians' and sociologists' experience of the American research university has less in common with that of natural scientists and engineers than it ever did, or with what Clark Kerr could have imagined in the 1960s when he wrote about that new institutional form he called the "multiversity."[100] Moreover, the increasing familiarity that academic scientists, engineers, and medical researchers now have with industry, its concerns, its rhythms and routines, is alien to almost all humanists and social scientists. The vocabulary historians and sociologists have to describe these things, and the sensibilities they have available to evaluate them, are, to put it bluntly, seriously disconnected from the lived experience of their technoscientific colleagues.

Much of contemporary industrial science, particularly in high-tech and biotech, is strongly marked by organizational flexibility and associated normative uncertainties. Chapters 5 and 6 have introduced these uncertainties and have suggested some consequences for the role of the individual researcher. But present-day reconfigurations of organizational forms have extended those uncertainties and the population of researchers who live with them. If Kenneth Mees, of the Eastman Kodak Research Laboratory in the 1910s and 1920s, was skeptical of the idea of a rule-book for how to organize industrial research, some contemporary start-up entrepreneurial companies scarcely know what a rule-book would be; others celebrate institutional spontaneity while venture capitalists sometimes look on in benign approval. Nevertheless, anything recognizable as an institution—and that includes a start-up drug discovery or wireless company—must have certain goals and certain routines for achieving those goals, however loosely defined both are and however contested both may be. How are such goals and routines established, made visible, and, to the extent they are, rendered legitimate to members of the institution?

Here, the answer remains broadly similar to the one described in connection with the institutions of industrial research a half-century and more ago: such authority is embodied by particular individuals. At Hybritech, it was cofounder Howard Birndorf who embodied not only the vision that monoclonal antibodies could be produced, but the structures and work rhythms involved in searching for them. As venture capitalist Brook Byers said, "In the birth of a new industry, the pace of how things are supposed to go is unclear. How fast are things supposed to happen? Nobody knows. So everybody figures they are supposed to happen about as fast as they're happening. Someone's got to build in a sense of urgency," and that somebody was Birndorf: "I would have to say he was the keeper of the sense of urgency."[101] It was Robert Noyce who embodied those goals and routines for Fairchild Semiconductors; Craig Venter for Celera Genomics and TIGR in their early days; Tom White for Cetus and Celera Diagnostics; Steve Jobs for Apple and then for Pixar; Larry Ellison for Oracle; Irwin Jacobs for QUALCOMM; and it is Leroy Hood who does so for the nonprofit Institute for Systems Biology in Seattle. In none of these cases, or in many other areas of high-tech, biotech, and organized nonprofit research, can patterns for success be taken off the shelf and impersonally incorporated in a manual.

The world of entrepreneurial science, and its relationships with charismatic authority, has not escaped the notice of social and economic theorists. Joseph Schumpeter's view of entrepreneurship had Weberian roots. Economic conduct was wholly rational, just on the condition that "things have time to hammer logic into men," but where such routines were unavailable, or where they were disrupted by the entrepreneurs themselves, institutional actions depended on embodied intuition.[102] What Schumpeter referred to as personal capitalism was supposed to follow the same historical trajectory as Weber's charismatic authority. But a few contemporary economists take a different view. Richard Langlois, a commentator on Schumpeter, has underlined the constitutive relationship between the role of late modern entrepreneurs and the charismatic nature of their authority: "The charismatic authority and coherent vision of . . . entrepreneurship remains an inevitable part of capitalism, however modern. For reasons that have to do with the nature of cognition and the structure of knowledge in organized society, some essential part of capitalism must always remain personal."[103] And against Schumpeter's contention that the entrepreneur was a figure of declining significance in late capitalism, Langlois points out that charismatic authority "solves a coordination problem in a situation of 'chaos' in which rights, roles and responsibilities are in flux," that is to say, using the sensibilities of this book, in circumstances of radical normative uncertainty.[104] But, far more importantly, participants in late modern entrepreneurial science themselves reflect upon the personal virtues of those individuals who embody the institution, giving it shape, direction, and legitimacy. "Walking the corridors"—the personal circulation of a leader around the organization—is understood, and talked about, as a way of making organizational goals and norms visible, and few commentaries on the world of the entrepreneurial start-up omit to mention the leader's circulating physical presence. The word used by both participants and commentators to describe this personally embodied leadership happens to be "charisma," however loosely their usage relates to the Weberian original.[105] What they mean to point out is embodied leadership rather than impersonal criteria, the moral and cognitive authority of a familiar person rather than of the rulebook or the organizational chart. "Charisma" is, in these ways, not just a protean vernacular usage; it is a consequential, reality-making usage. The charismatic nature of entrepreneurial action is widely recognized by participants: they use it to make sense of their world, to coordinate actions within it, to recognize legitimate conduct, and to help make a future. The world of late modern entrepreneurial science is at once the leading edge of capitalism and an ongoing set of experiments in charismatic authority.

Visions of the Future

UNCERTAINTY AND VIRTUE
IN THE WORLD OF HIGH-TECH
AND VENTURE CAPITAL

Inspiration in the field of science by no means plays any greater
role, as academic conceit fancies, than it does in the field of mastering
problems of practical life by a modern entrepreneur.

Max Weber, *Science as a Vocation*

SOMETHING VENTURED

Normative uncertainty has its varieties and its degrees. At one end, the
preparation of a Big Mac or the filing of a medical insurance claim is subject to very great organizational specificity and routine, though even here
there is no way exhaustively to spell out all the features of right action and
how they are to apply in all circumstances. At the other extreme are many
of the courses of action involved in the making of late modern technoscientific artifacts, techniques, and knowledges. Here, right actions can
rarely, if ever, be prescribed from an institutionalized template, and the
pertinence of any preexisting template is problematic. The closer you get
to the leading edges of technoscientific change, the greater the degree
of normative uncertainty: What is proper behavior for a scientist and a
manager in a start-up biotech company? How does one arrange inquiries
and lines of authority and responsibility within that sort of organization,
indeed what kind of organization *is* it? Will there be a market for a product that is not an improved razor blade—which people are reasonably
believed to want—but a disruptive technology that is a substantially new
thing in the world, that people at the time of an investment decision do
not *know* they want? The future is unpredictable in principle. As David
Hume showed in the eighteenth century, you cannot *prove* that the Sun
will rise tomorrow, or even that there will be death and taxes. There are,
however, certain sorts of activity in the late modern world that engage
with future uncertainties in highly focused, systematic, and consequential ways and in which those uncertainties are massive. It is in these sorts of
activities that people often become reflectively aware of the conditions

of uncertainty and strive to manage them. This chapter is concerned with the practicalities of uncertainty management in the worlds that bind scientific entrepreneurs to the sources of capital for turning entrepreneurial ideas into concrete realities. It is about "venture capital," about how venture capitalists (VCs) confront radical uncertainties, and, especially, about how the virtues of familiar people figure in the operation of what, by some accounts, is the most ruthlessly instrumental sector of late modern capitalism and late modern technoscience. How do VCs go about deciding which early-stage high-tech companies will be successes? How do they make up their minds to write a check for several million dollars, the outcome of which may be a drug that cures cancer, or nothing at all—a waste of investors' money? We will see that judgment in these worlds of leading-edge technoscience and finance often implicates knowledge of the virtues of familiar people. People and their virtues *matter*—and that mattering is absolutely central to the rationally calculative worlds where late modern finance meets technoscience.

The world occupied by VCs and the entrepreneurs seeking their support is not one in which distinctions between "science" and "technology," or, indeed, between doing science and doing business, are consequential actors' categories. Of course, every competent member of that world has occasions to parse the notions of "science," "technology," and "business." Every one of them is familiar with the stages that may link, at one end, more or less disinterested inquiry, and, at the other, a product on the market helping cure cancer and contributing to a company's bottom line; everyone understands that what's published in the journals is "science" and what's put in a business plan are some details that may or may not be "science"; and everyone knows that there are institutions called universities and institutions called companies, and that these differ in all sorts of pertinent respects, just as they resemble each other in other respects. However, they are also quite generally aware that these categories do not have species essences and little of their communal speech seems to depend upon making crucial distinctions between them. They are not, in general, very interested in such distinctions. Put the questions "What is the difference between science and technology, between pure research and applied research, between science and business?" to typical members of this world, and you will, as likely as not, get blank looks in reply. What VCs and entrepreneurs seem to understand is that one may want to do science in order to get a product into the world; one may realize that getting a product into the world involves the doing of science; one may think that getting products into the world permits more and better science to be done; and one may function in a world in which being a good scientist is signified by patents held and companies founded. Where is science? Where is technology? Where is business? Substantially, they are in the same place. This is late modern technoscience in a fully realized form. And it is also a world in which normative uncertainties flourish in uniquely intense and consequential forms. The world of VC-backed science and technology lies at the cutting edge of late modern change; it is future making; it is hard and instrumental; it is the engine of the capitalist economy and its ever-changing bases. So it is a perspicuous, if seemingly paradoxical, place to pose questions about patterns of familiarity and about the attribution of virtues to familiar people.

Gambling is a traditional activity, but it was not until fairly recently that there were such people as venture capitalists in the world. In all cultures and at all times people have staked money on inherently unknowable futures, hoping to benefit from outcomes they can do little, if anything, to affect. Sometimes gamblers are aware that they are simply exposing themselves to the fates, spinning Fortune's Wheel; sometimes they think that they have, or can get, an edge on the future—that they know something that others don't know about how things will turn out; sometimes too these people think that they have special expertise in predicting the future, that there are learned skills or aptitudes that allow them to pick winners. Such people include both bettors at a horse race and VCs, and occasionally both groups offer commentary on how to be successful in predicting radically uncertain futures.[1] And sometimes that commentary constitutes a more or less reliable account of aspects of how they actually do behave, how they decide where to place their wagers on radically uncertain futures. More rarely, interested observers can directly check such commentary against evidence deriving from the scenes in which judgment happens.

Venture capital is a form of private finance. Its practitioners tend to specialize in bringing new businesses into the world, financing start-up companies in high-tech and biotech, but they may also invest in other sorts of companies. Private finance is typically called upon when the business is too risky to be taken on by banks or traditional sources of public capital. It is estimated that there are about 3,500 VCs in the U.S., managing over $100 billion of assets (both numbers vary wildly from source to source). And if you are a VC, then your life is structured by the exercise of predicting what you know to be the highly, even impossibly, unpredictable, most especially when you are dealing with early-stage companies, doing what's called "seed investing" or even a "first-round."[2] You may appreciate that your previous experience will be of limited use, as each new venture presents its special challenges and problems — the more so if the venture is a step into the radically unknown.[3] Envisaging what the wireless communications future will be like and which companies will succeed in that future isn't much like opening a new McDonald's franchise. Venture capital is *such* an uncertain world that one highly successful VC described it as "a perpetual stroll into the fog," and another complained that his first fund was referred to by critics as "a sociology experiment."[4] But if you're a VC, that's what you do, and, if volumes of internal and external commentary are worth anything at all — and they may or may not be — some VCs are good at doing this impossible thing and others are not so good at it.

Spectacular successes are well advertised: particular investments multiplied by thirty, sixty, hundreds of times when a start-up company goes public or gets taken over by a larger company. But reliable statistics on these things are hard to come by, and even harder to interpret when you do come by them. (VC firms package their individual investments into "funds" whose assets have to be realized at a certain set time, typically five to ten years after launch. VC funds are private and largely unregulated. Firms are under no obligation to publish information about the performance of their funds, and most make their investors sign nondisclosure agreements [NDAs], a condition that has only recently been breached by university and state pension investors, under pressure from the leading Silicon Valley newspaper, the *San Jose Mercury News*, and the *Houston Chronicle*.)[5] Even the best VC firms acknowledge that successful investments — by most criteria of "success" — are far outnumbered by failures. Amy Radin, a Citygroup in-house VC, asserts that "we are building failure into the model."[6] One of the most respected VCs, Tom Perkins of Kleiner Perkins, once described one of his firm's first funds as "a barrel full of piss with a couple of cherries floating in it."[7] But the preferred metaphorical genre in the VC world is baseball: "home runs" (or even "grand slams") are what you're really looking for, hoping that the occasional home run will more than compensate for the many strikeouts.[8] Benno Schmidt, one of the founding fathers of venture capital, said that "we don't live so much on our batting average as our slugging average . . . We live off our extra-base hits."[9] Going for the big rewards necessarily implies a high failure rate. VCs are all in a high-risk business, but both individuals and VC firms naturally vary in style. As the San Diego biotech entrepreneur-turned-VC Ivor Royston puts it, "You can't get home runs without striking out. Babe Ruth taught us that . . . So you can either be a venture capitalist that hits singles all the time—very low-level risk—or you can take that big leap—go for the home run or something very—maybe, you know, a major paradigm shift, something very novel, extraordinary. Tremendous risk, but you're going to hit [the] ball out of the park. That's what I do."[10] (Recently, the baseball metaphor took a literal form when a group of private investors—including VCs "playing with their own money"—got together to finance an independent baseball league. Some of them viewed the investment as a bit of fun, while others stuck with a tempered form of the traditional VC ambition: "I know some people expect this thing to be a home run . . . But me, I'd settle for a ground-rule double.")[11]

A quoted rule of thumb is that three out of every ten VC investments are total losses; three carry on a marginal existence but present difficulties in extracting the original investment (often referred to as "the walking—or living—dead"); two give returns of 200–300%; and two yield more than ten times the original investment.[12] Horsley Bridge Partners, which manages private equity investments for institutions and pension funds, placed money in about sixty VC funds from 1985 to 1996, and of these funds' investments in 1,765 companies, 278 (16%) were write-offs, 330 (19%) were liquidated below cost, 685 (39%) returned profits of 100–500%, 136 (8%) of 1,000–2,500%, and only 89 (5%) more than 2,500%. So picking big winners—even during the fat years of the late 1990s—was an extremely difficult thing to do.[13] Some commentators say that VCs aim at "a reasonable possibility" of multiplying their original investment in a firm by five or ten times within five years. The Silicon Valley VC Jim Breyer (of Accel Partners) says that "we always go into a deal hoping to make ten times our money" over the course of the investment.[14] That means VCs are looking for a 20–40% annual return over the life of funds, though the major trade association's figures show a 26% average return to investors through the 1990s.[15] During the technology boom of the 1990s, a small number of VC firms yielded returns for their investors of over 100% a year: the celebrated Kleiner Perkins Caufield & Byers of Palo Alto made a 287% "internal rate of return" (IRR) on its 1996 fund, and Matrix Partners' 1997 fund returned an astonishing 516% a year, while by 2003 most funds invested at the height of the high-tech bubble were "underwater," showing negative returns, and, when they came to be liquidated, almost certainly lost much, if not most, of their investors' capital.[16] Venture capitalists do not have to "play with their own money," though many of them have some investments in their funds. They are the "general partners" and the wealthy individuals and institutions who invest in their funds (including pension funds and university endowments) are known as "limited partners." The limited partners expect high returns from the funds and, in that expectation, they pay high fees. VC firms take an annual management fee of 1.5 to 3% on funds committed, plus a "carry" (or percentage taken on total fund returns) of 20–35%, where the higher carry has been demanded only by such "star" VC firms as Kleiner Perkins, Sequoia, Accel, and Benchmark. A normal formula is known as "two and twenty"—a management fee equal to 2% of committed capital and 20% of net profits. For VCs, the carry represents their major material interest in funds' success, and the willingness of limited partners to pay such sizeable fees represents their expectation that these investments will radically outperform safer alternatives.[17] Being good at being a VC is not easy, and most—including many of those who exude confidence in their skills and boast of their achievements—are not good. Of course, actual results will look better or worse according to what fund you pick, and, as we all now know, what time period you select. During the boom years of dot-com investing in the late 1990s, VCs justified the commitment of large sums to companies that had no revenues, and sometimes not even a plausible business model, by invoking the tag "if you snooze you lose."[18] But the question with which this chapter is concerned is not how VCs make good rather than bad decisions; rather, it is about how they do whatever it is they do. What do VCs *think* they're doing when they *think* they're doing it pretty well? And how, so far as one can judge, do they *actually* confront the radical uncertainties of the world in which they've chosen to make their living?

The dot-com bust of spring 2000, with its abrupt and precipitous decline in high-tech share prices, did indeed have a massive effect on the environment in which VCs operated. 2000 marked a historic high point both in the number of deals done by American VCs and in the sums invested: 7,813 deals totaling over $104 billion. In 2001, the number of deals dropped by almost a half and the capital committed declined by more than 60%. A more gradual decline then followed, until a revival commencing in late 2003, which has continued to the time of this writing in the summer of 2007. Venture capital activity at present remains far below that of 2000, but is nevertheless above 1998 levels.[19] After the bubble burst, the "wash-out" rate of VCs has been very high, and many of them with what were deemed the right credentials and touted for greatness are now doing something else.[20] What is not evident, however, is that the dot-com bust effected any fundamental change in how VCs go about their business, how they decide to invest in technological futures. True, the volume of self-congratulatory internal commentary and external adulation abruptly declined after spring 2000, together with at least some of the "sex appeal" of the testosterone-charged VC way of life. Post-bubble, the substantial drying up of opportunities for VCs to cash out their investments in companies through initial public offerings (IPOs) meant that it became more expensive to commit funds to very early-stage enterprises. More and more, VCs looked to companies that were somewhat more mature and even to companies that had already generated revenue streams. As one spokesperson for the VC industry put it in 2006, "If venture capitalists are looking at two equally bright concepts—one is two guys and a dog sitting in a garage and the other is a working prototype—they're more likely to invest in the thing that's working."[21] Nevertheless, enthusiasm for investing in early-stage companies was reviving at the same time, with Web 2.0 and "green" energy deals leading the way.[22] By 2007, there were renewed concerns that VCs investing in these sorts of companies were again letting their hearts rule their heads, though commitments to homeland security and military technology firms were almost as large as to clean energy companies.[23] Some commentators even said that personal judgment was being at least partly replaced by rigid financial criteria: "After a period when 'epiphany replaced rigour,' the leading American venture firms are going back to basics."[24] A few VCs even shifted some or all of their attention to the less-uncertain M&A (mergers and acquisitions) field.

However, technological and market trends emerging in the last several years have gone some way to reverse the power dynamics in the entrepreneur-VC relationship and also to reintroduce a more welcoming attitude to high-risk investments. Web 2.0 participatory information-sharing ventures—e.g., MySpace, Facebook, YouTube, Flickr, Digg.com—are different from biotech in often having very low capital requirements and barriers to entry, while generating early revenue streams, meaning that VCs are sometimes queuing up to get barely postadolescent geeks to take their money.[25] The changing power relations involved in the rise of Web 2.0 enterprises is indexed by the "courtship" by Accel's famous VC Jim Breyer of Facebook founder Mark Zuckerberg. Breyer tried to buy the entrepreneur a drink at the Woodside Pub, but Zuckerberg was still not twenty-one: "I had a glass of pinot noir and he had a Sprite," Breyer said.[26] Once skeptical of repeating the mistakes of the 1990s, some VCs are now willingly embracing risk-reward ratios similar to those of the disavowed decade.[27] Moreover, the nature of VC investing means that the activity inevitably remains in the domain of very high uncertainty. Michael Moritz, of Sequoia Capital, funded Google, saying that "it's all too easy to identify the things that might go wrong with an investment . . . It's far more difficult to identify what might be possible . . . Safer isn't better."[28] And the slowing down of the IPO market in 2005 and 2006 induced several, previously conservative, VC firms to take on much more "technology risk"—investing large sums in companies without proven technologies—than they once had.[29] It is correct to note that many—not all—VC firms became, so to speak, more conservative post-bubble, but it would not be correct to conclude that the 2000 bust had any qualitative effect on how they make their investment decisions. VCs aimed to reduce their exposure to risk, but there is only so much that they can do in risk-reduction if they wish to remain venture capitalists: if you want investment vehicles that are much lower in risk, then you will be competing with banks in public capital markets—and you will not be able to attract capital that is willing or able to pay your very high fees. Radical uncertainty is therefore a *defining feature* of the VC world. VCs can and do try to reduce these uncertainties: one Southern California VC, talking to me early in 2003 and making no reference to post-2000 changes, wanted it understood that he did *not* welcome risk or see himself as a gambler:

> Venture capitalists are not paid to take risks. [This is a] common misperception. We're paid to make money. The fact that we have to take some risks to do it is hard, but our investors don't say, "Go see how many risks you can take." In fact, I'm in the business of mitigating risk. If I have five deals, all of which are equally interesting, I want the one with the least risk, with the most complete management team, with the most protected technology.[30]

The cost of reducing uncertainties is reducing potential for reward, and VCs intending to make money by investing in start-up companies have to accept high uncertainty as a defining feature of the world in which they choose to live.

PERFORMING THE FUTURE IN THE PRESENT

There is a vast literature on how to be a successful entrepreneur and a smaller, but still large, literature on how to be a successful VC. Both genres prescribe how the crucial encounter between VC and entrepreneur ought to go—how the entrepreneur can get investment capital from the VC and how the VC can pick a winning investment. This sort of commentary—mainly how-to manuals and participants' reflections on how they actually did it—is one sort of evidence used in this chapter, and there is also a considerable body of Silicon Valley- and Route 128-type high journalism, much of which is extraordinarily sensitive to the concerns about "people-mattering" that run through this book. But another source of evidence comes more directly. In the period from about 1999 to 2003, I spent time as an informal observer at a series of meetings in Southern California where early-stage high-tech and biotech entrepreneurs presented their "pitches" to a mixed audience of VCs, private "angel investors" (commonly "cashed-out" entrepreneurs investing their own money), intellectual property lawyers, accountants, potential business partners or licensors of the technology, and techies in related areas. And my understanding of *how* the future is predicted and made investable derives from reading, from talking with relevant participants, and, especially, from *watching* as people predicted technoscientific futures and others helped them do so. These are scenes in which interest is *performed*: entrepreneurs seek to *interest* investors in a particular vision of a technological future and investors look for the grounds on which they would allow themselves to get *interested* and, indeed, invested in that vision.[31]

What do such scenes and performances look like? How is a vision of the future performed in real time? The relevant scenes are a series of "Springboard" breakfast meetings, usually held at the Faculty Club of the University of California, San Diego, sponsored and hosted by CONNECT, a nonprofit, self-financing part of University Extension.[32] CONNECT was founded in 1985 and the Springboard program, designed to provide free assistance to start-up companies and entrepreneurs, has been going on since 1993. It is well regarded locally and nationally, having "graduated" about two hundred companies that had gone on to raise about half a billion dollars in capital by 2004, and has been widely imitated around the world.[33] There are typically twelve to sixteen people present—though one meeting in July 2002 drew about thirty—all invited by the CONNECT staff largely because they are thought to possess relevant expertise. There are almost always paper executive summaries of the business plan available to be picked up on entry, and participants leaf through this document while eating their muffins or bagels, occasionally referring to it during the presentation and subsequent commentary. (See figure 16.) What motivates these people to get up and on the freeways so early? Some have to get going well before 6 a.m., and this is not a scene in which many deals actually get done or money gets made. The professional term of art for the instrumental purpose of these meetings is "networking," though I've heard a lot of attendees say such noninstrumental things as "I

FIGURE 16. A meeting of the Platform program of the Massachusetts Technology Transfer Center (MTTC) in August 2007. MTTC is an organization within the University of Massachusetts President's Office in Boston, and the Platform is similar to UCSD CONNECT's Springboard program, both venues in which very early-stage entrepreneurs are coached and then "pitch" their company to an invited mixed audience of potential investors, service providers, and techies in similar lines of work. Here a postdoc electrical engineer at MIT is presenting an idea for a company based on a silicon chip designed to pre-concentrate biomolecules. (Photograph reproduced by permission of the Massachusetts Technology Transfer Center.)

want to help" and "it's usually interesting" or "fun." If you're a sociologist, what motivational language you want to apply is another issue, and the epilogue to this book offers some brief remarks about how one might interpret the form of life known as networking.

The enterprises pitching at the Springboard meetings are usually very early-stage companies, some not actually incorporated yet, all a long way from going public.[34] They're something between a dream in an entrepreneur's eye and a small-scale incorporated reality. Some entrepreneurs have not yet given up their day jobs; others have. They've all got sweat equity, and now they need infusions of capital to hire some crucial employees and to secure premises and plant. At the stage they've typically reached, they might need up from $1.5 to $3 million to go on with. That's

typically a VC-sized capital infusion, while bands of angel investors—
clubbed together for administrative convenience and sociability—tend
to function in the quarter-million to one-and-a-half million dollar range
and individual angels may put in as little as $15,000.[35] (VCs can be very
rude about the sophistication of individual angels and angel bands: sometimes they like to say that "angels get killed"; sometimes they character-
ize angels among the category known as "the three Fs: friends, family,
and fools.")[36] Some capital may have already come from second mort-
gages, some (literally) from friends and family, some—imprudently but
not uncommonly—from credit card debt.[37] A company presenting at
Springboard in June 2000, and founded in the previous year, had already secured a $500,000 capital infusion through the CEO's friendship
with a then still-buoyant dot-com toy-selling firm. A few Springboard-
presenting entrepreneurs may now be ready to pitch to VCs proper, possibly at one of the many venture fairs around the country, but many more
hope to be in that position quite soon.

These Springboard sessions are an edgy hybrid between a *rehearsal* for
a proper pitch to VCs or angel investors and the pitch itself. Presenters
have been elaborately prepped by CONNECT staff about how to make an
effective pitch—everything from pace and tone of voice, to the best use
of PowerPoint visuals, to how to respond in the following Q&A period:
"This is what you want to do if and when you make a pitch on Sand Hill
Road." Many of them genuinely believe that *this* pitch will result in VC
or angel investment, even though they're told that the chances are slim.
Certainly, the entrepreneurs all take the pitch very seriously and *act* as if
it's "the real thing." In fact, it's *not* unknown for deals to get done as a result
of the present pitch, and this pitch, after fine tuning, is usually the one
they go on to present to VCs, if indeed they get that far. The seriousness
of the occasion is signaled by the fact that the entrepreneurs (if male) are
almost always in suit and tie, while the audience is usually more casually
dressed. No one is asked to sign an NDA here: the presumption is that
everyone present is normally trustworthy—having been vouched for by
staff—and that key trade secrets are, in any case, not going to be revealed.
(At a similar meeting in Boston, the staff person in charge prefaced the
session by cuing those present that "it's all proprietary here," and wound
up by reminding them to "keep the information inside the room," but
nothing was offered to be signed and the tone was a redundant counsel
to use common sense.) It's understood in these circles that you've got to
take these sorts of risks to get the expected benefits. Tom White of Celera
Diagnostics, asked about the status and enforceability of confidentiality

agreements negotiated with academic collaborators, said, "It's always a calculated risk. You just have to make your own judgment, and you are careful as to the amount and degree of information that you provide, and you kind of assess how reliable the person is."[38] (Obliged by my former university's Human Research Protection Program to get signed informed consent forms from entrepreneurs I interviewed for material in the preceding chapter, some of them looked bemused at what they took to be these curious academic artifacts and tried to bat them away.)

As is the current norm, presentations are structured around Power-Point slides, somewhere between ten to fifteen slides for a fifteen-minute pitch. Unlike academic talks, the entrepreneurs almost never seem to have the slightest problem keeping to their allotted time, partly owing to the extensive coaching available to them about how to do the thing properly.[39] (When, exceptionally, one entrepreneur took three minutes longer than intended, he was gently criticized and strongly advised not to do that again.) "Rehearse, rehearse, rehearse!" one entrepreneurs' workshop manual preaches.[40] "They practice you to death," said a grateful biotech entrepreneur about one of the CONNECT coaching programs I attended.[41] It's well understood that VCs haven't got a lot of time, that your pitch is likely to be one of dozens that they hear in the course of a year, and that you'd better be concise. VCs are said to display status by visibly letting their attention wander if presenters go on too long; accordingly, entrepreneurs are coached to observe the maxim "less is more."[42] Investors aren't usually scientists, and presenters are advised to simplify the science as far as possible, reserving displays of enthusiasm for the business opportunities. "Stay away from techno-speak," one guide cautioned: you can always talk to the geeks one-on-one later.[43] At one pitch, the presenter (a biology professor) clearly got too "technical" for the potential investors, and was explicitly recommended in future to "dumb it down": he was told that he had been "way too technical"; you have to bring the presentation down to audience level; the audience mustn't get lost in scientific detail "or they tune out."[44] (Investors will want to know at the outset whether scientific claims have been peer-reviewed and published in reputable journals, but they needn't be given a science lecture at this stage: all the "technical stuff" and the legal status of intellectual property can, and probably will, be checked out later.)[45] Matters covered follow a fairly standard format. In fact, if you've got PowerPoint on your computer and you click on the Business Plan option, you'll have the standard format already laid out for you, in twelve slides, with the bullet points all specified. All you have to do is fill in the details. (Or you can just go to any number of on-line sites to see business plan specifications clearly outlined for the entrepreneur.)[46] What's the technology? What's its legal status? Who's the management and technical staff? What's the market? Who's the competition? What are the risks and rewards? How do you see revenues and profits developing over the next three to seven years? What's the present management team? What do you need now in the way of capital, staff, and related support?

Presentations and business plans that are manifestly sloppy, or that aren't done in due form, clearly don't count as good ideas, and, again, entrepreneurs are aware of, and many seek out, coaching assistance to enable them to do the job professionally.[47] "It's not what you say," a guidebook for entrepreneurs declares, "but *how you say it*—in your business plan and in your presentation." No one model is right for all purposes, the guidebook concedes, but there is an ABC of plan preparation: "Always Be Concise."[48] That much about the pitch is fairly rule governed, even ritualized, and sometimes VCs just ask you for your PowerPoint presentation, which they can review in the comfort of their offices before deciding whether to have you in personally for a one-on-one, or a meet-the-partners, session, lasting maybe an hour. (If you just turn your back on the audience and *read* your slides, a VC advises, "your audience will find you dispensable.")[49] Coaching manuals refer entrepreneurs to "the rule of seven" (or its several variants): no more than seven bullets per slide; no more than seven words per bullet; and they advise you how to stand (weight evenly balanced on both feet, no slouching); how to use your hands (avoiding such "wooden-speaker positions" as the "fig leaf," "the mortician," "the tight-rope walker," and "gunfighter" or "gorilla"); not to grab the podium ("the death grip"); to make sure you've emptied your pockets so that no one can hear your keys and coins rattling, how to move your eyes from person to person in the room ("three beats each, then move on—make eye-contact with everyone in the room"), but do not shift the gaze from right to left as if watching a ping-pong game.[50] Joshua Boger, the founder, president, and CEO of Vertex Pharmaceuticals, aimed for the illusion of "intimacy" in pitching to a roomful of potential investors, using a "variety of stage tricks to elicit such feigned intimacy—lowering his voice to make a point; speaking to the back row, which gave those in the forward rows the impression that he was speaking to them." His major device, Barry Werth records, "was a strenuously rehearsed spontaneity. As he spoke, Boger listened intently to his own words, then quickly anticipated the questions of those in the audience and tried to answer them matter-of-factly in the next sentence or two." As much as any

Renaissance courtier, Boger understood the art of artlessness: "You can't fake excitement and you can't fake sincerity," Boger noted with no evident intention of irony. "It can't be done . . . and to do it properly you have to practice."[51] Investors recurrently indicate that they're bored with presentations that are *too* smooth and *too* scripted. "Just tell me about it," they say. And a guide to biotech entrepreneurs advises against an overly "flashy presentation": "Too much glitz makes it look like you are masking a bad idea."[52]

INSTITUTIONALIZED SKEPTICISM AND ITS LIMITS

There are also some explicit criteria VCs are said to look for in deciding whether a specific investment is credible and attractive. There is paper involved and formal criteria have to be satisfied. Presentations, almost always, come together with a paper business plan, maybe thirty pages— with appendices pushing it up to perhaps one hundred—though the prefatory executive summary is usually no more than three pages and may be as short as a page and a half.[53] The business plan should look plausible and professional; it should not evidently contradict what potential investors competently know to be the facts of the matter; and it helps if it resembles, or can be made to resemble, past well-known success stories. Entrepreneurs are sometimes strongly cautioned not to deviate from industry norms in projecting sales and growth. It is reassuring if the present future can be made to resemble past futures. And there are some widely repeated folksy heuristics that VCs say they use in picking winners and that entrepreneurs are counseled to be aware of in pitching to VCs: among many examples, "look for companies selling aspirins rather than vitamins"; "there's no premium for complexity"; "set your sights higher than your sandbox"; "find markets the size of Texas"; "don't sell a dollar for 95 cents"; "don't invest in companies whose business you can't explain during an elevator ride" (the elevator pitch); "real entrepreneurs quit their day-jobs"; "don't let your mouth make a promise your ass can't keep".[54] The "Virtual CEO" and VC-broker Randy Komisar says that VCs want to know "three basic things": "Is it a big market? Can your product win over and defend a large share of that market? Can your team do the job?"[55] They want positive answers to all three, but they accept that every claim has, nevertheless, to be evaluated. And, in any case, the process called "due diligence" is designed, in principle, to allow investors to detect flies in the entrepreneurial ointment or misrepresentations of matters of fact.[56] In principle, what due diligence is intended to do is very hard to achieve: it's hard to verify empirical claims about past and present states of affairs and much harder to evaluate the plausibility of projections about the future. Due diligence, as routinely practiced, is an attempt to check out, for example, the credentials, character, and track record of the entrepreneurs; the likely size of the market; the security of intellectual property; the plausibility of the science and technology, as attested by relevant experts; and the existence and plans of potential competitors. It tries to identify risks rather than rewards.[57]

Yet, even here, legends circulate in Silicon Valley about appallingly amateurish and incompetent presentations, decided on over breakfast at Buck's in Woodside, Hobee's or Il Fornaio in Palo Alto, or the Konditorei in Portola Valley and sealed with a handshake, deals that turned out to make everyone involved disgustingly rich.[58] (Coffee and the coffeehouses in Silicon Valley are at least as important in VC deal making as they were for commercial relations in seventeenth-century London. A journalist noted that "to pass a day in the home of the tech industry doing meetings is to sit through one long round of cappuccinos and lattes.")[59] Gordon Moore tells an iconic story about how he and Bob Noyce got several millions in 1968 to start Intel from the VC Arthur Rock, the deal getting done "within an afternoon" and "without a business plan."[60] Familiarity here trumped formalism: Rock and Noyce had by then known each other for eleven years and used to go hiking and camping together. "Bob just called me on the phone," Rock said: "We'd been friends for a long time . . . Documents? There was practically nothing . . . We put out a page-and-a-half little circular, but I'd raised the money even before people saw it."[61] When software engineer Pierre Omidyar, the founder of eBay—before Google, the most successful Internet company of all—approached Sand Hill Road VCs late in 1996, he didn't have a PowerPoint presentation or even a business plan, and when he borrowed a VC's laptop to show the firm the eBay Web site, the server had gone down.[62] Even after the 2000 bubble burst, and VCs' vows henceforth to do things in a more deliberate manner, the pattern is repeated: a participant in a meeting between Facebook's Mark Zuckerberg and Bay area VC Peter Thiel said that when the entrepreneur "started telling us the Facebook story . . . it was pretty quickly apparent that this business was taking on aspects of an eBay or a Google, something with just an extraordinary growth rate. About eight minutes into the talk, it was clear to me that Peter was going to invest."[63] The firm quickly committed $12.2 million to Facebook, prompting comments from some observers of the VC scene that no cautionary lessons had been learned from the excesses of the late 1990s.[64]

The San Diego biotech entrepreneur (turned VC) Ivor Royston cele-
brates the role of intuition over due diligence, citing the example of Tom
Perkins, one of the VCs who backed his own start-up biotech company:

> Tom Perkins . . . was a very intuitive person. It's not like he had to do ex-
> tensive due diligence, you know. Once he got comfortable with the tech-
> nology, intuitively, and it made sense, and he got comfortable with the
> people, he was willing, basically, to bet on that, to bet on you . . . I admire
> that kind of thing. I think more and more people should, instead of doing
> extensive due diligence. Just trust their instincts, their gut, you know.[65]

"I don't see a correlation between due diligence and success," Royston
concluded. It's something you have to do for institutional reasons, but
Royston was skeptical of its informational role.[66] This was a sentiment
strongly endorsed by the San Diego VC Tim Wollaeger, who was CFO at
Hybritech in the 1980s. Wollaeger judged that "there is very little correlation between the amount of *Due Diligence* performed and the success of an
investment," and went on to enumerate a series of successful investments
he made within three days of hearing the initial proposal: "I do not believe
a great deal of Due Diligence will save you from bad investments. It is
better to get going on the companies that feel right to you."[67] And the
people who feel similarly right. Royston again:

> Oh, you do have to trust the people—I mean—that are presenting to
> you, and . . . we actually do significant reference checks on people, but,
> I mean, you know, when you meet somebody, . . . you should be able to
> tell whether these are good people to be partners with. It's like getting
> married. How do you describe to somebody who you want to get married
> to? And that's the same thing . . . You try to describe the characteristics
> of the person you want to marry, but, you know, falling in love and—it's
> the same—I can't tell you—I mean I have people sitting in front of me
> describing great science and interesting ideas, but I don't trust the guy,
> just from his—the way he talks, his mannerisms, how he talks to me. I've
> dismissed companies just because I don't like the guy.[68]

For these reasons, some VCs, when pressed about why some are successful in their investment choices and others—seemingly equally well-
qualified—are not, despair of identifying publicly visible criteria: "I think
what a lot of these guys [VC failures] learned, some the hard way, is that
you're a natural athlete or you're not . . . Some can do it, and some can't,
and like with athletes there's no way of telling until they take the field."[69]
Even some VCs setting greater store by the due diligence process freely

acknowledge its limitations: "I believe what due diligence does, it doesn't prevent you from making bad investments, it prevents you from making stupid investments, which are two different things. Stupid investments are the ones where you say, 'Boy if I just did a little checking I would have known that this guy was a flaming jerk.'"[70]

The forces of familiarity that operate so strongly in VCs' business worlds flow along some of the same channels as the rest of their lives. Work and play, so to speak, perhaps overlap more extensively than they do in the worlds of other professionals. Deals are done, or at least discussed, at restaurants and coffee shops (of course), on the golf course, on a sailing boat, on fishing trips to the Sierras, at baseball games. That's one way of thinking about why it's such a masculine world. Only 8% of the partner-level members of the National Venture Capital Association are women, and only a quarter of VC firms had women in management positions in 2000, but the female VCs who reflect on this state of affairs offer a wide range of complex and interesting explanations. The story most compatible with "discrimination" in VC firms draws an analogy between becoming a general partner and marriage. Carole Dressler, a headhunter for VC firms, says that "integrating a new partner is like marrying one person to six... You're dealing with close-knit organizations that depend on lots of camaraderie, and when you're talking about a group of males making decisions, they often take on male partners," and, she might have added, invest in male-led start-ups. (Only 6.5% of the firms in which VCs invested in 2006 were founded by women, even though women are estimated to own almost 30% of American companies and are founding new companies at twice the rate of men.) Yet the dearth of female VCs is only slightly worse than in the techie world in which VCs invest: in 2006, only 11% of U.S. computer engineering graduates were women. And another female VC attributes the shortfall among VCs to women's lack of presence in the career pools from which VCs are fished—not so much MBAs (and especially not in the West Coast VC scene), but people with entrepreneurial technology backgrounds: "The main reason there aren't more women in venture capital," observes Mobius's VC Heidi Roizen, "is that there aren't more women in the positions that lead to careers in tech investing, like founding a tech company or becoming a CEO of one."[71] Yet three caveats need to be made. First, patterns of familiarity in the VC-entrepreneur world are *not* solely those that operate to the detriment of either women, or, indeed, of minority groups. Although mountain biking and baseball might plausibly be regarded as typical sites of male bonding, other important informal channels include membership on the boards of voluntary associations and educational institutions, and, not uncommonly on the West Coast, churches—and these are not notably male domains. Second, while it is reasonable to think that male VCs are more comfortable forming business relationships with male partners and entrepreneurs, one might also expect discomfort across racial, religious, and ethnic boundaries, and yet this is an ethnically cosmopolitan world in which Indian, Korean, Israeli, Turkish, and Russian VCs and entrepreneurs flourish together. Finally, such solid evidence as there is of structural disadvantage flowing from patterns of familiarity is *not* incompatible with a shared *psychological* experience of extreme meritocracy. It is just very hard—though not totally unknown—in this world to hear talk of group discrimination. It is not part (or at most it is a very minor part) of members' vernacular—as it *is* for, say, members of academic sociology or history departments. Almost everybody in this world insists that "all cats are gray in the dark"—that good technology, good management, and good judgment of technology and management are all that matter. Women VCs have been notably voluble in rejecting the notion that they are discriminated against because they are women: VC Peg Wyant, for example, is well aware of the present state of affairs, and wishes it were otherwise, but is optimistic that "all of the components are in place for this to be a nonissue by the end of the decade."[72] And women VCs insist as vigorously as their male partners that their investment decisions are gender- and ethnicity-blind.[73] Gender and ethnicity are, indeed, elements in the patterns of familiarity that structure relations in the world inhabited by VCs and entrepreneurs, but they do not map unproblematically onto psychological states; they are not the only elements in such patterns; and they are cross-cut by considerations that work in ways whose effect on members of specific groups is impossible to determine.

Patterns of familiarity powerfully regulate access to VCs' attention. It is one thing to prepare a business plan or presentation in due form that testifies to inherently interesting technologies and opportunities; it is another to get the plan even looked at by a VC.[74] Jim Breyer of the elite VC firm Accel Partners says his firm gets 10,000 pitches a year, of which only several hundred get "seriously considered" and maybe twelve to fifteen invested in.[75] According to one overall estimate, as few as 25 of 1,000 submitted business plans ever result in a face-to-face meeting, but that face-time is vital since "once your idea has a face on it, you're already ahead of the other piles of paper."[76] A VC interviewed by Harvard Business School professors told them that "we got 300 approaches last year, 150 of those in writing; we met with about 15, then about 10 for a second meeting, and of those we did [a deal with] one or two."[77] It is a commonplace in the VC world that plans coming in over the transom (these days, as e-mail attachments) *never* get to the general partners' desks, and, in the rare cases that they do, they have to survive ruthless selection typically entrusted to lower-level administrative staff. One Sequoia Capital VC said he took home thirty business plans a night, but these already had gone through subordinates' triage.[78] The accountancy and consultancy firm Pricewater-houseCooper estimates that only 5% of all business plans received by VC firms are read—by anyone—beyond the executive summary.[79] A Southern California VC told me that he had almost never in his career looked at a business plan that came over the transom: "The unsolicited business plans; the stuff that comes over e-mail, 90% of it gets stopped by my office manager."[80] Another said that he did sometimes look at unsolicited plans, but had found that it was scarcely worth his while: "I have never invested in a single over-the-transom idea for the simple reason that none of them has met my investment standards . . . You're introduced by somebody you know, maybe somebody you trust, as opposed to somebody you don't know, who you don't trust."[81] PricewaterhouseCoopers similarly warns entrepreneurs not to cold-call VCs: "Introductions to venture capitalists through referral sources they respect improves the odds of securing financing," and the firm notes that it "can provide these introductions."[82]

A rule of thumb in this world mobilizes a vernacular version of social psychologist Stanley Milgram's work on the "small-world phenomenon"—the finding that two randomly selected people in the United States could be connected with each other through fewer than six intermediaries on average.[83] Entrepreneurs thinking of approaching a VC or angel investor are sometimes counseled to have in hand a social map connecting them with the investor through no more than "two degrees of separation"—someone who knows the entrepreneur who knows someone who the investor knows.[84] At a Springboard presentation in August 2000, one of the CONNECT staff cautioned the techie founders of a wireless mapping technology start-up to get themselves better networked: "Raising money is a contact sport. You want to know people who know people." Contacting a venture capitalist "cold"—without a prior introduction or shared node of familiarity—is sometimes seen in itself as a consequential mark of naivete: "Anyone whom I don't know who approaches me directly with a business plan shows me they haven't passed Entrepreneurship 101."[85] A handbook for entrepreneurs similarly cautions that "business plans that get financed are rarely those that just come in over the transom. 'It's not what you know, but who you know,' rings very true in the venture capital industry."[86] Robert Kunze adds that "viable business plans come to me almost entirely from referrals," from people he's familiar with who know the VC business: "This networking process is itself an informal investment screen."[87] "It is usually the quality of the references that determines whether a proposal becomes a deal," one commentator observed.[88] In 1978, Ivor Royston got an appointment with the VC Brook Byers without a business plan at all—a six-page business plan came later, at Byers's request. Royston got into Byers's office because Royston's wife, Colette, had once dated the VC, leading Hybritech's first president later to remark: "Now this is the way great companies get started. It's who you've dated, you know."[89] Another entrepreneur got into the presence of a Kleiner Perkins general partner because a member of his management team had many years previously dated the VC's sister.[90] Sometimes VCs feel themselves obliged to meet with an entrepreneur to whom they would not otherwise give time just because of what is owing to the introducer. So if a tech transfer officer at a respected institution asked for the favor—"'Look, I want you to really meet with one of my scientists. I think he's got something interesting'"—Ivor Royston said that he would feel obliged to grant a meeting: "We sometimes do it as a courtesy, you know, just to be politically correct."[91] Almost without exception, business plans need to come *introduced*—hand-carried, as it were, by VCs or entrepreneurs or institutional representatives with whom the prospective source of capital is already familiar. Familiarity with the personal characteristics of others, and with their known integrity and capacities of judgment, is what opens doors in this world. Without the advantages of familiarity, the satisfaction of formal criteria means almost nothing.

When you venture into the unknown you need all the familiarity you can get. The VC world is one in which the biggest prizes famously come from "out-of-the-box" thinking, from the totally unpredictable: not, as VCs say, "Faster, Better, Cheaper" but "Brave New World." There's a fine line, VCs remind themselves, between the high-tech bizarre and the brilliant, so you can't be certain that pitches (and the entrepreneurs that make them) that sound and look crazy really *are* crazy. One otherwise cynical VC records that he was not necessarily put off by entrepreneurs who said they were inspired by God or who wanted to make megabucks to give away to the Maharishi: "Anything legal that would motivate [him] to work 80 hours a week was fine with me . . . Anyone wanting to start up a semiconductor company in 1983 had to be a little nuts—along with the venture capitalist who backed them."[92] A Silicon Valley VC observed that "the best deals are always the craziest," and his colleague agreed that "if you want to go where the big wins are, be a little bit crazy . . . Conventional wisdom leads to conventional returns. If you're afraid of being fantastically wrong, you'll never be fantastically right."[93] (I am personally familiar with the CEO of a small drug discovery company who was vigorously cautioned against telling potential investors that Jesus helped write his business plan, but who nevertheless persisted in evangelizing potential investors as well as employees. That company, with its theologically unreconstructed CEO, recently raised $40 million in additional VC finance.)[94] Plausibility is not understood to be a very reliable guide to picking "the new new thing." VCs love to quote the alleged 1977 dictum of the founder of Digital Equipment Corporation that "there is no reason for any individual to have a computer in their home."[95]

FAMILIAR PEOPLE AND UNFAMILIAR FUTURES

At the California breakfast meetings that I attended, it's often pretty evident what is a polished presentation and what isn't, and polish is a goal in its own right, even if slick professionalism is no guarantee that you will attract angel or venture capital. So potential investors are looking at bits of paper and PowerPoint slides and seeing whether what's on them satisfies certain criteria, criteria that investors can and do articulate and write down — though it's remarkable the limited extent to which such criteria actually feature in VCs' commentary on their work. But they are also, and VCs are reflectively well aware of this, looking at a *performance* that's right in front of them. They're looking at *someone performing* a credible technological and commercial future. And VCs do have quite a lot to say about such things.

There's a proverb in VC circles — they're great ones for proverbs — that advises you to "bet on the jockey, not on the horse."[96] This proverb, like all proverbs in naturally occurring settings, has, of course, a ceteris paribus clause: you shouldn't bet on entrepreneurs just because they seem innovative, prudent, and professional; nevertheless, given that other features of the proposition are plausible, the decisive factor, sometimes more important than the promise of the technology or of the market, is individual and collective *management*, and this is particularly pertinent in early-stage investing when uncertainties are at their greatest. They're the people who are going to make it happen, and, if the entrepreneurs look good, then that's very important indeed. "More than anything else," an entrepreneur's guide says, "venture capitalists are judges of people."[97] So, as the late Harvard Business School guru General Georges Doriot said in a widely quoted remark, go for "an A-quality man with a B-quality project, but not the other way round."[98] And that's the thought informing Arthur Rock's claim that Intel was the only company he ever invested in that he "was absolutely, 100 percent sure would be a success, because of [Gordon] Moore and [Bob] Noyce." He had known them personally for such a long time, and he trusted them. "The lesson from Intel?" Rock rhetorically asks himself: "The necessity of having great management."[99] "I believe so strongly in people," Rock said, "that I think talking to the individual is much more important than finding out too much about what they want to do."[100]

That's one interpretation of what it means in VC circles to bet on the jockey. But betting on the jockey has got even greater scope, consequence, and interest for understanding how late-modern technoscientific futures are made credible and therefore *made*. Consider the scene in which the pitch is made: projections of the market and of revenues; projections of how the technology will be developed and manufactured; and projections about how likely competitors will appear and be dealt with—all these, however plausible, take the form of bits of paper or electronic bytes. In futures as radically unpredictable as these, anything may happen—and, as they say, usually does—to deflate the value of these bits and bytes. The so-called safe harbor statement required by the Securities and Exchange Commission since the Private Securities Litigation Reform Act of 1995, for example, absolutely *requires* entrepreneurs to use deflationary boilerplate language in approaching investors, language that so severely qualifies the truth of technical and commercial claims that these mandated disclaimers are much more skeptical than anything that could be said by an external critic. In effect, safe harbor legal/linguistic conventions formally tell you that you're not to take entrepreneurs' forward-looking statements as worth much more than the paper they're written on.[101] VCs are highly skilled at discounting the supposed facts, claims, and future projections in entrepreneurs' business plans. One VC's cynical rumination on the subject was titled "Lies, Damn Lies, and Marketing Lies." Entrepreneurs' Lie Number 10, for example: "We are the new paradigm!" VC: "Whatever the hell that means. If I had a dime for every company that has told me it was the new paradigm, I would be rich enough to pay Bill Gates's legal bill."[102]

There is, however, something in these scenes that is more durable than the bits and bytes, something on which you can plausibly rely: the entrepreneur in front of you, pitching these claims to investors—if he doesn't go under a proverbial bus or have a radical personality change—*will be there in the investable future*. What you see is very likely what you will get. If he seems innovative, prudent, and professional today, he will very likely be so a year or three in the future.[103] That's what's called a safe bet, or as safe as you get in this business.[104] And so it's also a safe bet that if problems arise with the business plan—and they always do—the entrepreneur in front of you now is the one who's going to fix those problems, insofar as they are fixable. Jockeys live longer than horses, and they're the ones that have got to steer the horse, to correct the errors of its ways, and to make adaptations to course conditions and the behavior of rival horses. One mark of an investable "great team" is, of course, a proven record of success, but so is past evidence of learning from failure. VCs are said to "attach a lot of importance to what they term 'scar tissue'—evidence that the person has learned from experience," a capacity to learn that can be projected into the uncertain future.[105] Greylock's Henry McCance glossed General Doriot's maxim about "A teams" and "B products" by noting some consequences of the extremely rapid rate of change in both technology and market opportunities: "Often what you invest in may evolve to something quite different over the length of a relationship. So if you have an A team, they will be able to internalize the changes, make the strategic calls that are necessary, and adapt to the changing environment that they're forced to live in."[106] Or, as the entrepreneur-turned-VC Ivor Royston says

> The idea is that if you invest in the right people and the technology doesn't work, then the people will find new technology, whereas if you have good technology and the wrong people, the technology can really flounder . . . So, yeah, you invest in people over technology.[107]

PRUDENCE AND PASSION

So what do you look for in a jockey? One thing might be a certain disengaged, cool, calculating, and wholly instrumental rationality. VCs are, after all, capitalists, and one thing that everybody knows about capitalism is that it's the bottom line, and only the bottom line, that matters. VCs worry that high-tech entrepreneurs sometimes don't know how to let go. Inevitably, VC investment means a dilution of the founders' share of equity and, almost certainly, of corporate control: in exchange for their infusion of capital, VC firms typically take preferred stock equity in the company or various combinations of equity and debt instruments. That's to say, they wind up owning the "founders' company"—at least temporarily. In addition, the VC firm will normally place some of their people on the company's board and will help search out management expertise (including CEOs, CFOs, and COOs), additional directors, and members of a scientific advisory board (SAB). The founders get the resources essential for their company to succeed, but it's now "their company" only in a historic or moral sense. That's why VCs sometimes want the entrepreneur to specify an "exit strategy," or to look forward to a "liquidity event," a few years down the line—just so they fully realize that what's their baby now won't, in all probability, be their baby for very long. Should a company go public—the dream of all VCs—the average entrepreneur winds up owning 3 to 10% of "their" company, and very likely will be eased out of any effective control.[108] VCs want entrepreneurs to appreciate the cliché that "it's better to have a small piece of a big pie than a big piece of a small pie, or no piece at all."[109] As an angel investor put it, one thing you're looking for in an entrepreneur who is the CEO of a start-up is a willingness to let go if it's deemed necessary: "Will the individual get out of the way when others perceive that he has not adjusted to the changing needs of the business?"[110] The CEO of a fax modem start-up company relates how he was bluntly told by a VC within hours of his pitch that "I don't fund companies with CEOs who need to learn on the job."[111] Of course, the reward for dilution, and for letting go, should be great piles of cash, and VCs want to be reassured that entrepreneurs understand the deal—that "cashing out" is what they're about. VCs tell stories about the disastrous consequences that have followed from entrepreneurs who refused to let the baby go. Especially after the dot-com crash, some investors say that they're looking more carefully for the personality type called "the serial entrepreneur," someone who makes manifest a commitment to cashing out and going on to build another company, the mythic paradigm of which is Jim Clark of Silicon Graphics, Netscape, and Healtheon (now WebMD).[112] And sometimes they say that they make this personality assessment "within seconds of shaking your hand."[113]

But cool rationality is not the only quality VCs look for in high-tech entrepreneurs. Indeed, very often they look for character traits pointing in the opposite direction entirely. VCs say that they're looking for entrepreneurs who display *passion, commitment, and vision.* Read the reflections of VCs, and these words, or their synonyms, stare out at you on practically every page. The words are designators for the personal virtues VCs are looking for in entrepreneurs. VCs want to see passion, vision, and commitment in entrepreneurs because building a high-tech company is a highly uncertain business: it's frustrating, it's emotionally draining, and it demands heroically hard work—24/7s for months and even years. You should, of course, have "Plan B," but there's a distinction between strategy and tactics. An investment analyst evaluating the COO of a Seattle biotech company described him as "smooth and competent . . . but does he have the stomach for the long haul?"[114] "Only passion will get you through the tough times," Randy Komisar told an entrepreneur whom he suspected of lacking passion.[115] Entrepreneurs understand that realizing their vision will probably demand the sacrifice of all other forms of personal fulfillment: no social life, no outside interests, a subsistence regime of Diet Pepsi and cold pizza.[116] Entrepreneurs (and the VCs who back them) are nature's optimists. Several months post-bubble, I attended a panel on "failure," also run by the staff of CONNECT. Scarcely a single "failed" entrepreneur presenting at this program neglected to quote the dictum that "what doesn't kill me makes me strong." They all had either dusted themselves off and started over again or were keenly looking forward to doing so soon. (Some VCs also responded to the bubble burst with optimism: the first and last PowerPoint slides in a 2002 presentation by the San Diego VC Bill Stensrud featured an explicit gesture to Nietzsche's version of the dictum—"What does not destroy me makes me stronger"—announcing that the most fertile conditions always followed a forest fire. "The smoke has cleared—we now know the rules; Small furry mammals will win."[117] The VC Ivor Royston told me in spring 2003 that this was "a good time to be a buyer. This is the best time to be doing venture capital now because the market's down."[118]

But even these optimists know the strong probability that all this sacrifice is likely to result in failure and that many failures will inevitably precede any success. That's one reason why VCs and angels like to see entrepreneurs' "skin in the game": they want the founders to put their own money in the business, as a mark of genuine commitment and seriousness of purpose.[119] And VCs may want entrepreneurs, especially academics, to *illustrate* commitment by showing that they really do *want* equity and financial returns. As one VC put it, "we would be leery if someone came in and—if a professor at the university came in and said, 'Oh, I don't want any stock.' To us, no stock means no commitment . . . No commitment to the company, even though the stock could be his ticket to, you know, financial success. We've actually had professors say, 'I don't want any stock because it's going to cause conflicts at the university,' what have you. But we wouldn't fund those people. Number 1, it suggests to us that they really want this company to support their research at the university . . . and they're not really committed. The commitment means having stock in the company, even if you're a professor." At the same time, signs that an entrepreneur might want to cash out too early are taken as warnings: "We are looking for people that are really entrepreneurial, who are willing to really put in the time and sweat to see the company's success. And we're looking for real commitment to the company, absolutely. Any sign that they're just going to cash out early, we would walk."[120]

Perhaps harder for some outside observers to credit, VCs and angel investors often say that good-bet entrepreneurs should not be, and often are not, motivated by the money, or even by "success," save as a stage in the drive for further "success."[121] Money *follows* from a passionate commitment to change the world (plus, of course, good technology and a good business model), and money is a socially useful *index* that one has in fact succeeded. During the 1990s glory days of Silicon Valley, monetary success was so ubiquitous that it lost its "nuancing" capacity; as Po Bronson says, it "didn't impress": "The way to stand out is to make something that has a big impact on the course of technology." "Find a way to make a difference," one VC observes, "and the rest will follow—personal success, promotions, financial gain."[122] According to one commentator, what impressed Benchmark Capital's Bob Kagle about eBay's Pierre Omidyar was that the entrepreneur "was consumed by the idea of community—every other sentence, he spoke about the eBay *community*, learning from the *community*, protecting the *community*. It was a passion similar to what, in Bob-speak, Kagle had for deals that brought out *the humanity*; that's what Kagle liked most of all, *the humanity*. The more Omidyar talked about his community vision, the more Kagle, as he put it, was 'lovin' him—this guy is *good people*.' And Omidyar felt the same about Kagle."[123] Years later, Omidyar has become a VC himself, his Omidyar Network announcing its commitment "to fostering individual self-empowerment on a global scale," funding both for-profits and nonprofits—including Third World microfinance institutions—that promote "equal access to information, resources and tools, the ability to connect to others with shared interests, and a sense of ownership over outcomes."[124] Another VC notes that the high-tech entrepreneurs his firm backed "were driven, not to make money, but to make their technology the best in the world... The entrepreneurs who came in here and said they were motivated to start a company because they wanted to get rich were not the ones that had the best ideas. The ones who came in and said that they had an idea to make the world better or the technology better were the ones you wanted to back. The money followed the ideas."[125] Or, as Komisar says he tells the

MBA classes he occasionally teaches, "it's the romance, not the finance that makes business worth pursuing."[126]

Robert Teitelman notes how early Silicon Valley entrepreneurs "carried the aggressive idealism of the 1960s into the new calling of business": utopian, commune-founding hippies and electronic or biotech visionaries—many of them were, after all, *the same people*.[127] Make no mistake about it: money is *very* important in this world. But VCs recurrently express wariness of the *sufficiency* of money motivation—both for themselves and for entrepreneurs.[128] And public displays of the personal virtues of passion, commitment, and vision are looked for as signs that entrepreneurs have a chance at success. Asked about the role of money motivation among entrepreneurs, a VC responded: "The guy who wants to do it for money? He's going to bail on you when the going gets tough and everyone's going to have tough going. These things are built by people who have a passion. You need people who want to change the world. Very few people [have it], you don't want guys who go, 'I'm doing this because I'm going to make a bunch of money.' That's a by-product."[129] Again, VCs occasionally think about themselves in the same terms. Brook Byers of Kleiner Perkins had no problem admitting that he got "paid and compensated well," but insisted that the true reward of his job was "to see the future. I get to see the future."[130] There is no reason to worry about the superficial soft-headedness: the vocabulary of virtue does indeed map onto that of monetary reward. Indeed, post-2000, some VCs have developed a special-purpose reactive public rhetoric: investments that some observers might ascribe to altruistic motives are carefully justified by bottom-line considerations. This is notably the case with the accelerating current funding of alternative (or "green") energy start-ups, and one Palo Alto VC's PowerPoint show invariably ends with a slide stressing "that to the extent his motivations are tinged green, it has to do with the color of money."[131] VC firms are not charities, and nothing they might say about "passion," "vision," and "changing the world" should be understood as an expectation that they, or the entrepreneurs in whom they invest, ought to lose money or have an insouciant disregard for capital growth and profit. The point at issue is how motives are parsed, stipulated, understood, and acted upon in the relevant communities.

Would-be high-tech entrepreneurs sometimes are told, and come to believe, that VCs respond to embodied displays of passion. Scientist-entrepreneurs, for example, are advised that they must "move away from the emotionally neutral language of research journals, opting for the language of desire used in marketing and business development circles." You should "use the active voice whenever possible, expressing your plans in strong, visionary language."[132] Coaching manuals advising how to perform an effective pitch remind entrepreneurs that the credibility of an uncertain future resides in the performer's *body* and its use to communicate pertinent emotions: "You are the personification of your business plan—you bring it to life." You should "let your energy be expressed through your hands"; you should use gestures to "communicate excitement"; you should actually "be excited—let your nervousness show as excitement." Remember that "your audience will only be as excited as you are!"[133] There is, as one VC acknowledged, "a large 'show biz' component to a financing presentation," and there are also risks to appearing overexcited: "Trivial perceptions can turn off a potential investor. For example, one investor turned down one of my companies, I later learned, because the presenter's facial flush was taken as proof the man was an alcoholic," when he was in truth suffering from an allergy.[134] And while assessing entrepreneurs' virtues and capacities from their presentational abilities is a radically imperfect gauge, it is nevertheless recognized as essential. As General Doriot said, "We have to judge a man and an idea," but it's the person who stands in front of you who speaks for the idea and the credibility of its future. In Erving Goffman's terms of art, "impressions" might be what's in your business plan, but VCs are also looking for "impressions given off."[135]

This is what VCs, and allied sorts of investors, say they do. It's also possible, in a limited way, to *see* the process in action. At one of the Springboard meetings I attended in August 2000, a clinical psychiatrist at the university medical school presented his plan for a genomics company. His company would identify the genes for a number of clinically recognized conditions—notably bipolar disorder—with a view to producing psychotropic drugs targeted on those genes. The presentation went well, and all the usual amazing visions of the future were sketched: references were made to a genomic "landrush" and analogies were offered between the yet-to-be-incorporated start-up and both Genset and Millennium, the former attracting $100 million in investment from Johnson & Johnson. Then, rather out of the blue, the psychiatrist volunteers to his tough-minded audience a story about what "really" motivates him: "This comes from the heart," he says. "I *see* these folk," meaning the mentally ill. "I see them suffer. That's very important to me." There's an audible murmur of approval; the room comes alive. Encouraged, the psychiatrist reminds the audience that we all know someone who's suffering, someone that his company could help. An angel investor seconds the sentiment: "You're a *psychiatrist*. That's key. That's profound. Be who you are. I found this a very *honest* presentation—which I think is key." Conversely, six weeks later a presentation for an automated optical screening start-up didn't go quite so well. Audience members not-so-gently criticized the entrepreneur for revenue projections that seemed too modest for the claimed technological potential. The CEO-founder was projecting only $3.5 million in sales three to four years into the future and gaining only 10% of the market for such instruments. That's "too long for too little," an angel investor and techie interjected. A VC noted that "nine out of ten times we take the numbers [projections of revenues and market share] down. But here—if this is as compelling as you say it is—if it's 'breakthrough technology'—why are your projections [so modest]" and why do you not project greater enthusiasm? "Don't fall back. Don't be so agreeable. If you really believe [that your automated screening technology] will replace the human eye and skilled pathologist, then *say so*." To which criticism, the entrepreneur lamely responded: "I underpromise"; he wanted to be "more modest" than is the norm in biotech, where great and global things are promised and rarely delivered. But still the investors pushed him to *perform* a more compelling future: "You're not going to attract VCs"; "I don't think you can sell us what *you* don't believe in . . . Be bold and be sincere." At a 2007 Boston Platform meeting, the academic electrical engineer pitching a silicon chip designed to preconcentrate biomolecules was mildly taken to task for not connecting the technology to "real-life examples," such as those involved in cancer treatment: "Get the Wow-factor in!"; "Link this to real patient needs." Performed passion and integrity visibly and audibly index future profit.[136]

If some VCs say they are looking for these virtues in entrepreneurs, some also ascribe the same virtues to themselves, saying that *they* want to change the world, and make it a better place, and that money is only an index of their success in doing so. They're not cold-blooded killers; they're as passionate, committed, and visionary as any entrepreneur. One Southern California VC firm summarizes its investment process as "Passionate Commitment—Rational Decisions."[137] The VC Tommy Davis described his feelings about investing in a potentially world-changing company like Genentech: "We hug each other and dance around. I don't know if we're going to make any money out of this, but if it works we'll have done a wonderful thing . . . [It's] so exciting I can hardly stand it."[138] Of course, there are different portrayals of VCs circulating in and around their world and different stories about how they come to make their judgments. Entrepreneurs—and especially the hordes of the disappointed and the resentful—sometimes say that VC stands for "vulture capitalist," and another name for them is "company nappers."[139] It is said that they take over founders' companies, not because they think that entrepreneurs really need management expertise but because they are control freaks. It is also said — and particularly in biotech — that VCs pigheadedly misunderstand the research process and set short-term "milestones" that distort scientific inquiry and hamper its ability ultimately to deliver the goods.[140] One Internet entrepreneur who got VC backing for his firm, and was then pushed out, put it this way: "Think about this: It's your idea, you write the business plan, you don't sleep for months, you talk all these people into joining you. [Then] all of a sudden, you get fired and they take your fucking stock away. It was like losing your kid."[141]

And then it is sometimes said that VCs' self-burnished image as charismatic and visionary judges of "the new new thing" is largely hype: for the most part, VCs are really just lemmings, playing follow-the-leader. They look around for deals that are attracting interest from star players and then try to hitch their carriage to a train that is already gathering speed. Many investments are syndicated — involving two or more VC firms — and, once a star "lead investor" takes a position, lesser VC firms all try to pile in. Then they take unjustified credit for judgments that were never theirs at all. One Southern California VC I spoke with has no problems whatsoever in acknowledging the importance of syndication and in certain aspects of the follow-the-leader syndrome:

> Absolutely, there's a lot of follow-the-leader, and syndication does have an aspect of having someone help you make the decision. Some of that is very rational . . . I don't know everything. I'm going to get someone else's opinion of it. And if saying, "everyone else thinks it's a bad idea therefore I'm going to say it's a good idea anyway and go forward" — occasionally that's a stroke of brilliance and insight but a lot of times it's just bullheaded stupidity . . . And syndication serves another function besides the decision making, which is the ongoing workload. So I think a lot of firms learn that doing it all by themselves, there's a lot of downside to it and it's not just that you might get [burned] . . . I think what bothers entrepreneurs is when someone [a VC] says I'm this bold, heroic person, when in fact they're not.[142]

Another VC conceded that many of his colleagues tried to mitigate risk by surrendering individual judgment, but maintained that these people were unworthy of their titles, and, indeed, his firm always *insisted* on being the lead investor in any deal. Clearly, the tactics of both VC firms and individual VCs vary enormously, and, equally clearly, entrepreneurs' charge of lemming-like behavior does pick out a substantial feature of that world. Yet, insofar as the charge sticks, the responsibility of making individual judgments about the future is never eliminated; it is just shifted from one part of the investment community to another. And the same applies to the grounds of individual decision making. If not *every* VC manages uncertainty through the resources of familiarity and the search for entrepreneurs' moral makeup, then the VC process itself manifestly does.

The VC world *is* hardheaded; it is often ruthless; and it is governed by bottom-line considerations. But none of these characterizations should be seen in necessary tension with the role of familiarity and assessments of personal virtue in the decision-making process. Consider, for example, the time frame in which VCs operate. From the point of view of much academic science, that time frame may seem unwarrantedly short, but from the point of view of other investment practices it is quite long. As one Bay Area VC put it, "In venture capital, only about 10 percent is making the investment. The other 90 percent is living with it."[143] VC investing is often about building relationships with entrepreneurs that may last up to ten years.[144] When VCs take management positions with the companies in which they invest—if not directly, then by proxy—they may be in personal contact with the management on a regular basis, and that is one reason why it has been an adage about VCs that they would never invest in companies more than a day's travel from their offices.[145] "It's very time consuming," one VC told me: "It's not a passive investment." The relationship between VCs and the entrepreneurs in whom they have invested can take many forms, and it is often both practically and morally fraught, but the instrumental importance of building and managing that relationship is universally acknowledged. The same VC recognized the naturalness of an affective relationship with funded entrepreneurs: "[It is] a long-term relationship . . . and I think [we] do become friends. [We] might go on trips together, might do fishing trips or things like that."[146] Another VC, equally attuned to the affective texture of the VC-entrepreneur association, was conscious of both positive and negative implications. On the plus side, familiarity was a powerful voucher of virtues and capacities that would otherwise call for extensive background checks and the satisfaction of formal criteria: "I just hired a guy at one of my companies and I know a limited amount about his technical skills, but I've known him personally for thirty years. I have no question about his honesty, his character, his open-mindedness, and all of that. You know? And I didn't make a single breakfast call." On the other hand, this *is* fundamentally a business relationship and there are risks to getting too close: "You're friendly but they're not necessarily truly close friends because at the end of the day they are people you work with. And it's hard. I fired two people who either before or during the process became friends, and it's hard. But do you [become friends]? Sure. I've got [friendly] with one of my former CEOs, her daughter, and so when she started up a new company I said, 'I'm not going on the board. I can't do it. We are too close. I cannot be objective about that.'"[147] And, apart from more personal considerations, there are several reasons why the fragile relationship between VCs and entrepreneurs might nevertheless take place on an affective field. First, though VCs do sometimes invest in what might be called "ordinary" businesses—restaurant and clothing chains, casinos, theme parks, and the like—the VC industry has come to be largely defined by the backing of high-tech and biotech enterprises. This means that relations between VCs and entrepreneurs commonly are between "status-equals"—both socially and educationally, though whether a Stanford Ph.D. in computer science or molecular biology outranks a Harvard MBA is moot. Second, VCs and high-tech entrepreneurs often recognize each other's aspirations and prized virtues: both—to use C. P. Snow's phrase again—consider themselves as having "the future in their bones," and, of course, in some instances VCs (like many angel investors) are themselves cashed-out entrepreneurs.[148] They regard themselves as elites driving the world forward, and VCs often derive their sense of self-esteem from the necessity of their role in transforming entrepreneurs' ideas into concrete realities. While VCs may look down their noses at entrepreneurs' lack of business acumen or management skills, their admiration for the scientists and engineers in whom they invest is often acknowledged and articulated. VCs consider investable entrepreneurs as serious people and they like to consider themselves as partners in serious enterprises.

Yet, whether or not the outcome of the VC-entrepreneur relationship is acknowledged as affective, there is more agreement on the significance of affect in deciding whether that relationship is going to happen at all. So VC John Fisher of Draper Fisher Jurvetson went so far as to say, "I don't care how good the concept is or the market is, if I don't like the guy I don't want to do the deal. Life is too short."[149] Expressions of that sentiment are, if not universal, at least common in the private investment world. "I must like the founders," said one angel investor, sharing his accumulated folksy wisdom; dealing with "good people" was his first

rule for success.[150] Most importantly, the resources of personal familiarity are acknowledged as a powerful source of relevant *information*, and,
again, among the most durable types of information that can be had in a
world of such radical uncertainties. Face-time, as it's called, is rich in information about the individual with whom one is dealing—information
about motivations, capacities, and constitutions. It is what's called "high-
bandwidth" interaction: the face-to-face domain offering, as one business
school commentator put it, "unusual capacity for interruption, repair,
feedback and learning," and in predicting the unpredictable, you need
all the help you can get.[151] One Silicon Valley VC, when asked by a *New
York Times* journalist why videoconferencing couldn't take the place of
face-to-face meetings, "scoffed at the suggestion of virtual meetings as
a feasible medium of establishing trust. He said that if the matter were
important—and human beings were involved—he believed that there
would never, ever be a replacement for face-to-face meetings."[152]

VCs *demand* face-time with their entrepreneurial supplicants, partly
to signal who's who in a power relationship, but partly to secure high-
bandwidth information about what's going on. When a Southern California start-up sent only its senior vice president to the East Coast to
meet with Highland Capital Partners, the VC Bob Davis was not best
pleased: "Why isn't [the CEO] here? Planes still fly from Irvine to Boston,
last I heard."[153] As two investors in early-stage companies note, "Angels
are aware that, ultimately, their assessment of a brand-new company
may come down to instincts about the character and competence of
the founder. And the best way to discover this is in face-to-face meetings, formal and informal."[154] A Southern California VC who wanted to
stress to me how much of his job involved "objective criteria" nevertheless acknowledged that "there's also a big component of just how is
the chemistry?"—a "chemistry" that could be realized only face-to-face.
"How do you react when I look at you and say, 'Steve, you look great. Now
probably sometime within the next year or two we're going to ask you to
step aside. How are you going to handle that? Let me ask you that.' And I
sit and I gauge how much you're lying to me and how much you're lying to
yourself? And so that's *not* objective." And so, like many VCs, he tried to
engineer situations of extended face-to-face interaction: "Part of it is, 'let's
go to a ballgame.' At my old firm, we didn't do it too often, but when we
could we'd get someone in a poker game. You'd learn a lot." So too could
the entrepreneur: "At my old firm, we had one partner; his hands would
shake when he was bluffing. You know? You knew this guy couldn't

bluff." Entrepreneurs were well advised to use close social interaction to take the measure of investors: "I know venture capitalists I would not have invest in my company because they're evil people, assholes. I mean truly, they are . . . Some because they're dumb as rocks and some of those dumb as rocks ones, they've got a Harvard MBA, but they're still dumb as rocks."[155] And the best way to find out who was evil, who was an asshole, and who was dumb was to spend some time with them. This is one of many reasons for the emphasis entrepreneurs and investors place on networking in general, but especially on those forms of networking that put relevant people into dense forms of interaction. "If you're sitting around the boardroom, around the table, it's all stiff," one Silicon Valley entrepreneur remarked. But "as soon as you're out mountain biking together, you're bonding in a way you just can't do otherwise." A participant in the intense geek-VC mountain-biking scene observed that "if you're not part of the peloton, you're not part of the deal."[156] "There are a lot of gear-heads out there," another Valley VC said. A typical bike ride in the Santa Cruz hills can take between one and four hours, plenty of time to talk about new investing and technological opportunities, even while denying that business is really happening: "We'd never admit that we're doing it for the networking . . . The people I ride with are basically tech execs, but that's not the main objective," he insists.[157] Bay Area VCs not keen on bicycles have been known to find similar potential for hair-down, personality-revealing interaction playing indoor court games. Brook Byers of Kleiner Perkins said that he only really discovered on the racquetball court that Howard Birndorf—one of the founders of La Jolla's Hybritech—had the entrepreneurial Right Stuff: "It was like a different person . . . It was like Howard unleashed. He was diving for balls, throwing himself against walls . . . He filled the room. That was the real Howard. I finally understood."[158] And, lacking opportunities for truly scenic mountain biking, an East Coast VC-techie scene features the invitation-only Nantucket Conference, a loosened-tie networking and family-fun event (visits to a whaling museum, a micro-brewery, artisanal chocolate and basket-weaving factories, lobster dinners) for VCs and entrepreneurs, where pitches and PowerPoint presentations are banned and an off-the-record policy is enforced.[159]

The whole object of action in this late modern world of future making is to be part of the deal, and being part of the deal means going along for the ride in the peloton. In so much external commentary on this world, there are two things that don't often seem to match up or make mutual sense: the ruthless instrumentalism and the clubbability, the bottom-line criteria of leading-edge capitalism and the moral texture of networking among familiar people. And yet it's been the theme of this chapter that there is no contradiction and not much tension between these two features of the VC-entrepreneur world. People matter; their personal constitutions matter; their virtues matter. And the reason they matter has to do with the radical uncertainty of these future-making practices. You need to know about the virtues of people because there is little else you can rely on that is so durable and so salient. While there is a clear link with the premodern modes of familiarity that some social historians and social theorists assure us is "lost," the reliance on familiarity and the personal virtues is no mere "survival" of premodernity.[160] Such things don't belong just, or even naturally, to the premodern "world we have lost"; they belong equally, or even especially, to the world of making the worlds to come.

FIGURE 17. A reception at the University of California, San Diego, 26 April 2004. The group at the front left includes an angel investor who was formerly CEO of a high-tech company, a university tech transfer officer, and an industrial liaison officer of the engineering school. Further back, a venture capitalist (far left) is talking to the director of the university's tech transfer office (second from left). Also in the picture are the dean of engineering, the chair of the bioengineering department, intellectual property attorneys, a senior executive of a large telecommunications company, a lawyer who helps start-up companies locate capital, and various members of the university's entrepreneurial support organizations. In attendance, but out of frame, are an "entrepreneur-in-residence" at a VC firm who, in her previous life, had been CEO of a communications and software company; the engineering school's corporate associates officer; one of the University of California's regents; the founder of a high-tech human resources company; and many other engineering and scientific professors and scientific officers at local high-tech and biotech companies. Racks of name tags are at the left; the food and drinks are on a table at the right. (Photograph by Melissa Jacobs and reproduced courtesy of the Jacobs School of Engineering, University of California, San Diego.)

The Way We
Live Now

EPILOGUE

We must go about our work and meet "the challenges of the day,"
both in our human relations and our vocation.

Max Weber, *Science as a Vocation*

It is a sunny late April day in San Diego. There is nothing remarkable about that—the city advertises itself as having the world's most perfect climate, and inhabitants profess themselves "stressed" when a cloud disrupts the perfect blue of the sky or when the temperature strays from the range between 70 and 80°F. Throughout the area, there are many groups of people gathered to enjoy this perfect day, and one group of about 200 is occupying a concrete deck overlooking the Pacific Ocean. The deck is attached to a stylish, famous-architect-designed building at the University of California, San Diego (UCSD), and the people have just come outside after listening to speeches celebrating the accomplishments and character of someone they have known for many years and who is about to leave the area for an opportunity on the East Coast (figure 17).

The people present are drawn largely from groups described in the last two chapters. They are scientists, engineers, and research physicians; high-tech and biotech entrepreneurs; CEOs, CSOs, and CTOs of start-up companies; venture capitalists and angel investors; intellectual property lawyers and service providers to the high-tech community; and academic administrators basking in pleasure—both at the perfection of the day, of course, and, especially, at the sight of all of these people assembled on the premises of a major public research university. It is a visible sign that the university is fulfilling one of its major acknowledged functions in a late modern economy, building bridges between knowledge making and wealth making, doing the sorts of things that make political and business leaders happy. It is, however, a collection of people that could be found on practically any day of the week, come together for any number of ostensible purposes, at any number of institutions or physical sites. This is late modern American technoscience collectively embodied, though whether technoscience is now happening is an interesting question. For some purposes, this might as well have been a holiday party. Many of these people make a lot of money, for themselves, for investors, and for the city and region whose tax rolls benefit from the high-paying jobs their activities produce. Billions of dollars of personal net worth are assembled here. Communities around the country, and around the world, compete with each other to get groups like this to live and work in their neighborhoods. These are the people who not only speak in the name of Reality but who transform that knowledge into artifacts, wealth, and power. They are among our culture's most authoritative people, and they are increasingly valued by the structures of civic power.

History weighs lightly on these people, if it does at all. They are late modernity's New Men and Women. It's not just that they usually know little history or that they read fewer history books than others of equivalent income and education. (That's possibly true, but more than anyone securely knows.) While they are avid consumers of anecdotes about past technoscientific and economic successes and failures, and while they clearly believe that much is to be learned from such anecdotes, the stories tend to have a very short shelf life. In San Diego, for example, there is deep folk consciousness about the origins of the local biotech and wireless industries, both of which go back to the 1970s, but, then, many of those involved in that history are still powerful and respected local players, so the anecdotal history is embodied, alive, and currently consequential. And, as was pointed out in chapter 7, in the particular case of the biological and biomedical sciences, players of a certain age—say in their 50s or 60s—often carry with them consciousness of immense historical changes in the relationship between science and commerce that occurred during their own careers. Such scientists quite commonly produce spontaneous histories of what it was to be a biologist then and what it's like now, of what the scientific virtues were and are, of what institutions best support those virtues. But for the most part these people—biomedical scientists as well as electrical engineers—tend to have so little use for history because they see what they're doing as almost wholly unprecedented. Generals and politicians often reckon that the past has lessons for them, because, to a degree, war is war and elections are elections, so the past may be a storehouse of useful examples. But where do you find relevant historical patterns for designing a personal computer, convincing investors that people will want to buy one, and then making a market for one? Where are the relevant historical patterns for monetizing a Web search company, for figuring out what teenagers will want to do with their cell phones and how much they will pay for them, for deciding the organizational form of a company running an online dating service? Return, however, to the scene on the Sun-drenched deck in San Diego: one of the things these people are doing that seems to them without much historical precedent is what they're doing right now.

If you asked them what they were doing, you might get several different answers: many would point to the formal purpose of the occasion—honoring a departing colleague. There's a good feeling about the scene. The colleague was much-liked and the speeches celebrating her work and character have been jokey, heartfelt, and, sometimes, emotional: a praise-poem was collectively composed and declaimed from a podium. While there are promises to stay in touch at a distance of several thousand miles, she will be greatly missed and everyone really knows that bonds of familiarity will inevitably be loosened and may eventually be broken. But others might say they were networking—making, maintaining, and reinforcing a wide range of contacts pertinent to the worlds in which they live and work. Honoring the departing colleague is the ostensible occasion for the scene, but almost any other reason to get these people together would serve a networking purpose. They are forming new social relationships, making sure that the ones they have will remain intact, and reminding relevant others that they are still in play—that they still remember them, are interested in them, want to know what's up with them. If you're not familiar with networking—and many people outside of the business and political worlds are only vaguely aware of it—then it's hard to explain how it's done, and, although it may seem easy to explain *why* it's done, that's not so straightforward either. If you set aside as much as you think you know about the scene in front of you, about what the people are like and what motivates them, and about how they relate to each other, then you might wind up with this sort of thin description of the behaviors on display.

The group does not network as a whole, but in ephemeral subgroups containing about three to six people. It's hard to describe how these subgroups form, but, as people drift into earshot space, they greet those with whom they're already familiar: there's a lot of bonhomie—two-hand (and up-the-arm) handshakes, pats on the shoulder, air kissing (and sometimes actual cheek kissing) between familiar women and between men and women. These are people who seem to be having a lot of fun, enjoying the day, enjoying the occasion, enjoying each other's company. There's a bit of effervescence in the atmosphere. Among people who already know each other, this is a first-name world as a matter of course: your second name is usually provided at first introduction and then rarely, if ever, used again. For someone used to stiff and incestuous academic conferences—at least in the humanities and social sciences—there seems to be an unusually high background level of civility. No one sidling up to a subgroup seems to be ignored, and there's always some sort of acknowledgment of the new presence, giving an opportunity for introductions. Either the person unknown to the group introduces themselves, or, if known to at least one of the members, the newcomer is introduced by the "at home" person: name and association or line of work. Everybody wears a big easy-to-read-at-a-distance name tag, with institutional affiliation, but that's nothing very particular to this sort of group. Introductions are accompanied by business cards. Cards are always at the ready, exchanged and pocketed, with only a casual glance, though their retrieval later, and likely entry into an electronic database or Rolodex file, is a big part of the story.

At this occasion, and at many other similar ones, waiters circulate with sparkling water, fruit juice, sodas, and wine, and most people either stand around with a glass in their hand or find some place on which to rest it while they talk. There is always food, and, if that is not also brought around by circulating waiters, then it's arrayed at a buffet somewhere towards the margins of the scene. The food plays a crucial role—and not just because it's usually pretty good Pacific Rim fusion stuff (bacon-wrapped scallops, grilled shrimp marinated in nuoc mam and lemongrass, satay kebabs), but also because it's a major way in which subgroups fragment and reconstitute themselves. You don't just leave a group without a word, and you don't preface your departure by saying "Look, there's Ellen; I'll see you later." A common way of separating yourself from one group, and joining another, is by way of a detour to the buffet: "I'm starving; I'm going to grab some of those scallops." It's little more than a guess, but the subgroups seem to stay more or less stable in their membership for no more than fifteen minutes, probably less. It's a semi-fluid social environment: during the hour or so the scene on the deck lasts, most individuals probably have extended face-to-face interaction with three or four groups and have some sort of interaction with most of the people present. There do not seem to be any singletons—disconsolately lurking at the margins—nor do dyads appear, except fleetingly. People accustomed to this sort of thing tell me that such occasions, and forms of interaction, are routine parts of their lives, occurring maybe twice a week. This world has got a more intense face-to-face dimension than that inhabited by the aca-demics who may write about it. Networking, of course, happens in more stable social forms. You can accept membership on various boards—local government, nonprofits, industry consortia—with the expectation that denser forms of networking will be part of the service and benefits. There is a lot of "volunteering" in this world; a lot of work done for free; a lot of professed concern with community welfare. (Some people say that this is more characteristic of the area, or of the West Coast, while others deny that there's any substantial regional difference in high-tech networking and forms of interaction.) Over the course of membership on such boards, lasting maybe several years, with meetings maybe monthly, one will prob-ably form ties with a small number of people also serving, and these ties will likely prove to be durable. Again, when asked why they accept mem-bership, sometimes people respond by saying that they're "interested" or that they "want to help out," but sometimes here too they say these are valuable opportunities for networking. There seems to be no special pat-tern to what account you get.

Just as there are several different ways in which participants give ac-counts of what they're doing, so there are several different repertoires in which the onlooker (participant-observer, friend of some of those present) can talk about such scenes, several different registers in which motivations can be ascribed and the behaviors made out as actions, several different ways in which such scenes belong to the currents of deep and recent history described in this book. Here is one: this is a scene, a form of social interaction, and an assemblage of persons and roles that never existed before in the history of the world. These people are doing what they're doing because the world in which they live and work is beset by the most radical uncertainties—uncertainties brought about partly by their own work (new forms of communication, new ways of being human) and partly by the conditions in which they can effectively do that work (how to fund, manage, and organize activities that have few or no rele-vant institutional precedents). They're in the business of technoscientific and economic future making—trying to discover drugs that will cure or alleviate cancer or wireless technologies that may become world stan-dards; starting up companies and managing their early growth; deciding which of them to join, back, or invest in; figuring out how structures of civic governance and institutions of higher education can best encour-age, adapt to, or, much more rarely, resist the futures that are being made by people like those assembled here. The marked openness of the forms of interaction, and the notable and diffuse civility characteristic of the scene, is a direct consequence of these radical uncertainties. Who knows which person, or even what *kind* of person, may or may not prove useful some years or months down the road? This is not like a conference of academic historians, say, where you understand with very great reliability who's likely to know some archives relevant to your work, to review your book, to have students to whom you might give a postdoc, to be on a panel of a grant-giving agency, to consider you for a job at a better university. If, however, you're trying to start up a digital optical-screening company, or to find a good investment in a drug-discovery company, it's a lot harder to circumscribe the set of people who may or may not eventually prove pertinent to the success of your enterprise. This has all the appearance of a meritocratic scene because everybody's money is equally good, and because there's no very secure way of telling from visible signs or stable background knowledge whose ideas, technology, or intellectual property may turn out to be valuable. Business is business, and whatever any individual may think about the virtues and vices of men, women, Jews, Muslims, Indians, Iranians, African-Americans, homosexuals, or merlot drinkers is not considered to have anything to do with the business at hand. (No need to idealize or romanticize the extent of this apparent egalitarianism: you can be cynical about its social realities and bases if you want, though it's not easy to be cynical about members' psychological dispositions; you can talk about it in instrumental terms if you prefer, even if academic humanists and social scientists tend to see workplace equity largely as a matter of principles and ideals. Or—less fashionably among external observers—you can take what it seems to be for what it is.) Well-networked people tend to do better than poorly-networked people, but the networks concerned don't seem to have much to do with traditional "old boy networks" based on religion, school, and parental wealth. The open-endedness and the evident civility of these scenes are the upshot of the radically uncertain businesses the participants are in. Best to keep everyone—or as many of them as possible—in play; best to make sure that as large a group as possible knows who you are and what you're up to; best to know who's who and what's what in that indefinitely large group; best to leave a good impression on as many members as possible. Information is currency, but it's hard to get a grip on it, running through your fingers like water.

That sort of story about networking is almost wholly instrumental: it accepts the evident civility of networking scenes, while offering a causal explanation of the phenomenon with reference to the rational maximization of possible material returns, to an econometric model of human behavior. It is an account emerging out of an Idea of Modernity, even Late Modernity. It makes out the people concerned to be hardheaded maximizers, and, if you offered an account like this to a social science journal, it would show that you too are head-headed. It is not solely an outsider's account: you can, indeed, get some participants to recognize it, and there are, of course, many books on the market that tell you how to network and that justify networking in roughly these terms, but, on the whole, it's not a story that emerges spontaneously from scenes like this. So here is another story about the same sort of scene: it is not so hardheaded, and it might have some difficulty getting published in the academic social science journals. This story starts by noting how much of the action belonging to this scene could count as a description of a premodern order. Instead of interpreting this scene as a distinctive instantiation of the Idea of Modernity, it recognizes much about it that belongs also to the alleged premodern World We Have Lost. Almost all the two hundred people present, and many more who could not be there but sent regrets, know each other by face, name, line of work, and institutional affiliation. The word "community" is much used in self-description, as in "the San Diego high-tech community," and members spontaneously and repeatedly remark on the fact that "everyone knows everybody else." They know a lot about each other, and not just who they work for and how much they are worth. Many seem also to know the names and affiliation of spouses, partners, and kids. As well as these networking events, they do a lot of lunch together and, for some tastes, too many frighteningly early breakfast meetings and evening receptions (with bacon-wrapped scallops and pinot grigio: not quite dinner and not quite not dinner). The texture of personal familiarity, the patterns of personal introductions and references, the awareness of the importance of friendship and the work devoted to maintaining and extending friendship circles, the exchange of cards, the observance of courtesies, the free rendering of services and the offering of thanks for those freely rendered services, and the informal awareness of exchange equity—all of these are capable of bearing an econometric account but, if taken at face value, look very like central features of the World We Have Lost.

Then there is the importance of personal reputation. Stories about people and their personal characteristics, their virtues and vices, travel around the community with remarkable speed and efficiency. This late modern community has many village-like characteristics: if, for example, you behave badly—misrepresent the status of your intellectual property to an angel investor, screw the entrepreneurs you invest in more than the industry norm, don't meet your milestones, or even if you acquire a reputation for workplace abuse (sexual or otherwise)—these things get around. If you're an entrepreneur, and you get assistance, and then you don't take your turn assisting other entrepreneurs, then that gets around too. It's not good for you to acquire a bad reputation. You can, if you like, gloss reputation in econometric instrumental terms, or you can take it at face value, just as it usually crops up in the local vernacular. The "how to network" book genre has its parallel in the late Renaissance and early modern "courtesy" (or "How to be a Gentleman") literature, but it's unlikely that many people at this California scene have actually read any such "how to" book: whatever they know how to do, they seem to have learned through direct modeling—seeing how people normally go on in this community, seeing what's approved and what isn't, then deciding whether, and on what terms, they want to be part of the community. Few, if any, of these people think of themselves, or each other, as moral paragons, though they seem to have focal awareness of the virtues of familiar people and their salience to patterns of practical life. Moreover, there is, in my experience, much more professed altruism attending their activities than is the norm in the average university history or social science department. It's not all that rare to hear people spontaneously say that they're trying to "make the world a better place" and that they're committed to wiping out some dread disease. How you interpret such professions is an issue, but it's not so easy to discern cynicism or to respond cynically when altruistic professions happen in front of you.

So we have two quite different accounts of what's going on here: who these people are, what their community is like, what makes them tick. The one is biased towards the instrumental and functional, the other to the phenomenological; the one ironical (things are not what they *seem*), the other concerned about *seeming* as part of the interaction order. But both accept as a matter of fact that crucial aspects of late modernity, and especially those activities that count as its leading edge, conduct their affairs importantly through resources of familiarity and through identifying the virtues and vices of familiar people. What these people do, they do on a moral field. This book has intermittently used both sorts of account to highlight the significance of the personal. I have not hesitated to refer the significance of personal familiarity and personal virtue to radical normative uncertainty obtaining in late modern technoscience, and, to that extent, I have offered an instrumental interpretation. But I do not see that such an interpretation explains anything about this world *better*, or that it is more *legitimate*, than an account that takes matters more at face value. Each of them seems to cover the same domain, and it is of no more than local academic interest—especially in the social sciences—that the first is preferred to the second. That is just a twist on an old fault line in the human sciences: objectivism versus subjectivism, legislative versus interpretative reason, "scientific/explanatory" versus "hermeneutic" models of what the social scientist can or should say about human action and culture. If the instrumental interpretation seems more hardheaded, and if the "face value" approach seems to romanticize late modern technoscience, that is little more than an artifact of the conventions of academic social science, of the state they've got themselves in; and chapters 4 through 6 gave examples of some tensions between official social scientific models of human behavior and those operative in the institutions themselves. I started out by saying that one of my major aims was to describe this world; I have no interest in either celebrating or condemning it. To say that some commentary on the late modern order has *got it badly wrong* is not the same thing as rendering moral or political judgment on people and institutions. One can criticize many aspects of late modern technoscience without being obliged to misrepresent what it's like to live and work in technoscientific worlds. And what it's like is something like a vocation.

The meeting on the California deck is about over. There's a slight chill in the air as the Sun sets and the marine cloud layer moves in. The group is dispersing and members of the community are going their separate ways. Tomorrow, the Sun will rise in the East, a fact that everyone here knows with great certainty, even though some may have heard that philosophers believe this incapable of proof. Tomorrow, these men and women will go back to their labs and offices: some of them will try to get evidence about whether certain chemical compounds obtained from marine organisms have pharmacological interest; some will be conducting due diligence on a possible investment opportunity; others will be deciding whether corporate resources should best be directed towards research on the possible genetic bases of bipolar disorder or of obesity. The outcomes of these efforts are radically uncertain: they are at the opposite epistemic pole to knowledge of the Sun rising tomorrow in the East. In between are the more or less durable characteristics, and the more or less stable virtues, of the people who speak on behalf of nature, technology, and the future.

Notes

PREFACE

1. Steven Shapin and Simon Schaffer, *Leviathan and the Air-Pump: Hobbes, Boyle, and the Experimental Life* (Princeton, NJ: Princeton University Press, 1985); Steven Shapin, *A Social History of Truth: Civility and Science in Seventeenth-Century England* (Chicago: University of Chicago Press, 1994). There was also a bit of that sort of thing towards the end of Shapin, *The Scientific Revolution* (Chicago: University of Chicago Press, 1996), and in my contributions to Christopher Lawrence and Steven Shapin, eds., *Science Incarnate: Historical Embodiments of Natural Knowledge* (Chicago: University of Chicago Press, 1998), introduction and ch. 1.

2. The usual gestures here are to Hans-Georg Gadamer, *Truth and Method*, 2nd rev. ed., trans. Joel Weinsheimer and Donald G. Marshall (New York: Crossroad, 1989) and his notion of the "fusion of horizons," but much is to said for the blunt Anglo-Saxon idiom of E. H. Carr, *What Is History?* (New York: Alfred A. Knopf, 1962).

3. Michael Oakeshott, "History and the Social Sciences," in Institute of Sociology, *The Social Sciences: Their Relations in Theory and in Teaching* (London: Le Play House Press, 1936), pp. 71–81, on pp. 74–75.

4. It cannot be said that academic usage is notably stable in such matters: Thorstein Veblen, for example, located the commencement of the "late modern" in the mid-eighteenth-century: "The Place of Science in Modern Civilization," *American Journal of Sociology* 11 (1906), 585–609, on p. 596.

5. Zygmunt Bauman, *Liquid Modernity* (Cambridge: Polity Press, 2000); idem, *Postmodern Ethics* (Oxford: Blackwell, 1993); idem, *Identity* (Oxford: Blackwell, 2004); Brian Heaphy and Jane Franklin, *Late Modernity and Social Change* (London: Routledge, 2004); Ulrich Beck, Anthony Giddens, and Scott Lash, *Reflexive Modernization: Politics, Tradition, and Aesthetics in the Modern Social Order* (Cambridge: Polity Press, 1994); Anthony Giddens, *Modernity and Self-Identity: Self and Society in the Late Modern Age* (Cambridge:

Polity Press, 1991); also John Urry, *Global Complexity* (Cambridge: Polity Press, 2002); John Fornas, *Cultural Theory and Late Modernity* (London: Sage, 1995).

6. Quoted in Robin Dougherty, "Between the Lines with Steven Pinker," *Boston Globe*, 26 December 2004, p. D9.

CHAPTER ONE

1. Bruno Latour, *Science in Action: How to Follow Scientists and Engineers through Society* (Cambridge, MA: Harvard University Press, 1987), p. 174.

2. The gesture is to Peter Laslett, *The World We Have Lost: Further Explored*, 3rd ed. (London: Routledge, 1983; orig. publ. 1965); idem, "The Face to Face Society," in idem, ed., *Philosophy, Politics and Society* (Oxford: Basil Blackwell, 1963), pp. 157–184; also Niklas Luhmann, *Trust and Power: Two Works*, trans. Howard Davis, John Raffan, and Kathryn Rooney; eds. Tom Burns and Gianfranco Poggi (Chichester: John Wiley, 1979); Anthony Giddens, *The Consequences of Modernity* (Stanford, CA: Stanford University Press, 1989), pp. 21–27, 79–85.

3. Max Weber, *The Protestant Ethic and the Spirit of Capitalism*, trans. Talcott Parsons (New York: Charles Scribner's, 1958; orig. publ. 1905), pp. 21–22; see also Zygmunt Bauman, *Postmodern Ethics* (Oxford: Basil Blackwell, 1993), p. 5.

4. Max Weber, *Economy and Society: An Outline of Interpretive Sociology*, 2 vols., eds. Gunther Roth and Claus Wittich, trans. Ephraim Fischoff et al. (Berkeley: University of California Press, 1978), Vol. II, pp. 1112, 1115.

5. Max Weber, "The Meaning of Discipline," in *From Max Weber: Essays in Sociology*, eds. H. H. Gerth and C. Wright Mills (London: Routledge, 1991), pp. 253–264, on p. 253.

6. Steven Shapin, *A Social History of Truth: Civility and Science in Seventeenth-Century England* (Chicago: University of Chicago Press, 1994), esp. chs. 1, 6; idem, "Cordelia's Love: Credibility and the Social Studies of Science," *Perspectives on Science* 3 (1995), 255–275.

7. Paul Rabinow, "American Moderns: On Science and Scientists," in idem, *Essays on the Anthropology of Reason* (Princeton, NJ: Princeton University Press, 1996), pp. 162–188, on p. 184 ("de-magification" is Rabinow's rendering of Weber's *Entzauberung*, commonly rendered as "disenchantment"); see also Rabinow, *French DNA: Trouble in Purgatory* (Chicago: University of Chicago Press, 1999), p. 11; Weber, "Science as a Vocation [1918]," in *From Max Weber*, pp. 129–156, on p. 149; Charles Thorpe, "Violence and the Scientific Vocation," *Theory, Culture & Society* 21 (2004), 59–84, esp. p. 63.

8. Bauman, *Postmodern Ethics*, p. 34; also Paul Rabinow and Talia Dan-Cohen, *A Machine to Make a Future: Biotech Chronicles* (Princeton, NJ: Princeton University Press, 2004), pp. 99–100.

9. Stephen Turner, "Charisma Reconsidered," *Journal of Classical Sociology* 3 (2003), 5–26, on pp. 23–24.

10. Paul Rabinow, *Making PCR: A Story of Biotechnology* (Chicago: University of Chicago Press, 1996), p. 17; see also idem, "Science as a Vocation: Truth versus Meaning,"

in idem, *Anthropos Today: Reflections on Modern Equipment* (Princeton, NJ: Princeton University Press, 2003), pp. 96–101; idem and Dan-Cohen, *A Machine to Make a Future: Biotech Chronicles*, esp. ch. 4. A more circumscribed engagement with Weber's essay in relation to present-day organizational realities is Edward J. Hackett, "Science as a Vocation in the 1990s: The Changing Organizational Culture of Academic Science," in *Degrees of Compromise: Industrial Interests and Academic Values*, eds. Jennifer Croissant and Sal Restivo (Albany: State University of New York Press, 2001), pp. 101–137. And, here and elsewhere, I owe much to the work of my former student Charles Thorpe, especially his treatment of charisma in *Oppenheimer: The Tragic Intellect* (Chicago: University of Chicago Press, 2006), ch. 1.

11. Rabinow, *French DNA*, p. 20.

12. Thorstein Veblen, *The Higher Learning in America: A Memorandum on the Conduct of Universities by Business Men* (New York: Sagamore Press, 1957; orig. publ. 1918), p. 5; see also pp. 55–56. On the origins of notions of "the personal equation" in natural scientific practice, and, especially, in astronomy, see Simon Schaffer, "Astronomers Mark Time: Discipline and the Personal Equation," *Science in Context* 2 (1988), 115–145.

13. Claude Bernard, *Introduction to the Study of Experimental Medicine*, trans. Henry Copley Greene (New York: Dover, 1957; orig. publ. 1865), pp. 40, 42–43. (On "Art is I . . .": Bernard was here alluding to a saying by Victor Hugo.) And see Lorraine J. Daston, "Objectivity and the Escape from Perspective," *Social Studies of Science* 22 (1992), 597–618, on pp. 609, 613; idem, "Fear & Loathing of the Imagination in Science," *Dædalus* 134, no. 4 (Fall 2005), 16–30, on p. 23.

14. Ernest Renan, *L'avenir de la science* (Paris: Calmann-Levy, 1890), p. 228, quoted in Daston, "Objectivity and the Escape from Perspective," p. 609.

15. Immanuel Kant, *The Critique of Judgment*, trans. James Creed Meredith (Oxford: Oxford University Press, 1952; orig. publ. 1790), pp. 168–170; see also Daston, "Fear & Loathing," p. 22.

16. Ralph Waldo Emerson, *Representative Men: Seven Lectures* (Boston: Houghton, Mifflin, 1903; orig. publ. 1849), p. 43.

17. Roland Barthes, "The Brain of Einstein," in idem, *Mythologies*, trans. Annette Lavers (New York: Noonday Press, 1975; orig. publ. 1957), pp. 68–70, on p. 69; also Michael Hagner, *Geniale Gehirne: Zur Geschichte der Elitegehirnforschung* (Munich: Deutscher Taschenbuch Verlag, 2007; orig. publ. 2004), pp. 293–302.

18. John Updike, "One Cheer for Literary Biography," *New York Review of Books* 46, no. 2 (4 February 1999) [accessed 18 February 2006 at http://www.nybooks.com/articles/article-preview?article_id=607]. For T. S. Eliot on a depersonalized view of the imaginative artist, see Eliot, "Tradition and the Individual Talent," in idem, *The Sacred Wood: Essays on Poetry and Criticism* (London: Faber and Faber, 1967; orig. publ. 1920), pp. 39–49, on pp. 44–45.

19. Jonathan Derbyshire, "Life Behind the Mind of a Philosopher," *Financial Times*, 11/12 February 2006, p. W4.

20. Søren Kierkegaard, *The Point of View for My Work as an Author*, trans. Walter Lowrie, ed. Benjamin Nelson (New York: Harper Torchbooks, 1962; orig. publ. 1859),

p. 44. This was, for Kierkegaard, a practical exercise in constituting himself as a *religious* author: "It was important for me to alter my personal mode of existence to correspond with the fact that I was making the transition to the statement of religious problems. I must have an existence-form corresponding with this sort of authorship" (ibid., p. 56). Ludwig Wittgenstein, who much admired Kierkegaard for just this reason, believed that there was no use "in studying philosophy if all it does for you is to enable you to talk with some plausibility about some abstruse questions of logic, etc., & if it does not improve your thinking about the important questions of everyday life": letter from Wittgenstein to Norman Malcolm, 12 November 1944, quoted in Malcolm, *Ludwig Wittgenstein: A Memoir* (Oxford: Oxford University Press, 1958), p. 39.

21. Henry Brougham, *Lives of the Men of Letters and Science, Who Flourished in the Time of George III*, 2 vols. (London: Charles Knight, 1845–1846), Vol. I, p. 510. (The context is a sketch of the life of the Scottish geometrician Robert Simson.)

22. Thomas Henry Huxley, "On the Method of Zadig [1880]," in idem, *Collected Essays, Vol. IV: Science and Hebrew Tradition* (New York: D. Appleton, 1900), pp. 1–23, on p. 2. Huxley was gesturing at Voltaire's 1747 "orientalist" philosophical tale *Zadig*, about a virtuoso of inference. On disembodiment as a theme in presentations of scientific and intellectual selves, see Steven Shapin, "The Philosopher and the Chicken: On the Dietetics of Disembodied Knowledge," in Lawrence and Shapin, *Science Incarnate*, pp. 21–50; Hélène Mialet, "Do Angels Have Bodies? Two Stories about Subjectivity in Science: The Cases of William X and Mister H," *Social Studies of Science* 29 (1999), 551–581; idem, "Reading Hawking's Presence: An Interview with a Self-Effacing Man," *Critical Inquiry* 29 (2003), 571–598.

23. John R. Baker, *The Scientific Life* (New York: Macmillan, 1943), pp. 36–37.

24. Derek J. de Solla Price, *Little Science, Big Science* (New York: Columbia University Press, 1963), p. 69.

25. André Gide, *Oscar Wilde: In Memoriam* (New York: Philosophical Library, 1949), p. 16.

26. A. N. Whitehead, *Science and the Modern World* (London: Scientific Book Club, 1946; orig. publ. 1926), p. 120.

27. David Hume, *Treatise of Human Nature* (Oxford: Oxford University Press, 1978; orig. publ. 1740), p. 469:

> In every system of morality, which I have hitherto met with, I have always remark'd, that the author proceeds for some time in the ordinary ways of reasoning, and establishes the being of a God, or makes observations concerning human affairs; when of a sudden I am surpriz'd to find, that instead of the usual copulations of propositions, is, and is not, I meet with no proposition that is not connected with an ought, or an ought not. This change is imperceptible; but is however, of the last consequence. For as this ought, or ought not, expresses some new relation or affirmation, 'tis necessary that it shou'd be observ'd and explain'd; and at the same time that a reason should be given; for what seems

altogether inconceivable, how this new relation can be a deduction from others, which are entirely different from it.

Also Henry Sidgwick, *Lectures on the Ethics of T. H. Green, Mr. Herbert Spencer and J. Martinear* (London: Macmillan, 1901), p. 145; and G. E. Moore, *Principia Ethica* (Cambridge: Cambridge University Press, 1903), p. 58. And see fine cultural historical treatment by Robert N. Proctor, *Value-Free Science? Purity and Power in Modern Knowledge* (Cambridge, MA: Harvard University Press, 1991), pp. 39–62, 134–136, 204–205.

28. Henri Poincaré, *The Value of Science* [1905], in idem, *The Value of Science: Essential Writings of Henri Poincaré* (New York: Modern Library, 2001), pp. 179–353, on p. 190. Poincaré's claim was just that the search for the two kinds of truth elicited the same sort of emotions and their outcomes generated the same sort of anxieties, not that it was possible to move logically from the one to the other.

29. Ludwig Wittgenstein, "Lecture on Ethics," quoted in Ray Monk, *Ludwig Wittgenstein: The Duty of Genius* (Harmondsworth: Penguin, 1990), p. 277.

30. Veblen, "The Place of Science in Modern Civilization," p. 600.

31. Weber, "Science as a Vocation [1918]," p. 142; see also Fritz Ringer, *Max Weber: An Intellectual Biography* (Chicago: University of Chicago Press, 2004), pp. 23–243 (for the contexts of these remarks).

32. Leo Tolstoy, "Modern Science [1898]," in idem, *Recollections & Essays*, trans. Aylmer Maude (London: Oxford University Press, 1937), pp. 176–187, on p. 186. Tolstoy implicitly described the historical process by which what Weber came to call *Zweckrationalität* had split from *Wertrationalität*:

> People must live, but in order to live they must know how to live. And men have always obtained this knowledge—well or ill—and in conformity with it have lived and progressed. And this knowledge of how men should live has—from the days of Moses, Solon, and Confucius—always been considered a science, the very essence of science. Only in our time has it come to be considered that the science telling us how to live is not a science at all, but that the only real science is experimental science—commencing with mathematics and ending in sociology. And a strange misunderstanding results.

Ibid., p. 178.

33. Note, however, that within social science Emile Durkheim described "the antithesis between science and ethics" as "that formidable argument with which the mystics of all times have wished to cloud human reason." He argued that a new "science of ethics" would not only "teach us to respect moral reality" but also give us "the means to improve it": Emile Durkheim, *Selected Writings*, ed. and trans. Anthony Giddens (Cambridge: Cambridge University Press, 1972), p. 121.

34. Albert Einstein, *Out of My Later Years* (New York: Philosophical Library, 1950), p. 114.

35. Ibid., pp. 21–22 (emphases in original). Signs that this formulation was becoming a commonplace among scientists by the 1930s are in William H. George, *The Scientist*

in Action: A Scientific Study of His Methods (London: Williams & Norgate, 1936), p. 64: "Whenever a scientist is making a statement of what is, for example, moral or immoral, he is not speaking as a scientist."

36. Einstein, "Physics and Reality [1936]," in idem, *Ideas and Opinions* (New York: Modern Library, 1994), pp. 318–356, on p. 318.

37. Einstein, *Out of My Later Years*, p. 114.

38. Harvard University, Committee on the Objectives of a General Education in a Free Society, *General Education in a Free Society, Report of the Harvard Committee* (Cambridge, MA: Harvard University Press, 1945), p. 59.

39. F. S. C. Northrop, "The Physical Sciences, Philosophy, and Human Values," in *Physical Science and Human Values*, ed. E. P. Wigner (Princeton, NJ: Princeton University Press, 1947), pp. 98–113, on p. 107; comments on Northrup by Henry Margenau in ibid., p. 115. For Margenau's disputes with Northrop, see also Henry Margenau, *Ethics & Science* (Princeton, NJ: D. Van Nostrand, 1964), pp. 14–30, 181–198, and, for the naturalistic fallacy, see ibid., pp. 98–137.

40. Karl K. Darrow, comment on Northrop, in Wigner, ed., *Physical Science and Human Values*, pp. 116–117.

41. Richard P. Feynman, *The Meaning of It All: Thoughts of a Citizen-Scientist* (Reading, MA: Perseus, 1998), pp. 16–17, 44. These were the John Danz Lectures given at the University of Washington in 1963.

42. Richard P. Feynman, *"Surely You're Joking, Mr. Feynman!": Adventures of a Curious Character* (New York: Bantam Books, 1985), p. 313; see also Steven Shapin, "Milk and Lemon," *London Review of Books* 27, no. 13 (7 July 2005), pp. 10–13.

43. The term is not meant to pick out a decline of morale but a reflective distinction between the possession of technical knowledge and personal virtue.

44. Weber, *Protestant Ethic*, p. 182.

45. Robert C. Dynes, "State of the Campus Address, March 22, 2001": http://orpheus.ucsd.edu/chancellor/state.html [accessed 10 July 2007]. Dynes was president of the University of California system from October 2003 until August 2007, when he resigned amidst scandals over undisclosed lucrative compensation packages for top administrators, reportedly saying that he would spend the lame duck portion of his presidency "advanc[ing] the University's research, development, and delivery portfolio in partnership with industry": http://www.latimes.com/news/local/la-me-dynes14aug14,1,3440034.story?track=rss; http://www.universityofcalifornia.edu/dynes/pressrelease.html [both accessed 21 August 2007].

46. Steven Shapin, "Ivory Trade," *London Review of Books*, 25, no. 17 (11 September 2003), pp. 15–19; cf. Derek Bok, *Universities in the Marketplace: The Commercialization of Higher Education* (Princeton, NJ: Princeton University Press, 2003); Masao Miyoshi, "Ivory Tower in Escrow," *boundary 2* 27 (2000), 7–50, esp. pp. 26–28. Diversity of role, function, and value in the research university was one of the lessons of the most prescient existing account of the institution, Clark Kerr's *The Uses of the University* (Cambridge, MA: Harvard University Press, 1963).

47. Here I owe much to the work of my former student Mark Peter Jones on biotech entrepreneurs, particularly his notion of ephemeral, contingently emerging, and institutionally variable "free spaces" of inquiry: Mark Peter Jones, "Biotech's Perfect Climate: The Hybritech Story," unpubl. Ph.D. thesis, University of California, San Diego, 2005.

CHAPTER TWO

1. Robert K. Merton, "The Normative Structure of Science," in idem, *The Sociology of Science: Theoretical and Empirical Investigations*, ed. Norman W. Storer (Chicago: University of Chicago Press, 1973; art. orig. publ. 1942), pp. 267–278, on pp. 275–276; see also idem, "Science and the Social Order" (1938), in ibid., pp. 254–266, on p. 259 (for a preliminary gesture at this idea). Merton returned to this stipulation time and again: see idem, "Priorities in Scientific Discovery" (1957), in ibid., pp. 286–324, on pp. 290–291; idem, *Sociological Ambivalence and Other Essays* (New York: Free Press, 1976), pp. 34–35; also Walter Hirsch, *Scientists in American Society* (New York: Random House, 1968), p. 91.

2. Merton had just joined the faculty at Columbia when the quoted article appeared. In 1939, there were fewer than a thousand sociologists in the U.S., but by 1957 there were 4,500: figures from Michael T. Kaufman's obituary, "Robert K. Merton, Versatile Sociologist and Father of the Focus Group, Dies at 92," *New York Times*, 24 February 2003, p. B7.

3. Robert K. Merton, *Social Theory and Social Structure*, rev. ed. (New York: Free Press, 1957), p. 532. (For Durkheim on social versus psychological facts, see Emile Durkheim, *Selected Writings*, ed. and trans. Anthony Giddens [Cambridge: Cambridge University Press, 1972], pp. 70–75.) Yet here and elsewhere (idem, "Forward," to Bernard Barber, *Science and the Social Order* [Glencoe, IL: Free Press, 1952], pp. 7–20, on p. 14), Merton warned that tracing the "connections between science and society" was *not* "to impugn the motives of scientists," which, so far as the sociologist was concerned, could be as high-minded as heroic legend stipulated. Barber himself (*Science and the Social Order*, pp. 59–60) followed Merton in distinguishing between "social factors"—as proper sociological explananda—and "the personal motives of individual working scientists."

4. Merton neither invented this tactic nor was his version its only Harvard articulation: see, notably, Talcott Parsons, "The Professions and Social Structure," *Social Forces* 17 (1939), 457–469.

5. Michel Foucault, "Truth and Power," in idem, *Power/Knowledge: Selected Interviews and Other Writings 1972–1977*, ed. Colin Gordon, trans. Gordon, Leo Marshall, John Mepham, and Kate Soper (New York: Pantheon, 1980), pp. 109–133, esp. pp. 126–129; see also Charles R. Thorpe, "J. Robert Oppenheimer and the Transformation of the Scientific Vocation," unpubl. Ph.D. thesis, University of California, San Diego, 2001, ch. 1; idem, "Violence and the Scientific Vocation," *Theory, Culture & Society* 21 (2004), 59–84, esp. pp. 60, 64–67; Paul Rabinow, *French Modern: Norms and Forms of the Social Environment* (Cambridge, MA: MIT Press, 1989), pp. 16, 251.

6. Karl R. Popper, "The Sociology of Knowledge," in idem, *The Open Society and Its Enemies, Vol. II: The High Tide of Prophecy: Hegel, Marx, and the Aftermath* (Princeton, NJ: Princeton University Press, 1966; orig. publ. 1945), pp. 212–223, on pp. 217–220. *The Open Society* was written in New Zealand between 1938 and 1943–that is, contemporaneously with Merton's essay.

7. Historians well understand that the word "scientist" was made up by the English mathematician and philosopher William Whewell in 1840, and that before then standard English designations included "natural philosopher" and "man of science," while other practitioners now swept up under the designation "scientist" included the "chemist," the "astronomer," the "natural historian," and, importantly, the "mathematician," etc. But few fully appreciate how long the usage of "scientist" was resisted, both within the communities of natural knowledge and elsewhere in the culture. "Man of science" or "scientific worker" were common usages, in texts where the word "scientist" never appeared, well into the twentieth century. In the mid-1880s, the president of the American Association for the Advancement of Science acknowledged that the term "scientist" was, in common usage, interchangeable with the "man of science" or "scientific man," but insisted that proper usage marked crucial differences:

> The word "scientist" is a coinage of the newspaper reporter, and, as ordinarily used, is very comprehensive. Webster defines a scientist as being "one learned in science, a savant,"—that is, a wise man,—and the word is often used in this sense. But the suggestion which the word conveys to my mind is rather that of one whom the public suppose to be a wise man, whether he is so or not, of one who claims to be scientific.

John S. Billings, "Scientific Men and Their Duties," *Science* 8, no. 201 (10 December 1886), 541–551, on p. 541. As late as 1920, the English classicist and professor of literature George Saintsbury recorded his distaste for the neologism, referring to "real men of science (who cannot be too carefully distinguished from 'scientists')": *Notes on a Cellar-Book* (London: Macmillan, 1963; orig. publ. 1920), p. 204n.

8. Steven Shapin, "The Image of the Man of Science," in *The Cambridge History of Science, Vol. 4: Eighteenth-Century Science*, ed. Roy Porter (Cambridge: Cambridge University Press, 2003), pp. 159–183; idem, "The Man of Science," in *The Cambridge History of Science, Vol. 3: Early Modern Science*, eds. Lorraine Daston and Katharine Park (Cambridge: Cambridge University Press, 2006), pp. 179–191. There is a large literature on the natural theological tradition and its moral uses: see, e.g., John Hedley Brooke, *Science and Religion: Some Historical Perspectives* (Cambridge: Cambridge University Press, 1991), esp. chs. 4–6; and, for aspects of the British case, see John Gascoigne, "From Bentley to the Victorians: The Rise and Fall of British Newtonian Natural Theology," *Science in Context* 2 (1988), 219–256.

9. For seventeenth-century references to the natural philosopher as nature's priest, see Harold Fisch, "The Scientist as Priest: A Note on Robert Boyle's Natural Theology," *Isis* 44 (1953), 252–265; Simon Schaffer, "Godly Men and Mechanical Philosophers: Souls and Spirits in Restoration Natural Philosophy," *Science in Context* 1 (1987), 55–85;

and Steven Shapin, *A Social History of Truth: Civility and Science in Seventeenth-Century England* (Chicago: University of Chicago Press, 1994), ch. 4.

10. Joseph Priestley, *The History and Present State of Electricity*, 2 vols., 3rd ed. (London: C. Bathurst and T. Lowndes, 1775), Vol. I, p. xxiii.

11. John F. W. Herschel, *A Preliminary Discourse on the Study of Natural Philosophy* (Chicago: University of Chicago Press, 1987; orig. publ. 1830), p. 16; see also Richard R. Yeo, *Defining Science: William Whewell, Natural Knowledge and Public Debate in Early Victorian Britain* (Cambridge: Cambridge University Press, 1993), pp. 119–120.

12. Richard Gregory, *Discovery, or the Spirit and Service of Science* (New York: Macmillan, 1928; orig. publ. 1916), p. 50. Gregory's book was popular, and editions continued to appear during the time Merton was a graduate student.

13. See fine studies by David A. Hollinger, "Inquiry and Uplift: Late Nineteenth-Century American Academics and the Moral Efficacy of Scientific Practice," in *The Authority of Experts: Studies in History and Theory*, ed. Thomas Haskell (Bloomington: Indiana University Press, 1984), pp. 142–156, esp. pp. 142–143, and, especially, Yeo, *Defining Science*.

14. Among several fine treatments of Scientific Naturalism, see especially Frank M. Turner, *Between Science and Religion: The Reaction to Scientific Naturalism in Late Victorian England* (New Haven, CT: Yale University Press, 1974).

15. Weber, "Science as a Vocation," in *From Max Weber: Essays in Sociology*, eds. H. H. Gerth and C. Wright Mills (London: Routledge, 1991), pp. 142–143.

16. An attribution of this much-quoted, but possibly apocryphal, remark is G. E. Hutchinson, "Homage to Santa Rosalia, or Why Are There So Many Kinds of Animals?" *American Naturalist* 93 (1959), 145–159, on p. 146n.

17. Galileo Galilei, "Letter to the Grand Duchess Christina [1615]," *Discoveries and Opinions of Galileo*, trans. Stillman Drake (Garden City, NY: Doubleday Anchor, 1957), pp. 173–216, on p. 200.

18. Isaac Newton, "Scholium," in idem, *Mathematical Principles of Natural Philosophy*, trans. Andrew Motte [1729] and trans. rev. by Florian Cajori (Berkeley: University of California Press, 1934), pp. 6–12, on p. 6.

19. Newton, "General Scholium," in idem, *Mathematical Principles of Natural Philosophy*, pp. 543–547, on p. 546.

20. Peter Dear, *Discipline and Experience: The Mathematical Way in the Scientific Revolution* (Chicago: University of Chicago Press, 1995); also Robert S. Westman, "The Astronomer's Role in the Sixteenth Century: A Preliminary Study," *History of Science* 18 (1980), 105–147.

21. Practitioners of mathematics and natural philosophy also differed contingently in their social standing, where the philosopher was more prestigious, and better rewarded, than the mathematician: Mario Biagioli, *Galileo, Courtier: The Practice of Science in the Culture of Absolutism* (Chicago: University of Chicago Press, 1993).

22. Thomas Henry Huxley, "On the Educational Value of the Natural History Sciences [1854]," in idem, *Collected Essays, Vol. III: Science and Education: Essays* (New York: D. Appleton, 1900), pp. 38–65, on p. 45.

23. Daniel G. Brinton, "The Character and Aims of Scientific Investigation," *Science* n.s. 1, no. 1 (4 January 1895), 3–4. Brinton was an archaeologist at the University of Pennsylvania who was, at the time of writing this essay, president of the American Association for the Advancement of Science.

24. William James, "What Pragmatism Means: Lecture Two," in idem, *Pragmatism: A New Name for an Old Way of Thinking* (Buffalo, NY: Prometheus Books, 1991; orig. publ. 1907), pp. 22–38, on pp. 34–35.

25. Peter Dear, *The Intelligibility of Nature: How Science Makes Sense of the World* (Chicago: University of Chicago Press, 2006).

26. See John Theodore Merz, *A History of European Thought in the Nineteenth Century*, 4 vols. (New York: Dover Publications, 1965; orig. publ. 1904–1912), Vol. III, pp. 578–586.

27. Henry A. Rowland, "The Highest Aim of the Physicist, Presidential Address Delivered at the Second Meeting of the Society, on October 28, 1899," *Bulletin of the American Physical Society* 1 (1899), 4–16, on p. 13; also in *Science* n.s. 10, no. 258 (8 December 1899), 825–833.

28. David Starr Jordan, "A Sage in Science," *Science* n.s. 9, no. 221 (14 April 1899), 529–532, on p. 530.

29. E. W. Scripture, "The Nature of Science and Its Relation to Philosophy," *Science* n.s. 1, no. 13 (29 March 1895), 350–352.

30. Conway MacMillan, "The Scientific Method and Modern Intellectual Life," *Science* n.s. 1, no. 20 (17 May 1895), 537–542, on p. 542.

31. E.g., Percy W. Bridgman, *The Logic of Modern Physics* (New York: Macmillan 1927), p. 5: "What do we mean by the length of an object? We evidently know what we mean by length if we can tell what the length of any and every object is, and for the physicist nothing more is required."

32. Albert Einstein, "Scientific Truth [1929]," in idem, *Ideas and Opinions* (New York: Modern Library, 1994; orig. publ. 1954), p. 286.

33. William O. Baker, "The Moral Un-Neutrality of Science," *Science* n.s. 133, no. 3448 (27 January 1961), 261–262, on p. 261.

34. C. P. Snow, "Address by Charles P. Snow [to Annual Meeting of American Association for the Advancement of Science]," *Science* n.s. 133, no. 3448 (27 January 1961), 256–259, on p. 257.

35. Giovanni Papini, "What Pragmatism Is Like," *Popular Science Monthly* 71, no. 4 (October 1907), 351–358 (trans. Katharine Royce), on pp. 353–354 (emphases in original).

36. See, notably, Alan W. Richardson, "Toward a History of Scientific Philosophy," *Perspectives on Science* 5 (1997), 418–451.

37. Hans Reichenbach, *The Rise of Scientific Philosophy* (Berkeley: University of California Press, 1951), pp. 53–54.

38. The upper case marks a distinction between Method—considered a formal statement of effective and universal means for making scientific knowledge—and

method—taken as just *what scientists do*. All scientific practice is *methodical*, while the regulative role of Method in the practice of science remains deeply problematic.

39. Especially J. E. McGuire and P. M. Rattansi, "Newton and the 'Pipes of Pan,'" *Notes and Records of the Royal Society* 21 (1966) 108–143; also Robert K. Merton, *On the Shoulders of Giants: A Shandean Postscript* (New York: Free Press, 1965).

40. Robert Boyle, "Some Considerations Touching the Usefulness of Experimental Natural Philosophy," in *The Works of the Honourable Robert Boyle*, ed. Thomas Birch, 2nd ed., 6 vols. (London: J. & F. Rivington, 1772; essay orig. publ. 1663), Vol. II, pp. 1–246, on pp. 61–62. Boyle professed sympathy with those "ancient heathens" who referred extraordinary skill in healing "to the gods, or godlike persons" (ibid., p. 200).

41. Robert Boyle, *An Epistolical Discourse . . . Inviting All True Lovers of Vertue and Mankind to a Free and Generous Communication* . . . , reprinted in Margaret E. Rowbottom, "The Earliest Published Writing of Robert Boyle," *Annals of Science* 6 (1950), 380–385, on p. 384.

42. Simon Schaffer, "Genius in Romantic Natural Philosophy," in *Romanticism and the Sciences*, eds. Andrew Cunningham and Nicholas Jardine (Cambridge: Cambridge University Press, 1990), pp. 82–98.

43. Richard R. Yeo, "Scientific Method and the Image of Science, 1831–1890," in *The Parliament of Science: The British Association for the Advancement of Science, 1831–1931*, eds. Roy MacLeod and Peter Collins (Northwood, Middlesex: Science Reviews, 1981), pp. 65–88; Jack Morrell and Arnold Thackray, *Gentlemen of Science: Early Years of the British Association for the Advancement of Science* (Oxford: Clarendon Press, 1981), esp. chs. 5, 8.

44. Roland Barthes, "The Brain of Einstein," in idem, *Mythologies*, trans. Annette Lavers (New York: Noonday Press, 1975; orig. publ. 1957), pp. 68–70, on pp. 68–69.

45. See, notably, J. B. Morrell, "The Chemist Breeders: The Research Schools of Liebig and Thomas Thomson," *Ambix* 19 (1972), 1–46.

46. Samuel Smiles, *Self-Help; with Illustrations of Conduct and Perseverance* (London: John Murray, 1880; orig. publ. 1859), pp. 94–96, 122, 131. For a bluff endorsement of Smiles's sentiments by an American executive in the 1920s, see Anon., "Why I Never Hire Brilliant Men," *American Magazine* (February 1924), 12, 13, 117–118, 121–122, on p. 121, drawing from the history of science (the lives of Newton and Darwin) the same lessons about dogged determination versus genius.

47. Quoted in Janet Browne, "The Natural Economy of Households: Charles Darwin's Account Books," typescript, 2007, from Francis Darwin, ed., *The Life and Letters of Charles Darwin*, 3 vols. (London: John Murray, 1887), Vol. III, p. 179.

48. E.g., Lorraine J. Daston, "Fear & Loathing of the Imagination in Science," *Dædalus* 134, no. 4 (Fall 2005), 16–30, esp. pp. 16–17, 28–30.

49. Steven Shapin, "'A Scholar and a Gentleman': The Problematic Identity of the Scientific Practitioner in Early Modern England," *History of Science* 29 (1991), 279–327; idem, "The Image of the Man of Science"; idem, "The Man of Science"; Stephen Gaukroger, *Descartes: An Intellectual Biography* (Oxford: Clarendon Press, 1995), p. 23 (for the remark by Descartes' father).

50. The motivating force of English Puritanism in the pursuit of natural knowledge was central to Merton's celebrated "thesis" of the 1930s: Robert K. Merton, *Science, Technology and Society in Seventeenth-Century England* (New York: H. Fertig, 1970; orig. publ. 1938).

51. For these *éloges*, see especially Charles B. Paul, *Science and Immortality: The Éloges of the Paris Academy of Sciences (1699–1791)* (Berkeley: University of California Press, 1980), on which the following paragraphs largely rely, and, for Georges Cuvier's *éloges* of the late eighteenth and early nineteenth centuries, see Dorinda Outram, "The Language of Natural Power: The Funeral *Éloges* of Georges Cuvier," *History of Science* 16 (1978), 153–178; also idem, *Georges Cuvier: Vocation, Science and Authority in Post-Revolutionary France* (Manchester: Manchester University Press, 1984), esp. pp. 66–67, 110–111. For important treatment of eighteenth- and early nineteenth-century debates over the virtue and mental capacities of Isaac Newton, see Richard Yeo, "Genius, Method and Morality: Images of Newton in Britain, 1760–1860," *Science in Context* 2 (1988), 257–284.

52. Adam Smith, *The Theory of Moral Sentiments*, eds. D. D. Raphael and A. L. Macfie (Oxford: Clarendon Press, 1976; orig. publ. 1759), p. 125.

53. Condorcet's *éloge* of Franklin (read 13 November 1790), quoted in Paul, *Science and Immortality*, p. 67.

54. Anon., "Philosopher," in *Encyclopedia: Selections*, eds. Denis Diderot and Jean le Rond d'Alembert, [this edition] eds. and trans. Nelly S. Hoyt and Thomas Cassirer (Indianapolis: Bobbs-Merrill, 1965; *Encyclopédie* orig. publ. 1751–1777), pp. 283–289, on p. 287. (The author was probably the grammarian César Chesneau Dumarsais, writing in the early 1750s.)

55. Immanuel Kant, "What Is Enlightenment?" in idem, *Foundations of the Metaphysics of Morals and What Is Enlightenment?*, 2nd ed., revised, trans. Lewis White Beck (New York: Macmillan, 1990; essay orig. publ. 1784), pp. 83–90, on pp. 84–86.

56. Charles Babbage, *Reflections on the Decline of Science in England, and on Some of Its Causes* (London: B. Fellowes, 1830), p. 23.

57. Roy Porter incisively noted "the lack of pressure to publish" bearing on gentlemen-geologists in the eighteenth century: "Gentlemen and Geology: The Emergence of a Scientific Career, 1660–1920," *Historical Journal* 21 (1978), 809–836, on p. 815. Indeed, gentlemen-amateurs often worried about the gentility of—as it was sometimes put—appearing in the character of an author. See also David P. Miller, "'My Favourite Studdys': Lord Bute as Naturalist," in *Lord Bute: Essays in Reinterpretation*, ed. Karl W. Schweizer (Leicester: Leicester University Press, 1988), pp. 213–239, on pp. 215, 218.

58. See Roger Hahn, "Scientific Careers in Eighteenth-Century France," in *The Emergence of Science in Western Europe*, ed. Maurice Crosland (London: Macmillan, 1975), pp. 127–138; also Paul, *Science and Immortality*, pp. 69–85, and Maurice Crosland, "The Development of a Professional Career in Science in France," in *The Emergence of Science in Western Europe*, ed. Crosland, pp. 139–159.

59. In these connections, see Peter Galison, *Einstein's Clocks, Poincaré's Maps: Empires of Time* (New York: W. W. Norton, 2003), esp. ch. 2.

60. "Tyndall's Philosophy: Its Inevitable End," anonymous letter to editor, *New York Times*, 6 September 1874, p. 1.

61. Edward Carpenter, *Modern Science: A Criticism* (Manchester: John Heywood, 1885), pp. 59–60.

62. Leo Tolstoy, "Modern Science [1898]," in idem, *Recollections & Essays*, trans. Aylmer Maude (London: Oxford University Press, 1937), pp. 176–187, on p. 184. For the collection of Carpenter essays that inspired Tolstoy's project, see Edward Carpenter, *Civilisation: Its Cause and Cure, and Other Essays* (London: Swan Sonnenschein, 1889).

63. Ralph Waldo Emerson, *The American Scholar . . . An Address Delivered before the φBK Society at Cambridge, August 1837* (New York: Laurentian Press, 1901), pp. 5, 24. Emerson definitely meant to include the man of science in his notion of the scholar, as shown here (pp. 34, 42) by his approving mentions of Newton, Flamsteed, Linnaeus, Cuvier, Davy, and Herschel.

64. David Starr Jordan, "Nature Study and Moral Culture," *Science* n.s. 4, no. 84 (7 August 1896), 149–156, on pp. 149–150, 152. At the time of writing, Jordan was president of Stanford.

65. Arthur D. Little, "The Fifth Estate," *Science* n.s. 60, no. 1553 (8 October 1924), 299–306, on pp. 299, 301.

66. See, notably, J. A. Secord, "The Discovery of a Vocation: Darwin's Early Geology," *British Journal for the History of Science* 24 (1991), 133–157; James R. Moore, "Darwin of Down: The Evolutionist as Squarson-Naturalist," in *The Darwinian Heritage*, ed. David Kohn (Princeton, NJ: Princeton University Press, 1985), pp. 435–481; Janet Browne, *Charles Darwin: Voyaging: A Biography* (Princeton, NJ: Princeton University Press, 1995), esp. pp. 228–229.

67. J. B. Morrell, "Individualism and the Structure of British Science in 1830," *Historical Studies in the Physical Sciences* 3 (1971), 183–204; see also Yeo, *Defining Science*, p. 117.

68. Babbage, *Reflections on the Decline of Science*, p. 18.

69. For Franklin, see Eugene C. Bingham, "Research in Colleges," *Science* n.s. 61, no. 1572 (13 February 1925), 174–176. Margaret Thatcher liked the Faraday story, but her science and university policies did not reflect a lesson learned: Margaret Thatcher, *The Path to Power* (New York: HarperCollins, 1995), p. 176.

70. See, among many examples, John J. Stevenson, "The Debt of the World to Pure Science," *Science* n.s. 7, no. 167 (11 March 1898), 325–334, on p. 332; see also Ronald Kline, "Construing 'Technology' as 'Applied Science': Public Rhetoric of Scientists and Engineers in the United States, 1880–1945," *Isis* 86 (1995), 194–221.

71. Babbage, *Reflections on the Decline of Science*, p. 17.

72. Alexis de Tocqueville, "Why the Americans are More Addicted to Practical Than to Theoretical Science," in idem, *Democracy in America*, trans. Henry Reeve, 2 vols. (New York: The Colonial Press, 1899) Vol. 2, ch. 10, on pp. 43, 46; see also R. H.

Shryock, "American Indifference to Basic Science During the 19th Century," *Archives internationales d'histoire des sciences* 28 (1948), 50–65.

73. George H. Daniels, "The Pure Science Ideal and Democratic Culture," *Science* n.s. 156, no. 3783 (30 June 1967), 1699–1705.

74. Henry S. Carhart, "The Educational and Industrial Value of Science," *Science* n.s. 1, no. 15 (12 April 1895), 393–402, on p. 399. Carhart was professor of physics at the University of Michigan.

75. Charles Richet, *The Natural History of a Savant*, trans. Oliver Lodge (London: J. M. Dent, 1927; orig. publ. 1923), pp. 4–5.

76. Russell Moseley, "From Avocation to Job: The Changing Nature of Scientific Practice," *Social Studies of Science* 9 (1979), 511–522.

77. Francis Galton, *English Men of Science: Their Nature and Nurture* (London: Macmillan, 1874), p. 260. For Galton's qualms about defining the characteristics of "men of science"—on grounds of their scarcity—see Yeo, *Defining Science*, pp. 116–117.

78. Billings, "Scientific Men and Their Duties [1886]," pp. 543–544.

79. Richard Hamer, "The Romantic and Idealistic Appeal of Physics," *Science* n.s. 61, no. 1570 (30 January 1925), 109–110, on p. 109.

80. Christopher Hamlin, "Scientific Method and Expert Witnessing: Victorian Perspectives on a Modern Problem," *Social Studies of Science* 16 (1986), 485–513, on pp. 491–492, 495–496. For an entry to American scientists' problems with expert witnessing in the late nineteenth century, see William P. Mason, "Expert Witnessing," *Science* n.s. 6, no. 137 (13 August 1897), 243–248.

81. Thorstein Veblen, *The Higher Learning in America: A Memorandum on the Conduct of Universities by Business Men* (New York: Sagamore Press, 1957; orig. publ. 1918), pp. 31, 43; see also Daniel J. Kevles, Jeffrey L. Sturchio, and P. Thomas Carroll, "The Sciences in America, Circa 1880," *Science* n.s. 209, no. 4452 (4 July 1980), 26–32, esp. pp. 30–32.

82. Weber, "Science as a Vocation," pp. 135, 144, 150–152.

CHAPTER THREE

1. H. L. Mencken, *Prejudices: Third Series* (New York: Alfred A. Knopf, 1922), p. 266; this remark first appeared in the *New York Evening Mail*, 25 March 1918.

2. William H. George, *The Scientist in Action: A Scientific Study of His Methods* (London: Williams & Norgate, 1936), p. 17. An American edition appeared in 1938.

3. David Lindsay Watson, *Scientists Are Human* (London: Watts & Co., 1938). This is the first essay or book title I have encountered using exactly this expression, or a cognate form; cf. A. V. Hill, "The Humanity of Science," in Sir John Boyd Orr et al., *What Science Stands For* (London: George Allen & Unwin, 1937), pp. 30–38. After the Second World War, the phrase became something of a cliché: see, for example, Jacob Bronowski, "Science Is Human," in *The Humanist Frame*, ed. Julian Huxley (New York: Harper & Brothers, 1961), pp. 83–94; H. N. Parton, *Science Is Human: Essays*, ed. M. H. Panckhurst (Dunedin, New Zealand: University of Otago Press, 1972); Glenn T. Seaborg, "The Scientist as a Human Being," in idem, *A Scientist Speaks Out: A Personal Perspective on*

Science, Society and Change (Singapore: World Scientific, 1996), pp. 89–96; orig. publ. in *Chemical and Engineering News* 42, no. 51 (21 December 1964), 60–62; George F. Kneller, *Science as a Human Endeavor* (New York: Columbia University Press, 1978); Stewart E. Perry, *The Human Nature of Science* (New York: Free Press, 1966); George Russell Harrison, *What Man May Be: The Human Side of Science* (New York: W. Morrow, 1956); Ralph E. Oesper, *The Human Side of Scientists* (Cincinnati, OH: University of Cincinnati Press, 1975).

4. Robert K. Merton, review of Watson, *Scientists Are Human*, *Isis* 31 (1940), 466–467.

5. Personal communication to author from Robert K. Merton (April 1996).

6. Watson, *Scientists Are Human*, pp. xiii–xiv.

7. Michael Polanyi, *Personal Knowledge: Towards a Post-Critical Philosophy* (Chicago: University of Chicago Press, 1958).

8. Watson, *Scientists Are Human*, pp. 45, 83.

9. Albert Einstein, "My First Impressions of the U.S.A. [1921]," in idem, *Ideas and Opinions* (New York: Modern Library; orig. publ. 1954), pp. 3–7, on p. 4. It would be artificial, here or elsewhere, to massage the evidently gendered language used to refer to scientific practitioners, and I will treat women's presences, absences, and the extent to which late modern technoscience was a masculine world in later chapters.

10. Albert Einstein, "Principles of Research [1918]," in idem, *Ideas and Opinions*, pp. 244–248, on pp. 244–245. These passages were prominently quoted by Watson, *Scientists Are Human*, pp. 35–36; see also Fritz Stern, *Einstein's German World* (Princeton, NJ: Princeton University Press, 1999), pp. 67–68.

11. Elizabeth Kay MacLeod, "Politics, Professionalism and the Organisation of Scientists: The Association of Scientific Workers, 1917–1942," unpubl. Ph.D. thesis, University of Sussex, 1975.

12. Anon., "The Encouragement of Basic Research," *Science* n.s. 61, no. 1567 (9 January 1925), 43–44.

13. J. D. Bernal, *The World, the Flesh & the Devil: An Inquiry into the Future of the Three Enemies of the Rational Soul* (Bloomington: Indiana University Press, 1969; orig. publ. 1929), p. 77.

14. J. D. Bernal, *Science in History* (London: Watts & Co., 1954), p. 918. (This statement is, indeed, from a postwar context, but it is an unusually assertive version of sentiments Bernal was expressing through the 1930s.)

15. Anon., "Scientists in Battle," *New York Times*, 21 May 1928, p. 20.

16. A. V. Hill, "The International Status and Obligations of Science [1933]," in idem, *The Ethical Dilemma of Science* (London: Scientific Book Guild, 1962), pp. 205–221.

17. Hill, "The Humanity of Science [1937]," pp. 30–31, 34.

18. David Brewster, *The Life of Sir Isaac Newton* (London: John Murray, 1831), p. 337. I am indebted here, and in following sections, to the fine scholarship and acute interpretations of Richard R. Yeo, especially his "Genius, Method and Morality: Images of Newton in Britain, 1760–1860," *Science in Context* 2 (1988), 257–284.

19. Augustus de Morgan, *Essays on the Life and Work of Newton*, ed. Philip E. B. Jourdain (Chicago: Open Court, 1914), pp. 119–182, on pp. 121–122. De Morgan's piece

originally appeared as Review of Brewster's "Memoirs of Newton," *North British Review* 23 (August 1855), 307–338. For the relevant added passages of the later Brewster biography, see Sir David Brewster, *The Life of Sir Isaac Newton*, rev. and ed. W. T. Lynn (Edinburgh: Gall & Inglis, [ca. 1875]; from 1855 edition), pp. 193–194. For a fine account of nineteenth-century controversies over how to treat Newton's life, see Rebekah Higgitt, "The Apple of Their Eye? Biographies of Isaac Newton, 1820–1870," unpubl. D.Sc. thesis, Imperial College, London, 2004; for de Morgan's response to French allegations that Newton had fraudulently claimed Pascal's work on universal gravitation as his own, see idem, "'Newton dépossédé!' The British Response to the Pascal Forgeries of 1867," *British Journal for the History of Science* 36 (2003), 437–453, on pp. 442–443, 446–448, 452–453. See also Paul Theerman, "Unaccustomed Role: The Scientist as Historical Biographer: Two Nineteenth-Century Portrayals of Newton," *Biography* 8 (1985), 145–162.

20. De Morgan, *Essays on the Life and Work of Newton*, pp. 124–125.

21. Ibid., p. 134.

22. Ibid., p. 182.

23. See debates on this question in Michael Shortland and Richard Yeo, eds., *Telling Lives in Science: Essays on Scientific Biography* (Cambridge: Cambridge University Press, 1996); notably, the recent advocacy of a return to moralizing modes by Thomas Söderqvist, "Existential Projects and Existential Choice in Science: Science Biography as an Edifying Genre," in ibid., pp. 45–84, and idem, *Science as Autobiography: The Troubled Life of Niels Jerne* (New Haven, CT: Yale University Press, 2003).

24. Thomas Babington Macaulay, "Lord Bacon," in idem, *Literary Essays Contributed to the Edinburgh Review* (Oxford: Oxford University Press, 1913), pp. 289–410 (orig. publ. in *Edinburgh Review* [July 1837]), on pp. 289–291; see also pp. 322–323. For the moral constitution of the man of science in the setting of nineteenth-century debates over methodology, see the fine Richard R. Yeo, *Defining Science: William Whewell, Natural Knowledge, and Public Debate in Early Victorian Britain* (Cambridge: Cambridge University Press, 1993).

25. Friedrich Nietzsche, "Beyond Good and Evil [1886]," in *The Philosophy of Nietzsche*, trans. Helen Zimmern (New York: Modern Library, 1954), pp. 369–616, on pp. 499–502.

26. Friedrich Nietzsche, "The Genealogy of Morals [1887]," in *The Philosophy of Nietzsche*, trans. Horace B. Samuel, pp. 617–807, on p. 740.

27. Nietzsche, "The Genealogy of Morals," p. 744. For asceticism as a trope in the identity of the scientist in the late nineteenth and early twentieth century, see Rebecca M. Herzig, *Suffering for Science: Reason and Sacrifice in Modern America* (New Brunswick, NJ: Rutgers University Press, 2005).

28. Julien Benda, *The Treason of the Intellectuals*, trans. Richard Aldington (New York: William Morrow & Company, 1928; orig. publ. 1927). The source of the epigraph was the French idealist philosopher Charles Bernard Renouvier (1815–1903).

29. Ibid., pp. 43–44, 61, 99 (emphases added); see also Charles Thorpe, "Violence and the Scientific Vocation," *Theory, Culture & Society* 21 (2004), 59–84, on pp. 64–65.

30. Benda, *The Treason of the Intellectuals*, pp. 107, 139–140, 158, 181.

31. Michel Foucault, "Truth and Power," in idem, *Power/Knowledge: Selected Interviews and Other Writings 1972–1977*, ed. Colin Gordon, trans. Gordon, Leo Marshall, John Mepham, and Kate Soper (New York: Pantheon, 1980), pp. 109–133, esp. pp. 126–129; see also Alvin W. Gouldner, *The Future of Intellectuals and the Rise of the New Class* (New York: Seabury Press, 1979), pp. 48–49 (for the "technical intelligentsia versus the "intellectuals"); Charles Thorpe, *Oppenheimer: The Tragic Intellect* (Chicago: University of Chicago Press, 2006), pp. 5–6.

32. George B. Schley, "Society's Need for Patents to University Research Workers," *Journal of Industrial and Engineering Chemistry* 29 (1937), 1319–1322, on p. 1320, and partly quoted in Grischa Metlay, "Reconsidering Renormalization: Stability and Change in 20th-Century Views of University Patents," *Social Studies of Science* 36 (2006), 565–597, on p. 577.

33. Daniel H. Kevles, *The Physicists: The History of a Scientific Community in Modern America* (New York: Alfred A. Knopf, 1977), pp. 99 (letter to Langmuir ca. 1911) and 101; for disapproval of the de-moralization of American physicists in this period, see notably Paul Forman, "Social Niche and Self-Image of the American Physicist," in *Proceedings of the International Conference on the Restructuring of the Physical Sciences in Europe and the United States 1945–1960*, eds. Michelangelo De Maria, Mario Grilli, and Fabio Sebastiani (Singapore: World Scientific Publishing, 1989), pp. 96–104.

34. C. S. Peirce, "The Scientific Attitude and Fallibilism [1896–1899]," in *The Philosophy of Peirce: Selected Writings*, ed. Justus Buchler (London: Routledge & Kegan Paul, 1940), pp. 42–59, on pp. 43–44.

35. [E. L. Youmans?], "Editor's Table: Official Science at Washington," *Popular Science Monthly* 27, no. 6 (October 1885), 844–847, on p. 846; cf. Kevles, *The Physicists*, p. 54.

36. J. R. Eastman, "The Relations of Science and the Scientific Citizen to the General Government," *Science* n.s. 5, no. 118 (2 April 1897), 525–531, on pp. 527–528.

37. G. A. Pearson, "Some Conditions for Effective Research," *Science* n.s. 60, no. 1543 (25 July 1924), 71–73, on pp. 71–72.

38. See, in this connection, George H. Daniels, "The Pure Science Ideal and Democratic Culture," *Science* n.s. 156, no. 3783 (30 June 1967), 1699–1705.

39. Roslynn D. Haynes, *From Faust to Strangelove: Representations of the Scientist in Western Literature* (Baltimore, MD: Johns Hopkins University Press, 1994), pp. 297–299.

40. Sinclair Lewis, *Arrowsmith* (New York: Harcourt Brace, 1925). That influence seems to have been especially strong among the atomic scientists who worked on the Manhattan Project in World War II. See, among many examples, R. R. Wilson, "My Fight against Team Research," in *The Twentieth-Century Sciences: Studies in the Biography of Ideas*, ed. Gerald Holton (New York: W. W. Norton & Company, 1972), pp. 468–479, on p. 468: "As a youth, I read *Arrowsmith* by Sinclair Lewis. That romantic idealization of a man dedicated to research made a deep impression upon me"; Victor F. Weisskopf, *The Joy of Insight: Passions of a Physicist* (New York: Basic Books, 1991), p. 320: "Surely it is

the task of artists and writers to bring the great ideas of our time to the public. I believe that Sinclair Lewis's *Arrowsmith* was the last great novel to describe the excitement of scientific research"; Glenn T. Seaborg, *Adventures in the Atomic Age: From Watts to Washington* (New York: Farrar, Straus and Giroux, 2001), p. 12. The influence of *Arrowsmith* on James Watson is mentioned in Errol C. Friedberg, *The Writing Life of James D. Watson* (Cold Spring Harbor, NY: Cold Spring Harbor Laboratory Press, 2005), pp. 5–7, and celebrated in James D. Watson, *Avoid Boring People: Lessons from a Life in Science* (New York: Alfred A. Knopf, 2007), p. 20. And see also Charles E. Rosenberg, "Martin Arrowsmith: The Scientist as Hero," *American Quarterly* 15 (1963), 447–458; Herzig, *Suffering for Science*, ch. 6.

41. Lewis, *Arrowsmith*, p. 136; see also Paul de Kruif, *Microbe Hunters* (New York: Harcourt, Brace & World, 1953; orig. publ. 1926), p. 3.

42. Quoted in Kevles, *The Physicists*, p. 185.

43. Thomas J. LeBlanc, review of *Arrowsmith*, *Science* n.s. 61, no. 1590 (19 June 1925), 632–634; also Kevles, *The Physicists*, p. 170.

44. Edwin Slosson, "Must Scientists Wear Whiskers?" *Independent* [New York], 28 November 1925, p. 601 (quoted in Kevles, *The Physicists*, p. 175).

45. Lewis, *Arrowsmith*, pp. 280–281.

46. Ibid., p. 279.

47. Anon., "What Is Reason For?" *Science* n.s. 62, no. 1595 (24 July 1925), 83–84, quoting from the *New York World*.

48. Quoted in Forman, "Social Niche and Self-Image of the American Physicist," pp. 97–98 (though Pupin is here erroneously called "Papin"); see also Ronald C. Tobey, *The American Ideology of National Science, 1919–1930* (Pittsburgh: University of Pittsburgh Press, 1971), esp. pp. 92–93, 154–155, 178–179.

49. The comment on Millikan's personality is in Anon., "Millikan Rays," *New York Times*, 12 November 1925, p. 24.

50. Arthur D. Little, "The Fifth Estate," *Science* n.s. 60, no. 1553 (8 October 1924), 299–306, on p. 304.

51. Both the Ehrlich and the Pasteur films were made by the same director, William Dieterle: see T. Hugh Crawford, "Screening Science: Pedagogy and Practice in William Dieterle's Film Biographies of Scientists," *Common Knowledge* 6 (1997), 52–68.

52. Anon., "Prof. Einstein Here, Explains Relativity," *New York Times*, 3 April 1921, p. 1.

53. Kevles, *The Physicists*, pp. 175–177.

54. Otto Nathan and Heinz Norden, eds., *Einstein on Peace* (New York: Simon and Schuster, 1960), p. 308.

55. Kevles, *The Physicists*, chs. 8–10; Daniel Lee Kleinman, *Politics on the Endless Frontier: Postwar Research Policy in the United States* (Durham, NC: Duke University Press, 1995), chs. 3–4; Tobey, *The American Ideology of National Science*, pp. 33–49; Daniel Charles, *Master Mind: The Rise and Fall of Fritz Haber, the Nobel Laureate Who Launched the Age of Chemical Warfare* (New York: HarperCollins, 2005).

56. Anon., "Crescat Scientia," *New York Times*, 26 January 1930, p. 54.

57. J. Robert Oppenheimer, "Physics in the Contemporary World [1947]," in idem, *The Open Mind* (New York: Simon and Schuster, 1955), pp. 81–102, on p. 88. The experience of guilt was, naturally, stronger among some American atomic scientists who had worked directly on the bomb, but it was also expressed by others, who had not. Max Born, for example, was referring to "Bomber" Harris and Lord Cherwell when he wrote, after the war, that "I had thought that a real scientist could not do base deeds. The scientists I admired and loved, like [James] Franck, Einstein, Rutherford, Planck, von Laue, seemed to confirm my belief. But Lindemann [Cherwell] did base things and opened the gates of hell for other men of his type, men efficient and clever, but not profound and wise, who later became leaders in science and its application to politics and war": Max Born, *My Life: Recollections of a Nobel Laureate* (London: Taylor & Francis, 1978), p. 262.

58. Among many sources, see Silvan S. Schweber, *In the Shadow of the Bomb: Bethe, Oppenheimer, and the Moral Responsibility of the Scientist* (Princeton, NJ: Princeton University Press, 2000); Mary Palevsky, *Atomic Fragments: A Daughter's Questions* (Berkeley: University of California Press, 2000); and my essay-review of both of these: Steven Shapin, "Don't Let That Crybaby in Here Again," *London Review of Books* 22, no. 17 (7 September 2000), 15–16; see also Alice Kimball Smith, *A Peril and a Hope: The Scientists' Movement in America, 1945–47* (Chicago: University of Chicago Press, 1965).

59. Erwin Chargaff, *Heraclitean Fire: Sketches from a Life before Nature* (New York: Rockefeller University Press, 1978), p. 136. See also Pnina Abir-Am, "From Biochemistry to Molecular Biology: DNA and the Acculturated Journey of the Critic of Science Erwin Chargaff," *History and Philosophy of Life Sciences* 2 (1980), 3–60, and, for "the pervasiveness of the modern integration between science and violence," Thorpe, "Violence and the Scientific Vocation," p. 60.

60. C. P. Snow, "Address by Charles P. Snow [to Annual Meeting of American Association for the Advancement of Science, 27 December 1960]," *Science* n.s. 133, no. 3448 (27 January 1961), 256–259 (emphasis added). This essay was subsequently published as "The Moral Un-Neutrality of Science," in Snow, *Public Affairs* (New York: Charles Scribner's, 1971), pp. 187–198), p. 256; see also Haynes, *From Faust to Strangelove*, pp. 246–263.

61. Lee A. DuBridge, "Scientists and Engineers: Quantity Plus Quality," *Science* n.s. 124, no. 3216 (17 August 1956), 299–304, on p. 301.

62. Alexander Haddow, "The Scientist as Citizen," *Bulletin of the Atomic Scientists* 12, no. 7 (September 1956), 245–252, on pp. 245, 252.

63. Excerpted from typescript "When Einstein Came Out with His World Revelations," quoted in Peter Galison, *Image and Logic: A Material Culture of Microphysics* (Chicago: University of Chicago Press, 1997), p. 282.

64. Stanislaw M. Ulam, *Adventures of a Mathematician* (New York: Charles Scribner's Sons, 1976), p. 151.

65. Steve J. Heims, *John Von Neumann and Norbert Wiener: From Mathematics to the Technologies of Life and Death* (Cambridge, MA: MIT Press, 1980), p. 362; see also p. 356: "Colleagues at Princeton caught something of von Neumann's style of adapting [to

different circumstances] in the story that made the rounds about him that he was not human but a demigod who 'had made a detailed study of humans and could imitate them perfectly.'"

66. Alvin W. Gouldner, "Cosmopolitans and Locals: Toward an Analysis of Latent Social Roles," *Administrative Science Quarterly* 2 (1957–1958), 281–306, 444–480.

67. Many of these issues are treated in Charles R. Thorpe and Steven Shapin, "Who Was J. Robert Oppenheimer? Charisma and Complex Organization," *Social Studies of Science* 30 (2000), 545–590; Thorpe, *Oppenheimer*, ch. 4.

68. Early in the career of the Manhattan Project, its director general Leslie Groves was particularly irritated by the abrasively anti-authoritarian, and anti-military, attitudes of the Hungarian emigré physicist Leo Szilard, and, according to one historian, "Groves seems to have attributed Szilard's brashness to the fact that he was a Jew": Richard Rhodes, *The Making of the Atomic Bomb* (New York: Simon and Schuster, 1986), p. 502; also Thomas P. Hughes, *American Genesis: A Century of Invention and Technological Enthusiasm 1870–1970* (New York: Viking, 1989), p. 397; Ellen W. Schrecker, *No Ivory Tower: McCarthyism and the Universities* (New York: Oxford University Press, 1986), pp. 133–134. Anti-Semitism was, of course, no great rarity in American military, business, governmental, and even academic circles, and the highly Jewish makeup of the Manhattan Project's scientific community, and especially of its physicists, is pertinent to emerging attitudes towards the scientist's character.

69. E. U. Condon, "Science and Security," *Science* n.s. 107, no. 2791 (25 June 1948), 659–665, on p. 661.

70. E. U. Condon, "The Duty of Dissent," *Science* n.s. 119, no. 3086 (19 February 1954), 227–228, on p. 228.

71. Conway Zirkle, "Our Splintered Learning and the Status of Scientists," *Science* n.s. 121, no. 3146 (15 April 1956), 513–519, on pp. 515–516.

72. Condon, "Science and Security," p. 661. For Condon's own troubles with the House Un-American Activities Committee, see Jessica Wang, "Science, Security, and the Cold War: The Case of E. U. Condon," *Isis* 83 (1992), 238–269; idem, *American Science in an Age of Anxiety: Scientists, Anticommunism and the Cold War* (Chapel Hill: University of North Carolina Press, 1999), pp. 130–147, 163–181; Schrecker, *No Ivory Tower*, pp. 141, 274–275. Fed up by the witch-hunting, and grotesquely accused by HUAC of being the "one of the weakest links" in national atomic security, Condon resigned as director of the National Bureau of Standards in 1951 and returned to industry, where he was research director at Corning Glass Company for several years. It was not until 1956 that Condon found universities (Washington University in St. Louis and, later, the University of Colorado) willing to stand up to the Congressional Red-baiters.

73. Walter Gellhorn, *Security, Loyalty, and Science* (Ithaca, NY: Cornell University Press, 1950), p. 120.

74. Phillip N. Powers, "Industrial Research Workers and Defense," in *Selection, Training, and Use of Personnel in Industrial Research*, Proceedings of the Second Annual Conference on Industrial Research June 1951, eds. David B. Hertz and Albert H. Rubenstein (New York: King's Crown Press, 1952), pp. 94–112, on p. 105.

75. Vannevar Bush, "To Make Our Security System Secure," *New York Times*, 20 March 1955, Sunday Magazine, pp. 9, 38, 42, 44, 47, on p. 38.

76. Melba Phillips, "Dangers Confronting American Science," *Science*, n.s. 116, no. 3017 (24 October 1952), 439–443, on pp. 441–442. The author was one of J. Robert Oppenheimer's first Ph.D. students. In the 1950s, she refused to testify before the McCarran Senate Internal Security Sub-Committee and was fired from her academic appointments at Brooklyn College and the Columbia University Radiation Laboratory; see also David Kaiser, "The Atomic Secret in Red Hands? American Suspicions of Theoretical Physicists During the Early Cold War," *Representations* 90 (Spring 2005), 28–60.

77. Quoted, from 1954 hearings, in Henry S. Hall, "Scientists and Politicians," in *The Sociology of Science*, eds. Bernard Barber and Walter Hirsch (New York: Free Press, 1962; art. orig. publ. in *Bulletin of the Atomic Scientists* 12, no. 2 [February 1956], pp. 46–52), pp. 269–287, on pp. 283–284.

78. Henry Margenau, David Bergamini, et al., *The Scientist (Life Science Library)* (New York: Time Incorporated, 1964), p. 35.

79. Lewis M. Terman, "Are Scientists Different?" *Scientific American* 192, no. 1 (January 1955), 25–29, on pp. 25, 29. For a later synopsis of psychological work on the question of personality differences, see Walter Hirsch, *Scientists in American Society* (New York: Random House, 1968), pp. 8–33.

80. United States Atomic Energy Commission, *In the Matter of J. Robert Oppenheimer: Transcript of Hearing before Personnel Security Board and Texts of Principal Documents and Letters. United States Atomic Energy Commission, May 27, 1954, through June 29, 1954*, ed. Philip M. Stern (Cambridge, MA: MIT Press, 1971), p. 17. This matter is well treated in Charles R. Thorpe, "Disciplining Experts: Scientific Authority and Liberal Democracy in the Oppenheimer Case," *Social Studies of Science* 32 (2002), 525–562; idem, "The Scientist in Mass Society: J. Robert Oppenheimer and the Postwar Liberal Imagination," in *Reappraising Oppenheimer: Centennial Studies and Reflections*, Berkeley Papers in the History of Science, Vol. 21, eds. Cathryn Carson and David A. Hollinger (Berkeley, CA: Office for History of Science and Technology, University of California, Berkeley, 2005), pp. 293–314; idem, *Oppenheimer*, ch. 7. For the wider significance of Jewish presence in contemporary American intellectual life, see David A. Hollinger, *Science, Jews, and Secular Culture: Studies in Mid-Twentieth-Century American Intellectual History* (Princeton, NJ: Princeton University Press, 1996), esp. pp. 3–4, 14–15 (for Oppenheimer as one of the iconic Jewish intellectuals).

81. United States Atomic Energy Commission, *In the Matter of J. Robert Oppenheimer*, p. 1016, quoted in Thorpe, "Disciplining Experts," p. 545.

82. Charles P. Curtis, *The Oppenheimer Case: The Trial of a Security System* (New York: Simon and Schuster, 1955), p. 152, quoted in Thorpe, "Disciplining Experts," p. 545.

83. J. Robert Oppenheimer, "Encouragement of Science," *Science* n.s. 111, no. 2885 (14 April 1950), 373–375, on p. 374.

84. Foucault, "Truth and Power," pp. 127–128.

85. Robert K. Merton, "The Machine, the Worker, and the Engineer," *Science* n.s. 105, no. 2717 (24 January 1947), 79–84, on p. 82. For Merton, the structural facts of differentiated labor, more than anything else, caused technical workers "to be indoctrinated with an ethical sense of limited responsibilities." The integration of scientists and engineers into "industrial bureaucracies" provided still another limit to a wider ethical sensibility.

86. Edward Teller, "Back to the Laboratories," *Bulletin of the Atomic Scientists* 6, no. 3 (March 1950), 71–72; see also Haynes, *From Faust to Strangelove*, p. 259.

87. Ralph E. Lapp, *The New Priesthood: The Scientific Elite and the Uses of Power* (New York: Harper & Row, 1965), pp. 114, 228. Note also the extreme heterogeneity of views among scientists about whether or not science was a unity, and what consequences flowed from its unity or disunity. C. P. Snow famously insisted in *The Two Cultures* (1959) that shared scientific sensibilities were almost too obvious to be described, while others saw disastrous consequences deriving from equally obvious disunity: Snow, *The Two Cultures and The Scientific Revolution*, the Rede Lecture 1959 (Cambridge: Cambridge University Press, 1959), pp. 9–10. See also the view of the Spanish psychiatrist Juan Lopez-Ibor: "Science no longer exists—it has been replaced by the sciences, and this dispersion of knowledge, this lack of a clear image of what is happening on earth, is one cause of today's human anguish": quoted in Margenau and Bergamini, *The Scientist*, p. 113.

88. James B. Conant, *On Understanding Science: An Historical Approach* (Oxford: Oxford University Press, 1947), pp. 7–8; see also elegant elaborations of similar views by the Columbia cultural historian Jacques Barzun, *Science: The Glorious Entertainment* (New York: Harper & Row, 1964), esp. pp. 61–80.

89. Quoted in Hall, "Scientists and Politicians," pp. 270–271.

90. Quoted in ibid., p. 272.

91. Kirtley F. Mather, "The Problem of Antiscientific Trends Today," *Science* n.s. 115, no. 2994 (16 May 1952), 533–537, on pp. 533–534.

92. Eric Larrabee, "Science, Poetry, and Politics," *Science* n.s. 117, no. 3042 (17 April 1953), 395–399, on pp. 398–399.

93. E. C. Stakman, "Science and Human Affairs," *Science* n.s. 113, no. 2928 (9 February 1951), 137–141, on pp. 138, 141. Other contemporary commentators reckoned that calls for a "moratorium" on scientific progress arose precisely because of scientists' attachment to the "ivory tower." Only if scientists were seen to be actively engaged with society could they win public trust: Reginald D. Manwell, "True Scientists," *Science* n.s. 118, no. 3067 (9 October 1953), 418–419, on p. 418.

94. Anthony Standen, *Science Is a Sacred Cow* (New York: E. P. Dutton, 1950), pp. 23–26. Albert Einstein is said to have loved this book: obituary of Anthony Standen, *New York Times*, 25 June 1993, p. B7.

95. Wadsworth Likely, "Scientists and Mobilization," *Science* n.s. 112, no. 2909 (29 September 1950), 349–351, on p. 352. Likely was an official with the nonprofit Washington organization Science Service, founded by the newspaper tycoon E. W. Scripps in

1921 to advance public appreciation of science: Tobey, *The American Ideology of National Science*, pp. 66–67.

96. Alan T. Waterman, "The Science of Producing Good Scientists," *New York Times Magazine*, 31 July 1955, pp. 9, 41, 43, on p. 43.

97. See, for example, Luis W. Alvarez, *Alvarez: Adventures of a Physicist* (New York: Basic Books, 1987), esp. ch. 8; Ulam, *Adventures of a Mathematician*, pp. 188–189; Heims, *John Von Neumann and Norbert Wiener*, pp. 357–360.

98. Jacob Bronowski, *Science and Human Values* (New York: Harper Torchbook, 1959; orig. publ. 1956), p. 75.

99. Edward Appleton, "Science for Its Own Sake," *Science* n.s. 119, no. 3082 (22 January 1954), 103–109, on pp. 105, 109.

100. J. Robert Oppenheimer, "Communication and Comprehension of Scientific Knowledge," in Melvin Calvin et al., *The Scientific Endeavor: Centennial Celebration of the National Academy of Sciences* (New York: Rockefeller University Press, 1965), pp. 271–279, on p. 272.

101. Oppenheimer, "Physics in the Contemporary World," pp. 92, 94. Note, however, Freeman Dyson's observation that Oppenheimer saw *himself* at Los Alamos, and afterwards, as a philosopher-king, the last one—"a man of wisdom who could get along with other men of wisdom who also had power": quoted in Palevsky, *Atomic Fragments*, p. 102.

102. Jacques Ellul, *The Technological Society*, trans. John Wilkinson, with an introduction by Robert K. Merton (New York: Vintage, 1964; orig. publ. 1954), pp. 435–436. There is little evidence that Ellul's work was much known in America before its translation, at which point it became—as it remains—a favorite text of the anti-technocratic liberal-left.

103. Detlev W. Bronk, "Science and Humanity," *Science* n.s. 109, no. 2837 (13 May 1949), 477–482, on p. 479.

104. Anon., "Attracting Young People into Science: Clinic Session Discussions," in *Selection, Training, and Use of Personnel in Industrial Research*, eds. Hertz and Rubenstein, pp. 182–187, on p. 183.

105. Glenn T. Seaborg, *The Creative Scientist: His Training & His Role* (Oak Ridge, TN: US Atomic Energy Commission, 1964), pp. 8–9; also idem, "How to Become a Scientist [1967]," in idem, *A Scientist Speaks Out: A Personal Perspective on Science, Society and Change* (Singapore: World Scientific, 1996), pp. 229–239, on pp. 235–236; see also Marcel LaFollette, *Making Science Our Own: Public Images of Science, 1910–1955* (Chicago: University of Chicago Press, 1990), pp. 71–73.

106. Seaborg, "The Scientist as a Human Being [1964]," in idem, *A Scientist Speaks Out*, pp. 89–96, on pp. 89, 94–95.

107. Seaborg, "The Role of Basic Research [1955]," in idem, *A Scientist Speaks Out*, pp. 1–11, on pp. 3–4.

108. Anon., "National Manpower Council," *Science* n.s. 117, no. 3049 (5 June 1953), 617–622, on p. 618.

109. Stephen B. Withey, "Public Opinion about Science and Scientists," *Public Opinion Quarterly* 23 (1959), 382–388, on p. 388.

110. Arthur S. Flemming, "Nation's Interest in Scientists and Engineers," *Scientific Monthly* 82, no. 6 (June 1956), 282–285, on p. 282; see also Haynes, *From Faust to Strangelove*, pp. 1–2.

111. Margaret Mead and Rhoda Métraux, "Image of the Scientist Among High-School Students: A Pilot Study," *Science* n.s. 126, no. 3270 (30 August 1957), 384–390, on pp. 384, 387, 389; see also David C. Beardslee and Donald D. O'Dowd, "The College-Student Image of the Scientist," ibid., n.s. 133, no. 3457 (31 March 1961), 997–1001; Margenau and Bergamini, *The Scientist*, pp. 30–31.

112. Donald E. Super and Paul B. Bachrach, *Scientific Careers and Vocational Development Theory: A Review, a Critique and Some Recommendations* (New York: Teacher's College, Columbia University Bureau of Publications, 1957), p. 1. The most representative studies of the psychology of eminent academic scientists in the period are by Anne Roe: "Personality and Vocation," *Transactions of the New York Academy of Sciences* 9 (1947), 257–267; *The Making of a Scientist* (New York: Dodd, Mead & Co, 1953); "Personal Problems and Science," in *Scientific Creativity: Its Recognition and Development: Selected Papers from the Proceedings of the First, Second, and Third University of Utah Conferences: "The Identification of Creative Scientific Talent,"* eds. Calvin W. Taylor and Frank Barron (New York: John Wiley & Sons, 1963), pp. 132–138; also Bernice T. Eiduson, *Scientists: Their Psychological World* (New York: Basic Books, 1962); idem and Linda Beckman, eds., *Science as a Career Choice: Theoretical and Empirical Studies* (New York: Russell Sage Foundation, 1973).

113. Anon., "Knowledge Is Power," *Time* 70, no. 21 (18 November 1957), 20–25, on p. 22.

114. Margenau and Bergamini, *The Scientist*, pp. 32–33 and p. 116 (for quotation).

115. Donald N. Michael, "Scientists through Adolescent Eyes: What We Need to Know, Why We Need to Know It," *Scientific Monthly* 84, no. 3 (March 1957), 135–140, on pp. 136–137. For Riesman, see David Riesman (with the assistance of Reuel Denney and Nathan Glazer), *The Lonely Crowd: A Study of the Changing American Character* (New Haven, CT: Yale University Press, 1950).

116. DuBridge, "Scientists and Engineers: Quantity Plus Quality," p. 302.

117. Luis W. Alvarez, "Berkeley in the 1930s," in *All in Our Time: The Reminiscences of Twelve Nuclear Pioneers*, ed. Jane Wilson (Chicago: Bulletin of the Atomic Scientists, 1975), pp. 10–21, on p. 15.

118. John Walsh, "A Conversation with Eugene Wigner," *Science* n.s. 181, no. 4099 (10 August 1973), 527–533, on p. 533; see also Wigner, "The Limits of Science," *Proceedings of the American Philosophical Society* 94, no. 5 (1950), 422–427.

119. Dwight D. Eisenhower, "Farewell Address [1961]," in *The Military-Industrial Complex*, ed. Carroll W. Pursell, Jr. (New York: Harper and Row, 1972), pp. 204–208. This speech was written by the president's administrative assistant Malcolm Moos, who went on to become president of the University of Minnesota, but there is solid evidence that Eisenhower specifically instructed Moos on the points quoted: see Herbert York,

personal communication to the author, quoted in Steven Shapin, "Megaton Man," *London Review of Books* 24, no. 8 (25 April 2002), 18–20. The fear of technocracy was a notable theme in Ralph Lapp's *The New Priesthood*, p. 3: "The danger is that a new priesthood of scientists may usurp the traditional roles of democratic decision-making."

120. Alvin M. Weinberg, "Impact of Large-Scale Science on the United States," in *Science and Society*, ed. Norman Kaplan (Chicago: Rand-McNally, 1965), pp. 551–559 (orig. publ. *Science* n.s. 134, no. 3473 [21 July 1961], 161–164), on pp. 551–553; see also idem, *Reflections on Big Science* (Cambridge, MA: MIT Press, 1967), esp. pp. 39–40. Specific complaints about "the new cult of 'bigness'" actually emerged at least as early as 1952, when a group of left-wing American scientists, including Linus Pauling, criticized the military source of much current funding, as well as "the overemphasis on, and the preoccupation with, mammoth installations and elaborate techniques" made possible by military support: Phillips, "Dangers Confronting American Science," p. 440.

121. Lapp, *The New Priesthood*, p. 14.

122. Norbert Wiener, "A Rebellious Scientist after Two Years," *Bulletin of the Atomic Scientists* 4 (November 1948), 338–339, and quoted in Lewis S. Feuer, *The Scientific Intellectual: The Psychological and Sociological Origins of Modern Science* (New York: Basic Books, 1963), p. 399.

123. Anon., "Scientist Scorns 'Ivory Tower' Life," *New York Times*, 13 March 1949, p. 62.

124. Norbert Wiener, *I Am a Mathematician: The Later Life of a Prodigy* (Garden City, NY: Doubleday, 1956), pp. 306–307, 359–360.

125. Einstein to a journalist from the magazine *The Reporter*, 18 November 1954, quoted in Otto Nathan and Heinz Norden, eds., *Einstein on Peace* (New York: Simon and Schuster, 1960), p. 613. This remark was very widely quoted by American social commentators: see, e.g., Feuer, *The Scientific Intellectual*, p. 400; C. Wright Mills, *The Power Elite* (New York: Oxford University Press, 1956), p. 217.

126. Quoted in Helen Dukas and Banesh Hoffmann, eds., *Albert Einstein: The Human Side: New Glimpses from the Archives* (Princeton, NJ: Princeton University Press, 1979), p. 19.

127. Peter Galison, "Bubble Chambers and the Experimental Workplace," in *Experiment and Observation in Modern Science*, eds. Peter Achinstein and Owen Hannaway (Cambridge, MA: MIT Press, 1985), pp. 309–373; idem, *Image and Logic*, ch. 5.

128. Percy W. Bridgman, *Reflections of a Physicist*, 2nd ed. (New York: Philosophical Library, 1955; orig. publ. 1950), pp. 294–300.

129. E.g., James Gleick, *Genius: The Life and Science of Richard Feynman* (New York: Pantheon Books, 1992); Richard Feynman, *No Ordinary Genius: The Illustrated Richard Feynman*, ed. Christopher Sykes (New York: W. W. Norton, 1994), p. 141.

130. Bentley Glass, "The Academic Scientist, 1940–1960," *Science* n.s. 132, no. 3427 (2 September 1960), 598–603, on p. 603.

131. Paul Weiss, "Experience and Experiment in Biology," *Science* n.s. 136, no. 3515 (11 May 1962), 468–471; see also Spencer Klaw, *The New Brahmins: Scientific Life in America*

(New York: William Morrow, 1968), pp. 255–256 (for approving remarks on Weiss's essay).

132. Chargaff, *Heraclitean Fire*, pp. 139–140.

133. Erwin Chargaff, "Amphisbaena," in idem, *Essays on Nucleic Acids* (Amsterdam: Elsevier, 1963), pp. 174–199, on p. 197.

134. Ibid., p. 175.

135. Erwin Chargaff, "A Few Remarks on Nucleic Acids, Decoding, and the Rest of the World," in idem, *Essays on Nucleic Acids*, pp. 161–173, on pp. 161, 163.

136. Chargaff, *Heraclitean Fire*, p. 138.

137. Ibid., p. 117.

138. Gunther S. Stent, "Philosophy: From Metaphysics to Language Philosophy," *Partisan Review* 64 (1997), 323–330 (comments in symposium). (I owe this reference to Charles Thorpe.) See also idem, "DNA," in *The Twentieth-Century Sciences: Studies in the Biography of Ideas*, ed. Gerald Holton (New York: W. W. Norton & Company, 1970), pp. 198–226.

139. Robert K. Merton, "The Normative Structure of Science," in idem, *The Sociology of Science*, ed. Norman W. Storer (Chicago: University of Chicago Press, 1973; art. orig. publ. 1942), pp. 267–278, on p. 276.

140. William Broad and Nicholas Wade, *Betrayers of the Truth: Fraud and Deceit in the Halls of Science* (New York: Simon and Schuster, 1982), p. 19; see also Judy Sarasohn, *Science on Trial: The Whistle-Blower, the Accused, and the Nobel Laureate* (New York: St. Martin's Press, 1993), p. 266, and Horace Freeland Judson, *The Great Betrayal: Fraud in Science* (New York: Harcourt Inc., 2004), ch. 5.

141. Broad and Wade, *Betrayers of the Truth*, pp. 60–61.

142. Judson, *The Great Betrayal*, esp. pp. 29–34, 40, 414 (quoting p. 40). Among the wilder charges in Judson's performance is the suggestion (p. 41) that what he calls "social constructionism" in the study of science arose from the 1970s because it described increasingly common strands of fraudulent science (not science in general) and that it may even have provided a license for fraud.

143. Charles E. Rosenberg, "Preface: Science in Play," in idem, *No Other Gods: Science and American Social Thought*, new and revised ed. (Baltimore, MD: Johns Hopkins University Press, 1997), pp. ix–xvi, on p. xi.

144. Richard Lewontin, "Dishonesty in Science," *New York Review of Books* 51, no. 18 (18 November 2004), 38–40, on p. 39.

145. E.g., Daniel S. Greenberg, *Science, Money, and Politics: Political Triumph and Ethical Erosion* (Chicago: University of Chicago Press, 2001), esp. pp. 348–349, 351.

146. Many of his columns for *Science* from the mid-1960s and 1970s were collected as Daniel S. Greenberg, *The Grant Swinger Papers* (Washington, DC: Science & Government Report, Inc., 1981).

147. Some years ago, I asked Greenberg how he had come to acquire his iconoclastic style of science reporting, not so common among his colleagues in the trade. He replied then (and has more recently confirmed my recollection) that he had started out as a police reporter, absorbing that profession's assumption that "the authorities were not

addicted to truthfulness, that the bastards were lying, lots, often maybe most of the time." He saw no problems in applying the crime reporter's cynicism straightforwardly to science politics. (Personal communication with author, 4 August 2004.)

148. Greenberg, *Science, Money, and Politics*, pp. 352–353.

149. Joel A. Snow, review of Daniel S. Greenberg, *The Politics of Pure Science*, *Bulletin of the Atomic Scientists* 24, no. 5 (May 1968), 34–36, on p. 34.

150. Robert M. Hutchins, *No Friendly Voice* (Chicago: University of Chicago Press, 1936), pp. 28, 155.

151. Robert M. Hutchins, "Science, Scientists, and Politics," in Hutchins et al., *Science, Scientists, and Politics*, an Occasional Paper on the Role of Science and Technology in the Free Society (Santa Barbara, CA: Center for the Study of Democratic Institutions, 1963), pp. 1–4, on pp. 1–2.

152. E.g., Stuart W. Leslie, *The Cold War and American Science: The Military-Industrial-Academic Complex at MIT and Stanford* (New York: Columbia University Press, 1994).

153. Feuer, *The Scientific Intellectual*, pp. 394, 396. Feuer (1913–2002) was a Harvard-educated sociologist who was teaching at Berkeley when this book appeared.

154. Lewis A. Coser, *Men of Ideas: A Sociologist's View* (New York: Free Press, 1965). Coser was active in New York left-wing Jewish intellectual circles after the war, later teaching sociology at Chicago, Berkeley, Brandeis, and SUNY Stony Brook. A public intellectual, he was one of the founders of *Dissent*.

155. Coser, *Men of Ideas*, pp. viii–x. And recall C. P. Snow's endorsement of the mathematician G. H. Hardy's complaint in the 1930s: "Have you noticed how the word 'intellectual' is used nowadays? There seems to be a new definition which certainly doesn't include Rutherford or Eddington or Dirac or Adrian or me. It does seem rather odd, don't y' know": Snow, *The Two Cultures*, p. 4.

156. Coser, *Men of Ideas*, p. viii; also p. 311.

157. Ibid., pp. 296, 298.

158. Ibid., p. 298.

159. Ibid., pp. 308, 311.

160. Snow, *The Two Cultures*, p. 10.

CHAPTER FOUR

1. E.g., David A. Hounshell, "The Evolution of Industrial Research in the United States," in *Engines of Innovation: U.S. Industrial Research at the End of an Era*, eds. Richard S. Rosenbloom and William J. Spencer (Cambridge, MA: Harvard University Press, 1996), pp. 13–85; John J. Beer, "Coal Tar Dye Manufacture and the Origins of the Modern Industrial Research Laboratory," *Isis* 49 (1958), 123–131; Leonard S. Reich, *The Making of American Industrial Research: Science and Business at GE and Bell, 1876–1926* (Cambridge: Cambridge University Press, 1985); John Rae, "The Application of Science to Industry," in *The Organization of Knowledge in Modern America, 1860–1920*, eds. Alexandra Oleson and John Voss (Baltimore, MD: Johns Hopkins University Press, 1979), pp. 249–268. Unless specifically indicated, the term "scientist" can be taken in these

connections to include academically qualified "engineers." Turn-of-the-century participants themselves tended to draw a distinction between so-called "Edisonian" empirical, "hit-or-miss" patterns of invention in the 1880s and 1890s and the more systematic applications of scientific knowledge represented by research in such companies as General Electric, AT&T, and Eastman Kodak.

2. John Mills, "Who Is the Research Man?" *Technology Review* 44 (1942), 451–452, 466, 468–469, esp. pp. 451–452; Ely Kahn, *The Problem Solvers: A History of Arthur D. Little, Inc.* (Boston: Little, Brown, 1986); Arnold Thackray, Jeffrey L. Sturchio, P. Thomas Carroll, and Robert Bud, *Chemistry in America, 1876–1976* (Dordrecht: D. Reidel, 1985); John Kenly Smith, Jr., "The Scientific Tradition in American Industrial Research," *Technology and Culture* 31 (1990), 121–131; Daniel J. Kevles, *The Physicists: The History of a Scientific Community in Modern America* (New York: Alfred A. Knopf, 1977), pp. 98–101, 148–149.

3. Charles L. Reese, "Scientific Ideals," *Science* n.s. 80, no. 2075 (5 October 1934), 299–303, esp. p. 301. (This was the presidential address of the American Chemical Society.)

4. Michael Aaron Dennis, "Accounting for Research: New Histories of Corporate Laboratories and the Social History of American Science," *Social Studies of Science* 17 (1987), 479–518, on p. 487. For Chandler, see Alfred D. Chandler, *The Visible Hand: The Managerial Revolution in American Business* (Cambridge, MA: Belknap Press of Harvard University Press, 1977).

5. Hounshell, "The Evolution of Industrial Research," p. 36. As Secretary of Commerce, the engineer Herbert Hoover enlisted in the cause, while stressing the dependence of commercial technology on basic science: Hoover, "The Vital Need for Greater Financial Support to Pure Science Research," *Mechanical Engineering* 48 (January 1926), 6–8.

6. Jonathan Liebenau, *Medical Science and Medical Industry: The Formation of the American Pharmaceutical Industry* (Baltimore, MD: Johns Hopkins University Press, 1987); John P. Swann, *Academic Scientists and the Pharmaceutical Industry: Cooperative Research in Twentieth-Century America* (Baltimore, MD: Johns Hopkins University Press, 1988). American industrial research and development (R&D) directed towards producing new drugs was an outgrowth of World War I, and, especially, of the seizure of German patent rights. In the early 1930s, the discovery of sulfa antibiotics gave additional impetus to industrial pharmaceutical research.

7. Kendall Birr, *Pioneering in Industrial Research: The Story of the General Electric Research Laboratory* (Washington, DC: Public Affairs Press, 1957), pp. 19–20.

8. George D. McLaughlin, "Research and Industry: Cooperation between Industry and University," *Scientific Monthly* 22, no. 4 (April 1926), 281–284.

9. For earlier sentiment about research and firm size, see Peter F. Drucker, "Management and the Professional Employee," *Harvard Business Review* 30, no. 3 (May–June 1952), 84–90, on p. 84; Elder, "Basic Research in Industry," p. 5.

10. See, notably, Harold J. Cook, *Matters of Exchange: Commerce, Medicine, and Science in the Dutch Golden Age* (New Haven, CT: Yale University Press, 2007).

11. Arthur D. Little, "Research: The Mother of Industry," *Scientific Monthly* 19, no. 2 (August 1924), 165–169, on p. 168.

12. Maurice Holland, with Henry F. Pringle, *Industrial Explorers* (New York: Harper & Brothers, 1928), p. 7.

13. Frank B. Jewett, "The Future of Industrial Research: The View of a Physicist," in Standard Oil Development Company, *The Future of Industrial Research: Papers and Discussion* (New York: Standard Oil Development Company, 1945), pp. 17–23, on p. 18.

14. S. C. Gilfillan, *The Sociology of Invention* (Cambridge, MA: MIT Press, 1970; orig. publ. 1935), pp. 53–54.

15. Francis Bello, "The World's Greatest Industrial Laboratory," *Fortune* (November 1958), 148–157, 208, 212, 214, 219–220, on p. 149. The aphorism echoes A. N. Whitehead, *Science and the Modern World* (London: Scientific Book Club, 1946; orig. publ. 1926), p. 120: "the greatest invention of the nineteenth century was the invention of the method of invention."

16. F. W. Preston, "Freedom of Research," *Scientific Monthly* 61, no. 6 (December 1945), 477–482, on p. 481.

17. Kettering's comments are from discussion in Standard Oil Development Company, *The Future of Industrial Research*, pp. 26–27.

18. Quoted in Bello, "The World's Greatest Industrial Laboratory," pp. 212, 214.

19. James Tait Elder, "Basic Research in Industry: Appraisal and Forecast," *Research Management* 6 (1963), 5–14, on pp. 6–7; also Sherman Kingsbury, "Organizing for Research," in *Handbook of Industrial Research Management*, ed. Carl Heyel (New York: Reinhold Publishing Corp., 1959), pp. 65–99, on pp. 71–72; P. G. Nutting, "Research and the Industries," *Scientific Monthly* 7, no. 2 (August 1918), 149–157, on p. 151: "Industrial research cannot be distinguished from 'pure' research, except that in one case it is scientific results that are the by-products, while, in the other it is the results of commercial interest which are regarded as incidental." For fine treatment of the later career of the so-called linear model causally relating pure and applied research, see Donald E. Stokes, *Pasteur's Quadrant: Basic Science and Technological Innovation* (Washington, DC: Brookings Institution Press, 1997), esp. chs. 1–2, and see the provocation by David Edgerton, " 'The Linear Model' Did Not Exist: Reflections on the History and Historiography of Science and Research in Industry in the Twentieth Century," in *The Science–Industry Nexus: History, Policy, Implications*, Nobel Symposium 123, eds. Karl Grandin, Nina Wormbs, and Sven Widmalm (Canton, MA: Science History Publications, 2004), pp. 31–57.

20. E.g., Vannevar Bush, *Science — The Endless Frontier: A Report to the President on a Program for Postwar Scientific Research*, National Science Foundation 40th Anniversary Edition (Washington, DC: National Science Foundation, 1990; orig. publ. 1945), p. 87 (for a 5% estimate). C. E. Kenneth Mees and John A. Leermakers, *The Organization of Industrial Scientific Research*, 2nd ed. (New York: McGraw-Hill, 1950), p. 44, estimated that a typical industrial research laboratory divided its labors between 10–20% fundamental research, 40–60% new product development, and 30–40% improvement of existing products or processes.

21. National Science Foundation, *National Patterns of Science and Technology Resources 1984* (Washington, DC: U.S. Government Printing Office, 1984), p. 33, as reproduced in Martin Kenney, *Biotechnology: The University-Industrial Complex* (New Haven, CT: Yale University Press, 1986), p. 35; also Ronald L. Geiger, "What Happened After Sputnik? Shaping University Research in the United States," *Minerva* 35 (1997), 349–367, on p. 360. The proportion of basic research done in universities rose swiftly (from about a quarter to about a half of the total) during subsequent decades, the effect of burgeoning Federal support of academic science in the post-Sputnik decades. Spending on the support of academic research quadrupled in the ten years after Sputnik (ibid., p. 365), but as late as 1984, industry was still doing about 20% of *all* American basic research.

22. Figures from Hounshell, "The Evolution of Industrial Research," p. 24; see also Bud, "Strategy in American Cancer Research," pp. 429–431.

23. Bush, *Science — The Endless Frontier*, Table I, p. 86; see also David M. Hart, *Forged Consensus: Science, Technology, and Economic Policy in the United States, 1921–1953* (Princeton, NJ: Princeton University Press, 1998), ch. 2.

24. Kevles, *The Physicists*, p. 273.

25. Figures assembled from various contemporary sources, including: National Resources Planning Board, *Research — A National Resource, II: Industrial Research*, Report of the National Research Council to the National Resources Planning Board, December 1940 (Washington, DC: Government Printing Office, 1941), esp. pp. 5–16; John R. Steelman, *Science and Public Policy: A Report to the President by John R. Steelman, Chairman, the President's Scientific Research Board*, 5 vols. (Washington, DC: Government Printing Office, 1947), Vol. I, p. 10; Mees and Leermakers, *The Organization of Industrial Scientific Research*, pp. 15–16; George Perazich, "Growth Rate of Industrial Research," *Science* n.s. 114, no. 2970 (30 November 1951), 3a; Yale Brozen, "The Economic Future of Research and Development," *Industrial Laboratories* 4, no. 12 (December 1953), 6–13, on pp. 6–7.

26. Arthur Gerstenfeld, *Effective Management of Research and Development* (Reading, MA: Addison-Wesley, 1970), pp. 16–18; F. Russell Bichowsky, *Industrial Research* (Brooklyn, NY: Chemical Publishing Co., 1942), p. iii; Don K. Price, *Government and Science: Their Dynamic Relation in American Democracy* (New York: New York University Press, 1954), p. 35.

27. For DuPont, see David A. Hounshell and John Kenly Smith, Jr., *Science and Corporate Strategy: DuPont R&D, 1902–1980* (Cambridge: Cambridge University Press, 1988), p. 287; for Bell Labs, Bello, "The World's Greatest Industrial Laboratory," p. 149.

28. National Science Foundation, Reviews of Data on Research and Development, NSF 59-46, no. 14, August 1959, pp. 1–2, as quoted in Simon Marcson, The Scientist in American Industry: Some Organizational Determinants in Manpower Utilization (New York: Harper & Brothers, 1960), p. 3. For formal attention to the problem of the professional in industry, see, notably, Drucker, "Management and the Professional Employee."

29. E.g., M. L. Tainter, "An Industrial View of Research Trends," *Science* n.s. 103, no. 2665 (25 January 1946), 95–99, on p. 97; Bush, *Science — The Endless Frontier*, p. 179; Ewan Clague, "Trends in Supply and Demand of Scientific Personnel," *Science* n.s. 107,

no. 2780 (9 April 1948), 355–360, on p. 356. For a survey of some academic consequences of Cold War anxieties about a specific shortage of physicists, see David Kaiser, "Cold War Requisitions, Scientific Manpower, and the Production of American Physicists after World War II," *Historical Studies in the Physical Sciences* 33 (2002), 131–159, and for continuity in the production of American physicists, see Spencer Weart, "The Physics Business in America, 1919–1940: A Statistical Reconnaissance," in *The Sciences in the American Context: New Perspectives*, ed. Nathan Reingold (Washington, DC: Smithsonian Institution Press, 1979), pp. 295–358.

30. Drucker, "Management and the Professional Employee," p. 84.

31. James Rowland Angell, "The Organization of Research," *Scientific Monthly* 11, no. 1 (July 1920), 25–42, on p. 30; C. E. Kenneth Mees, *The Organization of Industrial Scientific Research* (New York: McGraw-Hill, 1920), p. 104.

32. Charles V. Kidd, "The Federal Government and the Shortage of Scientific Personnel," *Science* n.s. 105, no. 2717 (24 January 1947), 84–88, on p. 86; also Clague, "Trends in Supply and Demand of Scientific Personnel," pp. 356–357.

33. Henry DeWolf Smyth, speaking before the American Association for the Advancement of Science in 1951, as quoted in Kaiser, "Cold War Requisitions," p. 138. Scientists such as I. I. Rabi were, however, disturbed by the stockpiling analogy, criticizing "the increasing tendency to treat science and the scientist as commodities" and to stockpile them as so much "tungsten or copper": quoted in Donald N. Michael, "Scientists through Adolescent Eyes: What We Need to Know, Why We Need to Know It," *Scientific Monthly* 84, no. 3 (March 1957), 135–140, on pp. 136–137.

34. Phillip N. Powers, "Industrial Research Workers and Defense," in *Selection, Training, and Use of Personnel in Industrial Research*, Proceedings of the Second Annual Conference on Industrial Research June 1951, eds. David B. Hertz and Albert H. Rubenstein (New York: King's Crown Press, 1952), pp. 94–112, on p. 98; see also Kevles, *The Physicists*, pp. 384–385.

35. Anon., "Researchers Encourage Students to Seek Careers in Science," *Industrial Laboratories* 9, no. 1 (January 1958), 28–29; see also Bentley Glass, "The Academic Scientist, 1940–1960," *Science* n.s. 132, no. 3427 (2 September 1960), 598–603, esp. pp. 598–599; Ralph M. Hower and Charles D. Orth III, *Managers and Scientists: Some Human Problems in Industrial Research Organizations* (Boston: Graduate School of Business Administration, Harvard University, 1963), p. 3: "The general public, especially since Sputnik I soared into orbit, has become increasingly aware that our survival as a nation depends upon the ability of scientists and engineers to solve theoretical and practical problems of tremendous scale and complexity."

36. E. M. Kipp, "Introduction of the Newly Graduated Scientist to Industrial Research," *Research Management* 3 (1960), 39–47, on p. 39. In the same year, a conference on *The Rate and Direction of Inventive Activity* mobilized economists to explore the possibilities for rationally managing the phenomenon: again, the salience of the enterprise was recognized as "the cold war and the growing awareness that our national security may depend on the output of our military research and development effort": Richard R. Nelson, introduction, in Universities-National Bureau Committee for Economic Research,

The Rate and Direction of Inventive Activity: Economic and Social Factors: A Conference of the Universities-National Bureau Committee for Economic Research and the Committee on Economic Growth of the Social Science Research Council (Princeton, NJ: Princeton University Press, 1962), pp. 3–16, on p. 4.

37. Henry L. Cox, "The Personal Approach in Dealing with Technical People," *Research Management* 6 (1963), 153–161, on p. 153; see also Arthur S. Flemming, "Nation's Interest in Scientists and Engineers," *Scientific Monthly* 82, no. 6 (June 1956), 282–285, on p. 283.

38. Thorstein Veblen, *The Higher Learning in America: A Memorandum on the Conduct of Universities by Business Men* (New York: Sagamore Press, 1957; orig. publ. 1918), pp. 23, 117.

39. Mees, *The Organization of Industrial Scientific Research*, pp. 103–104.

40. Robert Maynard Hutchins, *No Friendly Voice* (Chicago: University of Chicago Press, 1936), pp. 156–157. Hutchins thought that professors' salaries *should* be raised: "But they will never reach such heights that the professor will lose that fine impartiality with which he customarily regards the things of this world."

41. John Morris Weiss and Charles Raymond Downs, *The Technical Organization: Its Development and Administration* (New York: McGraw-Hill, 1924), p. 30.

42. Mills, "Who Is the Research Man?," p. 452.

43. Upton Sinclair, "Albert Einstein: As I Remember Him," *Saturday Review* (14 April 1956), 17–18, 56–59, on p. 17.

44. Quoting William H. Taliaferro, "Science in the Universities," *Science* n.s. 108, no. 2798 (13 August 1948), 145–148, on p. 148; see also Steelman, *Science and Public Policy*, Vol. III, pp. 142, 208; David Bendel Hertz, *The Theory and Practice of Industrial Research* (New York: McGraw-Hill, 1950), pp. 196–199; and Walter Hirsch, *Scientists in American Society* (New York: Random House, 1968), pp. 21–25.

45. Frederick J. Hammett, "Uncommitted Researchers," *Science* n.s. 117, no. 3029 (16 January 1953), 64.

46. Quoted in Harry S. Hall, "Scientists and Politicians," in *The Sociology of Science*, eds. Bernard Barber and Walter Hirsch (New York: Free Press, 1962), pp. 269–287, on pp. 275–276; art. orig. publ. in *Bulletin of the Atomic Scientists* 12, no. 2 (February 1956), 46–52.

47. National Manpower Council, *A Policy for Scientific and Professional Manpower: A Statement by the Council with Facts and Issues Prepared by the Research Staff* (New York: Columbia University Press, 1953), p. 72; also Anon., "National Manpower Council," *Science* n.s. 117, no. 3049 (5 June 1953), 617–622; Wadsworth Likely, "Scientists and Mobilization," *Science* n.s. 112, no. 2909 (29 September 1950), 349–352; Kevles, *The Physicists*, p. 388; Spencer Klaw, *The New Brahmins: Scientific Life in America* (New York: William Morrow & Co., 1968), pp. 75ff. Sidney W. Benson, "Sponsored Research," *Science* n.s. 116, no. 3009 (29 August 1952), 233, argued that academic research sponsored by industry could augment salaries and "so keep good men at academic posts."

48. Angell, "The Organization of Research," p. 29; see also Mills, "Who Is the Research Man?" p. 452.

49. John T. Walsh, preface to Paul Freedman, *The Principles of Scientific Research* (Washington, DC: Public Affairs Press, 1950; orig. publ. London: Macdonald, 1949), p. vii.

50. J. B. Kahn, Jr., " 'True' Scientists," *Science* n.s. 117, no. 3051 (19 June 1953), 697–698, on p. 697.

51. Hans Elias, " 'True' Scientists," *Science* n.s. 117, no. 3051 (19 June 1953), 698.

52. Glenn T. Seaborg, *Adventures in the Atomic Age: From Watts to Washington* (New York: Farrar, Straus and Giroux, 2001), pp. 131–132. Seaborg nevertheless elected to return to Berkeley at less than half the salary Chicago offered, though with promotion to full professor.

53. Kidd, "The Federal Government and the Shortage of Scientific Personnel," pp. 87–88.

54. Kevles, *The Physicists*, pp. 273–274.

55. U.S. Department of Labor, *Education, Employment and Earnings of American Men of Science*, Bulletin 1023 (Washington, DC: Government Printing Office, 1951), pp. 30–31; also Theresa R. Shapiro, "The Attitudes of Scientists toward Their Jobs," in *Human Relations in Industrial Research Management, Including Papers from the Sixth and Seventh Annual Conferences on Industrial Research: Columbia University, 1955 and 1956*, eds. Robert Teviot Livingstone and Stanley H. Milberg (New York: Columbia University Press, 1957), pp. 151–162, on pp. 157–158.

56. M. H. Trytten and Theresa R. Shapiro, "The Earnings of American Men of Science," *Science* n.s. 113, no. 2935 (30 March 1951), 345–347, on pp. 346–347. The survey documented a somewhat higher supplementary income for academics—whose nine-month contracts allowed them to consult and do summer-school teaching in the noncontracted months—but this added income did little to close the gap with industry.

57. *Fortune* editorial staff, "The Scientists," *Fortune* (October 1948), 106–112, 166, 168, 170, 173–174, 176, on pp. 108–109. This unsigned report was likely written (wholly or partly) by William H. Whyte, later famous for *The Organization Man* of 1956. For fine treatment of these themes, see David Kaiser, "The Postwar Suburbanization of American Physics," *American Quarterly* 56 (2004), 851–888. Five years later, the National Manpower Council's statistics put the median annual income of Ph.D. physicists at $6,400 for academics and $8,000 for both government and industrial employees. The median for the most senior university physicists was $7,500 while the median for the equivalent industrial physicists was $12,200: National Manpower Council, *A Policy for Scientific and Professional Manpower*, p. 192; see also David Kaiser, "Making Theory: Producing Physics and Physicists in Postwar America," unpubl. Ph.D. thesis, Harvard University, 2000, p. 91; idem, "The Postwar Suburbanization of American Physics," p. 869.

58. Anon., "Knowledge Is Power," *Time* 70, no. 21 (18 November 1957), 20–25, on p. 22.

59. Francis Bello, "The Young Scientists," in The Editors of *Fortune, The Mighty Force of Research* (New York: McGraw-Hill, 1956; art. orig. publ. 1953), pp. 21–39, on p. 35.

60. National Science Foundation, "Salaries and Selected Characteristics of American Scientists, 1966," *Reviews of Data on Science Resources*, no. 11 (Washington, DC: Government Printing Office, 1967), quoted in Hirsch, *Scientists in American Society*, pp. 4–5, 61.

61. Quoted in Kaiser, "The Postwar Suburbanization of American Physics," p. 869; idem, "Making Theory," p. 91.

62. Theresa R. Shapiro, "What Scientists Look For in Their Jobs," *Scientific Monthly* 76, no. 6 (June 1953), 335–340, on p. 339. Of all scientists interviewed in this study who changed jobs between 1939 and 1952, less than a quarter cited financial considerations, and about the same proportion mentioned "job interest" as decisive.

63. Powers, "Industrial Research Workers and Defense," pp. 95–96. Powers noted that defense-related industry competed with the Selective Service (the military draft) for skilled manpower, even before the outbreak of the Korean War. On physicists and the draft during the Korean War, see Kaiser, "Cold War Requisitions," especially pp. 144–146.

64. Charles Allen Thomas, "Creativity in Science: A Vital Human Resource, Part I," *Industrial Laboratories* 6, no. 10 (October 1955), 68–69, on p. 69. (Thomas had coordinated plutonium production for the Manhattan Project during World War II.)

65. National Resources Planning Board, *Research—A National Resource, II*: p. 9; also pp. 117–118; Mees and Leermakers, *The Organization of Industrial Scientific Research*, pp. 292, 298–299; Harold K. Work, "The University's Role in Training Research Workers," in *Selection, Training, and Use of Personnel*, eds. Hertz and Rubenstein, pp. 126–140, on p. 126.

66. F. W. Blair and N. Beverley Tucker, "Salary Policy," in *Research in Industry: Its Organization and Management*, ed. C. C. Furnas (New York: D. Van Nostrand, 1948), pp. 258–276, esp. pp. 259, 274; also Warner Eustis, "Personnel Policies and Personality Problems," in ibid., pp. 277–294; Mees and Leermakers, *The Organization of Industrial Scientific Research*, p. 297.

67. Weiss and Downs, *The Technical Organization*, pp. 31–32.

68. Mees, *The Organization of Industrial Scientific Research*, p. 103.

69. Charles D. Orth III, "The Optimum Climate for Industrial Research," in *Science and Society*, ed. Norman Kaplan (Chicago: Rand-McNally, 1965; art. orig. publ. 1959), pp. 194–210, on p. 201.

70. George Wise, *Willis R. Whitney: General Electric, and the Origins of U.S. Industrial Research* (New York: Columbia University Press, 1985), p. 62; Hounshell, "The Evolution of Industrial Research," p. 22.

71. Laurence A. Hawkins, *Adventure into the Unknown: The First Fifty Years of the General Electric Research Laboratory* (New York: William Morrow & Co., 1950), p. 4.

72. Reich, *The Making of American Industrial Research*, pp. 69–70, 75, 82–83, 92, 111, 126–127; Wise, *Willis R. Whitney*, pp. 119–126; Hounshell, "The Evolution of Industrial Research," pp. 22–23.

73. James B. Conant, *My Several Lives: Memoirs of a Social Inventor* (New York: Harper & Row, 1970), pp. 25–26. Hawkins (*Adventure into the Unknown*, p. 31) says that Whitney

offered Coolidge half-time at GE to continue his MIT research on electrical conduction in aqueous solutions; see also Patrick J. McGrath, *Scientists, Business, and the State, 1890–1960* (Chapel Hill: University of North Carolina Press, 2002), pp. 35 and 208 n. 6 (for comments about Conant's views on corporate research and its suitability for first-rate chemists).

74. Steelman, *Science and Public Policy*, Vol. III, pp. 142, 208; see also Hertz, *The Theory and Practice of Industrial Research*, pp. 196–199.

75. N. A. Shepard, "The Research Director's Job," in Furnas, ed., *Research in Industry*, pp. 56–70, on pp. 65–66.

76. Raymund L. Zwemer, "Incentives from the Viewpoint of a Scientist," in *Scientific Research: Its Administration and Organization*, eds. George P. Bush and Lowell H. Hattery (Washington, DC: American University Press, 1950), pp. 75–78, on pp. 75, 77.

77. Figures assembled from various contemporary sources cited in n. 25 above; also Darrell H. Voorhies, *The Co-ordination of Motive, Men and Money in Industrial Research, A Survey of Organization and Business Practice Conducted by the Department on Organization of the Standard Oil Company of California* (San Francisco: Standard Oil Company of California, 1946), pp. 3–5. In 1948, Ralph T. Cornwell, director of research at the Sylvania Division of the American Viscose Corporation, estimated the number of industry-employed "men in pursuit of new knowledge" at 100,000: "Professional Growth of the Research Man," in Furnas, ed., *Research in Industry*, pp. 295–307, on pp. 295–296.

78. *Fortune* editorial staff, "The Scientists," p. 109; National Science Foundation, *Scientific Personnel Resources: A Summary of Data on Supply Utilization and Training of Scientists and Engineers* (Washington, DC: Government Printing Office, 1955), p. 14 (Chart IV); quoted in William Kornhauser (with the assistance of Warren O. Hagstrom), *Scientists in Industry: Conflict and Accommodation* (Berkeley: University of California Press, 1962), p. 10; also National Manpower Council, *A Policy for Scientific and Professional Manpower*, p. 47.

79. Daniel J. Kevles, Jeffrey L. Sturchio, and P. Thomas Carroll, "The Sciences in America, Circa 1880," *Science* n.s. 209, no. 4452 (4 July 1980), 26–32, on p. 30 (for the 90% figure); Warren O. Hagstrom, *The Scientific Community* (Carbondale: Southern Illinois University Press, 1975; orig. publ. 1965), p. 38 (for the other figures).

80. Albert E. Hickey, Jr., "Basic Research: Should Industry Do More of It?" *Harvard Business Review* 36, no. 4 (July–August 1958), 115–122, on pp. 116–117; see also George P. Bush, "Principles of Administration in the Research Environment," in Bush and Hattery, eds., *Scientific Research: Its Administration and Organization*, pp. 161–183, on p. 164.

81. It was not until the 1960s that some academic social scientists began seriously to come to terms with this shift in institutional habitat: e.g., Hirsch, *Scientists in American Society*, p. 60.

82. Robert K. Merton, "The Normative Structure of Science," in idem, *The Sociology of Science*, ed. Norman W. Storer (Chicago: University of Chicago Press, 1973; art. orig. publ. 1942), pp. 267–278, on pp. 268, 270. For an extended account of what Merton meant by "socialization," and for an edgily defensive explanation that this was quite

different from the processes of "socialism," see Robert K. Merton, "Some Preliminaries to a Sociology of Medical Education," in *The Student-Physician: Introductory Studies in the Sociology of Medical Education*, eds. Merton, George G. Reader, and Patricia L. Kendall (Cambridge, MA: Harvard University Press, 1957), pp. 3–79, on pp. 41–43, 287–293 (endnotes to chapter).

83. See Steven Shapin, "Who Is the Industrial Scientist? Commentary from Academic Sociology and from the Shop-Floor in the United States, ca. 1900–ca. 1970," in *The Science—Industry Nexus* eds. Grandin, Wormbs, and Widmalm, pp. 337–363.

84. Merton, "The Normative Structure of Science," pp. 273, 275, 278. For parallel contemporary sentiments, see Mark A. May, "The Moral Code of Scientists," in *The Scientific Spirit and Democratic Faith* (New York: King's Crown Press, 1943), pp. 40–45, on p. 41: "All results of science, particularly those that have been published, are common property to men everywhere . . . Scientists who work for industries are in a somewhat different position because the firm that employs them does so with the idea of profiting by their discoveries"; see also Bernard Barber, *Science and the Social Order* (Glencoe, IL: Free Press, 1952), esp. pp. 131–133; Everett C. Hughes, *Men and Their Work* (Glencoe, IL: Free Press, 1958), p. 140: "Scientists chafe under secrecy . . . The great point in the scientist's code is full and honest reporting to his colleagues."

85. Merton, "The Normative Structure of Science," pp. 260, 276.

86. A partial exception to this generalization is the coedited volume: Robert K. Merton, George C. Reader, and Patricia L. Kendall, eds., *The Student-Physician: Introductory Studies in the Sociology of Medical Education* (Cambridge, MA: Harvard University Press, 1957), though Merton's own contribution to this volume—"Some Preliminaries to a Sociology of Medical Education" (pp. 3–79)—gives no clear indications of his own empirical familiarity with the medical practice or medical pedagogy.

87. Marcson, *The Scientist in American Industry*, pp. 5, 145; also idem, "The Professional Commitments of Scientists in Industry," *Research Management* 4 (1961), 271–275, on p. 271. Marcson's work was substantially based on "participant observation" and "semistructured depth interviews" with about one-fifth of the scientists in the research laboratory of an unnamed large electronics company, possibly RCA.

88. Richard S. Crog, "Ethics and Integrity in Personnel Relations," *Research Management* 7 (1964), 183–194, on p. 187.

89. Simon Marcson, "Role Adaptation of Scientists in Industrial Research," *IRE Transactions on Engineering Management*, EM-7, no. 4 (1960), 159–166, on p. 161.

90. Marcson, "Role Adaptation," p. 162; idem, *The Scientist in American Industry*, p. 148.

91. Marcson, "Professional Commitments of Scientists in Industry," p. 271.

92. Marcson, "Role Adaptation," pp. 163, 166.

93. E.g., Marcson, *The Scientist in American Industry*, p. 126.

94. William Kornhauser, "Strains and Accommodations in Industrial Research Organizations in the United States," *Minerva* 1 (1962), 30–42, on p. 30 (emphases in the original); see also idem, *Scientists in Industry*, ch. 1 (for an extended exposition of the "strain" thesis).

95. Kornhauser, *Scientists in Industry*, p. 41. Hagstrom soon developed this notion that the organizational forms of Big Science were transforming many scientists into "professional technicians," viewing their work as contract labor and being deflected from their proper commitment to professional norms: Warren O. Hagstrom, "Traditional versus Modern Forms of Scientific Teamwork," *Administrative Science Quarterly* 9 (1964), 241–263; idem, *The Scientific Community*, esp. pp. 149–152.

96. Kornhauser, "Strains and Accommodations," p. 32.

97. Kornhauser, *Scientists in Industry*, pp. 69–70, 73–74.

98. Kornhauser's empirical materials were more diverse than Marcson's. Kornhauser and his associate interviewed an unspecified number of research scientists, engineers, and managers in six industrial laboratories, a trade association laboratory, a government laboratory, and an independent research institute. It probably follows from this that Kornhauser's work better documents *variation* in patterns among organizations: "strains and accommodations between organization and profession vary according to the kind of organization": Kornhauser, *Scientists in Industry*, p. 82.

99. For other contributions to this genre, see, among many examples, William J. McEwen, "Position Conflict and Professional Orientation in a Research Organization," *Administrative Science Quarterly* 1 (1956), 208–224; Mark Abrahamson, "The Integration of Industrial Scientists," ibid. 9 (1964), 208–218; Todd La Porte, "Conditions of Strain and Accommodation in Industrial Research Organizations," ibid. 10 (1965), 21–38.

100. At least as early as 1920, Ordway Tead was offering "general principles" for the administration of human relations in industry which were founded not on "transient industrial conditions" but on solid "modern knowledge of human nature and its constituent elements." Tead was critical of F. W. Taylor for ignoring scientific psychology (as were later exponents of the so-called human relations school), but he was nevertheless confident that "scientific methods were applicable to problems of human relations": Ordway Tead and Henry C. Metcalf, *Personnel Administration: Its Principles and Practice* (New York: McGraw-Hill, 1920), pp. vii, 25, and, for criticism of Taylor, p. 256.

101. Edward H. Litchfield, "Notes on a General Theory of Administration," *Administrative Science Quarterly* 1 (1956), 3–29; also James D. Thompson, "On Building an Administrative Science," ibid., 102–111; Talcott Parsons, "Suggestions for a Sociological Approach to the Theory of Organizations (I) and (II)," ibid., 63–85, 225–239. For remarks on the early history and ambitions of *ASQ*, see Donald Palmer, "Taking Stock of the Criteria We Use to Evaluate One Another's Work: *ASQ* 50 Years Out," ibid. 51 (2006), 535–559, esp. pp. 537–538.

102. At about the same time, the editors of the proceedings of the Columbia University Industrial Research Conferences (both professors of industrial and management engineering at Columbia) said that their problem was to "study the job of research management, to find out what it has in common with the general management process and where it is unique . . . We are hopeful," they wrote, "that it is possible to begin work on a general theory of management, and it seems singularly appropriate to use the field of research as a point of reference." They lamented that at present there was

"substantially no objective core of knowledge [about research management] which can be passed on, that knowledge consists in the main of a number of rules of thumb and often unrelated and inconsistent prescriptions": Robert Teviot Livingstone and Stanley H. Milberg, preface to idem, eds., *Human Relations in Industrial Research Management*, pp. v–vii, on p. vi.

103. James D. Thompson, "Editor's Critique," *Administrative Science Quarterly* 1 (1956), 382–385, on pp. 382–383.

104. Herbert A. Shepard, "Nine Dilemmas in Industrial Research," *Administrative Science Quarterly* 1 (1956), 295–309, esp. pp. 295, 300. The paper was anthologized in Barber and Hirsch's (1962) *The Sociology of Science*, pp. 344–355, and was widely cited in the 1960s literature on scientists in organizations.

105. Shepard, "Nine Dilemmas in Industrial Research," p. 298. The Mertonian source is Robert K. Merton, "Patterns of Influence: Local and Cosmopolitan Influentials," in idem, *Social Theory and Social Structure*, revised ed. (New York: Free Press, 1957), pp. 387–420; orig. publ. as "Patterns of Influence," in *Communications Research, 1948–1949*, eds. Paul Lazarsfeld and Frank Stanton (New York: Harper and Brothers, 1949), pp. 189–202. Gouldner's systematic development of the contrast between "locals" and "cosmopolitans" (with no specific reference to industrial scientists) is Alvin W. Gouldner, "Cosmopolitans and Locals: Toward an Analysis of Latent Social Roles," *Administrative Science Quarterly* 2 (1957–1958), 281–306, 444–480. In addition to Shepard, the nomenclature and distinction of "locals" and "cosmopolitans" was soon picked up by sociologists studying a range of research organizations, e.g., Donald C. Pelz, "Some Social Factors Related to Performance in a Research Organization," *Administrative Science Quarterly* 1 (1956), 310–325; Robert W. Avery, "Enculturation in Industrial Research," *IRE Transactions on Engineering Management* EM-7, no. 1 (1960), 20–24; Norman W. Storer, "Science and Scientists in an Agricultural Research Organization: A Sociological Study," unpubl. Ph.D. thesis, Cornell University, 1961, pp. 98–99, 106–107, 170–171; Mark Abrahamson, "Cosmopolitanism, Dependence-Identification, and Geographical Mobility," *Administrative Science Quarterly* 10 (1965), 98–106. In 1963, the Merton student Barney Glaser felt obliged to synthesize the originally polar dichotomy between "local" and "cosmopolitan" scientists. Certain highly motivated scientists working in those uncommon organizations that happen to be committed to "the institutional goal of science" display hybrid "local-cosmopolitan" orientations: Barney G. Glaser, "The Local-Cosmopolitan Scientist," *American Journal of Sociology* 69 (1963), 249–259. But views that scientists in general were loyal principally to their profession and not to the organizations in which they happened to work were in lay as well as academic currency in the 1950s and 1960s, and probably trace back to classic tropes concerning the philosopher's relationship to the *polis*.

106. Charles Thorpe, "Disciplining Experts: Scientific Authority and Liberal Democracy in the Oppenheimer Case," *Social Studies of Science* 32 (2002), 525–562; idem, *Oppenheimer: The Tragic Intellect* (Chicago: University of Chicago Press, 2006), ch. 7.

107. Donald C. Pelz and Frank M. Andrews, *Scientists in Organizations: Productive Climates for Research and Development* (New York: John Wiley, 1966).

108. Anselm L. Strauss and Lee Rainwater, *The Professional Scientist: A Study of American Chemists* (Chicago: Aldine, 1962).

109. Kaplan, "Organization: Will It Choke or Promote the Growth of Science?"; Steven Box and Stephen Cotgrove, "Scientific Identity, Occupational Selection, and Role Strain," *British Journal of Sociology* 17 (1966), 20–28; Stephen Cotgrove and Steven Box, *Science, Industry and Society: Studies in the Sociology of Science* (London: George Allen and Unwin, 1970); Barry Barnes, "Making Out in Industrial Research," *Science Studies* 1 (1971), 157–175; also David Bloor and Celia Bloor, "Twenty Industrial Scientists: A Preliminary Exercise," in *Essays in the Sociology of Perception*, ed. Mary Douglas (London: Routledge & Kegan Paul, 1982), pp. 83–102.

110. William H. Whyte, Jr., *The Organization Man* (Garden City, NY: Doubleday Anchor, 1957; orig. publ. New York: Simon and Schuster, 1956). An academic journal like *ASQ* expressed mixed sentiments about *The Organization Man*: the sociologist reviewing it in 1957 enthusiastically endorsed its portrayal of creeping conformism, while some years later a contributor to *ASQ* airily dismissed the book as methodologically inadequate—"basically a literary work rather than a scientific treatise": William H. Form, review of *The Organization Man*, *Administrative Science Quarterly* 2 (1957), 124–126; John B. Miner, "Conformity among University Professors and Business Executives," ibid. 7 (1962), 96–109, on p. 97.

111. Norman Kaplan, "Introduction to Part III," *Science and Society*, ed. Kaplan, pp. 175–179, on p. 175; see also Barber and Hirsch, eds., *The Sociology of Science*. About three-quarters of the articles anthologized in the Kaplan collection, and about one quarter of those in the Barber and Hirsch volume, dealt with scientific organization and research management.

112. A. H. Hausrath, "Programs for Fuller Utilization of Present Resources of Scientific Personnel," *Science* n.s. 107, no. 2780 (9 April 1948), 360–363, on p. 362. Hausrath was director of scientific personnel for the Office of Naval Research.

113. Thomas Kuhn's celebrated paper on "The Essential Tension"—a precursor to key ideas in his seminal 1962 *The Structure of Scientific Revolutions*—was given in 1959 to a conference on scientific creativity that included contributions from officials in the Pentagon's Advanced Research Projects Agency and the Air Force Personnel and Training Research Center, as well as the Dow Chemical Company, and, while this was far from a technocratic performance, Kuhn was a commentator at conferences on scientific organization that were more frankly managerial in tone: *Scientific Creativity: Its Recognition and Development: Selected Papers from the Proceedings of the First, Second, and Third University of Utah Conferences: "The Identification of Creative Scientific Talent,"* eds. Calvin W. Taylor and Frank Barron (New York: John Wiley & Sons, 1963). For background to the military uses of human science during the Second World War, see Peter Buck, "Adjusting to Military Life: The Social Sciences Go to War, 1941–1950," in *Military Enterprise and Technological Change: Perspectives on the American Experience*, ed. Merritt Roe Smith (Cambridge, MA: MIT Press, 1985), pp. 205–252.

114. America's oldest university school of business was Pennsylvania's Wharton (1881), followed by schools at Chicago (1898), Dartmouth (1900), Harvard (1908), MIT

(1914), Columbia (1916), and Stanford (1925). Schools and institutes of industrial and labor relations were founded during and after World War II at, for example, Penn State, Cornell, and Illinois.

115. Hertz's most systematic work was *The Theory and Practice of Industrial Research* (1950), but throughout the 1950s he (and his major collaborator, Albert H. Rubenstein) produced a series of publications addressed to practical aspects of research management, with special reference to the enhancement of creativity.

116. Robert N. Anthony, *Management Controls in Industrial Research Organizations* (Boston: Graduate School of Business Administration, Harvard University, 1952), esp. pp. vii–viii, 26–28.

117. Hower and Orth, *Managers and Scientists*, pp. vii–viii, 7. The emerging, partially mathematized fields of "management science" and "operations research" also contributed to this practically oriented body of social science work through the 1950s and 1960s: Michael A. Fortun and Silvan S. Schweber, "Scientists and the Legacy of World War II: The Case of Operations Research (OR)," *Social Studies of Science* 23 (1993), 595–642, esp. pp. 620–625.

118. JoAnne Yates, *Control Through Communication: The Rise of System in American Management* (Baltimore, MD: Johns Hopkins University Press, 1989), esp. p. 256.

119. In the spring of 1947, the President's Scientific Research Board commissioned the National Opinion Research Center at the University of Denver to carry out a systematic, large-scale survey of scientists' views about "the satisfactions" of their work and the organizational conditions required for its successful prosecution, and the results were reported in Steelman, *Science and Public Policy*, Vol. III, Appendix III, pp. 205–252.

120. See, for example, the vocabulary of "strain" and "stress" in the Steelman Report, with special reference to the conflicts between scientists' desire for autonomy and free communication and the military's need for control and secrecy: Steelman, *Science and Public Policy*, Vol. III, pp. 33–35.

CHAPTER FIVE

1. Spencer Klaw, *The New Brahmins: Scientific Life in America* (New York: William Morrow & Co., 1968), p. 11.

2. Elmer W. Engstrom, "What Industry Requires of the Research Worker," in *Human Relations in Industrial Research, Including Papers from the Sixth and Seventh Annual Conferences on Industrial Research Management: Columbia University, 1955 and 1956*, eds. Robert Teviot Livingstone and Stanley H. Milberg (New York: Columbia University Press, 1957), pp. 69–79, on p. 69. (Engstrom was a radio engineer who had been a director of research at RCA since the 1940s, and in 1951—when his talk was delivered—was senior executive vice-president of that company.)

3. Carleton R. Ball, "Personnel, Personalities and Research," *Scientific Monthly* 23, no. 1 (July 1926), 33–45, on p. 35.

4. C. E. Kenneth Mees, "The Organization of Industrial Scientific Research," *Science* n.s. 43, no. 1118 (2 June 1916), 763–773, on p. 768.

5. Thomas S. Kuhn, *The Structure of Scientific Revolutions* (Chicago: University of Chicago Press, 1962).

6. John R. Steelman, *Science and Public Policy: A Report to the President by John R. Steelman, Chairman, The President's Scientific Research Board*, 5 vols. (Washington, DC: Government Printing Office, 1947), Vol. III (*Administration for Research*), pp. 28, 32.

7. Thornton Page, "Selecting the Research Team," in *Teamwork in Research*, eds. George P. Bush and Lowell H. Hattery (Washington, DC: American University Press, 1953), pp. 61–70, on pp. 62–32 (emphases added).

8. E.g., George P. Bush, "Principles of Administration in the Research Environment," in *Scientific Research: Its Administration and Organization*, eds. Bush and Lowell H. Hattery (Washington, DC: American University Press, 1950), pp. 161–183, on p. 165.

9. Carl Barus, "Research and Teaching," *Science* n.s. 57, no. 1476 (13 April 1923), 445–446, on p. 446.

10. James Phinney Baxter III, *Scientists against Time* (Boston: Little, Brown and Co., 1946), p. 12.

11. Vannevar Bush, January 26, 1945, in hearings before the Select Committee on Post-War Military Policy, 78th Congress, 2nd Session, Pursuant to House Resolution 465, pp. 244–245; as quoted in Baxter, *Scientists against Time*, p. 12.

12. E. J. Kahn, Jr., *The Problem Solvers: A History of Arthur D. Little, Inc.* (Boston: Little, Brown, 1986), on p. 43; also David F. Noble, *America by Design: Science, Technology, and the Rise of Corporate Capitalism* (New York: Alfred A. Knopf, 1977), pp. 124–125. The firm was originally founded in 1886 as Griffin & Little, performing chemical analyses of sugar, spices, metal ores, paper, ink, paints and stains, etc.

13. Mees, "The Organization of Industrial Scientific Research," p. 766; idem, *From Dry Plates to Ektachrome Film*, p. 50; Arthur W. Baum, "Doctor of the Darkroom," *Saturday Evening Post* (25 October 1947), 15–17, 47, 50, 52, on p. 17. (I thank Ray Curtin of the Eastman Kodak Company for this last reference.) See also D. E. H. Edgerton, "Industrial Research in the British Photographic Industry, 1879–1939," in *The Challenge of New Technology: Innovation in British Business Since 1850*, ed. Jonathan Liebenau (Aldershot: Gower, 1988), pp. 106–134, on p. 110. For an extended obituary, see Walter Clark, "Charles Edward Kenneth Mees: 1882–1960," *Biographical Memoirs of Fellows of the Royal Society* 7 (1961), 171–197.

14. C. E. Kenneth Mees, *The Organization of Industrial Scientific Research* (New York: McGraw-Hill, 1920), p. 102; also Reese Jenkins, *Images and Enterprise: Technology and the American Photographic Industry, 1839–1925* (Baltimore, MD: Johns Hopkins University Press, 1979), p. 308. Freedom from strict corporate accountability was a major reason why the fundamental research program that Westinghouse developed from the 1930s contributed, as one historian shows, "little to the bottom line": Thomas C. Lassman, "Industrial Research Transformed: Edward Condon at the Westinghouse Electric and Manufacturing Company, 1935–1942," *Technology and Culture* 44 (2003), 306–339, on pp. 309–310.

15. Mees, "The Organization of Industrial Scientific Research," p. 766. However, Mees freely acknowledged that the research laboratory of a photographic company was special in that, unlike other manufacturing firms, photography had no correlate

academic discipline to perform some of its inquiries into fundamental processes: as quoted in Maurice Holland, "Positive Negatives: C. E. Kenneth Mees, Director of Research Laboratory, Eastman Kodak Company," in idem, *Industrial Explorers* (New York: Harper & Brothers, 1928), pp. 224–239, on p. 229; see also C. E. Kenneth Mees and John A. Leermakers, *The Organization of Industrial Scientific Research*, 2nd ed. (New York: McGraw-Hill, 1950), p. 231 (note that this so-called second edition of Mees's 1920 book was an almost entirely different work).

16. Mees, "The Organization of Industrial Scientific Research," p. 768.

17. Mees and Leermakers, The Organization of Industrial Scientific Research, p. 237.

18. Quoted in T. A. Boyd, *Professional Amateur: The Biography of Charles Franklin Kettering* (New York: E. P. Dutton, 1957), p. 118.

19. Charles Kettering, "Head Lamp of Industry [speech delivered to U.S. Chamber of Congress, 1929]," in T. A. Boyd, ed., *Prophet of Progress: Selections from the Speeches of Charles F. Kettering* (New York: E. P. Dutton, 1961), pp. 77–89, on p. 79.

20. Darrell H. Voorhies, *The Co-ordination of Motive, Men and Money in Industrial Research, A Survey of Organization and Business Practice Conducted by the Department on Organization of the Standard Oil Company of California* (San Francisco: Standard Oil Company of California, 1946), pp. 53–54; Arthur Gerstenfeld, *Effective Management of Research and Development* (Reading, MA: Addison-Wesley, 1970), pp. 18–23 (for four-year payback and large- versus small-company figures); C. Wilson Randle, "Problems of Research and Development Management," *Harvard Business Review* 37, no. 1 (January–February 1959), 128–136, on p. 131 (for lack of formal evaluation criteria).

21. See, notably, Richard R. Nelson, "Uncertainty, Learning, and the Economics of Parallel Research and Development Efforts," *Review of Economics and Statistics* 43 (1961), 351–364; idem, "The Link between Science and Invention: The Case of the Transistor," in Universities-National Bureau Committee for Economic Research, *The Rate and Direction of Inventive Activity: Economic and Social Factors: A Conference of the Universities-National Bureau Committee for Economic Research and the Committee on Economic Growth of the Social Science Research Council* (Princeton, NJ: Princeton University Press, 1962), pp. 549–583, esp. pp. 550, 560, 562, 567–568; cf. David A. Hounshell, "The Medium Is the Message, or How Context Matters: The RAND Corporation Builds an Economics of Innovation, 1946–1962," in *Systems, Experts, and Computers: The Systems Approach in Management and Engineering, World War II and After*, eds. Thomas P. Hughes and Agatha C. Hughes (Cambridge, MA: MIT Press, 2000), pp. 255–310.

22. Richard R. Nelson, "The Economics of Invention: A Survey of the Literature," *Journal of Business* 32 (1959), 101–127, on p. 101. Nelson was decisively influenced here by the views of the Eastman Kodak's Kenneth Mees: see ibid., pp. 120–124; also idem, "The Simple Economics of Basic Scientific Research," *Journal of Political Economy* 67 (1959), 297–306, esp. pp. 302–304.

23. JoAnne Yates, *Control Through Communication: The Rise of System in American Management* (Baltimore, MD: Johns Hopkins University Press, 1989), p. 256.

24. Mees and Leermakers, *The Organization of Industrial Scientific Research*, p. 223 (Leermakers joined the Eastman Kodak Research Laboratory in 1934 and became its

director in 1964); Malcolm H. Hebb, with Miles J. Martin, "Free Inquiry in Industrial Research," *Research Management* 1 (1958), 67–83, on p. 67. Both Hebb and Martin were physicists who moved to the GE Research Laboratory from academic appointments sometime after 1955.

25. Among many examples, see Mees, *The Organization of Industrial Scientific Research*, pp. 20–21; also AT&T's John J. Carty, "Science and Business: An Address to the Chamber of Commerce of the United States, May 8, 1924," *Reprint and Circular Series of the National Research Council* (Washington, DC: National Research Council, 1929), pp. 1–2; John A. Leermakers, "Basic Research in Industry," *Industrial Laboratories* 2, no. 3 (March 1951), 2–3: "It is in the universities that true intellectual freedom, so essential to the development of the scientific attitude, is found."

26. See, for example, Rebecca S. Lowen, *Creating the Cold War University: The Transformation of Stanford* (Berkeley: University of California Press, 1997), esp. pp. 37–42; Rima D. Apple, "Patenting University Research: Henry Steenbock and the Wisconsin Alumni Research Foundation," *Isis* 80 (1989), 375–394; Henry Etzkowitz, *MIT and the Rise of Entrepreneurial Science* (London/New York: Routledge, 2002); Grischa Metlay, "Reconsidering Renormalization: Stability and Change in 20th-Century Views of University Patents," *Social Studies of Science* 36 (2006), 565–597; Daniel S. Greenberg, *Science for Sale: The Perils, Rewards, and Delusions of Campus Capitalism* (Chicago: University of Chicago Press, 2007).

27. Norman A. Shepard, "The Research Director's Job," in *Research in Industry: Its Organization and Management*, ed. C. C. Furnas (New York: D. Van Nostrand, 1948), pp. 56–70, on pp. 58–59; see also John A. Leermakers, "Basic Research in Industry," *Industrial Laboratories* 2, no. 3 (March 1951), 2–3.

28. John Morris and Charles Raymond Downs, *The Technical Organization: Its Development and Administration* (New York: McGraw-Hill, 1924), pp. 156–192, on p. 156.

29. Voorhies, *The Co-ordination of Motive, Men and Money in Industrial Research*, p. 4; and, for costing and evaluating research results, see pp. 60ff. This was a study commissioned by Dow, GE, US Rubber, Armour Research Foundation, and several oil companies.

30. Fred Olsen, "Evaluating the Results of Research," in Furnas, ed., *Research in Industry*, pp. 402–415, on p. 403; and see F. Russell Bichowsky, *Industrial Research* (Brooklyn, NY: Chemical Publishing Co., 1942), pp. 105–106.

31. Mees, *The Organization of Industrial Scientific Research*, pp. 127–129; see also S. C. Ogburn, "Research Management," *Industrial Laboratories* 2, no. 9 (September 1951), 6–9; Hertz, *Theory and Practice of Industrial Research*, pp. 255–284.

32. Mees, "The Organization of Industrial Scientific Research," p. 768.

33. Robert E. Wilson, "The Attitude of Management toward Research," *Chemical and Engineering News* 27, no. 5 (31 January 1949), 274–277, on p. 276.

34. Harold Gershinowitz, "Sustaining Creativity Against Organizational Pressures," *Research Management* 3 (1960), 49–56, on p. 53; see also Mees and Leermakers, *The Organization of Industrial Scientific Research*, pp. 224–226.

35. For industrial research as a "gamble," see, among very many examples, Mees, *The Organization of Industrial Scientific Research*, p. 132; Ogburn, "Research Management,"

p. 6; Olsen, "Evaluating the Results of Research," p. 410; Michael Polanyi, "Patent Reform," *Review of Economic Studies* 11, no. 2 (1944), 61–76, on p. 62; Henry L. Cox, "The Personal Approach in Dealing with Technical People," *Research Management* 6 (1963), 153–161, on p. 153; John Jewkes, David Sawers, and Richard Stillerman, *The Sources of Invention* (London: Macmillan, 1958), p. 129.

36. Quoted in Clark, "Charles Edward Kenneth Mees," pp. 181–182.

37. Kettering, "Head Lamp of Industry," p. 79.

38. Wilson, "The Attitude of Management toward Research," p. 276; see also Paul Freedman, *The Principles of Scientific Research* (Washington, DC: Public Affairs Press, 1950 [orig. publ. London: Macdonald, 1941]), pp. 202–203.

39. J. D. Bernal, "Fundamental and Applied Aspects of Research Problems," in *The Direction of Research Establishments: Proceedings of a Symposium held at the National Physical Laboratory [Teddington] on 26th, 27th & 28th September 1956* (New York: Philosophical Library, 1957), pp. I.1–17 and subsequent discussion pp. A.1–8 (chapters and commentary separately paginated), on p. A.1.

40. D. H. Killeffer, *The Genius of Industrial Research* (New York: Reinhold Publishing Corporation, 1948), p. 220.

41. Quoted in T. Brailsford Robertson, "The Cash Value of Scientific Research," *Scientific Monthly* 1, no. 2 (November 1915), 140–147, on p. 145.

42. Mees and Leermakers, *The Organization of Industrial Scientific Research*, p. 226.

43. Reich, *The Making of American Industrial Research*, pp. 107–108. Extremely complicated accounting procedures were adopted within GE to constitute the research laboratory as a "profit center," billing research costs to the division that requested the work and to the laboratory if unsolicited projects resulted in products that no one wanted. As Whitney once told Kenneth Mees in a private conversation, in a "really successful laboratory" the annual gross profit from new products introduced "as a result of research" should be "equal to the whole accumulated cost of the research laboratory up to date"—a remarkable rate of return that Whitney was wholly convinced of, but that was never formally acknowledged by even his notably research-friendly company: Mees and Leermakers, *The Organization of Industrial Scientific Research*, pp. 167–168.

44. E.g., Olsen, "Evaluating the Results of Research," in Furnas, ed., *Research in Industry*, pp. 402–415, on pp. 404, 407, 411; for "arbitrary" accounting standards in general, see Mees and Leermakers, *The Organization of Industrial Scientific Research*, p. 167.

45. J. M. McIlvain, "The Research Budget," in Furnas, ed., *Research in Industry*, pp. 145–158, on pp. 155–157.

46. Olsen, "Evaluating the Results of Research," p. 410.

47. Voorhies, *The Co-ordination of Motive, Men and Money in Industrial Research*, pp. 53–54. GE learned early on about the desirability of buffering its research budget from economic downturns: Reich, *The Making of American Industrial Research*, pp. 79, 92, 96; Kendall Birr, *Pioneering in Industrial Research: The Story of the General Electric Research Laboratory* (Washington, DC: Public Affairs Press, 1957), p. 54; for Charles Stine's insistence in the late 1920s that research at DuPont be protected from business

cycle changes in expenditure, see David A. Hounshell and John Kenly Smith, Jr., *Science and Corporate Strategy: DuPont R&D, 1902–1980* (Cambridge: Cambridge University Press, 1988), p. 225.

48. Wilson, "The Attitude of Management toward Research," p. 277.

49. Daniel H. Kevles, *The Physicists: The History of a Scientific Community in Modern America* (New York: Alfred A. Knopf, 1977), pp. 250, 273; also Russell Moseley, "From Avocation to Job: The Changing Nature of Scientific Practice," *Social Studies of Science* 9 (1979), 511–522, on pp. 519–520.

50. Edward U. Condon, "Recruitment and Selection of the Research Worker," in Bush and Hattery, eds., *Scientific Research: Its Administration and Organization*, pp. 61–64, on p. 61; also "The Westinghouse Research Fellowships," *Science* n.s. 86, no. 2244 (31 December 1937), 605–606; Clarence Zener, "Planning of Research Work at Westinghouse Electric," *IRE Transactions on Engineering Management* EM-3, no. 4 (1957), 94–96, on p. 95. The fellowships were continued until 1940, when war work put an end to them, and, after the war, as Condon said, there were too many high-paying positions available for physicists and there was little point in reviving the idea.

51. Frank B. Jewett, "Industrial Research," *Reprint and Circular Series of the National Research Council, #4* (Washington, DC: National Research Council, 1919), p. 7 (quoted in Kevles, *The Physicists*, p. 100).

52. Carty, "Science and Business" (quoted in Noble, *America by Design*, p. 115); see also Waldo H. Kliever, "Design of Research Projects and Programs," *Industrial Laboratories* 3, no. 10 (October 1952), 6–13, on p. 6; Patrick J. McGrath, *Scientists, Business, and the State, 1890–1960* (Chapel Hill: University of North Carolina Press, 2002), pp. 12–14.

53. James W. Hackett and B. L. Steierman, "The Organization of a Fundamental Research Effort at Owens-Illinois Glass Company," *Research Management* 6 (1963), 81–92, on pp. 85–86.

54. C. George Evans, *Supervising R&D Personnel* (New York: American Management Association, 1969), p. 24.

55. James B. Fisk, "Basic Research in Industrial Laboratories," in *Symposium on Basic Research, Sponsored by the National Academy of Sciences, the American Association for the Advancement of Science, and the Alfred P. Sloan Foundation*, ed. Dael Wolfle (Washington, DC: American Association for the Advancement of Science, 1959), pp. 159–167, on p. 163.

56. Fisk, quoted in Francis Bello, "The World's Greatest Industrial Laboratory," *Fortune* (November 1958), 148–157, 208, 212, 214, 219–220, on pp. 212, 214.

57. Fisk, "Basic Research in Industrial Laboratories," p. 164.

58. Ralph T. K. Cornwell, "Professional Growth of the Research Man," in Furnas, ed., *Research in Industry*, pp. 295–307, on pp. 295–296, 299; see also Mees, "The Organization of Industrial Scientific Research," p. 771; idem, *The Organization of Industrial Scientific Research*, pp. 100–101; see also similar views expressed by a vice president of Union Carbide: H. B. McClure, "External Communication of Research Results," in *Selection, Training, and Use of Personnel in Industrial Research*, Proceedings of the Second Annual Conference on Industrial Research June 1951, eds. David B. Hertz and Albert H. Rubenstein (New York: King's Crown Press, 1952), pp. 161–176, on p. 166.

59. Wilson, "The Attitude of Management toward Research," p. 277; see also H. A. Leedy, "Training of Young Researchers," in Hertz and Rubenstein, eds, *Selection, Training, and Use of Personnel in Industrial Research*, pp. 62–77, on p. 74.

60. Hounshell and Smith, *Science and Corporate Strategy*, pp. 300–301; for DuPont's concern for research secrecy, see also Yates, *Control Through Communication*, pp. 254–256.

61. Laurence A. Hawkins, "Does Patent Consciousness Interfere with Cooperation between Industrial and University Research Laboratories?" *Science* n.s. 105, no. 2726 (28 March 1947), 326–327, on p. 326; also "No 'Iron Curtain' Seen: G. E. Scientist Explains How the Patent System Helps All," *New York Times*, 29 March 1947, p. 23. Laurence Hawkins had been executive engineer in the GE Research Laboratory until 1945, having previously worked in the company's patent department, and he went on to say that those few industrialists who sought to "confine their research men in an ivory tower" would be acting very foolishly.

62. Quoted in Cornwell, "Professional Growth of the Research Man," p. 301; also, in criticism of the secrecy of the emerging national security State, E. U. Condon, "Science and Security," *Science* n.s. 107, no. 2791 (25 June 1948), 659–665, on p. 661.

63. See, for example, the vigorously pro-publication, and wholly pragmatic, views of a research manager at GE: R. W. Schmitt, "Why Publish Scientific Research from Industry?" *Research Management* 4 (1961), 31–41.

64. Hawkins, "Does Patent Consciousness Interfere with Cooperation?," p. 326.

65. Clark, "Charles Edward Kenneth Mees," p. 190.

66. Cornwell (Sylvania), "Professional Growth of the Research Man," p. 301; Wilson, "The Attitude of Management toward Research," p. 277; Schmitt (GE), "Why Publish Scientific Research from Industry?," p. 32. (Uniquely, in my experience, this last research manager offered evidence that he had read some works of Robert Merton, citing Merton's claims about the essential openness of genuine science while completely setting aside Merton's characterization of what industrial research was like: ibid., pp. 36, 39–40.) For a "must" policy on open publication, see N. Shepard, "How Can We Build Better Teamwork?," p. 805.

67. Hounshell and Smith, *Science and Corporate Strategy*, pp. 223, 301; Hounshell, "The Evolution of Industrial Research in the United States," in Richard S. Rosenbloom and William J. Spencer, eds., *Engines of Innovation: U.S. Industrial Research at the End of an Era* (Cambridge, MA: Harvard University Press, 1996), pp. 13–85, on pp. 39–42. The organic chemist Wallace Carothers had been recruited from Harvard by DuPont in 1928 to do fundamental research, the results of which were freely disseminated, but, when Carothers's group discovered polymers with evident implications for commercial synthetic fibers, his lab's agenda shifted in that direction, with corresponding adjustments in publication policies.

68. John A. Van Raalte, "Reflections on the Ph.D. Interview," *Research Management* 9 (1966), 307–317, on p. 310.

69. A notable contemporary exception is the software industry, where innovations are not commonly patented and where there are systematic efforts at maintaining

secrecy. Such behavior has, however, generated the vigorous "open-source" movement among software developers.

70. Cornwell, "Professional Growth of the Research Man," pp. 301–302.

71. Fisk, "Basic Research in Industrial Laboratories," p. 164.

72. Ralph Brown, "The Transistor as an Industrial Research Episode," *Scientific Monthly* 80, no. 1 (January 1955), 40–46, on p. 43.

73. Reich, *The Making of American Industrial Research*, p. 110. A 1961 survey of publication practices for basic research findings among 174 U.S. companies showed that 14% of them published "substantially all" basic research findings; 26% published "most"; 45% published "some"; and 16% published "none": National Science Foundation, *Publication of Basic Research Findings in Industry* (Washington, DC: Government Printing Office, 1961), pp. 11 ff., quoted in Hirsch, *Scientists in American Society*, p. 64.

74. Mees, "The Organization of Industrial Scientific Research," p. 771; idem, *The Organization of Industrial Scientific Research*, p. 100; see also Edgerton, "Industrial Research in the British Photographic Industry," p. 122.

75. Mees, *The Organization of Industrial Scientific Research*, p. 105 (emphasis added).

76. B. E. Noltingk, *The Art of Research: A Guide for the Graduate* (New York: Elsevier, 1965), p. 6.

77. An industrial research director interviewed by a GE personnel manager in the early 1950s remarked that "he was constantly amazed at the speed at which his scientists became interested in the welfare of the company." The personnel manager was aware that many commentators thought it impossible to get scientists to become "company conscious," but cited such evidence to argue that "this criticism is unsound": Lowell W. Steele, "Personnel Practices in Industrial Laboratories," *Personnel* 29 (1953), 469–476, on p. 471.

78. Mees and Leermakers, *The Organization of Industrial Scientific Research*, pp. 235–237; see also Norman Kaplan, "The Relation of Creativity to Sociological Variables in Research Organizations," in *Scientific Creativity: Its Recognition and Development: Selected Papers from the Proceedings of the First, Second, and Third University of Utah Conferences: "The Identification of Creative Scientific Talent,"* eds. Calvin W. Taylor and Frank Barron (New York: John Wiley & Sons, 1963), pp. 195–204, on pp. 198–200 (reporting that in several industrial laboratories only a small minority of scientists actually used their allocation of free research time—in one organization "never more than one-third," in another only 5–10%); Bichowsky, *Industrial Research*, p. 106 (for the advisability of keeping a "slush fund"—say, 5% of the research budget—for individual researchers, with the assent of group leaders, to do with pretty much as they pleased).

79. Mees, *The Organization of Industrial Scientific Research*, pp. 102–103.

80. Willard Dow, quoted in N. Shepard, "How Can We Build Better Teamwork?," p. 805; see also Bush, "Principles of Administration in the Research Environment," p. 174.

81. Francis Bello, "The Young Scientists," in the Editors of *Fortune*, *The Mighty Force of Research* (New York: McGraw-Hill, 1956; art. orig. publ. 1953), pp. 21–39, on p. 33.

82. Reich, *The Making of American Industrial Research*, p. 75 (for one-third); Birr, *Pioneering in Industrial Research*, p. 37, and Laurence A. Hawkins, *The Story of General Electric Research* ([Schenectady, NY]: General Electric Company, 1950), pp. 12–13 (for one-half).

83. Bush, "Principles of Administration in the Research Environment," p. 174.

84. Letter from Fieser to James B. Conant, 26 November 1927, quoted in Hounshell, "The Evolution of Industrial Research in the United States," on p. 27; see also Hounshell and Smith, *Science and Corporate Strategy*, pp. 228, 299. (Fieser ultimately turned down the DuPont position.) But see McGrath, *Science, Business, and the State*, pp. 35 and 208 n. 6, for dissent from Hounshell's sketch of Conant as a typical academic "snob" about industrial research. Conant did indeed maintain that "first-rate" students should wind up in universities rather than industrial laboratories, but he himself was heavily involved in industrial research and saw nothing awkward or illegitimate in these relationships.

85. Robert N. Anthony, *Management Controls in Industrial Research Organizations* (Boston: Graduate School of Business Administration, Harvard University, 1952), pp. 134–136.

86. E. Bright Wilson, Jr., *An Introduction to Scientific Research* (New York: McGraw-Hill, 1952), p. 5.

87. Hounshell, "The Evolution of Industrial Research," pp. 26–27. Evidently, the then-common practice of allowing industrial research workers free time could be unknown to academic social scientists critically commenting on the condition of scientists in industry during the 1950s. A Columbia professor of education, for example, wrote that such freedom was regarded by industrial research administrators as a "fantastic" and even laughable notion: Lyman Bryson, "Researchers in Industry," in Livingstone and Milberg, eds., *Human Relations in Industrial Research Management*, pp. 129–137, on p. 130.

88. Wilson, *Introduction to Scientific Research*, p. 5.

89. Bichowsky, *Industrial Research*, p. 102. Bichowsky had worked in the 1930s as a chemist for the Surface Combustion Corporation of Toledo, Ohio, and for the consulting division of the Dow Chemical Corporation, in Ann Arbor, Michigan; see also William Allen Hamor, "The Research Couplet: Research in Pure Science and Industrial Research," *Scientific Monthly* 6, no. 4 (April 1918), 319–330, on p. 330: academic research, Hamor wrote, "will always allure men of thought as contra-distinguished from men of action; and the real home of these investigators is the university because the time factor is there of secondary consideration."

90. Tom Burns, "Research, Development and Production: Problems of Conflict and Cooperation," in *Administering Research and Development: The Behavior of Scientists and Engineers in Organizations*, eds. Charles D. Orth III, Joseph C. Bailey, and Francis W. Wolek (Homewood, IL: Richard D. Irwin & Inc. and the Dorsey Press, 1964), pp. 112–129, on p. 122 (art. orig. publ. *IRE Transactions on Engineering Management* [March 1961], pp. 15–23); also on the physical location of the industrial research facility, see Mees, *The Organization of Industrial Scientific Research*, pp. 67–68; John J. Beer and David Lewis,

"Aspects of the Professionalization of Science," in *The Professions in America*, ed. Kenneth S. Lynn (Cambridge, MA: Houghton Mifflin, 1965), pp. 110–130, on pp. 113–114.

91. Bichowsky, *Industrial Research*, p. 103. Similar sentiments were expressed in 1948 by the director of research at Sylvania. "An effort should be made by the company to avoid, as much as possible, regimentation of the research worker. When the laboratory and the plant are under the same roof, this is not always possible": Cornwell, "Professional Growth of the Research Man," p. 306; and see Ogburn, "Research Management," p. 7: "Staff morale is greatly aided by a restful environment at a location where immediate plant or sales problems are not allowed to unduly hinder the true research function."

92. Howard E. Fritz and Douglas M. Beach (research managers at the B. F. Goodrich Co.), "The Modern Research Laboratory," in Furnas, ed., *Research in Industry*, pp. 308–340, on pp. 318–319.

93. Shepard, "How Can We Build Better Teamwork?," p. 806.

94. Steele, "Personnel Practices in Industrial Laboratories," p. 472.

95. Raymund L. Zwemer, "Incentives from the Viewpoint of a Scientist," in Bush and Hattery, eds., *Scientific Research*, pp. 75–78, on p. 75.

96. Freedman, *The Principles of Scientific Research*, p. 144.

97. D. G. Smellie, "Research Reports," in Furnas, ed., *Research in Industry*, pp. 159–181, on p. 161.

98. Warner Eustis, "Personnel Policies and Personality Problems," in Furnas, ed., *Research in Industry*, pp. 277–294, on p. 281. Eustis had been director of research and product development at the surgical dressings manufacturer, the Kendall Company, since 1929. For a detailed account of notebook and record-keeping in a routine industrial chemistry laboratory, see Weiss and Downs, *The Technical Organization*, pp. 94–108.

99. Quoted in Reich, *The Making of American Industrial Research*, p. 103.

100. R. G. Chollar, G. J. Wilson, and B. K. Green, "Creativity Techniques in Action," *Research Management* 1 (1958), 5–21, on pp. 10–11.

101. C. E. Kenneth Mees, *An Address to the Senior Staff of the Kodak Research Laboratories, November 9, 1955* (Rochester, NY: Kodak Research Laboratories, 1956), p. 28: "I learned from [Whitney] not how to run a laboratory but how not to run a laboratory. Dr. Whitney's policies were almost entirely negative, but they were very useful. He warned you against the things you shouldn't do." See also Mees, *From Dry Plates to Ektachrome Film*, p. 51; Reich, *The Making of American Industrial Research*, pp. 82, 99–100, 254, and 271 n.64; Jenkins, *Images and Enterprise*, pp. 310–311.

102. Clark, "Charles Edward Kenneth Mees," p. 181.

103. The closest thing I can find to an expression of role-conflict-through-academic-socialization from an industrial source is Lowell W. Steele, "Rewarding the Industrial Scientist: A Problem of Conflicting Values [1957]," in Livingstone and Milberg, eds., *Human Relations in Industrial Research Management*, pp. 163–175. But while Steele was indeed an executive at GE, he was not a natural scientist or engineer by training and seems to have had no direct shop-floor experience. Steele was an industrial personnel manager at the GE Research Laboratory, in charge of salary practices. He took a

Harvard M.B.A. in 1948, and later a MIT Ph.D. in economics and social science under the supervision of Herbert A. Shepard.

104. I. Gorog, "Successful Adjustment to Industrial Research: How Can University and Industry Assist," *Research Management* 9 (1966), 5–13, on pp. 7, 10–11; see also J. J. Tietjen, "The Transition from Graduate School to Industrial Research," ibid. 9 (1966), 109–113.

105. Horace A. Secrist, director of research at the Kendall Company, wrote that industrial management "often does not fully realize how great an adaptation the scientist does succeed in making when he moves from the university to the industrial laboratory," drawing attention here not to programmatic norms but to some mundane aspects of adjusting to *teamwork*: Secrist, "Motivating the Industrial Research Scientist," *Research Management* 3 (1960), 57–64, on pp. 58–59. A research manager at Sun Oil noted the importance of showing recruits—through organization charts—where they fit in the corporate scheme of things: E. M. Kipp, "Introduction of the Newly Graduated Recruit to Industrial Research," ibid., 39–47. The texture of industrial teamwork is discussed in detail in the next chapter.

106. E.g., Steelman, *Science and Public Policy*, Vol. III, p. 33.

107. Hawkins, *Adventure into the Unknown*, pp. 91–93 (for Willis Whitney's program of family socialization and its continuation as the GE "Whitney Club" after World War II).

108. Robert W. Avery, "Enculturation in Industrial Research," *IRE Transactions on Engineering Management* EM-7, no. 1 (1960), 20–24, on pp. 21–22.

109. See Burns, "Research, Development and Production," for a study of tensions between the research facility and production functions in British companies, and, for a participant's idyllic account of relations within the laboratory, see ibid., p. 123: " 'In the lab, we're very happy—sort of happy family relationships. The lab chief must select people on the grounds of getting on with others; they certainly do get on with everybody.' "

110. Steele, "Personnel Practices in Industrial Laboratories," p. 471.

111. Writing in 1914, Kenneth Mees gave reasons why universities were unlikely *ever* to become natural homes for scientific research: "For the last fifty years it has been assumed that the proper home for scientific research is the university, and that scientific discovery is one of the most important—if not the most important—function which a university can fulfill. In spite of this only a few of the American universities, which are admittedly among the best equipped and most energetic of the world, devote a very large portion of their energies to research work, while quite a number prefer to divert as little energy as possible from the business of teaching, which they regard as the primary function of the university": C. E. Kenneth Mees, "The Future of Scientific Research [editorial]," *Journal of Industrial and Engineering Chemistry* 6 (1914), 618–619, on p. 618.

112. E.g., Thorstein Veblen, *The Higher Learning in America: A Memorandum on the Conduct of Universities by Business Men* (New York: Sagamore Press, 1957; orig. publ. 1918); Robert E. Kohler, "The Ph.D. Machine: Building on the Collegiate Basis," *Isis* 81

(1990), 638–662; Laurence R. Veysey, *The Emergence of the American University* (Chicago: University of Chicago Press, 1965).

113. John W. Servos, "The Industrial Relations of Science: Chemical Engineering at MIT, 1900–1939," *Isis* 71 (1980), 531–549, esp. p. 532, for poor research resources at MIT in the early twentieth century.

114. J. J. Stevenson, "The Debt of the World to Pure Science," *Science* n.s. 7, no. 167 (11 March 1898), 325–334, on pp. 332–333.

115. Barus, "Research and Teaching [1923]," p. 446.

116. Philip Reichert, " 'Big Business Takes Over Research,' " *Science* n.s. 120, no. 3115 (10 September 1954), 434–435.

117. See Mees, "Discussion of Midgley," p. 48.

118. For measured criticism of Mees's minimalism, see Shepard, "How Can We Build Better Teamwork?," pp. 804–805.

119. E.g., Zener, "Planning of Research Work at Westinghouse Electric."

120. Marvin Smith (chief of engineering) to Westinghouse president Frank Merrick, 17 March 1937, quoted in Lassman, "Industrial Research Transformed," p. 313. The research manager selected was the physicist Edward U. Condon.

121. Norman Kaplan, "Organization: Will It Choke or Promote the Growth of Science?," in *The Management of Scientists*, ed. Karl Hill (Boston: Beacon Press, 1964), pp. 103–127, on pp. 103–105; also idem, "The Relation of Creativity to Sociological Variables in Research Organizations"; idem, "Some Organizational Factors Affecting Creativity," *IRE Transactions on Engineering Management* EM-7, no. 1 (1960), 24–30; idem, "The Role of the Research Administrator," *Administrative Science Quarterly* 4 (1959), 20–42; idem, "Research Administration and the Administrator: U.S.S.R. and U.S.," ibid. 6 (1961), 51–72.

122. Kaplan, "Organization: Will It Choke or Promote the Growth of Science?," p. 108.

123. Ibid., pp. 109, 114. As one British industrial research manager observed, the fact that academic chains of command are often less rigid than those "in other more highly organised establishments does not always prevent the exercise of considerable power that may be the more tyrannical because it is concealed": Noltingk, *The Art of Research*, p. 31. Noltingk worked at the Central Electricity Research Laboratories in Leatherhead.

CHAPTER SIX

1. Alvin M. Weinberg, "Impact of Large-Scale Science on the United States," in *Science and Society*, ed. Norman Kaplan (Chicago: Rand-McNally, 1965), pp. 551–559 (orig. publ. *Science* n.s. 134, no. 3473 [21 July 1961], 161–164). As historians have recently pointed out, the term "Big Science" had been used by the physicist Hans Bethe three years previously, and, while it is plausible that it was in some scientists' vernacular even before that, it was definitely Weinberg, and, later, the sociologist Derek Price, who gave it, so to speak, brand identity: James H. Capshew and Karen A. Rader, "Big

Science: Price to the Present," *Osiris* n.s. 7 (1992), 3–25, on p. 5 n.2; Derek J. de Solla Price, *Little Science, Big Science* (New York: Oxford University Press, 1963).

2. Dwight D. Eisenhower, "Farewell Address," in *The Military-Industrial Complex*, ed. Carroll W. Pursell, Jr. (New York: Harper and Row, 1972; address originally given 17 January 1961), pp. 204–208, on p. 207.

3. Alvin M. Weinberg, "Scientific Teams and Scientific Laboratories," in *The Twentieth-Century Sciences: Studies in the Biography of Ideas*, ed. Gerald Holton (New York: W. W. Norton, 1972), pp. 423–441, on p. 423: "Team science is characteristically conducted in the large multipurpose scientific laboratory, an institution that is predominantly a phenomenon of World War II and after."

4. E.g., Peter Galison, "The Many Faces of Big Science," in *Big Science: The Growth of Large-Scale Research*, eds. Peter Galison and Bruce Hevly (Stanford: Stanford University Press, 1992), pp. 1–17, on pp. 1–2; idem, *Image and Logic: A Material Culture of Microphysics* (Chicago: University of Chicago Press, 1997), pp. 239–311.

5. C. F. Kettering, "The Future of Science," *Science* n.s. 104, no. 2713 (27 December 1946), 609–614, on p. 609; see also Galison, *Image and Logic*, pp. 297–299 (for cooperation as a major lesson learned by academic scientists during the war).

6. Richard Goldschmidt, "Research and Politics," *Science* n.s. 109, no. 2827 (4 March 1949), 219–227, on pp. 219, 226.

7. See, e.g., R. F. Bud, "Strategy in American Cancer Research after World War II: A Case Study," *Social Studies of Science* 9 (1978), 425–459, esp. pp. 425–428.

8. P. G. Nutting, "Research and the Industries," *Scientific Monthly* 7, no. 2 (August 1918), 149–157, on p. 155; also Russell Moseley, "From Avocation to Job: The Changing Nature of Scientific Practice," *Social Studies of Science* 9 (1979), 511–522, on p. 520.

9. H. H. Whetzel, "Democratic Coordination of Scientific Efforts," *Science* n.s. 50, no. 1281 (18 July 1919), 50–55, on pp. 52, 55.

10. James Rowland Angell, "The Organization of Research," *Scientific Monthly* 11, no. 1 (July 1920), 25–42, on p. 33.

11. As quoted in David F. Noble, *America by Design: Science, Technology, and the Rise of Corporate Capitalism* (New York: Alfred A. Knopf, 1977), p. 110.

12. S. C. Gilfillan, *The Sociology of Invention* (Cambridge, MA: MIT Press, 1970; orig. publ. 1935), p. 116.

13. Frank B. Jewett, "The Future of Industrial Research: The View of a Physicist," in Standard Oil Development Company, *The Future of Industrial Research: Papers and Discussion* (New York: Standard Oil Development Company, 1945), pp. 17–23, on p. 18; Earl Place Stevenson, "Creative Technology," *Scientific Monthly* 76, no. 4 (April 1953), 203–206, on p. 203; see also Francis Bello, "The World's Greatest Industrial Laboratory," *Fortune* (November 1958), 148–157, 208, 212, 214, 219–220, on p. 149.

14. Sinclair Lewis, *Arrowsmith* (New York: Harcourt Brace, 1925). The allusion to *Arrowsmith*, and the quoted phrase, come from a skeptical meditation on the American mythology of scientific individualism and solitude by a scientific administrator writing just after World War II: M. J. Shear, "Teamwork in Scientific Research," *Personnel Administration* 9, no. 2 (November 1946), 3–8, 12, on p. 4.

15. Merle A. Tuve, "Is Science Too Big for the Scientist?" *Saturday Review* (6 June 1959), 48–52, on pp. 49–50.

16. Alvin M. Weinberg, *Reflections on Big Science* (Cambridge, MA: MIT Press, 1967), p. 43; see also Galison, *Image and Logic*, pp. 308–309; John Jewkes, David Sawers, and Richard Stillerman, *The Sources of Invention* (London: Macmillan, 1958), p. 162.

17. Robert R. Wilson, "My Fight against Team Research," in Holton, ed., *The Twentieth Century Sciences*, pp. 468–479, on p. 468.

18. Thomas Carlyle, "Signs of the Times [1829]," in idem, *Selected Writings*, ed. Alan Shelston (Harmondsworth: Penguin, 1971), pp. 61–85, on pp. 66–67.

19. Steven Shapin, "'The Mind Is Its Own Place': Science and Solitude in Seventeenth-Century England," *Science in Context* 4 (1991), 191–218.

20. John R. Baker, *The Scientific Life* (New York: Macmillan, 1943), pp. 33–34.

21. Vannevar Bush, "To Make Our Security System Secure," *New York Times*, 20 March 1955, Sunday Magazine, pp. 9, 38, 42, 44, 47, on pp. 9, 38.

22. Detlev W. Bronk, "The Role of Scientists in the Furtherance of Science," *Science* n.s. 119, no. 3086 (19 February 1954), 223-227, on p. 226; see also Bud, "Strategy in American Cancer Research," p. 429.

23. Curt P. Richter, "Free Research versus Design Research," *Science* n.s. 118, no. 3056 (24 July 1953), 91–93, on pp. 91–92.

24. William O. Baker, "The Moral Un-Neutrality of Science," *Science* n.s. 133, no. 3448 (27 January 1961), 262–262, on p. 262. This address was given in New York on 27 December 1960.

25. As quoted in Leonard Sayles and George Strauss, *Human Behavior in Organizations* (New York: Prentice Hall, 1966), p. 219.

26. R. D. Carmichael, "Individuality in Research," *Scientific Monthly* 9, no. 6 (December 1919), 514–525, on pp. 517, 521. Carmichael was a mathematician working at the University of Illinois. There is evident resonance here with Emerson's view of the English racial character: Ralph Waldo Emerson, *English Traits* (Boston: Houghton, Mifflin, 1903; orig. publ. 1876), esp. pp. 116–119.

27. Merle A. Tuve, "Development of the Section T Pattern of Research Organization," in *Teamwork in Research*, eds. George P. Bush and Lowell H. Hattery (Washington, DC: American University Press, 1953), pp. 135–142, on pp. 135–136; see also Michael Aaron Dennis, "'Our First Line of Defense': Two University Labs in the Postwar American State," *Isis* 85 (1994), 427–455, esp. pp. 430–44 (for Tuve at the wartime Applied Physics Laboratory).

28. William H. Taliaferro, "Science in the Universities," *Science* n.s. 108, no. 2798 (13 August 1948), 145–148, on p. 146.

29. L. Kowarski, "Team Work and Individual Work in Science [1962]," in *Science and Society*, ed. Kaplan, pp. 247–255, esp. p. 248; Walter Hirsch, *Scientists in American Society* (New York: Random House, 1968), pp. xi, 21, 92–93.

30. Erwin Chargaff, *Heraclitean Fire: Sketches from a Life before Nature* (New York: Rockefeller University Press, 1978), pp. 117, 121; and similar sentiments in Paul Weiss,

"Experience and Experiment in Biology," *Science* n.s. 136, no. 3515 (11 May 1962), 468–471.

31. Ralph E. Lapp, *The New Priesthood: The Scientific Elite and the Uses of Power* (New York: Harper & Row, 1965), p. 14.

32. Robert K. Merton, "The Normative Structure of Science," in idem, *The Sociology of Science: Theoretical and Empirical Investigations*, ed. Norman W. Storer (Chicago: University of Chicago Press, 1973; art. orig. publ. 1942), pp. 267–278, on pp. 275–276.

33. Bernard Barber, *Science and the Social Order* (Glencoe, IL: Free Press, 1952), p. 65.

34. Warren O. Hagstrom, "Traditional versus Modern Forms of Scientific Teamwork," *Administrative Science Quarterly* 9 (1964), 241–263, on p. 241.

35. Warren O. Hagstrom, *The Scientific Community* (Carbondale: Southern Illinois University Press, 1975; orig. publ. 1965), p. 111.

36. Quoted in Jewkes, Sawers, and Stillerman, *The Sources of Innovation*, p. 161.

37. Hagstrom, "Traditional versus Modern Forms of Scientific Teamwork," pp. 252–254. For Weber, see Max Weber, "Science as a Vocation," in *From Max Weber: Essays in Sociology*, eds. H. H. Gerth and C. Wright Mills (London: Routledge & Kegan Paul, 1991), pp. 129–156, on p. 131.

38. Hagstrom, *The Scientific Community*, pp. 153–154; see also Gerald M. Swatez, "The Social Organisation of a University Laboratory," *Minerva* 8 (1970), 36–58, esp. pp. 36–38 (for teamwork in academic high-energy physics, as informed by Hagstrom's views); and comments by the physicist Paul R. Zilsel, "The Mass Production of Knowledge," *Bulletin of the Atomic Scientists* 20, no. 4 (April 1964), 28–29, on p. 29: "the huge 'think factories' of our time are the equivalent of the Lancashire cotton mills of the industrial revolution. The scientists are many, and they are very busy producing staggering quantities of 'knowledge.' Their product, however, is increasingly taking on the character of a mere commodity; and their work takes on an alienated character of assembly line production, with no rhyme or reason or discernible relation to a meaningful whole."

39. Hagstrom, "Traditional versus Modern Forms of Scientific Teamwork," pp. 256–257; cf. Richard R. Nelson, "The Link between Science and Invention: The Case of the Transistor," in *The Rate and Direction of Inventive Activity: Economic and Social Factors: A Conference of the Universities-National Bureau Committee for Economic Research and the Committee on Economic Growth of the Social Science Research Council* (Princeton, NJ: Princeton University Press, 1962), pp. 549–583, on pp. 573–574.

40. William H. Whyte, Jr., *The Organization Man* (Garden City, NY: Doubleday Anchor, 1957; orig. publ. New York: Simon and Schuster, 1956). Note that Hagstrom importantly cited Whyte, as in Hagstrom, "Traditional versus Modern Forms of Scientific Teamwork," p. 246.

41. Whyte, *The Organization Man*, pp. 226–227. Of course, such general sentiments linking bureaucratic and hierarchical organization to the death (or "frustration") of science were not peculiar to Whyte. In Great Britain, for example, the Marxist scientist group, centering on J. D. Bernal, Joseph Needham, J. G. Crowther, and Lancelot Hogben, managed to combine enthusiastic approval of Soviet-style rational planning with

condemnation of capitalistic industrial secrecy and governmental laboratories' "over-organization" of scientific research: see, e.g., Bernard Lovell, *Science and Civilization* (London: Thomas Nelson, 1939), pp. 64–65.

42. Whyte, *The Organization Man*, pp. 207–208. For systematic psychological studies available in principle to support Whyte's claim about scientists' global "fierce independence," see, e.g., the problematic work of Anne Roe, *The Making of a Scientist* (New York: Dodd, Mead & Co., 1953).

43. Whyte, *Organization Man*, pp. 214, 230–232, 235; see also Kevles, *The Physicists*, p. 383.

44. Leonard Engel, "Get a Good Scientist ... *and Let Him Alone*," *Harper's Magazine* 208, no. 1244 (January 1954), 55–59, on p. 55.

45. Thomas Midgley, Jr., "The Future of Industrial Research: The View of a Chemist," in Standard Oil Development Company, *The Future of Industrial Research* pp. 30–46, esp. p. 42; also C. G. Suits, "The Engineer and the Fundamental Sciences," *Scientific Monthly* 76, no. 2 (February 1953), 90–99, esp. p. 92.

46. C. E. Kenneth Mees, *The Organization of Industrial Scientific Research* (New York: McGraw-Hill, 1920), p. 84.

47. Ralph Brown, "The Transistor as an Industrial Research Episode," *Scientific Monthly* 80, no. 1 (January 1955), 40–46, esp. p. 45; also Nelson, "The Link between Science and Invention: The Case of the Transistor," esp. pp. 578–579.

48. Thomas Midgley, quoted in Shear, "Teamwork in Scientific Research," p. 6. There *were* industrial problems that did *not* transgress disciplinary boundaries—though fewer and fewer as time went on—and here, Midgley allowed, the university-like form of "'head,' with assistants" worked quite well; see also John Morris Weiss and Charles Raymond Downs, *The Technical Organization: Its Development and Administration* (New York: McGraw-Hill, 1924), pp. 19–20.

49. Edward Teller, "The Work of Many People," *Science* n.s. 121, no. 3139 (25 February 1955), 267–275; see also David Kaiser, "The Atomic Secret in Red Hands? American Suspicions of Theoretical Physicists During the Early Cold War," *Representations* 90 (Spring 2005), 28–60, on pp. 31–33.

50. Stevenson, "Creative Technology," p. 204. The development of radar from the cavity magnetron was, Stevenson insisted, "not merely a story of mass production, but an example of enthusiastic creative teamwork": ibid., p. 203; see also Sir Harrie Massey, "Atomic Energy and the Development of Large Teams and Organizations," *Proceedings of the Royal Society of London A* 342 (1975), 491–497, on p. 494.

51. Morris G. Shepard, "Qualifications, Training, Aptitudes and Attitudes of Industrial Research Personnel," in *Research in Industry: Its Organization and Management*, ed. C. C. Furnas (New York: D. Van Nostrand, 1948), pp. 195–215, on p. 212.

52. See, notably, Thomas P. Hughes, *American Genesis: A Century of Invention and Technological Enthusiasm, 1870–1970* (New York: Viking, 1989), esp. ch. 8.

53. David Bendel Hertz, *The Theory and Practice of Industrial Research* (New York: McGraw-Hill, 1950), pp. 89, 188, 191.

54. F. Russell Bichowsky, *Industrial Research* (Brooklyn, NY: Chemical Publishing Co., 1942), p. 103.

55. Brown, "The Transistor as an Industrial Research Episode," p. 45.

56. Addison M. Rothrock, "Supervising and Training for Teamwork," in Bush and Hattery, eds., *Teamwork in Research*, pp. 82–90, on p. 83.

57. Stevenson, "Creative Technology," p. 204; see also Herbert A. Shepard, "Superiors and Subordinates in Research," in *The Direction of Research Establishments: Proceedings of a Symposium held at the National Physical Laboratory [Teddington] on 26th, 27th & 28th September 1956* (New York: Philosophical Library, 1957), pp. 1–13, on p. 12 (chapters separately paginated).

58. Peter Drucker, "Management and the Professional Employee," *Harvard Business Review* 30, no. 3 (May–June 1952), 84–90, on p. 85. (Drucker here was formally referring to professionals in general, but his examples were predominantly of scientific and technical employees.)

59. Francis Bello, "The Young Scientists," in the Editors of *Fortune, The Mighty Force of Research* (New York: McGraw-Hill, 1956), pp. 21–39, on p. 23. (This piece originally appeared in *Fortune* in 1953).

60. Josef Brozek and Ancel Keys, "General Aspects of Interdisciplinary Research in Experimental Human Biology," *Science* n.s. 100, no. 2606 (8 December 1944), pp. 507–512, on pp. 508–509; and quoted in Hertz, *The Theory and Practice of Industrial Research*, p. 187; see also Harold K. Work, "The University's Role in Training Research Workers," in *Selection, Training, and Use of Personnel in Industrial Research*, Proceedings of the Second Annual Conference on Industrial Research June 1951, eds. David B. Hertz and Albert H. Rubenstein (New York: King's Crown Press, 1952), pp. 126–140, esp. p. 134.

61. C. E. Kenneth Mees, "The Organization of Industrial Scientific Research," *Science* n.s. 43, no. 1118 (2 June 1916), 763–773, on p. 766; see also Jeffrey L. Sturchio, "Experimenting with Research: Kenneth Mees, Eastman Kodak, and the Challenges of Diversification," unpublished typescript paper delivered to the Hagley R&D Pioneers Conference, Wilmington, Delaware, 7 October 1985, pp. 11–12. For similar sentiments, see also Vernon Kellogg, "Isolation or Cooperation in Research," *Science* n.s. 63, no. 1626 (26 February 1926), 215–218; Angell, "The Organization of Research," pp. 35–36; Emil Ott, "The Team Approach in Research and Development," *Chemical and Engineering News* 28 (12 June 1950), 1994–1996.

62. Earl P. Stevenson, "The Frontiers of Industry," *Scientific Monthly* 22, no. 4 (April 1926), 285–288, on p. 285.

63. Kellogg, "Isolation or Cooperation in Research," p. 216 (for "not geniuses"); Weiss and Downs, *The Technical Organization*, pp. 3, 5–7 (for "gentlemen").

64. Albert W. Hull, "Selection and Training of Students for Industrial Research," *Science* n.s. 101, no. 2616 (16 February 1945), 157–160, on p. 158.

65. H. D. Arnold, director of research at Bell Labs, quoted in John Mills, "Who Is the Research Man?" *Technology Review* 44 (1942), 451–452, 466, 468–469, on p. 466.

66. Francis Bello, "Industrial Research: Geniuses Now Welcome," *Fortune* (January 1956), 96–99, 142, 144, 149–150. Bello, like Whyte, exempted the industrial research

"stars" (Bell, GE, Eastman Kodak) from his criticisms, and, while he reckoned that "there are still a good many laboratories where signs might still be posted: 'No geniuses wanted,'" he applauded big industry's willingness—perhaps, he thought, prompted by *Fortune*'s own arguments over the years—to develop "more flexibility in accommodating unusual personalities": ibid., p. 96. Two years earlier, a *Harper's Magazine* article had made a point of exempting Eastman Kodak, Bell Labs, and Merck from strictures against excessively organized research: Engel, "Get a Good Scientist . . . *and Let Him Alone*," p. 59.

67. Shear, "Teamwork in Scientific Research," p. 5.

68. Elmer W. Engstrom, "What Industry Requires of the Research Worker," in *Human Relations in Industrial Research Management, Including Papers from the Sixth and Seventh Annual Conferences on Industrial Research: Columbia University, 1955 and 1956*, eds. Robert Teviot Livingstone and Stanley H. Milberg (New York: Columbia University Press, 1957), pp. 69–79, esp. pp. 69, 74; also J. M. Morris, "Administration of Research in Industry," *Research Management* 5 (1962), 237–247, on pp. 239–240, for "at least six types of research minds," each of which had to be handled differently by management.

69. Emmett K. Carver, "Organization, Personalities, and Creative Thought," in Livingstone and Milberg, eds., *Human Relations in Industrial Research Management*, pp. 60–68, on p. 64. Carver was technical assistant to the general manager of the Kodak Park Works, Eastman Kodak Company. This view was not, however, uncontested: a history of wartime research organization celebrated scientific organization, but qualified that view by announcing that "one outstanding man will succeed where ten mediocrities will simply fumble": Baxter, *Scientists against Time*, p. 7.

70. Carleton R. Ball, "Personnel, Personalities and Research," *Scientific Monthly* 23, no. 1 (1926), 33–45, on p. 33. (Ball was an agronomist and a research administrator for the US Department of Agriculture.)

71. H. A. Leedy, "Training of Young Researchers," in Hertz and Rubenstein, eds., *Selection, Training, and Use of Personnel in Industrial Research*, pp. 62–77, on p. 64 (Leedy was director of the Armour Research Foundation); see also Merritt A. Williamson, "The Art of Interviewing for Technical Positions," *Research/Development* (formerly *Industrial Laboratories*) 11, no. 1 (1960), 17–24, esp. p. 17.

72. C. P. Haskins, "Characteristics of the Research Man and the Research Atmosphere," in Furnas, ed., *Research in Industry*, pp. 182–194, on p. 184.

73. Rothrock, "Supervising and Training for Teamwork," p. 89.

74. Earl Bartholomew of the Ethyl Research Laboratories (1935), quoted in Hattery, "Nature of Research Teamwork," in Bush and Hattery, eds., *Teamwork in Research*, pp. 3–9, on p. 8; see also Maurice Holland, with Henry F. Pringle, *Industrial Explorers* (New York: Harper & Brothers, 1928), p. 12 (on the virtue of "generosity").

75. Hull, "Selection and Training of Students for Industrial Research," pp. 157–158.

76. Roger Adams, R. C. Fuson, and C. S. Marvel, "The Graduate Training of Chemists," *Chemical and Engineering News* 28, no. 33 (14 August 1950), 2765–2767, on pp. 2766–2767.

77. Figures quoted in Hirsch, *Scientists in American Society*, pp. 4–5.

78. National Manpower Council, *A Policy for Scientific and Professional Manpower: A Statement by the Council with Facts and Issues Prepared by the Research Staff* (New York: Columbia University Press, 1953), p. 192.

79. See Laurence A. Hawkins, *Adventure into the Unknown: The First Fifty Years of the General Electric Research Laboratory* (New York: William Morrow & Co., 1950), pp. 96–97 (for three GE female researchers: Dr. Katharine Burr Blodgett (see figure 8); Dr. Dorothy Hall, an analytic chemist, who married a GE metallurgist and soon left to raise her two children; and Mrs. M. R. Andrews, who joined the laboratory in 1906, married a coworker, and remained with the laboratory after he unexpectedly died. Hawkins says there were others, but does not name them. Whitney was celebrated for his desire to bring women into the GE Research Laboratory: Holland, *Industrial Explorers*, p. 18.

80. E.g., Lee A. DuBridge, "Scientists and Engineers: Quantity Plus Quality," *Science* n.s. 124, no. 3216 (17 August 1956), 299–304.

81. Warner Eustis, "Personnel Policies and Personality Problems [1948]," in Furnas, ed., *Research in Industry*, pp. 277–294, on pp. 285–288. Eustis was research director at the Kendall Company. Eustis accepted that, "in general, the woman excels in detail work—compiling data, correlating facts, measuring minutiae, laboratory testing—but she frequently lacks the business, promotional point of view." Even so, Eustis insisted that "this generalization is becoming less true," and that there was no reason to denigrate either women's cognitive abilities or entrepreneurship (ibid., pp. 286–287).

82. Hawkins, *Adventure into the Unknown*, p. 97.

83. Thornton Page, "Selecting the Research Team," in Bush and Hattery, eds., *Teamwork in Research*, pp. 61–70, on p. 62 (for crew versus team); Edwin Frederick, "Planning Research for Team Attack," *Industrial Laboratories* 3, no. 3 (March 1952), 80–86; Hirsch, *Scientists in American Society*, pp. 116–117; E. Wight Bakke, "Teamwork in Industry," *Scientific Monthly* 66, no. 3 (March 1948) 213–220, esp. p. 214; Frederick Cottrell, "Scientists: Solo or Concerted?" in Bernard Barber and Walter Hirsch, eds., *The Sociology of Science* (New York: Free Press of Glencoe, 1962), pp. 388–393, on p. 391 (for opera); Angell, "The Organization of Research," p. 34 (for rejection of the military pattern); Holland, *Industrial Explorers*, p. 5 and Rothrock, "Supervising and Training for Teamwork," p. 391 (for two of the rare *defenses* of military-like organization); and Henry L. Cox, "The Personal Approach in Dealing with Technical People," *Research Management* 6 (1963), 153–161, on pp. 154–155 (for explicit contrast between the army platoon and the research team).

84. James B. Fisk, "Synthesis and Application of Scientific Knowledge for Human Use," in Melvin Calvin et al., *The Scientific Endeavor: Centennial Celebration of the National Academy of Sciences* (New York: Rockefeller University Press, 1965), pp. 293–302, on p. 296.

85. Angell, "The Organization of Research," p. 34.

86. See Mees's discussion of a piece by Thomas Midgley in Standard Oil Development Company, *The Future of Industrial Research*, pp. 47–48; cf. C. E. Kenneth Mees

and John A. Leermakers, *The Organization of Industrial Scientific Research*, 2nd ed. (New York: McGraw-Hill, 1950), p. 233.

87. James B. Fisk, "Basic Research in Industrial Laboratories," in *Symposium on Basic Research, Sponsored by the National Academy of Sciences, the American Association for the Advancement of Science, and the Alfred P. Sloan Foundation*, ed. Dael Wolfle (Washington, DC: American Association for the Advancement of Science, 1959), pp. 159–167, on p. 165.

88. Howard R. Tolley, "The Individualist and the Research Team," in Bush and Hattery, eds., *Teamwork in Research*, pp. 79–81.

89. Mees, *The Organization of Industrial Scientific Research*, pp. 80–83.

90. Ibid., pp. 83–84; C. E. Kenneth Mees, "The Production of Scientific Knowledge," *Journal of Industrial and Engineering Chemistry* 9 (1917), 1137–1141, on pp. 1138–1139.

91. Mees, *The Organization of Industrial Scientific Research*, p. 131.

92. Shear, "Teamwork in Scientific Research," p. 7.

93. Quoted in Theresa R. Shapiro, "What Scientists Look For in Their Jobs," *Scientific Monthly* 76, no. 6 (June 1953), 335–340, on p. 337.

94. Royden C. Sanders, Jr., "Interface Problems between Scientists and Others in Technically Oriented Companies," in *The Management of Scientists*, ed. Karl Hill (Boston: Beacon Press, 1964), pp. 75–86, on pp. 77–78.

95. Mees and Leermakers, *The Organization of Industrial Scientific Research*, p. 208; see also George P. Bush, "Teamwork in Research: A Commentary and Evaluation," in Bush and Hattery, eds., *Teamwork in Research*, pp. 171–186, on p. 179.

96. Clarence Zener, "Planning of Research Work at Westinghouse Electric," *IRE Transactions on Engineering Management* EM-3, no. 4 (1957), 94–96, on p. 95; see also Eugene p. Wigner, "The Limits of Science," *Proceedings of the American Philosophical Society* 94 (1950), 422–427, on p. 427; Weinberg, "Scientific Teams and Scientific Laboratories," p. 424; David E. Green, "Group Research," *Science* n.s. 119, no. 3092 (2 April 1954), pp. 444–445.

97. Ball, "Personnel, Personalities and Research," pp. 36–37: "Great personalities usually are men of a single idea which often becomes almost an obsession. Frequently they are unable to see the ramifying phases of the problem and its contacts with an obligation to other ideas, institutions or movements. This attitude makes for a narrow though perhaps brilliant development and not for a broad and comprehensive progress."

98. William C. Taylor, "Mogul vs. Mavericks: A Search for Heroes," *New York Times*, 9 May 2004, Business Section, p. 4; also James Surowiecki, *The Wisdom of Crowds: Why the Many Are Smarter Than the Few and How Collective Wisdom Shapes Business, Economies, Societies, and Nations* (New York: Doubleday, 2004).

99. For Bartlett on "social constructiveness," see David Bloor, "Whatever Happened to 'Social Constructiveness,'" in *Bartlett, Culture and Cognition*, ed. Akiko Saito (London: Psychology Press, 2000), pp. 194–215; idem, "F. C. Bartlett and the Origins of Social Constructionism in 1930s Cambridge," typescript of talk given at the Max

Planck Institute for the History of Science, Berlin, 2002. (I thank Bloor for drawing my attention to Bartlett's views in this connection.)

100. Sir Frederic C. Bartlett, *Remembering: A Study in Experimental and Social Psychology* (Cambridge: Cambridge University Press, 1932), pp. 276–277.

101. See, for instance, David A. Hollinger, "The Defence of Democracy and Robert K. Merton's Formulation of the Scientific Ethos," *Knowledge and Society*, Vol. 4, eds. Robert Alun Jones and Henrika Kuklick (Greenwich CT: JAI Press, 1983), pp. 1–15; idem, "Free Enterprise and Free Inquiry: The Emergence of Laissez-Faire Communitarianism in the Ideology of Science in the U.S.," *New Literary History* 21 (1990), 897–920; idem, "Science as a Weapon in *Kulturkämpfe* in the United States During and After World War II," *Isis* 86 (1995), 440–454.

102. Stevenson, "Creative Technology," p. 204; see also Bakke, "Teamwork in Industry," p. 220, for "the kind of free enterprise and democratic teamwork to which we are accustomed and to which we have committed our destiny as a nation."

103. Hawkins, *Adventure into the Unknown*, p. 91.

104. Walter Hirsch, "Knowledge for What?" *Bulletin of the Atomic Scientists* 21, no. 5 (May 1965), 28–31, on p. 28.

105. Michael Polanyi, "The Republic of Science: Its Political and Economic Theory," *Minerva* 1 (1962), 54–74, on p. 62; also idem, *Personal Knowledge: Towards a Post-Critical Philosophy* (Chicago: University of Chicago Press, 1958), pp. 174–184 (for remarks on the contrast between academic and industrial science). Note that strands of post–World War II commentary in the Federal government sided explicitly with Marxist pro-planners and argued against the anti-Marxist, anti-planning sentiments of the Society for Freedom in Science: see, for example, John R. Steelman, *Science and Public Policy: A Report to the President by John R. Steelman, Chairman, The President's Scientific Research Board*, 5 vols (Washington, DC: Government Printing Office, 1947), Vol. III, p. 130 (approvingly citing J. D. Bernal's *The Social Function of Science*).

106. Quoted again in Mees and Leermakers, *The Organization of Industrial Scientific Research*, p. 244, and also in L. A. Rogers, "What Constitutes Efficiency in Research?" *Journal of Bacteriology* 8 (1923), 197–213, on p. 206. (I have not been able to locate the source of this quotation in Little's publications.) Suspicion of the descriptive relevance of organization charts continued to be vigorously expressed by Arthur D. Little, Inc., management consultants well after World War II: see, e.g., Sherman Kingsbury, in association with Lawrence W. Bass and Warren C. Lothrop, "Organizing for Research," in *Handbook of Industrial Research Management*, ed. Carl Heyel (New York: Reinhold Publishing Corporation, 1959), pp. 65–99, on pp. 65, 77–78.

107. Mees, *The Organization of Industrial Scientific Research*, pp. 67–68.

108. C. E. Kenneth Mees, "Discussion [of Michael Polanyi, 'The Foundations of Freedom in Science']," in *Physical Science and Human Values*, ed. E. P. Wigner (Princeton, NJ: Princeton University Press, 1947), pp. 140–141 (emphases in original). In these connections, it should be noted that Mees had recently collaborated with the English zoologist John R. Baker on a semi-popular history of science. Baker had been a leading light, and a close associate of Polanyi, in the Society for Freedom in Science: see C. E.

Kenneth Mees, with the cooperation of John R. Baker, *The Path of Science* (London: Chapman & Hall, 1946).

109. Mees, "Discussion of Midgley," p. 48; cf. Mees and Leermakers, *The Organization of Industrial Scientific Research*, p. 233. Academic social scientists, skeptical of the very idea of freedom in industrial research, criticized Mees's "proverb" as "frequently not quite true": e.g., Lewis C. Mainzer, "Scientific Freedom in Government-Sponsored Research," *Journal of Politics* 23 (1961), 212–230, on pp. 222–223. Some research directors felt obliged to take a position—for, against, or qualifying—Mees's aphorisms: see, e.g., Norman A. Shepard (American Cyanamid), "How Can We Build Better Teamwork within Our Research Organizations?" *Chemical and Engineering News* 23, no. 9 (10 May 1945), 804–807, on p. 804.

110. Walter Clark, "Charles Edward Kenneth Mees: 1882–1960," *Biographical Memoirs of the Royal Society* 7 (1961), pp. 171–197, on p. 191.

111. C. E. Kenneth Mees, "Research and Business with Some Observations on Color Photography," *Vital Speeches of the Day* 2 (18 November 1935), 768–769; also idem, "Scope of Research Management," *Industrial and Engineering Chemistry* 24 (1932), 65–66. Mees's dicta are among the most widely quoted American pronouncements about research organization: see, e.g., Waldo H. Kliever, "Design of Research Projects and Programs," *Industrial Laboratories* 3, no. 10 (October 1952), 6–13, on p. 11; Nelson, "The Economics of Invention," p. 125; Rob Kaplan, *Science Says: A Collection of Quotations on the History, Meaning, and Practice of Science* (New York: Stonesong Press, 2001), pp. 105–106; Jewkes, Sawers, and Stillerman, *The Sources of Invention*, p. 138; and it is vigorously endorsed in Nelson's detailed study of the invention of the transistor at Bell Labs: "The Link between Science and Invention," pp. 571–572.

112. Mees, "Discussion of Midgley," p. 47.

113. Mees and Leermakers, *The Organization of Industrial Scientific Research*, pp. 20–21.

114. Ibid., pp. 234–235.

115. O. E. Buckley, quoted in Jewkes, Sawers, and Stillerman, *The Sources of Innovation*, pp. 137–138.

116. Kliever, "Design of Research Projects and Programs," p. 13.

117. Stevenson, "Creative Technology," p. 204. Stevenson was here quoting an unpublished speech in 1951 by R. E. Gibson (director of the Applied Physics Laboratory) to the Society for Personnel Administration.

118. E.g., Merritt L. Kastens, "Research—A Corporate Function," *Industrial Laboratories* 8, no. 10 (October 1957), 92–101; also D. L. Williams, *Planning of Research and Development* (New York: Wallace Clark, 1947), p. 5; Hertz, *The Theory and Practice of Industrial Research*, p. 202; Harvey Brooks, "Can Science Be Planned?" in idem, *The Government of Science* (Cambridge, MA: MIT Press, 1968), ch. 3, esp. pp. 59–72.

119. Steelman, *Science and Public Policy*, Vol. III, p. 28.

120. Bush, "Principles of Administration in the Research Environment," pp. 164–165.

121. Noble, *America by Design*, ch. 7, esp. p. 118: "As the industrial research laboratories grew in size, the role of the scientists within them came more and more to resemble

that of workmen on the production line and science became essentially a management problem."

122. Mees, "The Organization of Industrial Scientific Research," p. 768.

123. Mees, "Discussion of Midgley," p. 49; idem, *From Dry Plates to Ektachrome Film*, pp. 291–292; idem and Leermakers, *The Organization of Industrial Scientific Research*, p. 149; Clark, "Charles Edward Kenneth Mees," p. 189; Baum, "Doctor of the Darkroom," p. 47; Sturchio, "Experimenting with Research," pp. 18, 30 n. 22; Nelson, "The Simple Economics of Basic Research," p. 303 (where it became an iconic story about serendipitous discovery and the advantage that companies with a broad technological base have in exploiting unexpected discoveries). The scientist concerned was K. C. D. Hickman, and the new company founded to manufacture vitamins A and E using this technology was Distillation Products, Inc.; Mees's son Graham became the president: see "Kenneth Mees, Scientist, Dies; Ex-Research Head for Kodak," *New York Times*, 17 August 1960, p. 31; "Quantity Production of Vitamins," *New York Times*, 22 October 1944, p. E9. Similarly, Mees told a story (which appears in "Quantity Production of Vitamins") about a scientist at his laboratory who worked on a cellulose product that had no evident utility in photography but which turned out to be useful in surgical bandages. Eastman came to make this dressing commercially. Tennessee Eastman, derived from the laboratory's cellulose acetate work, operated plant Y-12 at Oak Ridge, Tennessee, during World War II, Mees providing many of the technicians for this plant from his own laboratory in Rochester.

124. Irving Langmuir, "The Growth of Particles in Smokes and Clouds and the Production of Snow from Supercooled Clouds," *Proceedings of the American Philosophical Society* 92 (July 1948), as quoted in Robert K. Merton and Elinor Barber, *The Travels and Adventures of Serendipity* (Princeton, NJ: Princeton University Press, 2004), pp. 190–191; see also pp. 69, 144–146, 173–174, 192.

125. Merton and Barber, *The Travels and Adventures of Serendipity*, pp. 144–146.

126. Brian Hindo, "Six Sigma: So Yesterday? In an Innovation Economy, It's No Longer a Cure-All," *BusinessWeek*, no. 4038 (11 June 2007), p. 11.

127. Anonymous letter writer in "Scrutinizing Six Sigma," ibid., no. 4041 (2 July 2007), pp. 90–91.

128. C. E. Kenneth Mees, *An Address to the Senior Staff of the Kodak Research Laboratories, November 9, 1955* (Rochester, NY: Kodak Research Laboratories, 1956), p. 28; Mees also learned something from the views of William Rintoul, the research manager at the Nobel Explosives Company: Clark, "Charles Edward Kenneth Mees," p. 181; Mees, *The Organization of Industrial Scientific Research*, pp. 68–69 (William Rintoul, private communication to Mees, 11 February 1919); Reese Jenkins, *Images and Enterprise: Technology and the American Photographic Industry, 1839–1925* (Baltimore, MD: Johns Hopkins University Press, 1979), pp. 310–311.

129. The hybrid nature of the research director—part scientist and part manager, partly a leader and partly in charge of an activity that could not be led—fascinated social scientists into the 1950s and 1960s, especially those responding to governmental concerns over scientific productivity in the Cold War years. For notable contributions

to this literature, see, for example, Norman Kaplan, "The Role of the Research Administrator," *Administrative Science Quarterly* 4 (1959), 20–42; idem, "Research Administration and the Administrator: U.S.S.R. and U.S.," ibid. 6 (1961), 51–72; Howard Baumgartel, "Leadership Style as a Variable in Research Administration," ibid. 2 (1957), 344–360; R. M. Cavanaugh, "Development of Managers: Training in a Research Division," ibid. 1 (1956), 373–381.

130. Mees, "The Organization of Industrial Scientific Research," p. 764.

131. Shepard, "How Can We Build Better Teamwork within Our Research Organizations?," pp. 804–805; idem, "The Research Director's Job," in Furnas, ed., *Research in Industry*, pp. 56–70, on pp. 56–57.

132. Shepard, "How Can We Build Better Teamwork within Our Research Organizations?," p. 805.

133. Mees and Leermakers, *The Organization of Industrial Scientific Research*, pp. 234–235.

134. Ibid., p. 235.

135. Ibid., pp. 122, 124; also Mees, *Address to the Senior Staff*, p. 28.

136. Mees, "Discussion of Midgley," p. 48.

137. Zener, "Planning of Research Work at Westinghouse Electric," p. 95.

138. Max Weber, *Economy and Society: An Outline of Interpretive Sociology*, 2 vols., eds. Gunter Roth and Claus Wittich, trans. Ephraim Fischoff et al. (Berkeley: University of California Press, 1978), Vol. II, p. 1115.

CHAPTER SEVEN

1. The work of the Austrian-American economist Joseph Schumpeter on the role of the entrepreneur as an agent of capitalist economic change is well known, and is mentioned briefly later in this chapter, but the genealogy of the English usage tracks back to the eighteenth-century Irish economist Richard Cantillon by way of John Stuart Mill and the twentieth-century American economist Frank H. Knight.

2. One need not look only at universities, and associated nonprofit research organizations, for the figure of the entrepreneurial scientist. Scientists employed by profit-making industrial organizations can become entrepreneurial by striking out on their own, founding their own companies and taking upon themselves responsibility for more of the commercializing process than they had within their former corporate home. Still other risk-taking and institutionally innovative scientists employed by large firms—those who "start a business within a business"—have become known as "intrapreneurs." For the American academic scientific entrepreneur from the late 1970s, see Daniel S. Greenberg, *Science for Sale: The Perils, Rewards, and Delusions of Campus Capitalism* (Chicago: University of Chicago Press, 2007), ch. 4.

3. E.g., Charles E. Rosenberg, "Science and Social Values in 19th-Century America: A Case Study in the Growth of Scientific Institutions," in *Science and Values: Patterns of Tradition and Change*, eds. Arnold W. Thackray and Everett Mendelsohn (New York:

Humanities Press, 1974), pp. 21–42; idem, "Rationalization and Reality in the Shaping of American Agricultural Research, 1875–1914," *Social Studies of Science* 7 (1977), 401–422.

4. Rima D. Apple, "Patenting University Research: Harry Steenbock and the Wisconsin Alumni Research Foundation," *Isis* 80 (1989), 374–394; David C. Mowery, Richard R. Nelson, Bhaven N. Sampat, and Arvids A. Ziedonis, *Ivory Tower and Industrial Innovation: University-Industry Technology Transfer Before and After the Bayh-Dole Act in the United States* (Stanford, CA: Stanford University Press, 2004), pp. 39–40, 58–65; Grischa Metlay, "Reconsidering Renormalization: Stability and Change in 20th-Century Views on University Patents," *Social Studies of Science* 36 (2006), 565–597.

5. Rebecca S. Lowen, *Creating the Cold War University: The Transformation of Stanford* (Berkeley: University of California Press, 1997), pp. 37–38, 114.

6. See also Henry Etzkowitz, "Knowledge as Property: The Massachusetts Institute of Technology and the Debate over Academic Patent Policy," *Minerva* 32 (1994), 383–421; idem, *MIT and the Rise of Entrepreneurial Science* (New York: Routledge, 2002); David Mowery and Bhaven Sampat, "University Patents and Patent Policy Debates in the USA, 1925–1980," *Industrial and Corporate Change* 10 (2001), 781–814; Charles Weiner, "Patenting and Academic Research: Historical Case Studies," *Science, Technology & Human Values* 12, no. 1 (Winter 1987), 50–62.

7. Mowery et al., *Ivory Tower and Industrial Innovation*, pp. 35–37.

8. Frederick Cottrell, "Patent Experience of the Research Corporation," *Transactions of the American Institute of Chemical Engineers* 28 (1932), 221–225, on p. 223 (quoted in Metlay, "Reconsidering Renormalization," p. 571).

9. Metlay, "Reconsidering Renormalization," p. 571.

10. Clark Kerr, *The Uses of the University* (Cambridge, MA: Harvard University Press, 1963), p. 20. Kerr was here ringing a change on a remark attributed to Robert M. Hutchins, president of the University of Chicago, who described the modern university "as a series of separate schools and departments held together by a central heating system."

11. Christopher Newfield, *Ivy and Industry: Business and the Making of the American University, 1880–1980* (Durham, NC: Duke University Press, 2003), pp. 168–170, 201 nn. 5–6. As Newfield summarizes, "While helping economic development was [recognized by the university as] a public service . . . , helping an industry with production and other business problems was not. Partnership with industry has to respect the university's primary commitment to basic research by contributing directly to basic research" (ibid., p. 169).

12. Ibid., pp. 171–173.

13. Figures quoted in Sheldon Krimsky, *Science in the Private Interest: Has the Lure of Profits Corrupted Biomedical Research?* (Lanham, MD: Rowman & Littlefield, 2003), p. 14.

14. Robert C. Dynes, "State of the Campus Address, March 22, 2001": http://orpheus.ucsd.edu/chancellor/state.html [accessed 10 July 2007].

15. Robert C. Conn, "University-Industry Interface in Engineering: The San Diego Experience," talk delivered at University of Southampton (U.K.), 3 April 2001, and accessed 27 July 2005 at http://www.engineering.soton.ac.uk/EEO/IMAGES/

southampton.pdf. Conn joined the San Diego Enterprise Partners venture capital firm in 2002, having previously cofounded a company of his own—Plasma & Materials Technologies, now Trikon Technologies.

16. The two most reliable current accounts of the origins and effects of the Bayh-Dole Act are Mowery et al., *Ivory Tower and Industrial Innovation*, and Greenberg, *Science for Sale*, pp. 53–70.

17. It turned out to be a disastrous decision, and by the late 1990s almost all of that investment was lost: Helen Epstein, "Crusader on the Charles," *New York Times Magazine*, 23 April 1989, pp. 26–28, 64–67, 74–76, 87, on p. 76; Diana B. Henriques, "Good Science, Bad Grades in Boston," *New York Times*, 3 May 1992, p. 110; Ronald Rosenberg, "Despite Losses, BU and Seragen Plow On," *Boston Globe*, 21 December 1997, P. G1; Masao Miyoshi, "Ivory Tower in Escrow," *boundary* 2 27 (2000), 7–50, esp. pp. 26–28.

18. Mowery et al., *Ivory Tower and Industrial Innovation*, pp. 101, 113, 159, 207 n.7.

19. See, for example, Anon., "Robert Swanson and Herbert Boyer: Giving Birth to Biotech," *BusinessWeek Online*, 18 October 2004 [accessed 14 August 2007 at www.businessweek.com/magazine/content/04-42/b3904017_mz072.htm].

20. The list of recent scholarly and journalistic writings complaining about these, and other aspects, of academic scientific entrepreneurship, and the commercialization of the American research university, is very large; the more notable examples include: Krimsky, *Science in the Private Interest*; Greenberg, *Science for Sale*; Martin Kenney, *Biotechnology: The University-Industrial Complex* (New Haven, CT: Yale University Press, 1986); Lawrence Soley, *Leasing the Ivory Tower* (Boston: South End Press, 1995); Sheila Slaughter and Larry L. Leslie, *Academic Capitalism: Politics, Policies, and the Entrepreneurial University* (Baltimore, MD: Johns Hopkins University Press, 1997); Jennifer Croissant and Sal Restivo, eds., *Degrees of Compromise: Industrial Interests and Academic Values* (Albany: State University of New York Press, 2001); Stanley Aronowitz, *The Knowledge Factory: Dismantling the Corporate University and Creating True Higher Learning* (Boston: Beacon Press, 2001); Philip Mirowski and Esther-Mirjam Sent, eds., *Science Bought and Sold: Essays in the Economics of Science* (Chicago: University of Chicago Press, 2002), esp. editors' introduction; Derek Bok, *Universities in the Marketplace: The Commercialization of Higher Education* (Princeton, NJ: Princeton University Press, 2003); Daniel L. Kleinman, *Impure Cultures: University Biology and the World of Commerce* (Madison: University of Wisconsin Press, 2003); Jennifer Washburn, *University, Inc.: The Corporate Corruption of Higher Education* (New York: Basic Books, 2005).

21. Anon., "Knowledge Is Power," *Time* 70, no. 21 (18 November 1957), 20–25, on p. 22. For morally charged historical work that describes and condemns the emergence of the fun-loving American physicist, see Paul Forman, "Social Niche and Self-Image of the American Physicist," in *Proceedings of the International Conference on the Restructuring of the Physical Sciences in Europe and the United States 1945–1960*, eds. Michelangelo De Maria, Mario Grilli, and Fabio Sebastiani (Singapore: World Scientific Publishing, 1989), pp. 96–104; see also concise and cogent remarks on this theme in Mary Jo Nye, *Before Big Science: The Pursuit of Modern Chemistry and Physics, 1800–1940* (Cambridge, MA: Harvard University Press, 1996), pp. 226–229.

22. Lewis S. Feuer, *The Scientific Intellectual: The Psychological and Sociological Origins of Modern Science* (New York: Basic Books, 1963); and, for sensitivity to asceticism in the development of American science, see Rebecca M. Herzig, *Suffering for Science: Reason and Sacrifice in Modern America* (New Brunswick, NJ: Rutgers University Press, 2005).

23. Robert L. Sinsheimer, "The Double Helix," *Science and Engineering*, September 1968, pp. 4–6; Philip Morrison, "The Human Factor in a Science First," *Life*, 1 March 1968, p. 8; Richard C. Lewontin, "'Honest Jim' Watson's Big Thriller about DNA," *Chicago Sunday Sun-Times*, 25 February 1968, pp. 1–2 (all reprinted in James D. Watson, *The Double Helix*, ed. Gunther S. Stent [New York: W. W. Norton, 1980; orig. publ. 1968], pp. 191–194, on p. 192 [Sinsheimer]; pp. 185–187, on p. 186 [Lewontin]; pp. 175–177, on p. 177 [Morrison]).

24. Watson, *The Double Helix*; and see also James D. Watson, *Genes, Girls and Gamow: After the Double Helix* (New York: Alfred A. Knopf, 2002).

25. James D. Watson, *Avoid Boring People: Lessons from a Life in Science* (New York: Alfred A. Knopf, 2007), pp. 193–194.

26. Richard P. Feynman, *"Surely You're Joking, Mr. Feynman!": Adventures of a Curious Character* (New York: Bantam Books, 1985), on pp. 157–158; also idem, *What Do YOU Care What Other People Think? Further Adventures of a Curious Character* (New York: W. W. Norton, 1988), *The Meaning of It All: Thoughts of a Citizen-Scientist* (Reading, MA: Perseus, 1998), *The Pleasure of Finding Things Out: The Best Short Works of Richard P. Feynman* (Cambridge, MA: Perseus, 1999).

27. Feynman to Tord Pramberg, 4 January 1967, in Richard P. Feynman, *Perfectly Reasonable Deviations from the Beaten Track: The Letters of Richard P. Feynman*, ed. Michelle Feynman (New York: Perseus, 2005), p. 230; see also Steven Shapin, "Milk and Lemon," *London Review of Books* 27, no. 13 (7 July 2005), pp. 10–13.

28. When he was a very young man, having just completed his first degree at MIT and before beginning his graduate work at Princeton, Feynman applied for employment at Bell Labs and was accepted: "In those days it was hard to find a job where you could be with other physicists," he wrote, saying he would have been very happy to work there: Feynman, *"Surely You're Joking, Mr. Feynman!,"* pp. 83–84.

29. John Sulston and Georgina Ferry, *The Common Thread: A Story of Science, Politics, Ethics, and the Human Genome* (Washington, DC: Joseph Henry Press, 2002), pp. 87–88.

30. The figure of the fun-loving, imperfectly socialized nerd is mythic in commentary on the worlds of hardware, software, and Internet innovation. For notable instances, see Linus Torvalds and David Diamond, *Just for Fun: The Story of an Accidental Revolutionary* (New York: HarperBusiness, 2001); Po Bronson, *The Nudist on the Late Shift* (New York: Random House, 1999); Robert X. Cringley, *Accidental Empires: How the Boys of Silicon Valley Make Their Millions, Battle Foreign Competition, and Still Can't Get a Date* (New York: HarperBusiness, 1996); and Michael Lewis, *The New New Thing: A Silicon Valley Story* (New York: W. W. Norton, 1997).

31. See, among many examples, J. Robert Moskin, "The Scientist as God and Devil," in idem, *Morality in America: A Report on Our Crisis of Immorality* (New York: Random House, 1966), pp. 57–71. Moskin was an editor for *Look* and *Collier's* magazines.

32. Philip J. Hilts, *Scientific Temperaments: Three Lives in Contemporary Science* (New York: Simon and Schuster, 1982), pp. 176–177. The relevant section of Hilts's book deals with the Harvard biologist and entrepreneur Mark Ptashne.

33. Tom Maniatis (a student of Mark Ptashne), quoted in ibid., p. 178.

34. A case could also be made for Leroy Hood, an inventor of the automatic DNA sequencer, founder of the Institute for Systems Biology in Seattle and of many companies, including Applied Biosystems and Amgen, but while Hood's entrepreneurial activities are legendary in the relevant technical communities, he is not yet a public icon of entrepreneurial science on the pattern of Venter and Mullis. For a fragmentary autobiography, see Hood, "My Life and Adventures Integrating Biology and Technology" [accessed 14 August 2007 at http://www.systemsbiology.org/download/2002Kyoto.pdf]. Other genomic entrepreneurs who have achieved some degree of public celebrity include Kári Stefánsson (of the controversial Icelandic genomics company deCode), Walter Gilbert (of Biogen and Harvard), Eric Lander (of MIT's Whitehead Center for Genome Research, the MIT/Harvard Broad Institute, and Millennium Pharmaceuticals), Joshua Boger (of Vertex Pharmaceuticals), and Daniel Cohen (of the Parisian Centre d'Etude du Polymorphisme Humain and Millennium).

35. The investors who financed TIGR simultaneously set up a for-profit company, Human Genome Sciences (HGS), to market any of TIGR's commercializable scientific discoveries. HGS was intended to have a six-month window during which it could market sequence knowledge before TIGR's researchers were free to publish. Craig Venter was given shares in HGS and almost immediately became very rich: Sulston and Ferry, *The Common Thread*, pp. 107–108, and, for Venter's own account, see J. Craig Venter, *A Life Decoded: My Genome: My Life* (New York: Allen Lane, Penguin Books, 2007), chs. 7–8.

36. "The Top 25 Managers," *BusinessWeek Online*, 8 January 2001 [accessed 10 July 2007 at http://www.businessweek.com/archives/2001/b3714001.arc.htm#B3714001]; "Voices of Innovation: Craig Venter," *BusinessWeek Online*, 11 October 2004 [accessed 10 July 2007 at http://www.businessweek.com/@@/Vq55v4cQJXo30RgA/magazine/content/04_41/b3903435.htm]; Michael D. Lemonick, "Gene Mapper: The Bad Boy of Science Has Jump-Started a Biological Revolution," *Time.com*, 17 December 2000 [accessed 5 August 2005 at http://www.time.com/time/poy2000/mag/venter.html]; Richard Preston, "The Genome Warrior: Craig Venter Has Grabbed the Lead in the Quest for Biology's Holy Grail," *New Yorker*, 12 June 2000, pp. 66–83; James Shreeve, *The Genome War: How Craig Venter Tried to Capture the Code of Life and Save the World* (New York: Alfred A. Knopf, 2004); Kevin Davies, *Cracking the Genome: Inside the Race to Unlock Human DNA: Craig Venter, Francis Collins, James Watson, and the Story of the Greatest Scientific Discovery of Our Time* (New York: Free Press, 2001); Ingrid Wickelgren, *The Gene Masters: How a New Breed of Scientific Entrepreneurs Raced for the Biggest Prize in Biology* (New York: Times Books, Henry Holt, 2002).

37. Preston, "The Genome Warrior."

38. Stu Borman, "Triumph Over Nature: Entrepreneurial Scientist J. Craig Venter Is On to New Ventures," *Chemical and Engineering News* 80, no. 33 (19 August 2002),

45–50 [accessed 14 August 2007 at http://pubs.acs.org/cen/coverstory/8033/print/8033craigventer.html]; Shreeve, *The Genome War*, pp. 68–75; Venter, *A Life Decoded*, ch. 2.

39. Lemonick, "Gene Mapper."

40. Shreeve, *The Genome War*, pp. 74–76, on p. 74.

41. James Shreeve, "Craig Venter's Epic Voyage to Redefine the Origin of the Species," *Wired*, issue 12.08 (August 2004): http://www.wired.com/wired/archive/12.08/venter.html [accessed 14 August 2007].

42. In a 2002 interview with an on-line trade publication, Venter claimed that the reason he set up the binary for-profit/nonprofit structures for completing the human genome sequence was "because I didn't want to be in a biotech company." Although Venter prospered mightily from the commercial success of Celera Genomics, and although he was the iconic figure of entrepreneurial genomics, he claims that the whole purpose of his enterprise "was to build the endowment of my foundation so I could go back and do science when it [the sequencing of the human genome] was done," and this, indeed, is a pretty accurate description of what Venter *did* do: "John Craig Venter Unvarnished," published on-line by BIO-IT World at http://www.bio-itworld.com/archive/111202/horizons_venter.html [accessed 14 August 2007]. For the NIH patent effort, see Sulston and Ferry, *The Common Thread*, pp. 86–88; Borman, "Triumph Over Nature."

43. Quoted in Preston, "The Genome Warrior" ; also Venter, *A Life Decoded*, esp. p. 259. Note, however, that it is one thing to make money by packaging and delivering information and another to seek to own that information.

44. Shreeve, "Craig Venter's Epic Voyage"; see also Anon., "The Journey of the Sorcerer," *Economist*, 2 December 2004 [accessed 10 July 2007 at http://economist.com/displaystory.cfm?story_id=S%27%29%28%3C%2EP1%5B%24%23%40%21L%0A].

45. Shreeve, "Craig Venter's Epic Voyage" ; also Venter, *A Life Decoded*, ch. 17.

46. http://www.sorcerer2expedition.org/version1/HTML/main.htm [accessed 5 August 2005].

47. Kary Mullis, *Dancing Naked in the Mind Field* (New York: Pantheon, 1998), ch. 3.

48. Emily Yoffe, "Is Kary Mullis God? (Or Just the Big Kahuna?) Nobel Prize Winner's New Life," *Esquire* 122, no. 1 (July 1994), pp. 68–75; Steven Shapin, "Nobel Savage," *London Review of Books* 21, no. 13 (1 July 1999), pp. 17–18.

49. Mullis, *Dancing Naked in the Mind Field*, pp. 196–197.

50. Ibid., pp. 7, 10–11, 139.

51. Kerr, *The Uses of the University*, pp. 90–91.

52. Robert Teitelman, *Gene Dreams: Wall Street, Academia, and the Rise of Biotechnology* (New York: Basic Books, 1989), pp. 187–188. For academic scientists contributing "social capital" — skill- and knowledge-bearing networks — to entrepreneurial companies, see Fiona Murray, "The Role of Academic Inventors in Entrepreneurial Firms: Sharing the Laboratory Life," *Research Policy* 33 (2004), 643–659 (invoking Merton's vocabulary of "locals" and "cosmopolitans").

53. But see Barry Werth, *The Billion-Dollar Molecule: One Company's Quest for the Perfect Drug* (New York: Simon and Schuster, 1994), pp. 70–73, for anxiety in the early days of the Cambridge biotech start-up Vertex Pharmaceuticals about the relationship between its experiments and those being done at Harvard, when it was mutually understood that the university's financial interests and those of the company were potentially in direct conflict.

54. E.g., Cyrus C. M. Mody, "Corporations, Universities, and Instrumental Communities: Commercializing Probe Microscopy, 1981–1996," *Technology and Culture* 47 (2006), 56–80, esp. pp. 61–62. For these purposes, I set aside not-for-profit, but substantially industry-funded, independent research institutions, of which the Scripps Research Institution and the Burnham Cancer Institute are prominent Southern California examples.

55. Theresa R. Shapiro, "What Scientists Look For in Their Jobs," *Scientific Monthly* 76, no. 6 (June 1953), 335–340, on p. 337.

56. I personally interviewed thirteen scientists and engineers; my then-assistant Charles Thorpe—now professor of sociology at the University of California, San Diego—interviewed a further thirteen and all interviews were transcribed. The interviews were conducted in 2002 in Southern California with support from the University of California Office of the President's Industry-University Cooperative Research Program. Conditions imposed on me by the local Human Subjects Committee required all interviewees to be identified by pseudonyms, even though most were perfectly willing to be named. The sections following also include views of named scientists and engineers deriving from published materials, and notes make clear which is which. For broadly complementary studies, see Jason Owen-Smith and Walter W. Powell, "Careers and Contradictions: Faculty Responses to the Transformation in Knowledge and Its Uses in the Life Sciences," *Research in the Sociology of Work* 10 (2001), 109–140; idem, "Standing on Shifting Terrain: Faculty Responses to the Transformation of Knowledge and Its Uses in the Life Sciences," *Science Studies* 15 (2002), 3–28.

57. Interviewed 31 July 2002.

58. This university's official policy prohibits professors from serving as executive officers of companies in which they have a commercial involvement, and O'Reilly does not so serve.

59. Interviewed on 9 and 13 August 2002.

60. As it transpired, he did *not* then leave the university but found happier academic employment in another department, outside the school of engineering, where he continues to provoke debate over the proper role of the university and the proper conduct of academic science. The letter of resignation, however, remained on his Web site until at least August 2005. In 2007, however, he changed his mind again and is now looking for nonacademic work.

61. Marvyn's Web-posted letter of resignation. Capitalizations are in the original. All quoted material, bar this excerpt from the resignation letter, come from an interview on 27 August 2002.

62. Interviewed 16 April 2002.

63. Interviewed 6 June 2002.

64. Interviewed 26 July 2002.

65. Joe Catanese, quoted in Paul Rabinow and Talia Dan-Cohen, *A Machine to Make a Future: Biotech Chronicles* (Princeton, NJ: Princeton University Press, 2005), p. 68.

66. Penni Crabtree, "A Legend in the Biotech World: Royston Is a Story That's Still Being Written," *San Diego Union-Tribune*, 14 December 2004 [accessed 10 July 2007]: http://www.signonsandiego.com/uniontrib/20041214/news_1b14royston. html. It should, however, be noted that Royston's departure may have had some connection with earlier investigations of possible abuse of his NIH grants (in which he was ultimately exonerated of all charges).

67. Arthur Kornberg, *For the Love of Enzymes: The Odyssey of a Biochemist* (Cambridge, MA: Harvard University Press, 1989), p. 294; see also idem, *The Golden Helix: Inside Biotech Ventures* (Sausalito, CA: University Science Books, 1995); also Paul Rabinow, *Making PCR: A Story of Biotechnology* (Chicago: University of Chicago Press, 1996), p. 29.

68. The protein chemist Gary David, quoted in Mark Peter Jones, "Biotech's Perfect Climate: The Hybritech Story," unpubl. Ph.D. thesis, University of California, San Diego, 2005, pp. 501–503. For systematic treatment of what they call "asymmetrical convergence" between the structures and cultures of academia and industry, with special reference to biotechnology, see Daniel Lee Kleinman and Steven P. Vallas, "Contradiction in Convergence: Universities and Industry in the Biotechnology Field," in *The New Political Sociology of Science: Institutions, Networks, and Power*, eds. Scott Frickel and Kelly Moore (Madison: University of Wisconsin Press, 2006), pp. 35–62. This fine article, not available to me at the time of writing, is broadly complementary to the sensibilities set out in this chapter.

69. Rabinow, *Making PCR*, pp. 43–44; idem and Talia Dan-Cohen, *A Machine to Make a Future: Biotech Chronicles* (Princeton, NJ: Princeton University Press, 2005), p. 68; Kenney, *Biotechnology: The University-Industrial Complex*, p. 18.

70. But see Marcia Angell, *The Truth About Drug Companies: How They Deceive Us and What to Do About It* (New York: Random House, 2004), for skepticism about the genuineness of Big Pharma's claims about research costs.

71. Interviewed 9 July 2002. As it turned out, Hawicke helped Yellowlees into a nonteaching, soft-money research scientist position in his group, and, at the time I talked to her, she was giving that a go before deciding whether, or how, to move into industry.

72. This sentiment was repeatedly pressed on me by a life scientist who had a joint appointment at a prestigious nonprofit research institute and a major research university and whose work was at the intersection of evolutionary theory and immunology. He said he had stopped submitting grant proposals and that his research was made possible only by "leakage" from the funding secured by his Nobel Prize–winning patron at the institute. For this scientist, and several others who spoke to me informally, scientific "fashion" severely constrained academic research freedom. Scientific freedom

was, of course, an ideal virtue, but the practical realities of academic research were much more problematic.

73. In recent years, there has been a notable rise in the numbers of university-based, soft-money-supported, nonprofessorial, and nonteaching "research scientists." So there is, increasingly, a set of reasons to do science at a university, without contributing in any way to its teaching functions. And as numbers of university "research scientists" increase, so there is an overlap between the institutional texture of academia and of nonprofit research institutes, whose scientific staff sometimes choose occasionally to teach at neighboring universities.

74. Interviewed 6 August 2002.

75. Werth, *The Billion-Dollar Molecule*, p. 228.

76. E.g., Luigi Corleone (interviewed 10 July 2002), a biologist and chief technology officer (CTO) at the small drug discovery company cofounded by O'Reilly. Corleone, perhaps wishing to be as paradoxical as possible—he has a company reputation for being an awkward customer—would not agree with *any* story about the differences between academic and industrial science. It all depended upon the particular lab you were in, the particular people and projects with which you were involved, the stage of your career, etc.

77. Interviewed 10 June 2002.

78. Interviewed 2 July 2002; see also Greenberg, *Science for Sale*, pp. 142–143.

79. Snide, and unsubstantiated, remarks by Lyons's students, colleagues, and, indeed, staff of the university's tech-transfer office insinuated that Lyons had all of his really good, patentable ideas on the one day a week that he consulted for a major wireless company.

80. Interviewed 31 May 2002; see also Greenberg, *Science for Sale*, pp. 82, 86.

81. Interviewed 25 September 2002.

82. Rabinow, *Making PCR*, p. 31; see also idem and Dan-Cohen, *A Machine to Make a Future*, p. 82.

83. Victoria Griffith, "Leader Picks Risky Way to Grow," *Financial Times*, 18 December 2003, p. 8.

84. Shapiro, "What Scientists Look For in Their Jobs," p. 337.

85. Interviewed 9 September 2002.

86. Werth, *The Billion-Dollar Molecule*, p. 23.

87. William C. Taylor, "Mogul vs. Mavericks: A Search for Heroes," *New York Times*, 9 May 2004, Business Section, p. 4; also Ben Elgin, "Managing Google's Idea Factory," *BusinessWeek*, 3 October 2005, pp. 88–90.

88. Victoria Griffith, "'If the Science Looks Good, I'll Go After It,'" *Financial Times*, 11 February 2005, p. 9.

89. Grant Fjermedal, *Magic Bullets* (New York: Macmillan, 1984), pp. 118–119.

90. Interviewed 2 July 2002.

91. Interviewed 10 July 2002.

92. Mullis, *Dancing Naked in the Mind Field*, pp. 39–40.

93. Rabinow, *Making PCR*, p. 80.

94. Kornberg, *The Golden Helix*, pp. 32, 129.

95. See, among many examples, Teitelman, *Gene Dreams*, esp. p. 169.

96. Fjermedal, *Magic Bullets*, p. 100.

97. Quoted in ibid., pp. 128–129.

98. Andrew Jack and Lisa Urquhart, "Change of Culture: How Big Pharma Is Picking the Best of Biotech as a Sector Starts to Mature," *Financial Times*, 12 January 2006, p. 13.

99. Paul Rabinow, *French DNA: Trouble in Purgatory* (Chicago: University of Chicago Press, 1999), p. 13; see also idem, *Making PCR*, pp. 57, 62; idem and Dan-Cohen, *A Machine to Make a Future*, pp. 34, 66, 97.

100. Kerr, *The Uses of the University*, esp. ch. 1.

101. Quoted in Fjermedal, *Magic Bullets*, p. 102. For Robert Oppenheimer's personal embodiment of time-discipline at wartime Los Alamos, see Charles Thorpe, "Against Time: Scheduling, Momentum, and the Moral Order at Wartime Los Alamos," *Journal of Historical Sociology* 17 (2004), 31–55.

102. Joseph A. Schumpeter, *The Theory of Economic Development* (Cambridge, MA: Harvard University Press, 1934), pp. 80, 85; see also Michel Crozier, *The Bureaucratic Phenomenon* (Chicago: University of Chicago Press, 1964), esp. p. 196; and cf. Alfred Chandler on "personal capitalism": Alfred D. Chandler, Jr., *The Visible Hand: The Managerial Revolution in American Business* (Cambridge, MA: Belknap Press of Harvard University Press, 1977).

103. Richard N. Langlois, "Schumpeter and Personal Capitalism," in *Microfoundations of Economic Growth: A Schumpeterian Perspective*, eds. Gunnar Eliasson and Christopher Green (Ann Arbor: University of Michigan Press, 1998), pp. 57–82, on p. 77.

104. Richard N. Langlois, "Personal Capitalism as Charismatic Authority: The Organizational Economics of a Weberian Concept," *Industrial and Corporate Change* 7 (1998), 195–213, on p. 206; also J. S. Coleman, "Rational Organization," *Rationality and Society* 2 (1990), 94–105.

105. For charismatic leadership in creative organizations in general, see Warren Bennis and Patricia Ward Biederman, *Organizing Genius: The Secrets of Creative Collaboration* (Reading, MA: Addison-Wesley, 1997); and Charles Thorpe and Steven Shapin, "Who Was J. Robert Oppenheimer? Charisma and Complex Organization," *Social Studies of Science* 30 (2000), 545–590.

CHAPTER EIGHT

1. See a vitally important, and very funny, commentary on selecting the winners of "classic" horse races cowritten by the political philosopher Michael Oakeshott: Guy Griffith and Michael Oakeshott, *A New Guide to the Derby: How to Pick the Winner*, 2nd ed. (London: Faber and Faber, 1947; orig. publ. 1936).

2. For a history of American venture capital, see Martha Louise Reiner, "A History of Venture Capital Organizations in the United States," unpubl. Ph.D. thesis, University

of California, Berkeley, 1989; also John W. Wilson, *The New Venturers: Inside the High-Stakes World of Venture Capital* (Reading, MA: Addison-Wesley, 1985), ch. 2. The role of venture capital in the origins of biotech is nicely summarized in Mark Peter Jones, "Biotech's Perfect Climate: The Hybritech Story," unpubl. Ph.D. thesis, University of California, San Diego, 2005, pp. 309–326. The term "venture capital" was probably coined by Jock Whitney and his associate Benno Schmidt to designate the investment practices of J. H. Whitney & Co., a risk investment vehicle created after World War II: Wilson, *The New Venturers*, p. 17.

3. Nitin Nohria, "Information and Search in the Creation of New Business Ventures: The Case of the 128 Venture Group," in *Networks and Organizations: Structure, Form, and Action*, eds. Nitin Nohria and Robert G. Eccles (Boston: Harvard Business School Press, 1992), pp. 240–261, on p. 242.

4. For "fog," see Michael Moritz (of Sequoia Capital), quoted in Richard Waters, "Start-Ups Are 'A Perpetual Stroll into the Fog,'" *Financial Times*, 16 February 2005, p. 9; for "sociology experiments," see Robert Kagle (of Benchmark Capital), in *Done Deals: Venture Capitalists Tell Their Stories*, ed. Udayan Gupta (Boston: Harvard Business School Press, 2000), pp. 15–28, on p. 20; see also David A. Kaplan, *The Silicon Boys and Their Valley of Dreams* (New York: William Morrow & Co., 1999), pp. 155–184.

5. Richard Waters, "Technology Boom Venture Capital Leaders Named," *Financial Times*, 18 February 2003, p. 18; idem, "Calpers Data Shine Light on VC and Buy-Out World," ibid., 5 May 2003, p. 21; Tom Stein, "Exposed! Just When They Thought Life Couldn't Get Worse, Venture Capitalists May Have to Divulge Their Best-Kept Secrets," *Red Herring*, 23 January 2003 [accessed 6 June 2004 at http://www.redherring.com/vc/2003/01/exposed012303.html].

6. Quoted in "Amy Radin: The Venture Capitalist," *Business Week Special Supplement "Inside Innovation,"* 19 June 2006, p. 26.

7. Tom Perkins to Ted Greene (of the San Diego biotech company Hybritech), as quoted in Jones, "Biotech's Perfect Climate," p. 525. (In this case, the "cherries" were Tandem Computer and Genentech, so the "barrel of piss" turned out to be very profitable indeed.)

8. Geoffrey Yang (of Redpoint Ventures), in Gupta, *Done Deals*, pp. 73–82, on p. 78; Mitch Kapor, in ibid., pp. 83–91, on p. 89. Paul Wythes (of Sutter Hill Ventures), in ibid., pp. 151–163, on p. 159, concurs that his firm too is in the home run business, but is exceptional in admitting that "we also want to hit a lot of doubles and triples—even an occasional single."

9. Quoted in Barry Werth, *The Billion-Dollar Molecule: One Company's Quest for the Perfect Drug* (New York: Simon and Schuster, 1994), p. 89.

10. As interviewed by author, 30 April 2003. The venture capitalists I talked to were not asked to, and did not, sign informed consent forms, nor did they ask for anonymity or confidentiality.

11. Gary Rivlin, "Root, Root, Root for the Start-Up," *New York Times*, 9 July 2006, pp. C1, C4, on p. C4.

12. Martin Kenney and Richard Florida, "Venture Capital in Silicon Valley: Fueling New Firm Formation," in *Understanding Silicon Valley: The Anatomy of an Entrepreneurial Region*, ed. Kenney (Stanford, CA: Stanford University Press, 2000), pp. 98–123, on pp. 101–102; see also Jeffrey Zygmont, *The VC Way: Investment Secrets from the Wizards of Venture Capital* (Cambridge, MA: Perseus, 2001), pp. 23, 34; Charles H. Ferguson, *High St@kes, No Prisoners: A Winner's Tale of Greed and Glory in the Internet Wars* (New York: Random House, 1999), p. 73; Robert J. Kunze, *Nothing Ventured: The Perils and Payoffs of the Great American Venture Capital Game* (New York: Harper-Business, 1990), pp. 19, 219–220. There is, however, no great consensus about these metrics.

13. Karen Southwick, *The King-Makers: Venture Capital and the Money Behind the Net* (New York: John Wiley, 2001), p. 102.

14. Quoted in Gary Rivlin, "Billion-Dollar Baby Dot-Coms? Uh-Oh, Not Again," *New York Times*, 2 September 2005, p. C7.

15. Jerry Kaplan, *Startup: A Silicon Valley Adventure* (Boston: Houghton Mifflin, 1995), p. 22; Wythes, in Gupta, *Done Deals*, p. 159; Dennis, in ibid., 188. For figures from the National Venture Capital Association (NVCA), see Fred Wainwright, "Riding on Angels' Wings," *Financial Times*, 15 August 2003, p. 7.

16. Richard Waters, "Technology Boom Venture Capital Leaders Name"; also idem, "Silicon Valley Adjusts Its Expectations," *Financial Times*, 20 February 2003, p. 8. VC firms use two complementary measures of how their funds are doing: the "internal rate of return" (IRR) and the "realization ratio." The IRR is a portfolio's annualized profit going back to investors, taking into account both the amount of money invested and the length of time it has been invested, less the VC firm's charges. The IRR is the most popular metric in the industry, though it is a metric that may vary wildly according to the year in which it is calculated and whether or not a particular fund has a big hit. The realization ratio is the ratio of the sum of cumulative distributions and remaining portfolio value to paid-in capital, expressed, for example, as 2x, 3x, etc: see, for example, Southwick, *The King-Makers*, pp. 61–63, 112–113.

17. Post-bubble, some VCs were predicting that both management fees and carry would decrease: Bill Stensrud (managing director of Enterprise Partners Venture Capital), "Venture Capital: Still Adventurous?," presentation to UCSD Economics Roundtable, 18 June 2002 (response to audience question).

18. Gary Rivlin, "Venture Capital Rediscovers the Consumer Internet," *New York Times*, 10 June 2005, p. C7.

19. Figures from *The Money Tree™ Report* of PricewaterhouseCoopers http://www. pwcmoneytree.com/moneytree/index.jsp [accessed 27 July 2006]. In 2006, venture capitalists invested $25.5 billion in 3,416 deals, the highest level of activity since 2001, when the figure was $40.7 billion: http://www.pwcmoneytree.com/exhibits/06Q4MT_Press_Release.pdf [accessed 26 January 2007].

20. Gary Rivlin, "So You Want to Be a Venture Capitalist," *New York Times*, 22 May 2005, section 3, pp. 1, 9.

21. Matt Richtel, "Venture Capital Moves Out of the Garage," *New York Times*, 28 July 2006, p. C6, quoting John S. Taylor of the National Venture Capital Association; see also "Once Burnt," *Economist*, 27 November 2004, p. 16.

22. https://www.pwcmoneytree.com/MTPublic/ns/moneytree/filesource/exhibits/2007Q2_MoneyTree_Release.pdf [accessed 7 August 2007].

23. Matthew L. Wald, "Venture Capital Rushes Into Alternative Energy," *New York Times*, 30 April 2007, p. C7; Richard Waters, "Bubble 2.0? A Silicon Valley Investment Boom Heads for a Shake-Out," *Financial Times*, 1 May 2007, p. 7; Chris Hughes, "VC Funding for Homeland Security Up by Third," ibid., 27 August 2007, p. 11.

24. "Once Burnt," pp. 16–17.

25. Sarah Lacy and Jessi Hempel, "Valley Boys," *BusinessWeek*, 14 August 2006, pp. 40–47, esp. p. 44; also Robert Weisman, "Where's Web 2.0?" *Boston Globe*, 7 August 2006, pp. D1, D4; Graham Bowley, "Start-Ups May Be Lovelier the Second Time Around," *Financial Times*, 9–10 September 2006, p. W5.

26. Matt Marshall, "VCs Backing 'Magic' of Youth," *San Jose Mercury News*, 12 February 2006: http://www.mercurynews.com/vcsurvey/ci_5186421 [accessed 19 July 2007].

27. Gary Rivlin, "Skeptics Take Another Look at Social Sites," *New York Times*, 9 May 2005, p. C1.

28. "Michael J. Moritz, Sequoia Capital," *BusinessWeek Online*, 29 September 2003: www.businessweek.com/magazine/content/03_39/b3851611.htm [accessed 14 January 2005]. For the extent of post-bubble risk aversion, see, for example, Linda Himelstein, "Venture Capital's Dilemma," *BusinessWeek Online*, 19 September 2003: www.businessweek.com/smallbiz/content/sep2003/sb20030919_4952_sb020.htm [accessed 14 January 2005].

29. Timothy J. Mullaney, "New Enterprise Piles on the Risk," *BusinessWeek*, 4 September 2006, p. 70.

30. Dave Titus (of Windward Ventures), as interviewed by author 30 January 2003.

31. Bruno Latour displays a nice sensibility to the active processes of generating and displacing interests, while my intention here is to show just how interest is *performed* and to draw attention to the concrete technical and *moral* scenes in which such performances occur. The virtues of familiar people *matter* in my account in a way in which they do not in Latour's: Latour, *Science in Action: How to Follow Scientists and Engineers through Society* (Cambridge, MA: Harvard University Press, 1987), pp. 108–121; idem, *The Pasteurization of France*, trans. Alan Sheridan and John Law (Cambridge, MA: Harvard University Press, 1988), pp. 65–67.

32. http://www.connect.org/about/index.htm [accessed 31 July 2006]. As a condition of participation, I promised not to name any companies or individuals concerned. More recently, I attended several very similar meetings at the Massachusetts Technology Transfer Center's "Platform" program in Boston.

33. http://www.connect.org/programs/springboard/index.htm [accessed 31 July 2006]; Andrea Siedsma, "Springboard Injects Success into San Diego Tech Firms," *UCSD CONNECT Newsletter*, Issue 15–11 (15 June 2004) [accessed 21 July 2004].

34. The pitch scene has become so culturally familiar and dramatically resonant that it features in a "reality" television program—*Dragons' Den*–which originated in Japan and is now screened in Britain, Australia, and several other countries: http://www.bbc.co.uk/dragonsden/ [accessed 27 July 2006]. For present purposes, I lump angels and VCs proper together as sources of private capital—using "VC" to designate both—since both sorts of investors equally require to be pitched and persuaded of a credible future.

35. Wainwright, "Riding on Angels' Wings," p. 7. The Federal Securities and Exchange Commission requires that individual angel investors have a $200,000 annual income or $1,000,000 in assets, and some bands of angels further require that members invest in a certain number of deals a year—in the case of Tech Coast Angels, two $25,000 deals. In 2004, angels provided $22 billion in U.S. venture funding compared to VCs' $20.4 billion. Members of angel bands sometimes explicitly say they do it partly for the "social fun," and a number of the more prominent bands of angels are "not-for-profit" ventures: Chris Nuttall, "Halos Burn Brighter in a Group for Business Angels," *Financial Times*, 13 April 2005, p. 9.

36. Nuttall, "Halos Burn Brighter."

37. Peter Kolchinsky, *The Entrepreneur's Guide to a Biotech Startup*, 3rd ed. (published on-line by Evelexa BioResources, accessed 24 May 2002 from www.evelexa.com), pp. 64–65. Other sources of capital—even before angel funds—include Federal SBIRs (Small Business Innovation Research program grants, instituted in 1982) if the entrepreneurs can manage their way through the forms and government bureaucracy. A more recent option is the STTR (Small Business Technology Transfer program) to encourage joint ventures between small business and nonprofit research laboratories. Both offer phase I grants of up to $100,000 and phase II of $500,000 to $750,000.

38. On real-world attitudes to confidentiality agreements in biotech, see Paul Rabinow and Talia Dan-Cohen, *A Machine to Make a Future: Biotech Chronicles* (Princeton, NJ: Princeton University Press, 2004), pp. 138–140, on p. 140.

39. The celebrated Band of Angels meeting in Los Altos, California, requires ten-minute pitches: John May and Cal Simmons, *Every Business Needs an Angel: Getting the Money You Need to Make Your Business Grow* (New York: Crown Business, 2001), p. 63. At UCSD CONNECT's Financial Forum (in which somewhat later-stage companies present to investors) presentations are pared down to six minutes, and at other events entrepreneurs are given one to two minutes each to do their "elevator pitch," just to pique investors' interest enough to prompt later contacts. For the elevator pitch, see Southwick, *The King-Makers*, pp. 118–119, and May and Simmons, *Every Business Needs an Angel*, pp. 75–76: "Public-speaking experts say that the most critical part of your presentation is the first sentence, which should grab and captivate. Rehearse it . . . Practice in front of a mirror to check your facial expression and body language." A Southern California group of angels sponsors a "Fast Pitch" (or "PitchFest") Competition, in which entrepreneurs are required to deliver fifty-second business presentations, the winners' prize being, at one time, a free two-hour coaching session and, on another occasion, $10,000 in cash and a chance to present their plan to an assembled audience

of VCs: "Angel Angles," *UCSD CONNECT Newsletter*, 23 April 2002 and 29 July 2003. A professional coach counsels entrepreneurs to "prepare and practice various length elevator pitches—60 seconds, 2 minutes, 5 minutes, etc. The shorter, the harder and the better": Diane West, "Secrets 2Connect with Your Audience," typescript summary of seminar on "Investor Presentations," 9 October 2000.

40. National Collegiate Inventors & Innovators Alliance, *Invention to Venture: Workshops in Technology and Entrepreneurship: Participants Guide* 2006 (Amherst, MA: NCIIA, 2006), p. 42.

41. Dr. Phil Moheno (of SanRx Pharmaceuticals), quoted in Andrea Siedsma, "Springboard Injects Success Into San Diego Tech Firms," *UCSD CONNECT Newsletter*, Issue 15-1 (15 June 2004) [accessed on-line 13 June 2004: www.connect.org].

42. West, "Secrets 2Connect."

43. National Collegiate Inventors & Innovators Alliance, *Invention to Venture*, p. 42.

44. UCSD Springboard, 30 July 2002.

45. A Springboard session (10 September 2002) rather came apart when an IP lawyer expressed doubts that the presenter had given adequate evidence of proper peer review and of the legal status of patents.

46. See, e.g., Marco Rochat and Keith Arundale, *Three Keys to Obtaining Venture Capital*, pp. 5–10: http://www.altassets.net/pdfs/3keys-13-11-02.pdf [accessed 1 August 2006]; "Business Plan Basics": http://www.vcfodder.com/?PAGE=188 [accessed 18 July 2007].

47. May and Simmons, *Every Business Needs an Angel*, pp. 70–71, 75: "You don't want a [band of angels] manager's first impression to be of someone who can't string together two coherent ideas without tripping"; "Make sure your presentation is smooth." There are many sources of professional advice and coaching available—both free and expensive—for entrepreneurs preparing pitches. See also Thomas Leech, *How to Prepare, Stage, & Deliver Winning Presentations*, 2nd ed. (New York: American Management Associations, 1993), whose Amazon.com Web page guides you to such "closely related" works as Dale Carnegie's *How to Win Friends and Influence People*. Its author is a professional consultant whose other works include *Say It Like Shakespeare & Other Winning Communication Tools* (New York: McGraw-Hill, 2001): http://winning-presentations.com/book.html [accessed 31 July 2006].

48. Jerry L. Kalman, "The Entrepreneur's Gestalt: Ten Tips for the Fund Raiser," in *Entrepreneur's Handbook for Raising Capital*, ed. Guy J. Ianuzzi (San Diego, CA: Mentus, Inc., 2000), pp. A4–A11, on pp. A7–A8.

49. Peter Kolchinsky, "Director's Note: Insider's Guide to Pitching," *Evelexa BioResources*, 7 August 2002, on-line at http://www.evelexa.com/archives/080702.cfm [accessed 6 March 2003].

50. Leech, *How to Prepare, Stage, & Deliver Winning Presentations*, pp. 149, 208–219; UCSD CONNECT, the UCSD Program in Technology and Entrepreneurship, *Entrepreneur's Anthology* (La Jolla: University of California, San Diego, 1997), pp. 20–23. More ascetic variants include "the rule of five": no more than five lines per slide and no more than five words per line. Another tip given to entrepreneurs is to take

a 3×5 index card and a child's fat wax crayon: what you can write clearly and legibly on that card is what should be on the PowerPoint slide. For concise directions for making "the powerful pitch," see Ray Smilor, *Daring Visionaries: How Entrepreneurs Build Companies, Inspire Allegiance, and Create Wealth* (Holbrook, MA: Adams Media Corp., 2001), pp. 116–121.

51. Werth, *The Billion-Dollar Molecule*, p. 98.

52. E.g., Komisar, *The Monk and the Riddle*, p. 13; Kolchinsky, *The Entrepreneur's Guide*, p. 70.

53. Steve Harmon, *Zero Gravity: Riding Venture Capital from High-Tech Start-Up to Breakout IPO* (Princeton, NJ: Bloomberg Press, 1999), pp. 63–64.

54. Ruthann Quindlen, *Confessions of a Venture Capitalist: Inside the High-Stakes World of Start-up Financing* (New York: Warner Books, 2000), pp. 41, 49, 61, 67, 89, 104, 137. Against these explicit heuristics for picking winners is a Zen maxim said to be popular among VCs: "Those who know, do not talk. Those who talk, do not know" (Kaplan, *The Silicon Boys*, p. 157).

55. Randy Komisar (with Kent Lineback), *The Monk and the Riddle: The Education of a Silicon Valley Entrepreneur* (Boston: Harvard Business School Press, 2000), p. 30.

56. For the due diligence process in venture capital, see Kunze, *Nothing Ventured*, pp. 30–40; Jones, "Biotech's Perfect Climate," pp. 418–424 (for a specific case); and, in angel investing, see May and Simmons, *Every Business Needs an Angel*, ch. 5. The term was taken over from bankers' procedures for determining creditworthiness.

57. Mark Van Osnabrugge and Robert J. Robinson, *Angel Investing: Matching Start-Up Funds with Start-Up Companies* (San Francisco: Jossey-Bass, 2000), pp. 151–153.

58. Quindlen, *Confessions of a Venture Capitalist*, p. 114; Wythes, in Gupta, *Done Deals*, p. 155; Kaplan, *The Silicon Boys*, pp. 160–161. After the high-tech stock-market collapse of spring 2000, commentators noted how the action shifted to Starbucks and Peet's coffee shops, "the meeting places of the Valley's new dispossessed," where lonely geeks prepare resumés and the more optimistic develop new business plans, waiting for the upturn: Richard Waters, "The Chips Are Down," *Financial Times*, 14 December 2002, FT Weekend Section, p. 1.

59. Waters, "The Chips Are Down," p. 1.

60. Gordon Moore, speaking on "Global Business," BBC World Service, 3 July 2002.

61. Quoted in Tim Jackson, *Inside Intel: Andy Grove and the Rise of the World's Most Powerful Chip Company* (New York: Plume, 1998), p. 22; also Gupta, *Done Deals*, pp. 140, 144–145.

62. Randall E. Stross, *Eboys: The First Inside Account of Venture Capitalists at Work* (New York: Crown Business, 2000), pp. 24–28. The continual retelling of that, and similar, stories, sends a lesson about the limits of professionalism in this world; e.g., Paul Abrahams and Thorold Barker, "Ebay, the Flea Market that Spanned the Globe," *Financial Times*, 11 January 2002, p. 16.

63. Matt Cohler, quoted in John Cassedy, "Me Media: How Hanging Out on the Internet Became Big Business," *New Yorker* (15 May 2006), pp. 50–59, on p. 53.

64. Gary Rivlin, "A Few Signs of Froth Do Not a Bubble Make," *New York Times*, 19 May 2006, p. C7.

65. Quoted in Jones, "Biotech's Perfect Climate," p. 528.

66. As interviewed by author, 30 April 2003; see also Southwick, *The King-Makers*, pp. 34–35.

67. Tim Wollaeger (of Kingsbury Associates, San Diego), "Due Diligence," typescript of talk given at University of California, San Diego, 7 September 2000.

68. As interviewed by author, 30 April 2003.

69. Sanford Robertson (former chairman of Robertson, Stephens & Company investment bank), quoted in Rivlin, "So You Want to Be a Venture Capitalist," p. 9.

70. Dave Titus (of Windward Ventures), as interviewed by author 30 January 2003. Almost needless to say, literature produced by business school academics generally takes a more optimistic view of the role and value of formal due diligence practices: see, for example, the views of two Harvard Business School professors: Van Osnabrugge and Robinson, *Angel Investing*, p. 151: "Absolutely vital to making a sound investment, due diligence verifies any business opportunities that survive the initial screening stage."

71. Constance Loizos, "Dearth of Women in VC Ranks Traced to Less Interest in Tech," *San Jose Mercury News*, 11 May 2007: http://www.siliconvalley.com/vcsurvey/ci_5873541 [accessed 19 July 2007].

72. Quoted in Theresa Forsman, "The Surge in Women VCs," *BusinessWeek*, 4 October 2000: http://www.businessweek.com/smallbiz/content/oct2000/sb2000104_097.htm [accessed 19 July 2007].

73. Candida Brush et al., *Gatekeepers of Venture Growth: A Diana Project Report on the Role and Participation of Women in the Venture Capital Industry* (Kansas City, MO: Kauffman Foundation, 2007): http://www.kauffman.org/items.cfm?itemID=416 [accessed 19 July 2007].

74. Kunze, *Nothing Ventured*, p. 24.

75. Gary Rivlin, "Getting In on the Next Little Thing," *New York Times*, 20 September 2005, pp. E1, E11, on E11.

76. Harmon, *Zero Gravity*, pp. 61, 63.

77. Van Osnabrugge and Robinson, *Angel Investing*, p. 146.

78. Southwick, *The King-Makers*, p. 191.

79. Marco Rochat and Keith Arundale, *Three Keys to Obtaining Venture Capital*, p. 5: http://www.altassets.net/pdfs/3keys-13-11-02.pdf [accessed 19 July 2007]; Jones, "Biotech's Perfect Climate," p. 406.

80. Dave Titus (of Windward Ventures), as interviewed by author, 30 January 2003.

81. Ivor Royston (of Forward Ventures), as interviewed by author 30 April 2003; see also Van Osnabrugge and Robinson, *Angel Investing*, p. 148, noting that referrals by VCs' and angels' personal friends and business associates tend to be more trusted that those by attorneys, accountants, and bankers.

82. Rochat and Arundale, *Three Keys to Obtaining Venture Capital*, p. 2.

83. Stanley Milgram, "The Small-World Problem," *Psychology Today* 1 (1967), 61–67; and see Duncan J. Watts, *Six Degrees: The Science of a Connected Age* (New York: W. W. Norton, 2003).

84. Smilor, *Daring Visionaries*, pp. 155–157.

85. View attributed to VCs in general in Stross, *Eboys*, p. 25; also Kolchinsky, *The Entrepreneur's Guide*, pp. 65, 68; Joseph R. Mancuso, *How to Prepare and Present a Business Plan* (New York: Prentice Hall, 1983), pp. 6, 23.

86. Robert Kibble, "Making Your Dreams a Reality," in *Entrepreneur's Handbook*, pp. A12–A17, on p. A17; see also Harmon, *Zero Gravity*, p. 62.

87. Kunze, *Nothing Ventured*, p. 27.

88. Wilson, *The New Venturers*, p. 7.

89. Ted Greene, quoted in Jones, "Biotech's Perfect Climate," pp. 400–401; James Sterngold, "The Johnny Appleseed of a Biotechnology Forest," *New York Times*, 18 August 1996, p. F1.

90. Southwick, *The King-Makers*, p. 193.

91. As interviewed by author, 30 April 2003.

92. Kunze, *Nothing Ventured*, p. 135.

93. Southwick, *The King-Makers*, p. 129.

94. For divinely inspired business plans, see Wilson, *The New Venturers*, p. 7.

95. Yang, in Gupta, *Done Deals*, pp. 80–81; Quindlen, *Confessions of a Venture Capitalist*, pp. 100–103, 157, 159–160; Komisar, *The Monk and the Riddle*, p. 36. (Sometimes the claim is attributed to Ken Olson of DEC, sometimes, in a different version, to Thomas Watson of IBM, but it turns out to be an urban legend, and there's no solid evidence that either ever said it: it's just a cautionary tale about the perils of certainty.)

96. UCSD CONNECT, *Entrepreneur's Anthology*, p. 17. On proverbs in early and late modern settings, see Steven Shapin, "Proverbial Economies: How Certain Social and Linguistic Features of Common Sense May Throw Light on More Prestigious Bodies of Knowledge, Science for Example," *Social Studies of Science* 31 (2001), 731–769, esp. pp. 755–758.

97. Kolchinsky, *The Entrepreneur's Guide*, p. 70.

98. Quoted in David Amis and Howard H. Stevenson, *Winning Angels: The 7 Fundamentals of Early Stage Investing* (London: Financial Times/Prentice Hall, 2001), pp. 80–81; see also Zygmont, *The VC Way*, p. 71. Doriot (1899–1987) was one of the original gurus of American venture capital, professor of industrial management at the Harvard Business School, and an early president of American Research and Development: Henry Etzkowitz, "The Invention of the Venture Capital Firm: American Research and Development (ARD)," in idem, *MIT and the Rise of Entrepreneurial Science* (London/New York: Routledge, 2002), pp. 89–101, esp. p. 92.

99. Quoted in Gupta, *Done Deals*, pp. 144–145. Rock is repeatedly quoted as saying "I invest in people, not technology": see, e.g., ibid., p. 147.

100. Quoted in Wilson, *The New Venturers*, pp. 35–36; see also Reid Dennis, in Gupta, *Done Deals*, pp. 188–189 ("If I got to the point where I believed in the people, then I made

the investment"); Komisar, *The Monk and the Riddle*, p. 36; and Quindlen, *Confessions of a Venture Capitalist*, pp. 34–35.

101. See Rabinow and Dan-Cohen, *A Machine to Make a Future: Biotech Chronicles*, pp. 38–39. An online investors' guide defines the safe harbor as "a legal provision to reduce or eliminate liability as long as good faith is demonstrated"; it "protect[s] management from liability for making financial projections and forecasts made in good faith": http://www.investopedia.com/terms/s/safeharbor.asp [accessed 21 December 2004].

102. Howard Anderson, "Lies Entrepreneurs Tell Venture Capitalists (And Vice Versa)," http://tenonline.org/art/0009.html [accessed 10 January 2005].

103. See endorsement of this view in the case of financial markets: Donald MacKenzie, "Empty Cookie Jar," *London Review of Books*, 22 May 2003. http://www.lrb.co.uk/v25/n10/mack01_./html [accessed 19 July 2007]: "In a world in which technologies and economic circumstances change rapidly, personal virtues may be the most stable things around."

104. Udayan Gupta, introduction, idem, *Done Deals*, pp. 1–12, on pp. 6–7. Here Gupta summarizes the opinions of Georges Doriot: "venture capital depend[s] on bets placed on individuals," though Doriot cautioned that the problem of judging how individuals with proven track records will behave in an uncertain organizational future is very difficult. Track records of entrepreneurial success are as good an index as you can expect of future success, but, even here, it is widely understood that it is an imperfect predictor. Of course, in ventures where "intellectual capital" is vital—and that includes almost all high-tech—people just *are* central, and it is recognized that their minds come packaged together with their personalities: see, e.g., Ann Winblad (of Hummer Winblad Venture Partners) in ibid., pp. 61–72, on p. 67. The VC Andy Marcuvitz of Matrix Capital took the view that "the financials" in entrepreneurs' presentations were useful mainly to test whether they had "thought seriously" about opportunities and costs. "We don't invest of the basis of the numbers . . . , nobody can predict how something like this will turn out. We invest if we like the team and the business idea": Ferguson, *High St@kes, No Prisoners*, p. 76.

105. Paul Tyrrell, "The Bruises of the Bandwagon," *Financial Times*, 25 April 2005, p. 10.

106. Quoted in Zygmont, *The VC Way*, p. 71.

107. Quoted in Jones, "Biotech's Perfect Climate," p. 511. Jones shows how Hybritech's second business plan—containing detailed projections of technological and market futures—served mainly as a vehicle for persuading investors that key company people understood the rules of the game.

108. Kunze, *Nothing Ventured*, pp. 19, 66, 136, 144, 155, 208, 222–223, 228 (for a roll call of unsuitable founders he had to fire); Emily Barker, "The Bullet-Proof Business Plan," *Inc.* (October 2001), pp. 102–104, on p. 104. Liquidity events include IPOs (initial public offerings, or "going public"), merging or being acquired by another company, and going broke and having whatever assets remain liquidated. The time horizon in

pre-Internet days was around five to seven years; during the dot-com boom VCs were looking for deals they could exit in eighteen months; now it's more in the three- to five-year range or even longer.

109. Jones, "Biotech's Perfect Climate," p. 557.

110. Charles Matthews, president of the San Diego Tech Coast Angels, in *UCSD CONNECT Newsletter*, 27 August 2002.

111. Michael Lutz, "The Entrepreneur/Investor Dilemma," *UCSD CONNECT Newsletter*, 29 July 2003.

112. Michael Lewis, *The New New Thing: A Silicon Valley Story* (New York: W. W. Norton, 2000), p. 80: "[Clark] wanted to create *the* company that invented the future. Once he'd done that, he wanted to do it again and again and again." Once you acquire this kind of reputation, you have VCs chasing after *you* rather than the other way round.

113. Paul B. Brown, "Seven Steps to Heaven: It Pays to Think Like an Angel Investor," *Inc.* (October 2001), pp. 75–81, on p. 76; cf. Quindlen, *Confessions of a Venture Capitalist*, pp. 49–52 (for a cautionary tale of an entrepreneur who, disastrously, refused to let his company be acquired).

114. Robert Teitelman, *Gene Dreams: Wall Street, Academia, and the Rise of Biotechnology* (New York: Basic Books, 1989), p. 161; see also Ferguson, *High St@kes, No Prisoners*, pp. 83–84.

115. Komisar, *The Monk and the Riddle*, p. 93.

116. See notably Po Bronson, *The Nudist on the Late Shift and Other True Tales of Silicon Valley* (New York: Random House, 1999); also Kunze, *Nothing Ventured*, p. 212.

117. Bill Stensrud: "Venture Capital: Still Adventurous?," presentation to UCSD Economics Roundtable, 18 June 2002; http://www.epvc.com/pdfs/news_ep_010314. pdf [accessed 31 July 2006].

118. Interviewed by author 30 April 2003.

119. Smilor, *Daring Visionaries*, pp. 113–114; Jim Pinto, "Pinto Perspectives on 'Angel' Investing," *UCSD CONNECT Newsletter*, 11 February 2003 [accessed online 1 August 2006 at http://www.jimpinto.com/writings/angelinvesting.html].

120. Ivor Royston, as interviewed by author, 30 April 2003.

121. Looking for the marks of an investable entrepreneur, an angel asked "Does the entrepreneur-CEO have . . . the personal drive to continue on a success path, as success is not an end in itself but merely a precursor to and suggestion of further success? A positive assessment lends credibility to a belief that the individual has the persistence to succeed—because that is so often what it takes to be a successful entrepreneur." You want to look for people who strive for success but are never satisfied with what they achieved: Charles Matthew, president, San Diego Tech Coast Angels, in *UCSD CONNECT Newsletter*, 27 August 2002.

122. E.g., Bronson, *The Nudist on the Late Shift*, pp. 219, 242; also Komisar, *The Monk and the Riddle*, pp. 48–49, advising a dot-com entrepreneur who professed solely a desire to get rich: "The money's never there until it's there. There must be something more, a purpose that will sustain you when things look bleakest . . . I can't muster the

energy for a company whose founders never hope to accomplish anything more than making some bucks."

123. Stross, *Eboys*, p. 28.

124. http://www.omidyar.net/about.php [accessed 20 November 2007]; also Connie Bruck, "Millions for Millions," *New Yorker*, 30 October 2006, pp. 62–73, esp. pp. 67–68.

125. Sanford Robertson (of Robertson Stephens), in Gupta, *Done Deals*, p. 134. Charles Waite (of Greylock Management) noted that there once was "a kind of missionary quality in venture capital years ago," which he attributed partly to the influence of General Doriot. Entrepreneurs were on a "mission" and VCs backing them shared in that "missionary zeal": in ibid., p. 231.

126. Komisar, *The Monk and the Riddle*, p. 93.

127. Teitelman, *Gene Dreams*, p. 22.

128. Of course, at the highest levels of Silicon Valley entrepreneurship, competition to have the most money was a way of gauging who was Head Boy, and it was *hugely* important to be Head Boy. After Jim Clark had accumulated $3 billion—more than he knew what to do with—he concluded that he now had to have more money than Oracle's Larry Ellison: Lewis, *The New New Thing*, pp. 259–260. On Larry Ellison, greed, toys, and boys, see, e.g., Kaplan, *The Silicon Boys*, pp. 119–154, esp. 120: "Others use him as a benchmark"; Mike Wilson, *The Difference Between God and Larry Ellison: Inside Oracle Corporation* (New York: William Morrow, 1997).

129. Dave Titus (of Windward Ventures), as interviewed by the author, 30 January 2003.

130. Quoted in Grant Fjermedal, *Magic Bullets* (New York: Macmillan, 1984), p. 99.

131. Gary Rivlin, "Seed Money for Green Ventures," *New York Times*, 22 June 2005, p. C1, C5, on p. C1. The VC quoted is Ira Ehrenpreis of Technology Partners. U.S. VC investments in "clean energy" have increased from 1.2% in 2000 to 2.6% in 2004; see also James C. Collins and Jerry I. Porras, *Built to Last: Successful Habits of Visionary Companies* (New York: HarperBusiness, 1997), ch. 3, for "pragmatic idealism."

132. Chris Edwards (a science management consultant and founding editor of *Nature Biotechnology*), "If You're So Smart, Why Aren't You Rich?" *HMS Beagle*: http://biomednet.com/hmsbeagle/34/labres/adapt.htm (posted 10 July 1998 and accessed 28 January 2003).

133. UCSD CONNECT, *Entrepreneur's Anthology*, p. 20.

134. Kunze, *Nothing Ventured*, pp. 227–228.

135. Erving Goffman, *The Presentation of Self in Everyday Life* (Garden City, NY: Anchor Books, 1959), pp. 13–27.

136. For a practical guide to entrepreneurship that insists on the fundamental importance of these capacities and gestures, see Smilor, *Daring Visionaries*, esp. chs. 1–2, pp. 16, 46–47. See also Leech, *How to Prepare, Stage, & Deliver Winning Presentations*, pp. 206–207: "character counts"; "speak from the heart"; "believe in yourself." Nevertheless, and, of course, not necessarily because of lessons learned from this encounter, the optical screening company later turned out to be one of the UCSD program's

significant success stories. In November 2005, the company closed on $17 million in financing with a consortium of VC firms, and its products are doing well in the marketplace, while the genomics company has come to nothing.

137. Prospectus for Forward Ventures V (a life sciences venture fund launched by Forward Ventures of San Diego in February 2003). A staff member of the first venture capital firm, American Research and Development, founded in the early post–World War II years, recalls the visionary commitment that motivated its officers: "'Let's reshape the world; let's change its ills and shortcomings.' So American Research and Development had a very high tone and intent and integrity over and beyond making money": William Congleton, interviewed in 1986 and quoted in Etzkowitz, "The Invention of the Venture Capital Firm," p. 95.

138. Quoted in Wilson, *The New Venturers*, p. 59.

139. See Kaplan, *The Silicon Boys*, p. 158 (quoting Jim Clark on John Doerr and Kleiner Perkins); Jones, "Biotech's Perfect Climate," pp. 428 (for "company nappers"). The rapacious behavior of "vulture capitalists" is often instantiated in the well-known story of Sequoia Capital's dismissal of Cisco Systems' founders: Zygmont, *The VC Way*, pp. 24–29.

140. See, for example, Arthur Kornberg, *The Golden Helix: Inside Biotech Ventures* (Sausalito, CA: University Science Books, 1995), pp. 29, 186, 213, 250–251.

141. Eric Greenberg, of Viant, quoted in Stross, *Eboys*, p. 64; also Kolchinsky, *The Entrepreneur's Guide*, p. 68.

142. Dave Titus, interviewed by author, 30 January 2003.

143. Bill Burnham (of Mobius Venture Capital), quoted in Julie Landry, "Something Ventured: Wall Streeters Break Down on Sand Hill Road," *Red Herring*, 21 August 2002 [accessed 5 March 2003 at http://www.redherring.com/vc/2002/08/vc082102.htm].

144. Jones, "Biotech's Perfect Climate," pp. 428–434; Antonio Gledson de Carvalho, Charles W. Calomiris, and João Amaro de Matos, *Venture Capital as Human Resource Management*, Working Paper 11250 (Cambridge, MA: National Bureau of Economic Research, 2005).

145. Scott Kirsner, "Will You Ever Catch Up?" *Boston Globe*, 22 July 2007, pp. D1, D5, on p. D5.

146. Ivor Royston, interviewed by author, 30 April 2003.

147. Dave Titus, interviewed by author, 30 January 2003.

148. C. P. Snow, *The Two Cultures and The Scientific Revolution*, the Rede Lecture 1959 (Cambridge: Cambridge University Press, 1959), p. 10. For Google technologists becoming VCs, and then funding other Web entrepreneurs, see Miguel Helft, "A Post-Google Fraternity of Investors," *New York Times*, 28 December 2007, pp. C1, C6.

149. Quoted in Zygmont, *The VC Way*, p. 64.

150. Pinto, "Pinto Perspectives on 'Angel' Investing."

151. Nohria, "Information and Search," pp. 242–243; see also Paul DiMaggio, "Nadel's Paradox Revisited: Relational and Cultural Aspects of Social Structure," in Nohria and Eccles, *Networks and Organizations*, pp. 118–142, and Nohria and Eccles, "Face-to-Face: Making Network Organizations Work," in idem, eds., *Networks and Organizations*,

pp. 288–308, on pp. 292–293. On networking, economic activity, and patterns of familiarity, see also Walter W. Powell and Laurel Smith-Doerr, "Networks and Economic Life," in *Handbook of Economic Sociology*, eds. Neil J. Smelser and Richard Swedberg (Princeton, NJ: Princeton University Press, 1994), pp. 368–402.

152. Randall Stross, "It's Not Who You Know; It's Where You Are," *New York Times*, 22 October 2006, Business Section, p. 3 (interviewing Allen Morgan of the Sand Hill Road Mayfield Fund).

153. Robert Weisman, "Trading Places," *Boston Globe*, 10 May 2004, pp. C1, C4.

154. May and Simmons, *Every Business Needs an Angel*, p. 111: "We know an angel who claims he never invests in a company until he's had the founder drive him somewhere . . . His theory is that you can tell how someone manages the world by the way he or she drives. We also know angels who go to quirky lengths to play golf or tennis with someone, again in the belief that how people conduct themselves in a competitive situation reveals volumes about their values and attitudes." They tell about how one angel manages to smuggle an unlabeled professional psychologist into these face-to-face meetings, but most investors evidently trust themselves to make the necessary personality assessments.

155. Dave Titus, as interviewed by author, 30 January 2003.

156. Alex Williams, "Wheels and Deals in Silicon Valley," *New York Times*, 4 December 2005, section 9, pp. 1, 6 (quoting Internet entrepreneur Scott Milener and think-tank director Paul Saffo).

157. Deepak Kamra (of Canaan Partners), quoted in Kevin Allison, "Take the Road to Success on Two Wheels," *Financial Times*, 21 May 2007, p. 12.

158. Sterngold, "Johnny Appleseed."

159. http://www.nantucketconference.com/about.html [accessed 21 May 2007].

160. Peter Laslett, "The Face to Face Society," in idem, ed., *Philosophy, Politics and Society* (Oxford: Basil Blackwell, 1963), pp. 157–184; idem, *The World We Have Lost: Further Explored*, 3rd ed. (London: Routledge, 1983; orig. publ. 1965); Niklas Luhmann, *Trust and Power: Two Works*, eds. Tom Burns and Gianfranco Poggi; trans. Howard Davis, John Raffan, and Kathryn Rooney (Chichester: John Wiley, 1979); idem, "Familiarity, Confidence, Trust: Problems and Alternatives," in *Trust: Making and Breaking Cooperative Relations*, ed. Diego Gambetta (Oxford: Basil Blackwell, 1989), pp. 21–27, 79–85.

Bibliography

This is a consolidated bibliography, making no distinction between primary and secondary sources. It omits almost all references to newspapers, most ephemeral magazine pieces, most in-house business publications, and almost all Web site citations. These are all given in full when cited in the text.

Abir-Am, Pnina, 1980. "From Biochemistry to Molecular Biology: DNA and the Acculturated Journey of the Critic of Science Erwin Chargaff." *History and Philosophy of Life Sciences* 2: 3–60.

Abrahamson, Mark, 1964. "The Integration of Industrial Scientists." *Administrative Science Quarterly* 9: 208–218.

Abrahamson, Mark, 1965. "Cosmopolitanism, Dependence-Identification, and Geographical Mobility." *Administrative Science Quarterly* 10: 98–106.

Adams, Roger, R. C. Fuson, and C. S. Marvel, 1950. "The Graduate Training of Chemists." *Chemical and Engineering News* 28, no. 33 (14 August): 2765–2767.

Alvarez, Luis W., 1975. "Berkeley in the 1930s." Pp. 10–21 in Jane Wilson, ed., *All in Our Time: The Reminiscences of Twelve Nuclear Pioneers*. Chicago: Bulletin of the Atomic Scientists.

Alvarez, Luis W., 1987. *Alvarez: Adventures of a Physicist*. New York: Basic Books.

Amis, David, and Howard H. Stevenson, 2001. *Winning Angels: The 7 Fundamentals of Early Stage Investing*. London: Financial Times/Prentice Hall.

Angell, James Rowland, 1920. "The Organization of Research." *Scientific Monthly* 11, no. 1 (July): 25–42.

Angell, Marcia, 2004. *The Truth About Drug Companies: How They Deceive Us and What to Do About It*. New York: Random House.

Anon., 1924. "Why I Never Hire Brilliant Men." *American Magazine* (February): 12, 13, 117–118, 121–122.

Anon., 1925a. "The Encouragement of Basic Research." *Science* n.s. 61, no. 1567 (9 January): 43–44.

Anon., 1925b. "What Is Reason For?" *Science* n.s. 62, no. 1595 (24 July): 83–84.

Anon., 1937. "The Westinghouse Research Fellowships." *Science* n.s. 86, no. 2244 (31 December): 605–606.

Anon., 1952. "Attracting Young People into Science: Clinic Session Discussions." In Hertz and Rubenstein 1952, pp. 182–187.

Anon., 1953. "National Manpower Council." *Science* ns. 117, no. 3049 (5 June): 617–622.

Anon., 1957a. *The Direction of Research Establishments: Proceedings of a Symposium held at the National Physical Laboratory [Teddington] on 26th, 27th & 28th September 1956.* New York: Philosophical Library.

Anon., 1957b. "Knowledge Is Power." *Time* 70, no. 21 (18 November): 20–25.

Anon., 1958. "Researchers Encourage Students to Seek Careers in Science." *Industrial Laboratories* 9, no. 1 (January): 28–29.

Anon., 1965. "Philosopher." Pp. 283–289 in Denis Diderot and Jean le Rond d'Alembert, eds., *Encyclopedia: Selections,* [this ed.] ed. and trans. Nelly S. Hoyt and Thomas Cassirer. Indianapolis: Bobbs-Merrill.

Anthony, Robert N., 1952. *Management Controls in Industrial Research Organizations.* Boston: Graduate School of Business Administration, Harvard University.

Apple, Rima D., 1989. "Patenting University Research: Harry Steenbock and the Wisconsin Alumni Research Foundation." *Isis* 80: 374–394.

Appleton, Edward, 1954. "Science for Its Own Sake." *Science* n.s. 119, no. 3082 (22 January): 103–109.

Aronowitz, Stanley, 2001. *The Knowledge Factory: Dismantling the Corporate University and Creating True Higher Learning.* Boston: Beacon Press.

Avery, Robert W., 1960. "Enculturation in Industrial Research." *IRE Transactions on Engineering Management* EM-7, no. 1: 20–24.

Babbage, Charles, 1830. *Reflections on the Decline of Science in England, and on Some of Its Causes.* London: B. Fellowes.

Baker, John R., 1943. *The Scientific Life.* New York: Macmillan.

Baker, William O., 1961. "The Moral Un-Neutrality of Science." *Science* n.s. 133, no. 3448 (27 January): 261–262.

Bakke, E. Wight, 1948. "Teamwork in Industry." *Scientific Monthly* 66, no. 3 (March): 213–220.

Ball, Carleton R., 1926. "Personnel, Personalities and Research." *Scientific Monthly* 23, no. 1 (July): 33–45.

Barber, Bernard, 1952. *Science and the Social Order.* Glencoe, IL: Free Press.

Barber, Bernard, and Walter Hirsch, eds., 1962. *The Sociology of Science.* New York: Free Press of Glencoe.

Barker, Emily, 2001. "The Bullet-Proof Business Plan." *Inc.* (October): 102–104.

Barnes, Barry, 1971. "Making Out in Industrial Research." *Science Studies* 1: 157–175.

Barthes, Roland, [1957] 1975. "The Brain of Einstein." Pp. 68–70 in idem, *Mythologies,* trans. Annette Lavers. New York: Noonday Press.

Bartlett, Frederic C., 1932. *Remembering: A Study in Experimental and Social Psychology.* Cambridge: Cambridge University Press.

Barus, Carl, 1923. "Research and Teaching." *Science* n.s. 57, no. 1476 (13 April): 445–446.

Barzun, Jacques, 1964. *Science: The Glorious Entertainment.* New York: Harper & Row.

Baum, Arthur W., 1947. "Doctor of the Darkroom." *Saturday Evening Post* (25 October): 15–17, 47, 50, 52.

Bauman, Zygmunt, 1993. *Postmodern Ethics.* Oxford: Basil Blackwell.

Bauman, Zygmunt, 2000. *Liquid Modernity.* Cambridge: Polity Press.

Bauman, Zygmunt, 2004. *Identity.* Oxford: Basil Blackwell.

Baumgartel, Howard, 1957. "Leadership Style as a Variable in Research Administration." *Administrative Science Quarterly* 2: 344–360.

Baxter, James Phinney, III, 1946. *Scientists against Time.* Boston: Little, Brown.

Beardslee, David C., and Donald D. O'Dowd, 1961. "The College-Student Image of the Scientist." *Science* n.s. 133, no. 3457 (31 March): 997–1001.

Beck, Ulrich, Anthony Giddens, and Scott Lash, 1994. *Reflexive Modernization: Politics, Tradition, and Aesthetics in the Modern Social Order.* Cambridge: Polity Press.

Beer, John J., 1958. "Coal Tar Dye Manufacture and the Origins of the Modern Industrial Research Laboratory." *Isis* 49: 123–131.

Beer, John J., and David Lewis, 1965. "Aspects of the Professionalization of Science." Pp. 110–130 in Kenneth S. Lynn, ed., *The Professions in America.* Cambridge, MA: Houghton Mifflin.

Bello, Francis, [1953] 1956. "The Young Scientists." Pp. 21–39 in the editors of *Fortune*, eds., *The Mighty Force of Research.* New York: McGraw-Hill.

Bello, Francis. 1958. "The World's Greatest Industrial Laboratory." *Fortune* (November): 148–157, 208, 212, 214, 219–220.

Benda, Julien, [1927] 1928. *The Treason of the Intellectuals*, trans. Richard Aldington. New York: William Morrow & Co.

Bennis, Warren, and Patricia Ward Biederman, 1997. *Organizing Genius: The Secrets of Creative Collaboration.* Reading, MA: Addison-Wesley.

Benson, Sidney W., 1952. "Sponsored Research." *Science* n.s. 116, no. 3009 (29 August): 233.

Bernal, J. D., 1954. *Science in History.* London: Watts & Co.

Bernal, J. D., 1957. "Fundamental and Applied Aspects of Research Problems." In Anon. 1957a, pp. I.1–17.

Bernal, J. D., [1929] 1969. *The World, the Flesh & the Devil: An Inquiry into the Future of the Three Enemies of the Rational Soul.* Bloomington: Indiana University Press.

Bernard, Claude, [1865] 1927. *Introduction to the Study of Experimental Medicine*, trans. Henry Copley Greene. New York: Dover.

Biagioli, Mario, 1993. *Galileo, Courtier: The Practice of Science in the Culture of Absolutism.* Chicago: University of Chicago Press.

Bichowsky, F. Russell, 1942. *Industrial Research.* Brooklyn, NY: Chemical Publishing Co.

Billings, John S., 1886. "Scientific Men and Their Duties." *Science* 8, no. 201 (10 December): 541–551.

Bingham, Eugene C., 1925. "Research in Colleges." *Science* n.s. 61, no. 1572 (13 February): 174–176.

Birr, Kendall, 1957. *Pioneering in Industrial Research: The Story of the General Electric Research Laboratory.* Washington, DC: Public Affairs Press.

Blair, F. W., and N. Beverley Tucker, 1948. "Salary Policy." In Furnas 1948, pp. 258–276.

Bloor, David, 2000. "Whatever Happened to 'Social Constructiveness'?" Pp. 194–215 in Akiko Saito, ed., *Bartlett, Culture and Cognition.* London: Psychology Press.

Bloor, David, 2002. "F. C. Bartlett and the Origins of Social Constructionism in 1930s Cambridge." Typescript of talk given at the Max Planck Institute for the History of Science, Berlin.

Bloor, David, and Celia Bloor, 1982. "Twenty Industrial Scientists: A Preliminary Exercise." Pp. 83–102 in Mary Douglas, ed., *Essays in the Sociology of Perception.* London: Routledge & Kegan Paul.

Bok, Derek, 2003. *Universities in the Marketplace: The Commercialization of Higher Education.* Princeton, NJ: Princeton University Press.

Borman, Stu, 2002. "Triumph Over Nature: Entrepreneurial Scientist J. Craig Venter Is On to New Ventures." *Chemical and Engineering News* 80, no. 33 (19 August): 45–50.

Born, Max, 1978. *My Life: Recollections of a Nobel Laureate.* London: Taylor & Francis.

Box, Steven, and Stephen Cotgrove, 1966. "Scientific Identity, Occupational Selection, and Role Strain." *British Journal of Sociology* 17: 20–28.

Boyd, T. A., 1957. *Professional Amateur: The Biography of Charles Franklin Kettering.* New York: E. P. Dutton.

Boyle, Robert, 1655. *An Epistolical Discourse . . . Inviting All True Lovers of Vertue and Mankind to a Free and Generous Communication . . .*, reprinted in Margaret E. Rowbottom, "The Earliest Published Writing of Robert Boyle." *Annals of Science* 6 (1950): 380–385.

Boyle, Robert, [1663] 1772. "Some Considerations Touching the Usefulness of Experimental Natural Philosophy." Pp. 1–246, vol. II in *The Works of the Honourable Robert Boyle*, ed. Thomas Birch, 2nd ed., 6 vols. London: J. & F. Rivington.

Brewster, David, 1831. *The Life of Sir Isaac Newton.* London: John Murray.

Brewster, David, 1855. *The Life of Sir Isaac Newton*, rev. and ed. W. T. Lynn. Edinburgh: Gall & Inglis.

Bridgman, Percy W., 1927. *The Logic of Modern Physics.* New York: Macmillan.

Bridgman, Percy W., [1950] 1955. *Reflections of a Physicist*, 2nd ed. New York: Philosophical Library.

Brinton, Daniel G., 1895. "The Character and Aims of Scientific Investigation." *Science* n.s. 1, no. 1 (4 January): 3–4.

Broad, William, and Nicholas Wade, 1982. *Betrayers of the Truth: Fraud and Deceit in the Halls of Science.* New York: Simon and Schuster.

Bronk, Detlev W., 1949. "Science and Humanity." *Science* n.s. 109, no. 2837 (13 May): 477–482.

Bronk, Detlev W., 1954. "The Role of Scientists in the Furtherance of Science." *Science* n.s. 119, no. 3086 (19 February): 223–227.

Bronowski, Jacob, [1956] 1959. *Science and Human Values*. New York: Harper Torchbook.

Bronowski, Jacob, 1961. "Science Is Human." Pp. 83–94 in Julian Huxley, ed., *The Humanist Frame*. New York: Harper & Brothers.

Bronson, Po, 1999. *The Nudist on the Late Shift and Other True Tales of Silicon Valley*. New York: Random House.

Brooke, John Hedley, 1991. *Science and Religion: Some Historical Perspectives*. Cambridge: Cambridge University Press.

Brooks, Harvey, 1968. *The Government of Science*. Cambridge, MA: MIT Press.

Brougham, Henry, 1845–1846. *Lives of the Men of Letters and Science, Who Flourished in the Time of George III*, 2 vols. London: Charles Knight.

Brown, Paul B., 2001. "Seven Steps to Heaven: It Pays to Think Like an Angel Investor." *Inc.* (October): 75–81.

Brown, Ralph, 1955. "The Transistor as an Industrial Research Episode." *Scientific Monthly* 80, no. 1 (January): 40–46.

Browne, Janet, 1995. *Charles Darwin: Voyaging: A Biography*. Princeton, NJ: Princeton University Press.

Brozek, Josef, and Ancel Keys, 1944. "General Aspects of Interdisciplinary Research in Experimental Human Biology." *Science* n.s. 100, no. 2606 (8 December): 507–512.

Brozen, Yale, 1953. "The Economic Future of Research and Development." *Industrial Laboratories* 4, no. 12 (December): 6–13.

Bruck, Connie, 2006. "Millions for Millions." *New Yorker*, 30 October, pp. 62–73.

Bryson, Lyman, 1957. "Researchers in Industry." In Livingstone and Milberg 1957, pp. 129–137.

Buck, Peter, 1985. "Adjusting to Military Life: The Social Sciences Go to War, 1941–1950." Pp. 205–252 in Merritt Roe Smith, ed., *Military Enterprise and Technological Change: Perspectives on the American Experience*. Cambridge, MA: MIT Press.

Bud, R. F., 1978. "Strategy in American Cancer Research after World War II: A Case Study." *Social Studies of Science* 9: 425–459.

Burns, Tom, 1964. "Research, Development and Production: Problems of Conflict and Cooperation." Pp. 112–129 in Charles D. Orth III, Joseph C. Bailey, and Francis W. Wolek, eds., *Administering Research and Development: The Behavior of Scientists and Engineers in Organizations*. Homewood, IL: Richard D. Irwin & Inc. and the Dorsey Press.

Bush, George P., 1950. "Principles of Administration in the Research Environment." In Bush and Hattery 1950, pp. 161–183.

Bush, George P., 1953. "Teamwork in Research: A Commentary and Evaluation." In Bush and Hattery 1953, pp. 171–186.

Bush, George P., and Lowell H. Hattery, eds., 1950. *Scientific Research: Its Administration and Organization.* Washington, DC: American University Press.

Bush, George P., and Lowell H. Hattery, eds., 1953. *Teamwork in Research.* Washington, DC: American University Press.

Bush, Vannevar, [1945] 1990. *Science — The Endless Frontier: A Report to the President on a Program for Postwar Scientific Research,* National Science Foundation 40th Anniversary Edition. Washington, DC: National Science Foundation.

Calvin, Melvin, et al., 1965. *The Scientific Endeavor: Centennial Celebration of the National Academy of Sciences.* New York: Rockefeller University Press.

Capshew, James H., and Karen A. Rader, 1992. "Big Science: Price to the Present." *Osiris* n.s. 7: 3–25.

Carhart, Henry S., 1895. "The Educational and Industrial Value of Science." *Science* n.s. 1, no. 15 (12 April): 393–402.

Carlyle, Thomas, [1829] 1971. "Signs of the Times." Pp. 61–85 in Alan Shelston, ed., *Selected Writings.* Harmondsworth: Penguin.

Carmichael, R. D., 1919. "Individuality in Research." *Scientific Monthly* 9, no. 6 (December): 514–525.

Carpenter, Edward, 1885. *Modern Science: A Criticism.* Manchester: John Heywood.

Carpenter, Edward, 1889. *Civilisation: Its Cause and Cure, and Other Essays.* London: Swan Sonnenschein.

Carr, E. H., 1962. *What Is History?* New York: Alfred A. Knopf.

Carty, John J., 1929. "Science and Business: An Address to the Chamber of Commerce of the United States, May 8, 1924." Pp. 1–2 in *Reprint and Circular Series of the National Research Council.* Washington, DC: National Research Council.

Carvalho, Antonio Gledson de, Charles W. Calomiris, and João Amaro de Matos, 2005. *Venture Capital as Human Resource Management,* Working Paper 11250. Cambridge, MA: National Bureau of Economic Research.

Carver, Emmett K., 1957. "Organization, Personalities, and Creative Thought." In Livingstone and Milberg 1957, pp. 60–68.

Cassedy, John, 2006. "Me Media: How Hanging Out on the Internet Became Big Business." *New Yorker,* 15 May, pp. 50–59.

Cavanaugh, R. M., 1956. "Development of Managers: Training in a Research Division." *Administrative Science Quarterly* 1: 373–381.

Chandler, Alfred D., Jr., 1977. *The Visible Hand: The Managerial Revolution in American Business.* Cambridge, MA: Belknap Press of Harvard University Press.

Chargaff, Erwin, 1963a. *Essays on Nucleic Acids.* Amsterdam: Elsevier.

Chargaff, Erwin, 1963b. "A Few Remarks on Nucleic Acids, Decoding, and the Rest of the World." In Chargaff 1963a, pp. 161–173.

Chargaff, Erwin, 1963c. "Amphisbaena." In Chargaff 1963a, pp. 174–199.

Chargaff, Erwin, 1978. *Heraclitean Fire: Sketches from a Life before Nature.* New York: Rockefeller University Press.

Charles, Daniel, 2005. *Master Mind: The Rise and Fall of Fritz Haber, the Nobel Laureate Who Launched the Age of Chemical Warfare.* New York: HarperCollins.

Chollar, R. G., G. J. Wilson, and B. K. Green, 1958. "Creativity Techniques in Action." *Research Management* 1: 5–21.

Clague, Ewan, 1948. "Trends in Supply and Demand of Scientific Personnel." *Science* n.s. 107, no. 2780 (9 April): 355–360.

Clark, Walter, 1961. "Charles Edward Kenneth Mees: 1882–1960." *Biographical Memoirs of Fellows of the Royal Society* 7: 171–197.

Coleman, J. S., 1990. "Rational Organization." *Rationality and Society* 2: 94–105.

Collins, James C., and Jerry I. Porras, 1997. *Built to Last: Successful Habits of Visionary Companies.* New York: HarperBusiness.

Conant, James B., 1947. *On Understanding Science: An Historical Approach.* Oxford: Oxford University Press.

Conant, James B., 1970. *My Several Lives: Memoirs of a Social Inventor.* New York: Harper & Row.

Condon, Edward U., 1948. "Science and Security." *Science* n.s. 107, no. 2791 (25 June): 659–665.

Condon, Edward U., 1950. "Recruitment and Selection of the Research Worker." In Bush and Hattery 1950, pp. 61–64.

Condon, Edward U., 1954. "The Duty of Dissent." *Science* 119, no. 3086 (19 February): 227–228.

Cook, Harold J., 2007. *Matters of Exchange: Commerce, Medicine, and Science in the Dutch Golden Age.* New Haven, CT: Yale University Press.

Cornwell, Ralph T. K., 1948. "Professional Growth of the Research Man." In Furnas 1948, pp. 295–307.

Coser, Lewis A., 1965. *Men of Ideas: A Sociologist's View.* New York: Free Press.

Cotgrove, Stephen, and Steven Box, 1970. *Science, Industry and Society: Studies in the Sociology of Science.* London: George Allen and Unwin.

Cottrell, Frederick, 1932. "Patent Experience of the Research Corporation." *Transactions of the American Institute of Chemical Engineers* 28: 221–225.

Cottrell, Frederick, [1960] 1962. "Scientists: Solo or Concerted?" In Barber and Hirsch 1962, pp. 388–393.

Cox, Henry L., 1963. "The Personal Approach in Dealing with Technical People." *Research Management* 6: 153–161.

Crawford, T. Hugh, 1997. "Screening Science: Pedagogy and Practice in William Dieterle's Film Biographies of Scientists." *Common Knowledge* 6: 52–68.

Cringley, Robert X., 1996. *Accidental Empires: How the Boys of Silicon Valley Make Their Millions, Battle Foreign Competition, and Still Can't Get a Date.* New York: HarperBusiness.

Crog, Richard S., 1964. "Ethics and Integrity in Personnel Relations." *Research Management* 7: 183–194.

Crosland, Maurice, ed., 1975a. *The Emergence of Science in Western Europe.* London: Macmillan.

Crosland, Maurice, 1975b. "The Development of a Professional Career in Science in France." In Crosland 1975a, pp. 139–159.

Crozier, Michel, 1964. *The Bureaucratic Phenomenon*. Chicago: University of Chicago Press.

Curtis, Charles P., 1955. *The Oppenheimer Case: The Trial of a Security System*. New York: Simon and Schuster.

Daniels, George H., 1967. "The Pure Science Ideal and Democratic Culture." *Science* n.s. 156, no. 3783 (30 June): 1699–1705.

Daston, Lorraine J., 1992. "Objectivity and the Escape from Perspective." *Social Studies of Science* 22: 597–618.

Daston, Lorraine J., 2005. "Fear & Loathing of the Imagination in Science." *Dædalus* 134, no. 4 (Fall): 16–30.

Davies, Kevin, 2001. *Cracking the Genome: Inside the Race to Unlock Human DNA: Craig Venter, Francis Collins, James Watson, and the Story of the Greatest Scientific Discovery of Our Time*. New York: Free Press.

de Morgan, Augustus, [1855] 1914. *Essays on the Life and Work of Newton*, ed. Philip E. B. Jourdain. Chicago: Open Court.

Dear, Peter, 1995. *Discipline and Experience: The Mathematical Way in the Scientific Revolution*. Chicago: University of Chicago Press.

Dear, Peter, 2006. *The Intelligibility of Nature: How Science Makes Sense of the World*. Chicago: University of Chicago Press.

Dennis, Michael Aaron, 1987. "Accounting for Research: New Histories of Corporate Laboratories and the Social History of American Science." *Social Studies of Science* 17: 479–518.

Dennis, Michael Aaron, 1994. " 'Our First Line of Defense': Two University Labs in the Postwar American State." *Isis* 85: 427–455.

DiMaggio, Paul, 1992. "Nadel's Paradox Revisited: Relational and Cultural Aspects of Social Structure." In Nohria and Eccles 1992a, pp. 118–142.

Drucker, Peter F., 1952. "Management and the Professional Employee." *Harvard Business Review* 30, no. 3 (May–June): 84–90.

DuBridge, Lee A., 1956. "Scientists and Engineers: Quantity Plus Quality." *Science* n.s. 124, no. 3216 (17 August): 299–304.

Dukas, Helen, and Banesh Hoffmann, eds., 1979. *Albert Einstein: The Human Side: New Glimpses from the Archives*. Princeton, NJ: Princeton University Press.

Durkheim, Emile, 1972. *Selected Writings*, ed. and trans. Anthony Giddens. Cambridge: Cambridge University Press.

Eastman, J. R., 1897. "The Relations of Science and the Scientific Citizen to the General Government." *Science* n.s. 5, no. 118 (2 April): 525–531.

Edgerton, David, 1988. "Industrial Research in the British Photographic Industry, 1879–1939." Pp. 106–134 in Jonathan Liebenau, ed., *The Challenge of New Technology: Innovation in British Business Since 1850*. Aldershot: Gower.

Edgerton, David, 2004. "'The Linear Model' Did Not Exist: Reflections on the History and Historiography of Science and Research in Industry in the Twentieth Century." In Grandin, Wormbs, and Widmalm 2004, pp. 31–57.

Eiduson, Bernice T., 1962. *Scientists: Their Psychological World*. New York: Basic Books.

Eiduson, Bernice T., and Linda Beckman, eds., 1973. *Science as a Career Choice: Theoretical and Empirical Studies*. New York: Russell Sage Foundation.

Einstein, Albert, [1954] 1994a. *Ideas and Opinions*. New York: Modern Library.

Einstein, Albert, 1994b. "Physics and Reality [1936]." In Einstein 1994a, pp. 318–356.

Einstein, Albert, 1994c. "Scientific Truth [1929]." In Einstein 1994a, p. 286.

Einstein, Albert, 1994d. "Principles of Research [1918]." In Einstein 1994a, pp. 244–248.

Einstein, Albert, 1994e. "My First Impressions of the U.S.A. [1921]." In Einstein 1994a, pp. 3–7.

Einstein, Albert, 1950. *Out of My Later Years*. New York: Philosophical Library.

Eisenhower, Dwight D. [1961] 1972. "Farewell Address." Pp. 204–208 in Carroll W. Pursell, Jr., ed., *The Military-Industrial Complex*. New York: Harper and Row.

Elder, James Tait, 1963. "Basic Research in Industry: Appraisal and Forecast." *Research Management* 6: 5–14.

Elias, Hans, 1953. " 'True' Scientists." *Science* n.s. 117, no. 3051 (19 June): 698.

Eliot, T. S., [1920] 1967. "Tradition and the Individual Talent." Pp. 39–49 in idem, *The Sacred Wood: Essays on Poetry and Criticism*. London: Faber and Faber.

Ellul, Jacques, [1954] 1964. *The Technological Society*, trans. John Wilkinson, with an introduction by Robert K. Merton. New York: Vintage.

Emerson, Ralph Waldo, 1901. *The American Scholar . . . An Address Delivered before the φBK Society at Cambridge, August 1837*. New York: Laurentian Press.

Emerson, Ralph Waldo, [1849] 1903a. *Representative Men: Seven Lectures*. Boston: Houghton, Mifflin.

Emerson, Ralph Waldo, [1876] 1903b. *English Traits*. Boston: Houghton, Mifflin.

Engel, Leonard, 1954. "Get a Good Scientist . . . and Let Him Alone." *Harper's Magazine* 208, no. 1244 (January): 55–59.

Engstrom, Elmer W., 1957. "What Industry Requires of the Research Worker [1951, 1953]." In Livingstone and Milberg 1957, pp. 69–79.

Etzkowitz, Henry, 1994. "Knowledge as Property: The Massachusetts Institute of Technology and the Debate over Academic Patent Policy." *Minerva* 32: 383–421.

Etzkowitz, Henry, 2002a. *MIT and the Rise of Entrepreneurial Science*. New York: Routledge.

Etzkowitz, Henry, 2002b. "The Invention of the Venture Capital Firm: American Research and Development (ARD)." In Etzkowitz 2002a, pp. 89–101.

Eustis, Warner, 1948. "Personnel Policies and Personality Problems." In Furnas 1948, pp. 277–294.

Evans, C. George, 1969. *Supervising R&D Personnel*. New York: American Management Association.

Ferguson, Charles H., 1999. *High St@kes, No Prisoners: A Winner's Tale of Greed and Glory in the Internet Wars*. New York: Random House.

Feuer, Lewis S., 1963. *The Scientific Intellectual: The Psychological and Sociological Origins of Modern Science*. New York: Basic Books.

Feynman, Richard P., 1985. *"Surely You're Joking, Mr. Feynman!": Adventures of a Curious Character.* New York: Bantam Books.

Feynman, Richard P., 1988. *What Do YOU Care What Other People Think? Further Adventures of a Curious Character.* New York: W. W. Norton.

Feynman, Richard P., 1994. *No Ordinary Genius: The Illustrated Richard Feynman,* ed. Christopher Sykes. New York: W. W. Norton.

Feynman, Richard P., 1998. *The Meaning of It All: Thoughts of a Citizen-Scientist.* Reading, MA: Perseus.

Feynman, Richard P., 1999. *The Pleasure of Finding Things Out: The Best Short Works of Richard P. Feynman.* Cambridge, MA: Perseus.

Feynman, Richard P., 2005. *Perfectly Reasonable Deviations from the Beaten Track: The Letters of Richard P. Feynman,* ed. Michelle Feynman. New York: Perseus.

Fisch, Harold, 1953. "The Scientist as Priest: A Note on Robert Boyle's Natural Theology." *Isis* 44: 252–265.

Fisk, James B., 1959. "Basic Research in Industrial Laboratories." Pp. 159–167 in Dael Wolfle, ed., *Symposium on Basic Research, Sponsored by the National Academy of Sciences, the American Association for the Advancement of Science, and the Alfred P. Sloan Foundation.* Washington, DC: American Association for the Advancement of Science.

Fisk, James B., 1965. "Synthesis and Application of Scientific Knowledge for Human Use." In Calvin et al. 1965, pp. 293–302.

Fjermedal, Grant, 1984. *Magic Bullets.* New York: Macmillan.

Flemming, Arthur S., 1956. "Nation's Interest in Scientists and Engineers." *Scientific Monthly* 82, no. 6 (June): 282–285.

Form, William H., 1957. Review of *The Organization Man. Administrative Science Quarterly* 2: 124–126.

Forman, Paul, 1989. "Social Niche and Self-Image of the American Physicist." Pp. 96–104 in Michelangelo de Maria, Mario Grilli, and Fabio Sebastiani, eds., *Proceedings of the International Conference on the Restructuring of the Physical Sciences in Europe and the United States 1945–1960.* Singapore: World Scientific Publishing.

Fornas, John, 1995. *Cultural Theory and Late Modernity.* London: Sage.

Fortun, Michael A., and Silvan S. Schweber, 1993. "Scientists and the Legacy of World War II: The Case of Operations Research (OR)." *Social Studies of Science* 23: 595–642.

Fortune editorial staff, 1948. "The Scientists." *Fortune* (October): 106–112, 166, 168, 170, 173–174, 176.

Foucault, Michel, 1980. "Truth and Power." Pp. 109–133 in Colin Gordon, ed., *Power/Knowledge: Selected Interviews and Other Writings 1972–1977,* trans. Gordon, Leo Marshall, John Mepham, and Kate Soper. New York: Pantheon.

Frederick, Edwin, 1952. "Planning Research for Team Attack." *Industrial Laboratories* 3, no. 3 (March): 80–86.

Freedman, Paul, [1941] 1950. *The Principles of Scientific Research.* Washington, DC: Public Affairs Press.

Friedberg, Errol C., 2005. *The Writing Life of James D. Watson*. Cold Spring Harbor, NY: Cold Spring Harbor Laboratory Press.

Fritz, Howard E., and Douglas M. Beach, 1948. "The Modern Research Laboratory." In Furnas 1948, pp. 308–340.

Furnas, C. C., ed., 1948. *Research in Industry: Its Organization and Management*. New York: D. Van Nostrand.

Gadamer, Hans-Georg, 1989. *Truth and Method*, 2nd rev. ed., trans. Joel Weinsheimer and Donald G. Marshall. New York: Crossroad.

Galilei, Galileo, [1615] 1957. "Letter to the Grand Duchess Christina." Pp. 173–216 in *Discoveries and Opinions of Galileo*, trans. Stillman Drake. Garden City, NY: Doubleday Anchor.

Galison, Peter, 1985. "Bubble Chambers and the Experimental Workplace." Pp. 309–373 in Peter Achinstein and Owen Hannaway, eds., *Experiment and Observation in Modern Science*. Cambridge, MA: MIT Press.

Galison, Peter, 1992. "The Many Faces of Big Science." Pp. 1–17 in Peter Galison and Bruce Hevly, eds., *Big Science: The Growth of Large-Scale Research*. Stanford: Stanford University Press.

Galison, Peter, 1997. *Image and Logic: A Material Culture of Microphysics*. Chicago: University of Chicago Press.

Galison, Peter, 2003. *Einstein's Clocks, Poincaré's Maps: Empires of Time*. New York: W. W. Norton.

Galton, Francis, 1874. *English Men of Science: Their Nature and Nurture*. London: Macmillan.

Gascoigne, John, 1988. "From Bentley to the Victorians: The Rise and Fall of British Newtonian Natural Theology." *Science in Context* 2: 219–256.

Gaukroger, Stephen, 1995. *Descartes: An Intellectual Biography*. Oxford: Clarendon Press.

Geiger, Ronald L., 1997. "What Happened After Sputnik? Shaping University Research in the United States." *Minerva* 35: 349–367.

Gellhorn, Walter, 1950. *Security, Loyalty, and Science*. Ithaca, NY: Cornell University Press.

George, William H., 1936. *The Scientist in Action: A Scientific Study of His Methods*. London: Williams & Norgate.

Gershinowitz, Harold, 1960. "Sustaining Creativity Against Organizational Pressures." *Research Management* 3: 49–56.

Gerstenfeld, Arthur, 1970. *Effective Management of Research and Development*. Reading, MA: Addison-Wesley.

Giddens, Anthony, 1989. *The Consequences of Modernity*. Stanford, CA: Stanford University Press.

Giddens, Anthony, 1991. *Modernity and Self-Identity: Self and Society in the Late Modern Age*. Cambridge: Polity Press.

Gide, André, 1949. *Oscar Wilde: In Memoriam*. New York: Philosophical Library.

Gilfillan, S. C., [1935] 1970. *The Sociology of Invention*. Cambridge, MA: MIT Press.

Glaser, Barney G., 1963. "The Local-Cosmopolitan Scientist." *American Journal of Sociology* 69: 249–259.

Glaser, Barney G., 1964. *Organizational Scientists: Their Professional Careers.* Indianapolis: Bobbs-Merrill.

Glass, Bentley, 1960. "The Academic Scientist, 1940–1960." *Science* n.s. 132, no. 3427 (2 September): 598–603.

Gleick, James, 1992. *Genius: The Life and Science of Richard Feynman.* New York: Pantheon Books.

Goffman, Erving, 1959. *The Presentation of Self in Everyday Life.* Garden City, NY: Anchor Books.

Goldschmidt, Richard, 1949. "Research and Politics." *Science* n.s. 109, no. 2827 (4 March): 219–227.

Gorog, I., 1966. "Successful Adjustment to Industrial Research: How Can University and Industry Assist." *Research Management* 9: 5–13.

Gouldner, Alvin W., 1957–1958. "Cosmopolitans and Locals: Toward an Analysis of Latent Social Roles." *Administrative Science Quarterly* 2: 281–306, 444–480.

Gouldner, Alvin W., 1979. *The Future of Intellectuals and the Rise of the New Class.* New York: Seabury Press.

Grandin, Karl, Nina Wormbs, and Sven Widmalm, eds., 2004. *The Science—Industry Nexus: History, Policy, Implications*, Nobel Symposium 123. Canton, MA: Science History Publications.

Green, David E., 1954. "Group Research." *Science* n.s. 119, no. 3092 (2 April): 444–445.

Greenberg, Daniel S., 1981. *The Grant Swinger Papers.* Washington, DC: Science & Government Report.

Greenberg, Daniel S., 2001. *Science, Money, and Politics: Political Triumph and Ethical Erosion.* Chicago: University of Chicago Press.

Greenberg, Daniel S., 2007. *Science for Sale: The Perils, Rewards, and Delusions of Campus Capitalism.* Chicago: University of Chicago Press.

Gregory, Richard, [1916] 1928. *Discovery, or the Spirit and Service of Science.* New York: Macmillan.

Griffith, Guy, and Michael Oakeshott, [1936] 1947. *A New Guide to the Derby: How to Pick the Winner*, 2nd ed. London: Faber and Faber.

Gupta, Udayan, ed., 2000. *Done Deals: Venture Capitalists Tell Their Stories.* Boston: Harvard Business School Press.

Hackett, James W., and B. L. Steierman, 1963. "The Organization of a Fundamental Research Effort at Owens-Illinois Glass Company." *Research Management* 6: 81–92.

Hackett, Edward J., 2001. "Science as a Vocation in the 1990s: The Changing Organizational Culture of Academic Science." Pp. 101–137 in Jennifer Croissant and Sal Restivo, eds., *Degrees of Compromise: Industrial Interests and Academic Values.* Albany: State University of New York Press.

Haddow, Alexander, 1956. "The Scientist as Citizen." *Bulletin of the Atomic Scientists* 12, no. 7 (September): 245–252.

Hagner, Michael, [2004] 2007. *Geniale Gehirne: Zur Geschichte der Elitegehirnforschung.* Munich: Deutscher Taschenbuch Verlag.

Hagstrom, Warren O., 1964. "Traditional versus Modern Forms of Scientific Teamwork." *Administrative Science Quarterly* 9: 241–263.

Hagstrom, Warren O., [1965] 1975. *The Scientific Community.* Carbondale: Southern Illinois University Press.

Hahn, Roger, 1975. "Scientific Careers in Eighteenth-Century France." In Crosland 1975a, pp. 127–138.

Hall, Henry S., [1956] 1962. "Scientists and Politicians." In Barber and Hirsch 1962, pp. 269–287.

Hamer, Richard, 1925. "The Romantic and Idealistic Appeal of Physics." *Science* n.s. 61, no. 1570 (30 January): 109–110.

Hamlin, Christopher, 1986. "Scientific Method and Expert Witnessing: Victorian Perspectives on a Modern Problem." *Social Studies of Science* 16: 485–513.

Hammett, Frederick J., 1953. "Uncommitted Researchers." *Science* n.s. 117, no. 3029 (16 January): 64.

Hamor, William Allen, 1918. "The Research Couplet: Research in Pure Science and Industrial Research." *Scientific Monthly* 6, no. 4 (April): 319–330.

Harmon, Steve, 1999. *Zero Gravity: Riding Venture Capital from High-Tech Start-Up to Breakout IPO.* Princeton, NJ: Bloomberg Press.

Harrison, George Russell 1956. *What Man May Be: The Human Side of Science.* New York: William Morrow & Co.

Hart, David M., 1998. *Forged Consensus: Science, Technology, and Economic Policy in the United States, 1921–1953.* Princeton, NJ: Princeton University Press.

Harvard University, Committee on the Objectives of a General Education in a Free Society, 1945. *General Education in a Free Society, Report of the Harvard Committee.* Cambridge, MA: Harvard University Press.

Haskins, C. P., 1948. "Characteristics of the Research Man and the Research Atmosphere." In Furnas 1948, pp. 182–194.

Hattery, Lowell H., 1953. "Nature of Research Teamwork." In Bush and Hattery 1953, pp. 3–9.

Hausrath, A. H., 1948. "Programs for Fuller Utilization of Present Resources of Scientific Personnel." *Science* n.s. 107, no. 2780 (9 April): 360–363.

Hawkins, Laurence A., 1947. "Does Patent Consciousness Interfere with Cooperation between Industrial and University Research Laboratories?" *Science* n.s. 105, no. 2726 (28 March): 326–327.

Hawkins, Laurence A., 1950a. *The Story of General Electric Research.* [Schenectady, NY]: General Electric Company.

Hawkins, Laurence A., 1950b. *Adventure into the Unknown: The First Fifty Years of the General Electric Research Laboratory.* New York: William Morrow & Co.

Haynes, Roslynn D., 1994. *From Faust to Strangelove: Representations of the Scientist in Western Literature.* Baltimore, MD: Johns Hopkins University Press.

Heaphy, Brian, and Jane Franklin, 2004. *Late Modernity and Social Change*. London: Routledge.

Hebb, Malcolm H., with Miles J. Martin, 1958. "Free Inquiry in Industrial Research." *Research Management* 1: 67–83.

Heims, Steve J., 1980. *John Von Neumann and Norbert Wiener: From Mathematics to the Technologies of Life and Death*: Cambridge, MA: MIT Press.

Herschel, John F. W., [1830] 1987. *A Preliminary Discourse on the Study of Natural Philosophy*. Chicago: University of Chicago Press.

Hertz, David B., 1950. *The Theory and Practice of Industrial Research*. New York: McGraw-Hill.

Hertz, David B., and Albert H. Rubenstein, eds., 1952. *Selection, Training, and Use of Personnel in Industrial Research*, Proceedings of the Second Annual Conference on Industrial Research June 1951. New York: King's Crown Press.

Hertz, David B., and Albert H. Rubenstein, 1953. *Team Research*. Boston and New York: Eastern Technical Publications.

Herzig, Rebecca M., 2005. *Suffering for Science: Reason and Sacrifice in Modern America*. New Brunswick, NJ: Rutgers University Press.

Hickey, Albert E., Jr., 1958. "Basic Research: Should Industry Do More of It?" *Harvard Business Review* 36, no. 4 (July–August): 115–122.

Higgitt, Rebekah, 2003. "'Newton dépossédé!' The British Response to the Pascal Forgeries of 1867." *British Journal for the History of Science* 36: 437–453.

Higgitt, Rebekah, 2004. "The Apple of Their Eye? Biographies of Isaac Newton, 1820–1870." Unpubl. D.Sc. thesis, Imperial College, London.

Hill, A. V., 1937. "The Humanity of Science." Pp. 30–38 in John Boyd Orr et al., *What Science Stands For*. London: George Allen & Unwin.

Hill, A. V., [1933] 1962. "The International Status and Obligations of Science." Pp. 205–221 in idem, *The Ethical Dilemma of Science*. London: Scientific Book Guild.

Hill, Karl, ed., 1964. *The Management of Scientists*. Boston: Beacon Press.

Hilts, Philip J., 1982. *Scientific Temperaments: Three Lives in Contemporary Science*. New York: Simon and Schuster.

Hirsch, Walter, 1965. "Knowledge for What?" *Bulletin of the Atomic Scientists* 21, no. 5 (May): 28–31.

Hirsch, Walter, 1968. *Scientists in American Society*. New York: Random House.

Holland, Maurice, with Henry F. Pringle, 1928. *Industrial Explorers*. New York: Harper & Brothers.

Hollinger, David A., 1983. "The Defence of Democracy and Robert K. Merton's Formulation of the Scientific Ethos." Pp. 1–15 in Robert Alun Jones and Henrika Kuklick, eds., *Knowledge and Society*, vol. 4. Greenwich, CT: JAI Press.

Hollinger, David A., 1984. "Inquiry and Uplift: Late Nineteenth-Century American Academics and the Moral Efficacy of Scientific Practice." Pp. 142–156 in Thomas Haskell, ed., *The Authority of Experts: Studies in History and Theory*. Bloomington: Indiana University Press.

Hollinger, David A., 1990. "Free Enterprise and Free Inquiry: The Emergence of Laissez-Faire Communitarianism in the Ideology of Science in the U.S." *New Literary History* 21: 897–920.

Hollinger, David A., 1995. "Science as a Weapon in *Kulturkämpfe* in the United States During and After World War II." *Isis* 86: 440–454.

Hollinger, David A., 1996. *Science, Jews, and Secular Culture: Studies in Mid-Twentieth-Century American Intellectual History*. Princeton, NJ: Princeton University Press.

Holton, Gerald, ed., 1972. *The Twentieth-Century Sciences: Studies in the Biography of Ideas*. New York: W. W. Norton.

Hoover, Herbert, 1926. "The Vital Need for Greater Financial Support to Pure Scientific Research." *Mechanical Engineering* 48 (January): 6–8.

Hounshell, David A., 1996. "The Evolution of Industrial Research in the United States." Pp. 13–85 in Richard S. Rosenbloom and William J. Spencer, eds., *Engines of Innovation: U.S. Industrial Research at the End of an Era*. Cambridge, MA: Harvard University Press.

Hounshell, David A., 2000. "The Medium is the Message, or How Context Matters: The RAND Corporation Builds an Economics of Innovation, 1946–1962." Pp. 255–310 in Thomas P. Hughes and Agatha C. Hughes, eds., *Systems, Experts, and Computers: The Systems Approach in Management and Engineering, World War II and After*. Cambridge, MA: MIT Press.

Hounshell, David A., and John Kenly Smith, Jr., 1988. *Science and Corporate Strategy: DuPont R&D, 1902–1980*. Cambridge: Cambridge University Press.

Hower, Ralph M., and Charles D. Orth III, 1963. *Managers and Scientists: Some Human Problems in Industrial Research Organizations*. Boston: Graduate School of Business Administration, Harvard University.

Hughes, Everett C., 1958. *Men and Their Work*. Glencoe, IL: Free Press.

Hughes, Thomas P., 1989. *American Genesis: A Century of Invention and Technological Enthusiasm, 1870–1970*. New York: Viking.

Hull, Albert W., 1945. "Selection and Training of Students for Industrial Research." *Science* n.s. 101, no. 2616 (16 February): 157–160.

Hume, David, [1740] 1978. *Treatise of Human Nature*. Oxford: Oxford University Press.

Hutchins, Robert Maynard, 1936. *No Friendly Voice*. Chicago: University of Chicago Press.

Hutchins, Robert Maynard, 1963. "Science, Scientists, and Politics." Pp. 1–4 in Hutchins et al., *Science, Scientists, and Politics*, An Occasional Paper on the Role of Science and Technology in the Free Society. Santa Barbara, CA: Center for the Study of Democratic Institutions.

Hutchinson, G. E., 1959. "Homage to Santa Rosalia, or Why Are There So Many Kinds of Animals?" *American Naturalist* 93: 145–159.

Huxley, Thomas Henry [1854] 1900a. "On the Educational Value of the Natural History Sciences." Pp. 38–65 in idem, *Collected Essays, Vol. III. Science and Education: Essays*. New York: D. Appleton.

Huxley, Thomas Henry, [1880] 1900b. "On the Method of Zadig." Pp. 1–23 in idem, *Collected Essays, Vol. IV. Science and Hebrew Tradition*. New York: D. Appleton.

Ianuzzi, Guy J., ed., 2000. *Entrepreneur's Handbook for Raising Capital*. San Diego, CA: Mentus.

Jackson, Tim, 1998. *Inside Intel: Andy Grove and the Rise of the World's Most Powerful Chip Company*. New York: Plume.

James, William, [1907] 1991. "What Pragmatism Means. Lecture Two." Pp. 22–38 in idem, *Pragmatism: A New Name for an Old Way of Thinking*. Buffalo, NY: Prometheus Books.

Jenkins, Reese, 1979. *Images and Enterprise: Technology and the American Photographic Industry, 1839–1925*. Baltimore, MD: Johns Hopkins University Press.

Jewett, Frank B., 1919. "Industrial Research." *Reprint and Circular Series of the National Research Council, #4*. Washington, DC: National Research Council.

Jewett, Frank B., 1945. "The Future of Industrial Research: The View of a Physicist." In Standard Oil Development Company 1945, pp. 17–23.

Jewkes, John, David Sawers, and Richard Stillerman, 1958. *The Sources of Invention*. London: Macmillan.

Jones, Mark Peter, 2005. "Biotech's Perfect Climate: The Hybritech Story." Unpubl. Ph.D. thesis, University of California, San Diego.

Jordan, David Starr, 1896. "Nature Study and Moral Culture." *Science* n.s. 4, no. 84 (7 August): 149–156.

Jordan, David Starr, 1899. "A Sage in Science." *Science* n.s. 9, no. 221 (14 April): 529–532.

Judson, Horace Freeland, 2004. *The Great Betrayal: Fraud in Science*. New York: Harcourt Inc.

Kahn, Ely J., Jr., 1986. *The Problem Solvers: A History of Arthur D. Little, Inc.* Boston: Little, Brown.

Kahn, J. B., Jr., 1953. "'True' Scientists." *Science* n.s. 117, no. 3051 (19 June): 697–698.

Kaiser, David, 2000. "Making Theory: Producing Physics and Physicists in Postwar America." Unpubl. Ph.D. thesis, Harvard University.

Kaiser, David, 2002. "Cold War Requisitions, Scientific Manpower, and the Production of American Physicists after World War II." *Historical Studies in the Physical Sciences* 33: 131–159.

Kaiser, David, 2004. "The Postwar Suburbanization of American Physics." *American Quarterly* 56: 851–888.

Kaiser, David, 2005. "The Atomic Secret in Red Hands? American Suspicions of Theoretical Physicists During the Early Cold War." *Representations* 90 (Spring): 28–60.

Kalman, Jerry L., 2000. "The Entrepreneur's Gestalt: Ten Tips for the Fund Raiser." In Ianuzzi 2000, pp. A4–A11.

Kant, Immanuel, [1790] 1952. *The Critique of Judgment*, trans. James Creed Meredith. Oxford: Oxford University Press.

Kant, Immanuel, [1784] 1990. "What Is Enlightenment?" Pp. 83–90 in idem, *Foundations of the Metaphysics of Morals and What Is Enlightenment?*, 2nd rev. ed., trans. Lewis White Beck. New York: Macmillan.

Kaplan, David A., 1999. *The Silicon Boys and Their Valley of Dreams*. New York: William Morrow & Co.

Kaplan, Jerry, 1995. *Startup: A Silicon Valley Adventure*. Boston: Houghton Mifflin.

Kaplan, Norman, 1959. "The Role of the Research Administrator." *Administrative Science Quarterly* 4: 20–42.

Kaplan, Norman, 1960. "Some Organizational Factors Affecting Creativity." *IRE Transactions on Engineering Management* EM-7, no. 1: 24–30.

Kaplan, Norman, 1961. "Research Administration and the Administrator: U.S.S.R. and U.S." *Administrative Science Quarterly* 6: 51–72.

Kaplan, Norman, 1963. "The Relation of Creativity to Sociological Variables in Research Organizations." In Taylor and Barron 1963, pp. 195–204.

Kaplan, Norman, 1964. "Organization: Will It Choke or Promote the Growth of Science?" In Hill 1964, pp. 103–127.

Kaplan, Norman, ed., 1965a. *Science and Society*. Chicago: Rand-McNally.

Kaplan, Norman, 1965b. "Introduction to Part III." In Kaplan 1965a, pp. 175–179.

Kaplan, Rob, 2001. *Science Says: A Collection of Quotations on the History, Meaning, and Practice of Science*. New York: Stonesong Press.

Kastens, Merritt L., 1957. "Research — A Corporate Function." *Industrial Laboratories* 8, no. 10 (October): 92–101.

Kellogg, Vernon, 1926. "Isolation or Cooperation in Research." *Science* n.s. 63, no. 1626 (26 February): 215–218.

Kenney, Martin, 1986. *Biotechnology: The University-Industrial Complex*. New Haven, CT: Yale University Press.

Kenney, Martin, and Richard Florida, 2000. "Venture Capital in Silicon Valley: Fueling New Firm Formation." Pp. 98–123 in Martin Kenney, ed., *Understanding Silicon Valley: The Anatomy of an Entrepreneurial Region*. Stanford, CA: Stanford University Press.

Kerr, Clark, 1963. *The Uses of the University*. Cambridge, MA: Harvard University Press.

Kettering, Charles, 1946. "The Future of Science." *Science* n.s. 104, no. 2713 (27 December): 609–614.

Kettering, Charles, [1929] 1961. "Head Lamp of Industry [speech delivered to U.S. Chamber of Congress, 1929]." Pp. 77–89 in T. A. Boyd, ed., *Prophet of Progress: Selections from the Speeches of Charles F. Kettering*. New York: E. P. Dutton.

Kevles, Daniel J., 1977. *The Physicists: The History of a Scientific Community in Modern America*. New York: Alfred A. Knopf.

Kevles, Daniel J., Jeffrey L. Sturchio, and P. Thomas Carroll, 1980. "The Sciences in America, Circa 1880." *Science* n.s. 209, no. 4452 (4 July): 26–32.

Kibble, Robert, 2000. "Making Your Dreams a Reality." In Ianuzzi 2000, pp. A12–A17.

Kidd, Charles V., 1947. "The Federal Government and the Shortage of Scientific Personnel." *Science* n.s. 105, no. 2717 (24 January): 84–88.

Kierkegaard, Søren, [1859] 1962. *The Point of View for My Work as an Author*, trans. Walter Lowrie, ed. Benjamin Nelson. New York: Harper Torchbooks.

Killeffer, D. H., 1948. *The Genius of Industrial Research*. New York: Reinhold Publishing Corporation.

Kingsbury, Sherman, in association with Lawrence W. Bass and Warren C. Lothrop, 1959. "Organizing for Research." Pp. 65–99 in Carl Heyel, ed., *Handbook of Industrial Research Management*. New York: Reinhold Publishing Corporation.

Kipp, E. M., 1960. "Introduction of the Newly Graduated Scientist to Industrial Research." *Research Management* 3: 39–47.

Klaw, Spencer, 1968. *The New Brahmins: Scientific Life in America*. New York: William Morrow & Co.

Kleinman, Daniel Lee, 1995. *Politics on the Endless Frontier: Postwar Research Policy in the United States*. Durham, NC: Duke University Press.

Kleinman, Daniel Lee, 2003. *Impure Cultures: University Biology and the World of Commerce*. Madison: University of Wisconsin Press.

Kleinman, Daniel Lee, and Steven P. Vallas, 2006. "Contradiction in Convergence: Universities and Industry in the Biotechnology Field." Pp. 35–62 in *The New Political Sociology of Science: Institutions, Networks, and Power*, eds. Scott Frickel and Kelly Moore. Madison: University of Wisconsin Press.

Kliever, Waldo H., 1952. "Design of Research Projects and Programs." *Industrial Laboratories* 3, no. 10 (October): 6–13.

Kline, Ronald, 1995. "Construing 'Technology' as 'Applied Science': Public Rhetoric of Scientists and Engineers in the United States, 1880–1945." *Isis* 86: 194–221.

Kneller, George F., 1978. *Science as a Human Endeavor*. New York: Columbia University Press.

Kohler, Robert E., 1990. "The Ph.D. Machine: Building on the Collegiate Basis." *Isis* 81: 638–662.

Kolchinsky, Peter, 2002. *The Entrepreneur's Guide to a Biotech Startup*, 3rd ed. [Published on-line by Evelexa BioResources, accessed 24 May 2002from www.evelexa.com.]

Komisar, Randy, with Kent Lineback, 2000. *The Monk and the Riddle: The Education of a Silicon Valley Entrepreneur*. Boston: Harvard Business School Press.

Kornberg, Arthur, 1989. *For the Love of Enzymes: The Odyssey of a Biochemist*. Cambridge, MA: Harvard University Press.

Kornberg, Arthur, 1995. *The Golden Helix: Inside Biotech Ventures*. Sausalito, CA: University Science Books.

Kornhauser, William, 1962. "Strains and Accommodations in Industrial Research Organizations in the United States." *Minerva* 1: 30–42.

Kornhauser, William, with the assistance of Warren O. Hagstrom, 1962. *Scientists in Industry: Conflict and Accommodation*. Berkeley: University of California Press.

Kowarski, L., [1962] 1965. "Team Work and Individual Work in Science." In Kaplan 1965a, pp. 247–255.

Krimsky, Sheldon, 2003. *Science in the Private Interest: Has the Lure of Profits Corrupted Biomedical Research?* Lanham, MD: Rowman & Littlefield.

Kruif, Paul de, [1926] 1953. *Microbe Hunters.* New York: Harcourt, Brace & World.

Kuhn, Thomas S., 1962. *The Structure of Scientific Revolutions.* Chicago: University of Chicago Press.

Kunze, Robert J., 1990. *Nothing Ventured: The Perils and Payoffs of the Great American Venture Capital Game.* New York: HarperBusiness.

La Porte, Todd, 1965. "Conditions of Strain and Accommodation in Industrial Research Organizations." *Administrative Science Quarterly* 10: 21–38.

Lacy, Sarah, and Jessi Hempel, 2006. "Valley Boys." *BusinessWeek*, 14 August, pp. 40–47.

LaFollette, Marcel, 1990. *Making Science Our Own: Public Images of Science, 1910–1955.* Chicago: University of Chicago Press.

Langlois, Richard N., 1998a. "Personal Capitalism as Charismatic Authority: The Organizational Economics of a Weberian Concept." *Industrial and Corporate Change* 7: 195–213.

Langlois, Richard N., 1998b. "Schumpeter and Personal Capitalism." Pp. 57–82 in Gunnar Eliasson and Christopher Green, eds., *Microfoundations of Economic Growth: A Schumpeterian Perspective.* Ann Arbor: University of Michigan Press.

Lapp, Ralph E., 1965. *The New Priesthood: The Scientific Elite and the Uses of Power.* New York: Harper & Row.

Larrabee, Eric, 1953. "Science, Poetry, and Politics." *Science* n.s. 117, no. 3042 (17 April): 395–399.

Laslett, Peter, 1963. "The Face to Face Society." Pp. 157–184 in idem, ed., *Philosophy, Politics and Society.* Oxford: Basil Blackwell.

Laslett, Peter, [1965] 1983. *The World We Have Lost: Further Explored*, 3rd ed. London: Routledge.

Lassman, Thomas C., 2003. "Industrial Research Transformed: Edward Condon at the Westinghouse Electric and Manufacturing Company, 1935–1942." *Technology and Culture* 44: 306–339.

Latour, Bruno, 1987. *Science in Action: How to Follow Scientists and Engineers through Society.* Cambridge, MA: Harvard University Press.

Latour, Bruno, 1988. *The Pasteurization of France*, trans. Alan Sheridan and John Law. Cambridge, MA: Harvard University Press.

Lawrence, Christopher, and Steven Shapin, eds., 1998. *Science Incarnate: Historical Embodiments of Natural Knowledge.* Chicago: University of Chicago Press.

LeBlanc, Thomas J., 1925. Review of *Arrowsmith. Science* n.s. 61, no. 1590 (19 June): 632–634.

Leech, Thomas, 1993. *How to Prepare, Stage, & Deliver Winning Presentations*, 2nd ed. New York: American Management Associations.

Leedy, H. A., 1952. "Training of Young Researchers." In Hertz and Rubenstein 1952, pp. 62–77.

Leermakers, John A., 1951. "Basic Research in Industry." *Industrial Laboratories* 2, no. 3 (March): 2–3.

Lemonick, Michael D. "Gene Mapper: The Bad-Boy of Science Has Jump-Started a Biological Revolution." *Time.com*, 17 December 2000.

Leslie, Stuart W., 1994. *The Cold War and American Science: The Military-Industrial-Academic Complex at MIT and Stanford*. New York: Columbia University Press.

Lewis, Michael, 2000. *The New New Thing: A Silicon Valley Story*. New York: W. W. Norton.

Lewis, Sinclair, 1925. *Arrowsmith*. New York: Harcourt Brace.

Lewontin, Richard, 2004. "Dishonesty in Science." *New York Review of Books* 51, no. 18 (18 November): 38–40.

Liebenau, Jonathan, 1987. *Medical Science and Medical Industry: The Formation of the American Pharmaceutical Industry*. Baltimore, MD: Johns Hopkins University Press.

Likely, Wadsworth, 1950. "Scientists and Mobilization." *Science* n.s. 112, no. 2909 (29 September): 349–351.

Litchfield, Edward H., 1956. "Notes on a General Theory of Administration." *Administrative Science Quarterly* 1: 3–29.

Little, Arthur D., 1924a. "Research: The Mother of Industry." *Scientific Monthly* 19, no. 2 (August): 165–169.

Little, Arthur D., 1924b. "The Fifth Estate." *Science* n.s. 60, no. 1553 (8 October): 299–306.

Livingstone, Robert Teviot, and Stanley H. Milberg, 1957. *Human Relations in Industrial Research Management, Including Papers from the Sixth and Seventh Annual Conferences on Industrial Research: Columbia University, 1955 and 1956*. New York: Columbia University Press.

Lovell, Bernard, 1939. *Science and Civilization*. London: Thomas Nelson.

Lowen, Rebecca S., 1997. *Creating the Cold War University: The Transformation of Stanford*. Berkeley: University of California Press.

Luhmann, Niklas, 1979. *Trust and Power: Two Works*, eds. Tom Burns and Gianfranco Poggi, trans. Howard Davis, John Raffan, and Kathryn Rooney. Chichester: John Wiley.

Luhmann, Niklas, 1989. "Familiarity, Confidence, Trust: Problems and Alternatives." Pp. 21–27 in Diego Gambetta, ed., *Trust: Making and Breaking Cooperative Relations*. Oxford: Basil Blackwell.

Macaulay, Thomas Babington, [1837] 1913. "Lord Bacon." Pp. 289–410 in idem, *Literary Essays Contributed to the Edinburgh Review*. Oxford: Oxford University Press.

MacKenzie, Donald, 2003. "Empty Cookie Jar." *London Review of Books*, 22 May: http://www.lrb.co.uk/v25/n10/mack_01./html [accessed 19 July 2007].

MacLeod, Elizabeth Kay, 1995. "Politics, Professionalism and the Organisation of Scientists: The Association of Scientific Workers, 1917–1942." Unpubl. Ph.D. thesis, University of Sussex.

MacMillan, Conway, 1895. "The Scientific Method and Modern Intellectual Life." *Science* n.s. 1, no. 20 (17 May): 537–542.

Mainzer, Lewis C., 1961. "Scientific Freedom in Government-Sponsored Research." *Journal of Politics* 23: 212–230.

Malcolm, Norman, 1958. *Ludwig Wittgenstein: A Memoir*. Oxford: Oxford University Press.

Mancuso, Joseph R., 1983. *How to Prepare and Present a Business Plan*. New York: Prentice Hall.

Manwell, Reginald D., 1953. "True Scientists." *Science* n.s. 118, no. 3067 (9 October): 418–419.

Marcson, Simon, 1960a. *The Scientist in American Industry: Some Organizational Determinants in Manpower Utilization*. New York: Harper & Brothers.

Marcson, Simon, 1960b. "Role Adaptation of Scientists in Industrial Research." *IRE Transactions on Engineering Management* EM-7, no. 4: 159–166.

Marcson, Simon, 1961. "The Professional Commitments of Scientists in Industry." *Research Management* 4: 271–275.

Margenau, Henry, 1964. *Ethics & Science*. Princeton, NJ: D. Van Nostrand.

Margenau, Henry, et al., 1964. *The Scientist (Life Science Library)*. New York: Time.

Mason, William P., 1897. "Expert Witnessing." *Science* n.s. 6, no. 137 (13 August): 243–248.

Massey, Harrie, 1975. "Atomic Energy and the Development of Large Teams and Organizations." *Proceedings of the Royal Society of London A* 342: 491–497.

Mather, Kirtley F., 1952. "The Problem of Antiscientific Trends Today." *Science* n.s. 115, no. 2994 (16 May): 533–537.

May, John, and Cal Simmons, 2001. *Every Business Needs an Angel: Getting the Money You Need to Make Your Business Grow*. New York: Crown Business.

May, Mark A., 1943. "The Moral Code of Scientists." Pp. 40–45 in idem, *The Scientific Spirit and Democratic Faith*. New York: King's Crown Press.

McClure, H. B., 1952. "External Communication of Research Results." In Hertz and Rubenstein 1952, pp. 161–176.

McEwen, William J., 1956. "Position Conflict and Professional Orientation in a Research Organization." *Administrative Science Quarterly* 1: 208–224.

McGrath, Patrick J., 2002. *Scientists, Business, and the State, 1890–1960*. Chapel Hill: University of North Carolina Press.

McGuire, J. E., and P. M. Rattansi, 1966. "Newton and the 'Pipes of Pan.'" *Notes and Records of the Royal Society* 21: 108–143.

McIlvain, J. M., 1948. "The Research Budget." In Furnas 1948, pp. 145–158.

McLaughlin, George D., 1926. "Research and Industry: Cooperation between Industry and University." *Scientific Monthly* 22, no. 4 (April): 281–284.

Mead, Margaret, and Rhoda Métraux, 1957. "Image of the Scientist Among High-School Students: A Pilot Study." *Science* n.s. 126, no. 3270 (30 August): 384–390.

Mees, C. E. Kenneth, 1914. "The Future of Scientific Research [editorial]." *Journal of Industrial and Engineering Chemistry* 6: 618–619.

Mees, C. E. Kenneth, 1916. "The Organization of Industrial Scientific Research." *Science* n.s. 43, no. 1118 (2 June): 763–773.

Mees, C. E. Kenneth, 1917. "The Production of Scientific Knowledge." *Journal of Industrial and Engineering Chemistry* 9: 1137–1141.

Mees, C. E. Kenneth, 1920. *The Organization of Industrial Scientific Research*. New York: McGraw-Hill.

Mees, C. E. Kenneth, 1932. "Scope of Research Management." *Industrial and Engineering Chemistry* 24: 65–66.

Mees, C. E. Kenneth, 1935. "Research and Business with Some Observations on Color Photography." *Vital Speeches of the Day* 2 (18 November): 768–769.

Mees, C. E. Kenneth, 1945. "Discussion of Midgley." In Standard Oil Development Company 1945, pp. 46–49.

Mees, C. E. Kenneth, with the cooperation of John R. Baker, 1946. *The Path of Science*. London: Chapman & Hall.

Mees, C. E. Kenneth, 1947. "Discussion [of Michael Polanyi, 'The Foundations of Freedom in Science']." In Wigner 1947, pp. 140–141.

Mees, C. E. Kenneth, 1956. *An Address to the Senior Staff of the Kodak Research Laboratories, November 9, 1955*. Rochester, NY: Kodak Research Laboratories.

Mees, C. E. Kenneth, 1961. *From Dry Plates to Ektachrome Film: A Story of Photographic Research*. New York: Ziff-Davis.

Mees, C. E. Kenneth, and John A. Leermakers, 1950. *The Organization of Industrial Scientific Research*, 2nd ed. New York: McGraw-Hill.

Mencken, H. L., 1922. *Prejudices: Third Series*. New York: Alfred A. Knopf.

Merton, Robert K., 1940. Review of Watson, *Scientists Are Human*. *Isis* 31: 466–467.

Merton, Robert K., 1947. "The Machine, the Worker, and the Engineer." *Science* n.s. 105, no. 2717 (24 January): 79–84.

Merton, Robert K., 1957a. *Social Theory and Social Structure*, rev. ed. New York: Free Press.

Merton, Robert K., [1949] 1957b. "Patterns of Influence: Local and Cosmopolitan Influentials." In Merton 1957a, pp. 189–202.

Merton, Robert K., 1957c. "Some Preliminaries to a Sociology of Medical Education." In Merton, Reader, and Kendall 1957, pp. 3–79.

Merton, Robert K., 1965. *On the Shoulders of Giants: A Shandean Postscript*. New York: Free Press.

Merton, Robert K., [1938] 1970. *Science, Economy and Society in Seventeenth-Century England*. New York: Harper.

Merton, Robert K., 1973a. *The Sociology of Science: Theoretical and Empirical Investigations*, ed. Norman W. Storer. Chicago: University of Chicago Press.

Merton, Robert K., [1938] 1973b. "Science and the Social Order." In Merton 1973a, pp. 254–266.

Merton, Robert K., [1942] 1973c. "The Normative Structure of Science." In Merton 1973a, pp. 267–278.

Merton, Robert K., [1957] 1973d. "Priorities in Scientific Discovery." In Merton 1973a, pp. 286–324.

Merton, Robert K., 1976. *Sociological Ambivalence and Other Essays*. New York: Free Press.

Merton, Robert K., and Elinor Barber, 2004. *The Travels and Adventures of Serendipity: A Study in Semiological Semantics and the Sociology of Science*. Princeton, NJ: Princeton University Press.

Merton, Robert K., George C. Reader, and Patricia L. Kendall, eds., 1957. *The Student-Physician: Introductory Studies in the Sociology of Medical Education*. Cambridge, MA: Harvard University Press.

Merz, John Theodore, [1904–1912] 1965. *A History of European Thought in the Nineteenth Century*, 4 vols. New York: Dover Publications.

Metlay, Grischa, 2006. "Reconsidering Renormalization: Stability and Change in 20th-Century Views on University Patents." *Social Studies of Science* 36: 565–597.

Mialet, Hélène, 1999. "Do Angels Have Bodies? Two Stories about Subjectivity in Science: The Cases of William X and Mister H." *Social Studies of Science* 29: 551–581.

Mialet, Hélène, 2003. "Reading Hawking's Presence: An Interview with a Self-Effacing Man." *Critical Inquiry* 29: 571–598.

Michael, Donald N., 1957. "Scientists through Adolescent Eyes: What We Need to Know, Why We Need to Know It." *Scientific Monthly* 84, no. 3 (March): 135–140.

Midgley, Thomas, Jr., 1945. "The Future of Industrial Research: The View of a Chemist." In Standard Oil Development Company 1945, pp. 30–46.

Milgram, Stanley, 1967. "The Small-World Problem." *Psychology Today* 1: 61–67.

Miller, David P., 1988. "'My Favourite Studdys': Lord Bute as Naturalist." Pp. 213–239 in Karl W. Schweizer, ed., *Lord Bute: Essays in Reinterpretation*. Leicester: Leicester University Press.

Mills, C. Wright, 1956. *The Power Elite*. New York: Oxford University Press.

Mills, John, 1942. "Who Is the Research Man?" *Technology Review* 44: 451–452, 466, 468–469.

Miner, John B., 1962. "Conformity among University Professors and Business Executives." *Administrative Science Quarterly* 7: 96–109.

Mirowski, Philip, and Esther-Mirjam Sent, eds., 2002. *Science Bought and Sold: Essays in the Economics of Science*. Chicago: University of Chicago Press.

Miyoshi, Masao, 2000. "Ivory Tower in Escrow." *boundary 2* 27: 7–50.

Mody, Cyrus C. M., 2006. "Corporations, Universities, and Instrumental Communities: Commercializing Probe Microscopy, 1981–1996." *Technology and Culture* 47: 56–80.

Monk, Ray, 1991. *Ludwig Wittgenstein: The Duty of Genius*. Harmondsworth: Penguin.

Moore, James R., 1985. "Darwin of Down: The Evolutionist as Squarson-Naturalist." Pp. 435–481 in David Kohn, ed., *The Darwinian Heritage*. Princeton, NJ: Princeton University Press.

Moore, G. E., 1903. *Principia Ethica*. Cambridge: Cambridge University Press.

Morrell, Jack, and Arnold Thackray, 1981. *Gentlemen of Science: Early Years of the British Association for the Advancement of Science.* Oxford: Clarendon Press.

Morrell, J. B., 1971. "Individualism and the Structure of British Science in 1830." *Historical Studies in the Physical Sciences* 3: 183–204.

Morrell, J. B., 1972. "The Chemist Breeders: The Research Schools of Liebig and Thomas Thomson." *Ambix* 19: 1–46.

Morris, J. M., 1962. "Administration of Research in Industry." *Research Management* 5: 237–247.

Morris, John, and Charles Raymond Downs, 1924. *The Technical Organization: Its Development and Administration.* New York: McGraw-Hill.

Moseley, Russell, 1979. "From Avocation to Job: The Changing Nature of Scientific Practice." *Social Studies of Science* 9: 511–522.

Moskin, J. Robert, 1966. "The Scientist as God and Devil." Pp. 57–71 in idem, *Morality in America: A Report on Our Crisis of Immorality.* New York: Random House.

Mowery, David, Richard, R. Nelson, Bhaven N. Sampat, and Arvids A. Ziedonis, 2004. *Ivory Tower and Industrial Innovation: University-Industry Technology Transfer Before and After the Bayh-Dole Act in the United States.* Stanford, CA: Stanford University Press.

Mowery, David, and Bhaven Sampat, 2001. "University Patents and Patent Policy Debates in the USA, 1925–1980." *Industrial and Corporate Change* 10: 781–814.

Mullis, Kary, 1998. *Dancing Naked in the Mind Field.* New York: Pantheon.

Murray, Fiona, 2004. "The Role of Academic Inventors in Entrepreneurial Firms: Sharing the Laboratory Life." *Research Policy* 33: 643–659.

Nathan, Otto, and Heinz Norden, eds., 1960. *Einstein on Peace.* New York: Simon and Schuster.

National Collegiate Inventors & Innovators Alliance, 2006. *Invention to Venture: Workshops in Technology and Entrepreneurship: Participants Guide 2006.* Amherst, MA: NCIIA.

National Manpower Council, 1953. *A Policy for Scientific and Professional Manpower: A Statement by the Council with Facts and Issues Prepared by the Research Staff.* New York: Columbia University Press.

National Resources Planning Board, 1941. *Research—A National Resource. II. Industrial Research*, Report of the National Research Council to the National Resources Planning Board, December 1940. Washington, DC: Government Printing Office.

National Science Foundation, 1955. *Scientific Personnel Resources: A Summary of Data on Supply Utilization and Training of Scientists and Engineers.* Washington, DC: Government Printing Office.

National Science Foundation, 1959. *Reviews of Data on Research and Development*, NSF 59-46, no. 14, August. Washington, DC: Government Printing Office.

National Science Foundation, 1961. *Publication of Basic Research Findings in Industry.* Washington, DC: Government Printing Office.

National Science Foundation, 1967. "Salaries and Selected Characteristics of American Scientists, 1966." *Reviews of Data on Science Resources,* no. 11. Washington, DC: Government Printing Office.

National Science Foundation, 1984. *National Patterns of Science and Technology Resources 1984.* Washington, DC: Government Printing Office.

Nelson, Richard R., 1959a. "The Economics of Invention: A Survey of the Literature." *Journal of Business* 32: 101–127.

Nelson, Richard R., 1959b. "The Simple Economics of Basic Scientific Research." *Journal of Political Economy* 67: 297–306.

Nelson, Richard R., 1961. "Uncertainty, Learning, and the Economics of Parallel Research and Development Efforts." *Review of Economics and Statistics* 43: 351–364.

Nelson, Richard R., 1962a. Introduction. In Universities-National Bureau 1962, pp. 3–16.

Nelson, Richard R., 1962b. "The Link between Science and Invention: The Case of the Transistor." In Universities-National Bureau 1962, pp. 549–583.

Newfield, Christopher, 2003. *Ivy and Industry: Business and the Making of the American University, 1880–1980.* Durham, NC: Duke University Press.

Newton, Isaac, [1729] 1934. "Scholium." Pp. 6–12 in idem, *Mathematical Principles of Natural Philosophy,* trans. Andrew Motte [1729] and trans. rev. by Florian Cajori. Berkeley: University of California Press.

Nietzsche, Friedrich, 1954a. *The Philosophy of Nietzsche,* trans. Horace B. Samuel. New York: Modern Library.

Nietzsche, Friedrich, [1886] 1954b. "Beyond Good and Evil." In Nietzsche 1954a, pp. 369–616.

Nietzsche, Friedrich, [1887] 1954c. "The Genealogy of Morals." In Nietzsche 1954a, pp. 617–807.

Noble, David F., 1977. *America by Design: Science, Technology, and the Rise of Corporate Capitalism.* New York: Alfred A. Knopf.

Nohria, Nitin, 1992. "Information and Search in the Creation of New Business Ventures: The Case of the 128 Venture Group." In Nohria and Eccles 1992a, pp. 240–261.

Nohria, Nitin, and Robert G. Eccles, eds., 1992a. *Networks and Organizations: Structure, Form, and Action.* Boston: Harvard Business School Press.

Nohria, Nitin, and Robert G. Eccles, 1992b. "Face-to-Face: Making Network Organizations Work." In Nohria and Eccles 1992a, pp. 288–308.

Noltingk, B. E., 1965. *The Art of Research: A Guide for the Graduate.* New York: Elsevier.

Northrop, F. S. C., 1947. "The Physical Sciences, Philosophy, and Human Values." In Wigner 1947, pp. 98–113.

Nutting, P. G., 1918. "Research and the Industries." *Scientific Monthly* 7, no. 2 (August): 149–157.

Nye, Mary Jo, 1996. *Before Big Science: The Pursuit of Modern Chemistry and Physics, 1800–1940.* Cambridge, MA: Harvard University Press.

Oakeshott, Michael, 1936. "History and the Social Sciences." Pp. 71–81 in Institute of Sociology, *The Social Sciences: Their Relations in Theory and in Teaching*. London: Le Play House Press.

Oesper, Ralph E., 1975. *The Human Side of Scientists*. Cincinnati, OH: University of Cincinnati Press.

Ogburn, S. C., 1951. "Research Management." *Industrial Laboratories* 2, no. 9 (September): 6–9.

Olsen, Fred, 1948. "Evaluating the Results of Research." In Furnas 1948, pp. 402–415.

Oppenheimer, J. Robert, 1950. "Encouragement of Science." *Science* n.s. 111, no. 2885 (14 April): 373–375.

Oppenheimer, J. Robert, [1947] 1955. "Physics in the Contemporary World." Pp. 81–102 in idem, *The Open Mind*. New York: Simon and Schuster.

Oppenheimer, J. Robert, 1965. "Communication and Comprehension of Scientific Knowledge." In Calvin et al. 1965, pp. 271–279.

Orth, Charles D., III, 1965. "The Optimum Climate for Industrial Research." In Kaplan 1965a, pp. 194–210.

Ott, Emil, 1950. "The Team Approach in Research and Development." *Chemical and Engineering News* 28, no. 12 (12 June): 1994–1996.

Outram, Dorinda, 1978. "The Language of Natural Power: The Funeral *Éloges* of Georges Cuvier." *History of Science* 16: 153–178.

Outram, Dorinda, 1984. *Georges Cuvier: Vocation, Science and Authority in Post-Revolutionary France*. Manchester: Manchester University Press.

Owen-Smith, Jason, and Walter W. Powell, 2001. "Careers and Contradictions: Faculty Responses to the Transformation in Knowledge and Its Uses in the Life Sciences." *Research in the Sociology of Work* 10: 109–140.

Owen-Smith, Jason, and Walter W. Powell, 2002. "Standing on Shifting Terrain: Faculty Responses to the Transformation of Knowledge and Its Uses in the Life Sciences." *Science Studies* 15: 3–28.

Page, Thornton, 1953. "Selecting the Research Team." In Bush and Hattery 1953, pp. 61–70.

Palevsky, Mary, 2000. *Atomic Fragments: A Daughter's Questions*. Berkeley: University of California Press.

Palmer, Donald, 2006. "Taking Stock of the Criteria We Use to Evaluate One Another's Work: *ASQ* 50 Years Out." *Administrative Science Quarterly* 51: 535–559.

Papini, Giovanni, 1907. "What Pragmatism Is Like." *Popular Science Monthly* 71, no. 4 (October): 351–358.

Parsons, Talcott, 1939. "The Professions and Social Structure." *Social Forces* 17: 457–469.

Parsons, Talcott, 1956. "Suggestions for a Sociological Approach to the Theory of Organizations (I) and (II)." *Administrative Science Quarterly* 1: 63–85, 225–239.

Parton, H. N., 1972. *Science Is Human: Essays*, ed. M. H. Panckhurst. Dunedin., New Zealand: University of Otago Press.

Paul, Charles B., 1980. *Science and Immortality: The Éloges of the Paris Academy of Sciences (1699–1791)*. Berkeley: University of California Press.

Pearson, G. A., 1924. "Some Conditions for Effective Research." *Science* n.s. 60, no. 1543 (25 July): 71–73.

Peirce, C. S., [1896–1899] 1940. "The Scientific Attitude and Fallibilism." Pp. 42–59 in Justus Buchler, ed., *The Philosophy of Peirce: Selected Writings*. London: Routledge & Kegan Paul.

Pelz, Donald C., 1956. "Some Social Factors Related to Performance in a Research Organization." *Administrative Science Quarterly* 1: 310–325.

Pelz, Donald C., and Frank M. Andrews, 1966. *Scientists in Organizations: Productive Climates for Research and Development*. New York: John Wiley.

Perazich, George, 1951. "Growth Rate of Industrial Research." *Science* n.s. 114, no. 2970 (30 November): 3a.

Perry, Stewart E., 1966. *The Human Nature of Science*. New York: Free Press.

Phillips, Melba, 1952. "Dangers Confronting American Science." *Science* n.s. 116, no. 3017 (24 October): 439–443.

Pinto, Jim. "Pinto Perspectives on 'Angel' Investing." *UCSD CONNECT Newsletter*, 11 February 2003 [accessed on-line 1 August 2006 at http://www.jimpinto.com/writings/angelinvesting.html].

Poincaré, Henri, [1905] 2001. *The Value of Science*. Pp. 179–353 in idem, *The Value of Science: Essential Writings of Henri Poincaré*, series ed. Stephen Jay Gould. New York: Modern Library.

Polanyi, Michael, 1944. "Patent Reform." *Review of Economic Studies* 11, no. 2: 61–76.

Polanyi, Michael, 1958. *Personal Knowledge: Towards a Post-Critical Philosophy*. Chicago: University of Chicago Press.

Polanyi, Michael, 1962. "The Republic of Science: Its Political and Economic Theory." *Minerva* 1: 54–74.

Popper, Karl R., [1945] 1966. "The Sociology of Knowledge." Pp. 212–223 in idem, *The Open Society and Its Enemies. Volume II: The High Tide of Prophecy: Hegel, Marx, and the Aftermath*. Princeton, NJ: Princeton University Press.

Porter, Roy, 1978. "Gentlemen and Geology: The Emergence of a Scientific Career, 1660–1920." *Historical Journal* 21: 809–836.

Powell, Walter W., and Laurel Smith-Doerr, 1994. "Networks and Economic Life." Pp. 368–402 in *Handbook of Economic Sociology*, eds. Neil J. Smelser and Richard Swedberg. Princeton, NJ: Princeton University Press.

Powers, Phillip N., 1952. "Industrial Research Workers and Defense." In Hertz and Rubenstein 1952, pp. 94–112.

Preston, F. W., 1945. "Freedom of Research." *Scientific Monthly* 61, no. 6 (December): 477–482.

Preston, Richard, 2000. "The Genome Warrior: Craig Venter Has Grabbed the Lead in the Quest for Biology's Holy Grail." *New Yorker*, 12 June, pp. 66–83.

Price, Derek J. de Solla, 1963. *Little Science, Big Science*. New York: Columbia University Press.

Price, Don K., 1954. *Government and Science: Their Dynamic Relation in American Democracy*. New York: New York University Press.

Priestley, Joseph, 1775. *The History and Present State of Electricity*, 2 vols., 3rd ed. London: C. Bathurst and T. Lowndes.

Proctor, Robert N., 1991. *Value-Free Science? Purity and Power in Modern Knowledge*. Cambridge, MA: Harvard University Press.

Quindlen, Ruthann, 2000. *Confessions of a Venture Capitalist: Inside the High-Stakes World of Start-up Financing*. New York: Warner Books.

Rabinow, Paul, 1989. *French Modern: Norms and Forms of the Social Environment*. Cambridge, MA: MIT Press.

Rabinow, Paul, 1996a. *Making PCR: A Story of Biotechnology*. Chicago: University of Chicago Press.

Rabinow, Paul, 1996b. "American Moderns: On Science and Scientists." Pp. 162–188 in idem, *Essays on the Anthropology of Reason*. Princeton, NJ: Princeton University Press.

Rabinow, Paul, 1999. *French DNA: Trouble in Purgatory*. Chicago: University of Chicago Press.

Rabinow, Paul, 2003. "Science as a Vocation: Truth versus Meaning." Pp. 96–101 in idem, *Anthropos Today: Reflections on Modern Equipment*. Princeton, NJ: Princeton University Press.

Rabinow, Paul, and Talia Dan-Cohen, 2005. *A Machine to Make a Future: Biotech Chronicles*. Princeton, NJ: Princeton University Press.

Rae, John, 1979. "The Application of Science to Industry." Pp. 249–268 in Alexandra Oleson and John Voss, eds., *The Organization of Knowledge in Modern America, 1860–1920*. Baltimore, MD: Johns Hopkins University Press.

Randle, C. Wilson, 1959. "Problems of Research and Development Management." *Harvard Business Review* 37, no. 1 (January–February): 128–136.

Reese, Charles L., 1934. "Scientific Ideals." *Science* n.s. 80, no. 2075 (5 October): 299–303.

Reich, Leonard S., 1985. *The Making of American Industrial Research: Science and Business at GE and Bell, 1876–1926*. Cambridge: Cambridge University Press.

Reichenbach, Hans, 1951. *The Rise of Scientific Philosophy*. Berkeley: University of California Press.

Reichert, Philip, 1954. "'Big Business Takes Over Research.'" *Science* n.s. 120, no. 3115 (10 September): 434–435.

Reiner, Martha Louise, 1989. "A History of Venture Capital Organizations in the United States." Unpubl. Ph.D. thesis, University of California, Berkeley.

Renan, Ernest, 1890. *L'avenir de la science*. Paris: Calmann-Levy.

Rhodes, Richard, 1986. *The Making of the Atomic Bomb*. New York: Simon and Schuster.

Richardson, Alan W., 1997. "Toward a History of Scientific Philosophy." *Perspectives on Science* 5: 418–451.

Richet, Charles, [1923] 1927. *The Natural History of a Savant*, trans. Oliver Lodge. London: J. M. Dent.

Richter, Curt P., 1953. "Free Research versus Design Research." *Science* n.s. 118, no. 3056 (24 July): 91–93.

Riesman, David, with the assistance of Reuel Denney and Nathan Glazer, 1950. *The Lonely Crowd: A Study of the Changing American Character*. New Haven, CT: Yale University Press.

Ringer, Fritz, 2004. *Max Weber: An Intellectual Biography*. Chicago: University of Chicago Press.

Rivlin, Gary. "So You Want to Be a Venture Capitalist." *New York Times*, 22 May 2005 section 3, pp. 1, 9.

Robertson, T. Brailsford, 1915. "The Cash Value of Scientific Research." *Scientific Monthly* 1, no. 2 (November): 140–147.

Rochat, Marco, and Keith Arundale. *Three Keys to Obtaining Venture Capital*. http://www.altassets.net/pdfs/3keys-13–11–02.pdf [accessed 1 August 2006].

Roe, Anne, 1947. "Personality and Vocation." *Transactions of the New York Academy of Sciences* 9 (1947): 257–267.

Roe, Anne, 1953. *The Making of a Scientist*. New York: Dodd, Mead & Co.

Roe, Anne, 1963. "Personal Problems and Science." In Taylor and Barron 1963, pp. 132–138.

Rogers, L. A., 1923. "What Constitutes Efficiency in Research?" *Journal of Bacteriology* 8: 197–213.

Rosenberg, Charles E., 1963. "Martin Arrowsmith: The Scientist as Hero." *American Quarterly* 15: 447–458.

Rosenberg, Charles E., 1974. "Science and Social Values in 19th-Century America: A Case Study in the Growth of Scientific Institutions." Pp. 21–42 in Arnold W. Thackray and Everett Mendelsohn, eds., *Science and Values: Patterns of Tradition and Change*. New York: Humanities Press.

Rosenberg, Charles E., 1977. "Rationalization and Reality in the Shaping of American Agricultural Research, 1875–1914." *Social Studies of Science* 7: 401–422.

Rosenberg, Charles E., 1997. "Preface: Science in Play." Pp. ix–xvi in idem, *No Other Gods: Science and American Social Thought*, new rev. ed. Baltimore, MD: Johns Hopkins University Press.

Rothrock, Addison M., 1953. "Supervising and Training for Teamwork." In Bush and Hattery 1953, pp. 82–90.

Rowland, Henry A., 1899. "The Highest Aim of the Physicist, Presidential Address Delivered at the Second Meeting of the Society, on October 28, 1899." *Bulletin of the American Physical Society* 1, 4–16. (Also in *Science* n.s. 10, no. 258 [8 December 1899]: 825–833.)

Saintsbury, George, [1920] 1963. *Notes on a Cellar-Book*. London: Macmillan.

Sanders, Royden C., Jr., 1964. "Interface Problems between Scientists and Others in Technically Oriented Companies." In Hill 1964, pp. 75–86.

Sarasohn, Judy, 1993. *Science on Trial: The Whistle-Blower, the Accused, and the Nobel Laureate*. New York: St. Martin's Press.

Sayles, Leonard, and George Strauss, 1966. *Human Behavior in Organizations*. New York: Prentice Hall.

Schaffer, Simon, 1987. "Godly Men and Mechanical Philosophers: Souls and Spirits in Restoration Natural Philosophy." *Science in Context* 1: 55–85.

Schaffer, Simon, 1988. "Astronomers Mark Time: Discipline and the Personal Equation." *Science in Context* 2: 115–145.

Schaffer, Simon, 1990. "Genius in Romantic Natural Philosophy." Pp. 82–98 in Andrew Cunningham and Nicholas Jardine, eds., *Romanticism and the Sciences*. Cambridge: Cambridge University Press.

Schley, George B., 1937. "Society's Need for Patents to University Research Workers." *Journal of Industrial and Engineering Chemistry* 29: 1319–1322.

Schmitt, R. W., 1961. "Why Publish Scientific Research from Industry?" *Research Management* 4: 31–41.

Schrecker, Ellen W., 1986. *No Ivory Tower: McCarthyism and the Universities*. New York: Oxford University Press.

Schumpeter, Joseph A., 1934. *The Theory of Economic Development*. Cambridge, MA: Harvard University Press.

Schweber, Silvan S., 2000. *In the Shadow of the Bomb: Bethe, Oppenheimer, and the Moral Responsibility of the Scientist*. Princeton, NJ: Princeton University Press.

Scripture, E. W., 1895. "The Nature of Science and Its Relation to Philosophy." *Science* n.s. 1, no. 13 (29 March): 350–352.

Seaborg, Glenn T., 1964. *The Creative Scientist: His Training & His Role*. Oak Ridge, TN: U.S. Atomic Energy Commission.

Seaborg, Glenn T., 1996a. *A Scientist Speaks Out: A Personal Perspective on Science, Society and Change*. Singapore: World Scientific.

Seaborg, Glenn T., [1967] 1996b. "How to Become a Scientist." In Seaborg 1996a, pp. 229–239.

Seaborg, Glenn T., [1964] 1996c. "The Scientist as a Human Being." In Seaborg 1996a, pp. 89–96.

Seaborg, Glenn T., [1955] 1996d. "The Role of Basic Research." In Seaborg 1996a, pp. 1–11.

Seaborg, Glenn T., 2001. *Adventures in the Atomic Age: From Watts to Washington*. New York: Farrar, Straus and Giroux.

Secord, J. A., 1991. "The Discovery of a Vocation: Darwin's Early Geology." *British Journal for the History of Science* 24: 133–157.

Secrist, Horace A., 1960. "Motivating the Industrial Research Scientist." *Research Management* 3: 57–64.

Servos, John W., 1980. "The Industrial Relations of Science: Chemical Engineering at MIT, 1900–1939." *Isis* 71: 531–549.

Shapin, Steven, 1991a. "'A Scholar and a Gentleman': The Problematic Identity of the Scientific Practitioner in Early Modern England." *History of Science* 29: 279–327.

Shapin, Steven, 1991b. "'The Mind Is Its Own Place': Science and Solitude in Seventeenth-Century England." *Science in Context* 4: 191–218.

Shapin, Steven, 1994. *A Social History of Truth: Civility and Science in Seventeenth-Century England*. Chicago: University of Chicago Press.

Shapin, Steven, 1995. "Cordelia's Love: Credibility and the Social Studies of Science." *Perspectives on Science* 3: 255–275.

Shapin, Steven, 1996. *The Scientific Revolution*. Chicago: University of Chicago Press.

Shapin, Steven, 1998. "The Philosopher and the Chicken: On the Dietetics of Disembodied Knowledge." In Lawrence and Shapin 1998, pp. 21–50.

Shapin, Steven, 1999. "Nobel Savage." *London Review of Books* 21, no. 13 (1 July): 17–18.

Shapin, Steven, 2000a. "Trust Me." *London Review of Books* 22, no. 9 (27 April): 15–17.

Shapin, Steven, 2000b. "Don't Let That Crybaby in Here Again." *London Review of Books* 22, no. 17 (7 September): 15–16.

Shapin, Steven, 2001. "Proverbial Economies: How Certain Social and Linguistic Features of Common Sense May Throw Light on More Prestigious Bodies of Knowledge, Science for Example." *Social Studies of Science* 31: 731–769.

Shapin, Steven, 2002. "Megaton Man." *London Review of Books* 24, no. 8 (25 April): 18–20.

Shapin, Steven, 2003a. "The Image of the Man of Science." Pp. 159–183 in Roy Porter, ed., *The Cambridge History of Science: Vol. 4. Eighteenth-Century Science*. Cambridge: Cambridge University Press.

Shapin, Steven, 2003b. "Ivory Trade." *London Review of Books* 25, no. 17 (11 September): 15–19.

Shapin, Steven, 2004. "Who Is the Industrial Scientist? Commentary from Academic Sociology and from the Shop-Floor in the United States, ca. 1900–ca. 1970." In Grandin, Wormbs, and Widmalm 2004, pp. 337–363.

Shapin, Steven, 2005. "Milk and Lemon." *London Review of Books* 27, no. 13 (7 July): 10–13.

Shapin, Steven, 2006. "The Man of Science." Pp. 179–191 in Lorraine Daston and Katharine Park, eds., *The Cambridge History of Science. Vol. 3: Early Modern Science*. Cambridge: Cambridge University Press.

Shapin, Steven, and Simon Schaffer, 1985. *Leviathan and the Air-Pump: Hobbes, Boyle, and the Experimental Life*. Princeton, NJ: Princeton University Press.

Shapiro, Theresa R., 1953. "What Scientists Look for in Their Jobs." *Scientific Monthly* 76, no. 6 (June): 335–340.

Shapiro, Theresa R., 1957. "The Attitudes of Scientists toward Their Jobs." In Livingstone and Milberg 1957, pp. 151–162.

Shear, M. J., 1946. "Teamwork in Scientific Research." *Personnel Administration* 9, no. 2 (November): 3–8, 12.

Shepard, Herbert A., 1956. "Nine Dilemmas in Industrial Research." *Administrative Science Quarterly* 1: 295–309.

Shepard, Herbert A., 1957. "Superiors and Subordinates in Research." Pp. 1–13 in *The Direction of Research Establishments: Proceedings of a Symposium held at the National*

Physical Laboratory [Teddington] on 26th, 27th & 28th September 1956. New York: Philosophical Library.

Shepard, Morris G., 1948. "Qualifications, Training, Aptitudes and Attitudes of Industrial Research Personnel." In Furnas 1948, pp. 195–215.

Shepard, Norman A., 1945. "How Can We Build Better Teamwork within Our Research Organizations?" *Chemical and Engineering News* 23, no. 9 (10 May): 804–807.

Shepard, Norman A., 1948. "The Research Director's Job." In Furnas 1948, pp. 56–70.

Shortland, Michael, and Richard Yeo, eds., 1996. *Telling Lives in Science: Essays on Scientific Biography.* Cambridge: Cambridge University Press.

Shreeve, James. "Craig Venter's Epic Voyage to Redefine the Origin of the Species." *Wired*, issue 12.08 (August 2004):http://www.wired.com/wired/archive/12.08/venter.html [accessed 14 August 2007].

Shreeve, James, 2004. *The Genome War: How Craig Venter Tried to Capture the Code of Life and Save the World.* New York: Alfred A. Knopf.

Shryock, R. H., 1948. "American Indifference to Basic Science During the 19th Century." *Archives internationales d'histoire des sciences* 28: 50–65.

Sidgwick, Henry, 1901. *Lectures on the Ethics of T. H. Green, Mr. Herbert Spencer and J. Martinear.* London: Macmillan.

Sinclair, Upton, 1956. "Albert Einstein: As I Remember Him." *Saturday Review* (14 April): 17–18, 56–59.

Sinsheimer, Robert L., 1968. "The Double Helix." *Science and Engineering*, September: 4–6.

Slaughter, Sheila, and Larry L. Leslie, 1997. *Academic Capitalism: Politics, Policies, and the Entrepreneurial University.* Baltimore, MD: Johns Hopkins University Press.

Smellie, D. G., 1948. "Research Reports." In Furnas 1948, pp. 159–181.

Smiles, Samuel, [1859] 1880. *Self-Help; with Illustrations of Conduct and Perseverance.* London: John Murray.

Smilor, Ray, 2001. *Daring Visionaries: How Entrepreneurs Build Companies, Inspire Allegiance, and Create Wealth.* Holbrook, MA: Adams Media.

Smith, Adam, [1759] 1976. *The Theory of Moral Sentiments*, eds. D. D. Raphael and A. L. Macfie. Oxford: Clarendon Press.

Smith, Alice Kimball, 1965. *A Peril and a Hope: The Scientists' Movement in America, 1945–47.* Chicago: University of Chicago Press.

Smith, John Kenly, Jr., 1990. "The Scientific Tradition in American Industrial Research." *Technology and Culture* 31: 121–131.

Snow, Joel A., 1968. Review of Daniel S. Greenberg, *The Politics of Pure Science. Bulletin of the Atomic Scientists* 24, no. 5 (May): 34–36.

Snow, C. P., 1959. *The Two Cultures and the Scientific Revolution*, the Rede Lecture 1959. Cambridge: Cambridge University Press.

Snow, C. P., 1961. "Address by Charles P. Snow [to Annual Meeting of American Association for the Advancement of Science, 27 December 1960]." *Science* n.s.

133, no. 3448 (27 January): 256–259. (Subsequently published as "The Moral Un-Neutrality of Science," in Snow, *Public Affairs* [New York: Charles Scribner's, 1971], pp. 187–198.)

Söderqvist, Thomas, 1996. "Existential Projects and Existential Choice in Science: Science Biography as an Edifying Genre." In Shortland and Yeo 1996, pp. 45–84.

Söderqvist, Thomas, 2003. *Science as Autobiography: The Troubled Life of Niels Jerne.* New Haven, CT: Yale University Press.

Soley, Lawrence, 1995. *Leasing the Ivory Tower.* Boston: South End Press.

Southwick, Karen, 2001. *The King-Makers: Venture Capital and the Money Behind the Net.* New York: John Wiley.

Stakman, E. C., 1951. "Science and Human Affairs." *Science* n.s. 113, no. 2928 (9 February): 137–141.

Standard Oil Development Company, 1945. *The Future of Industrial Research: Papers and Discussion.* New York: Standard Oil Development Company.

Standen, Anthony, 1950. *Science is a Sacred Cow.* New York: E. P. Dutton.

Steele, Lowell W., 1953. "Personnel Practices in Industrial Laboratories." *Personnel* 29: 469–476.

Steele, Lowell W., 1957. "Rewarding the Industrial Scientist: A Problem of Conflicting Values." In Livingstone and Milberg 1957, pp. 163–175.

Steelman, John R., 1947. *Science and Public Policy: A Report to the President by John R. Steelman, Chairman, The President's Scientific Research Board,* 5 vols. Washington, DC: Government Printing Office.

Stent, Gunther S., 1970. "DNA." In Holton 1972, pp. 198–226.

Stent, Gunther S., 1997. "Philosophy: From Metaphysics to Language Philosophy." *Partisan Review* 64: 323–330.

Stern, Fritz, 1999. *Einstein's German World.* Princeton, NJ: Princeton University Press.

Stevenson, Earl Place, 1926. "The Frontiers of Industry." *Scientific Monthly* 22, no. 4 (April): 285–288.

Stevenson, Earl Place, 1953. "Creative Technology." *Scientific Monthly* 76, no. 4 (April): 203–206.

Stevenson, John J., 1898. "The Debt of the World to Pure Science." *Science* n.s. 7, no. 167 (11 March): 325–334.

Stewart, Nathaniel, 1953. "Executive Talent in Industrial Laboratories." *Industrial Laboratories* 4, no. 10 (October): 7–12.

Stokes, Donald E., 1997. *Pasteur's Quadrant: Basic Science and Technological Innovation.* Washington, DC: Brookings Institution Press.

Storer, Norman W., 1961. "Science and Scientists in an Agricultural Research Organization: A Sociological Study." Unpubl. Ph.D. thesis, Cornell University.

Strauss, Anselm L., and Lee Rainwater, 1962. *The Professional Scientist: A Study of American Chemists.* Chicago: Aldine.

Stross, Randall E., 2000. *Eboys: The First Inside Account of Venture Capitalists at Work.* New York: Crown Business.

Sturchio, Jeffrey L., 1985. "Experimenting with Research: Kenneth Mees, Eastman Kodak, and the Challenges of Diversification." Unpublished typescript paper delivered to the Hagley R&D Pioneers Conference, Wilmington, Delaware, 7 October.

Suits, C. G., 1953. "The Engineer and the Fundamental Sciences." *Scientific Monthly* 76, no. 2 (February): 90–99.

Sulston, John, and Georgina Ferry, 2002. *The Common Thread: A Story of Science, Politics, Ethics, and the Human Genome*. Washington, DC: Joseph Henry Press.

Super, Donald E., and Paul B. Bachrach, 1957. *Scientific Careers and Vocational Development Theory: A Review, a Critique and Some Recommendations*. New York: Teacher's College, Columbia University Bureau of Publications.

Surowiecki, James, 2004. *The Wisdom of Crowds: Why the Many Are Smarter Than the Few and How Collective Wisdom Shapes Business, Economies, Societies, and Nations*. New York: Doubleday.

Swann, John P., 1988. *Academic Scientists and the Pharmaceutical Industry: Cooperative Research in Twentieth-Century America*. Baltimore, MD: Johns Hopkins University Press.

Swatez, Gerald M., 1970. "The Social Organisation of a University Laboratory." *Minerva* 8: 36–58.

Tainter, M. L., 1946. "An Industrial View of Research Trends." *Science* n.s. 103, no. 2665 (25 January): 95–99.

Taliaferro, William H., 1948. "Science in the Universities." *Science* n.s. 108, no. 2798 (13 August): 145–148.

Taylor, Calvin W., and Frank Barron, eds., 1963. *Scientific Creativity: Its Recognition and Development: Selected Papers from the Proceedings of the First, Second, and Third University of Utah Conferences: "The Identification of Creative Scientific Talent."* New York: John Wiley & Sons.

Tead, Ordway, and Henry C. Metcalf, 1920. *Personnel Administration: Its Principles and Practice*. New York: McGraw-Hill.

Teitelman, Robert, 1989. *Gene Dreams: Wall Street, Academia, and the Rise of Biotechnology*. New York: Basic Books.

Teller, Edward, 1950. "Back to the Laboratories." *Bulletin of the Atomic Scientists* 6, no. 3 (March): 71–72.

Teller, Edward, 1955. "The Work of Many People." *Science* n.s. 121, no. 3139 (25 February): 267–275.

Terman, Lewis M., 1955. "Are Scientists Different?" *Scientific American* 192, no. 1 (January): 25–29.

Thackray, Arnold, Jeffrey L. Sturchio, P. Thomas Carroll, and Robert Bud, 1985. *Chemistry in America, 1876–1976*. Dordrecht: D. Reidel.

Thatcher, Margaret, 1995. *The Path to Power*. New York: HarperCollins.

Theerman, Paul, 1985. "Unaccustomed Role: The Scientist as Historical Biographer: Two Nineteenth-Century Portrayals of Newton." *Biography* 8: 145–162.

Thomas, Charles Allen, 1955a. "Creativity in Science: A Vital Human Resource, Part I." *Industrial Laboratories* 6, no. 10 (October): 68–69.

Thomas, Charles Allen, 1955b. "Creativity in Science: A Vital Human Resource, Part II." *Industrial Laboratories* 6, no. 11 (November): 16–19.

Thompson, James D., 1956a. "Editor's Critique." *Administrative Science Quarterly* 1: 382–385.

Thompson, James D., 1956b. "On Building an Administrative Science." *Administrative Science Quarterly* 1: 102–111.

Thorpe, Charles R., 2001. "J. Robert Oppenheimer and the Transformation of the Scientific Vocation." Unpubl. Ph.D. thesis, University of California, San Diego.

Thorpe, Charles R., 2002. "Disciplining Experts: Scientific Authority and Liberal Democracy in the Oppenheimer Case." *Social Studies of Science* 32: 525–562.

Thorpe, Charles R., 2004a. "Violence and the Scientific Vocation." *Theory, Culture & Society* 21: 59–84.

Thorpe, Charles R., 2004b. "Against Time: Scheduling, Momentum, and the Moral Order at Wartime Los Alamos." *Journal of Historical Sociology* 17: 31–55.

Thorpe, Charles R., 2005. "The Scientist in Mass Society: J. Robert Oppenheimer and the Postwar Liberal Imagination." Pp. 293–314 in Cathryn Carson and David A. Hollinger, eds., *Reappraising Oppenheimer: Centennial Studies and Reflections*, Berkeley Papers in the History of Science, Vol. 21. Berkeley, CA: Office for History of Science and Technology, University of California, Berkeley.

Thorpe, Charles R., 2006. *Oppenheimer: The Tragic Intellect*. Chicago: University of Chicago Press.

Thorpe, Charles, and Steven Shapin, 2000. "Who Was J. Robert Oppenheimer? Charisma and Complex Organization." *Social Studies of Science* 30: 545–590.

Tietjen, J. J., 1966. "The Transition from Graduate School to Industrial Research." *Research Management* 9: 109–113.

Tobey, Ronald C., 1971. *The American Ideology of National Science, 1919–1930*. Pittsburgh, PA: University of Pittsburgh Press.

Tocqueville, Alexis de, 1899. "Why the Americans Are More Addicted to Practical Than to Theoretical Science." In idem, *Democracy in America*, trans. Henry Reeve, 2 vols. New York: The Colonial Press.

Tolley, Howard R., 1953. "The Individualist and the Research Team." In Bush and Hattery 1953, pp. 79–81.

Tolstoy, Leo, [1898] 1937. "Modern Science." Pp. 176–187 in idem, *Recollections & Essays*, trans. Aylmer Maude. London: Oxford University Press.

Torvalds, Linus, and David Diamond, 2001. *Just for Fun: The Story of an Accidental Revolutionary*. New York: HarperBusiness.

Trytten, M. H., and Theresa R. Shapiro, 1951. "The Earnings of American Men of Science." *Science* 113, no. 2935 (30 March): 345–347.

Turner, Frank M., 1974. *Between Science and Religion: The Reaction to Scientific Naturalism in Late Victorian England*. New Haven, CT: Yale University Press.

Turner, Stephen, 2003. "Charisma Reconsidered." *Journal of Classical Sociology* 3: 5–26.

Tuve, Merle A., 1953. "Development of the Section T Pattern of Research Organization." In Bush and Hattery 1953, pp. 135–142.

Tuve, Merle A., 1959. "Is Science Too Big for the Scientist?" *Saturday Review* (6 June): 48–52.

UCSD CONNECT, 1997. *Entrepreneur's Anthology*. La Jolla: University of California, San Diego.

U.S. Department of Labor, 1951. *Education, Employment and Earnings of American Men of Science*, Bulletin 1023. Washington, DC: Government Printing Office.

Ulam, Stanislaw M., 1976. *Adventures of a Mathematician*. New York: Charles Scribner's Sons.

United States Atomic Energy Commission, 1971. *In the Matter of J. Robert Oppenheimer: Transcript of Hearing before Personnel Security Board and Texts of Principal Documents and Letters. United States Atomic Energy Commission, May 27, 1954, through June 29, 1954*, ed. Philip M. Stern. Cambridge, MA: MIT Press.

Universities—National Bureau Committee for Economic Research, 1962. *The Rate and Direction of Inventive Activity: Economic and Social Factors. A Conference of the Universities-National Bureau Committee for Economic Research and the Committee on Economic Growth of the Social Science Research Council*. Princeton, NJ: Princeton University Press.

Urry, John, 2002. *Global Complexity*. Cambridge: Polity Press.

Van Osnabrugge, Mark, and Robert J. Robinson, 2000. *Angel Investing: Matching Start-Up Funds with Start-Up Companies*. San Francisco: Jossey-Bass.

Van Raalte, John A., 1966. "Reflections on the Ph.D. Interview." *Research Management* 9: 307–317.

Veblen, Thorstein, 1906. "The Place of Science in Modern Civilization." *American Journal of Sociology* 11: 585–609.

Veblen, Thorstein, [1918] 1957. *The Higher Learning in America: A Memorandum on the Conduct of Universities by Business Men*. New York: Sagamore Press.

Venter, J. Craig, 2007. *A Life Decoded: My Genome: My Life*. New York: Allen Lane, Penguin Books.

Veysey, Laurence R., 1965. *The Emergence of the American University*. Chicago: University of Chicago Press.

Vollmer, Howard M., 1965. *Work Activities and Attitudes of Scientists and Research Managers*. Menlo Park, CA: Stanford Research Institute.

Voorhies, Darrell H., 1946. *The Co-ordination of Motive, Men and Money in Industrial Research, A Survey of Organization and Business Practice Conducted by the Department on Organization of the Standard Oil Company of California*. San Francisco: Standard Oil Company of California.

Walsh, John, 1973. "A Conversation with Eugene Wigner." *Science* n.s. 181, no. 4099 (10 August): 527–533.

Wang, Jessica, 1992. "Science, Security, and the Cold War: The Case of E. U. Condon." *Isis* 83: 238–269.

Wang, Jessica, 1999. *American Science in an Age of Anxiety: Scientists, Anticommunism and the Cold War*. Chapel Hill: University of North Carolina Press.

Washburn, Jennifer, 2005. *University, Inc.: The Corporate Corruption of Higher Education*. New York: Basic Books.

Watson, David Lindsay, 1938. *Scientists Are Human*. London: Watts.

Watson, James D., [1968] 1980. *The Double Helix*, ed. Gunther S. Stent. New York: W. W. Norton.

Watson, James D., 2002. *Genes, Girls and Gamow: After the Double Helix*. New York: Alfred A. Knopf.

Watson, James D., 2007. *Avoid Boring People: Lessons from a Life in Science*. New York: Alfred A. Knopf.

Watts, Duncan J., 2003. *Six Degrees: The Science of a Connected Age*. New York: W. W. Norton.

Weart, Spencer, 1979. "The Physics Business in America, 1919–1940: A Statistical Reconnaissance." Pp. 295–358 in Nathan Reingold, ed., *The Sciences in the American Context: New Perspectives*. Washington, DC: Smithsonian Institution Press.

Weber, Max, [1905] 1958. *The Protestant Ethic and the Spirit of Capitalism*, trans. Talcott Parsons. New York: Charles Scribner's.

Weber, Max, 1978. *Economy and Society: An Outline of Interpretive Sociology*, 2 vols., eds. Gunther Roth and Claus Wittich, trans. Ephraim Fischoff et al. Berkeley: University of California Press.

Weber, Max, 1991a. *From Max Weber: Essays in Sociology*, eds. H. H. Gerth and C. Wright Mills. London: Routledge.

Weber, Max, 1991b. "The Meaning of Discipline." In Weber 1991a, pp. 253–264.

Weber, Max, 1991c. "Science as a Vocation." In Weber 1991a, pp. 129–156.

Weinberg, Alvin M., [1961] 1965. "Impact of Large-Scale Science on the United States." In Kaplan 1965a, pp. 551–559.

Weinberg, Alvin M., 1967. *Reflections on Big Science*. Cambridge, MA: MIT Press.

Weinberg, Alvin M., 1972. "Scientific Teams and Scientific Laboratories." In Holton 1972, pp. 423–441.

Weiner, Charles, 1987. "Patenting and Academic Research: Historical Case Studies." *Science, Technology & Human Values* 12, no. 1 (Winter): 50–62.

Weiss, John Morris and Charles Raymond Downs, 1924. *The Technical Organization: Its Development and Administration*. New York: McGraw-Hill.

Weiss, Paul, 1962. "Experience and Experiment in Biology." *Science* n.s. 136, no. 3515 (11 May): 468–471.

Weisskopf, Victor F., 1991. *The Joy of Insight: Passions of a Physicist*. New York: Basic Books.

Werth, Barry, 1994. *The Billion-Dollar Molecule: One Company's Quest for the Perfect Drug*. New York: Simon and Schuster.

West, Diane, 2000. "Secrets 2Connect with Your Audience." Typescript summary of seminar on "Investor Presentations," 9 October.

Westman, Robert S., 1980. "The Astronomer's Role in the Sixteenth Century: A Preliminary Study." *History of Science* 18: 105–147.

Whetzel, H. H., 1919. "Democratic Coordination of Scientific Efforts." *Science* n.s. 50, no. 1281 (18 July): 50–55.

Whitehead, A. N., [1926] 1946. *Science and the Modern World.* London: The Scientific Book Club.

Whyte, William H., Jr. [1956] 1957. *The Organization Man.* Garden City, NY: Doubleday Anchor.

Wickelgren, Ingrid, 2002. *The Gene Masters: How a New Breed of Scientific Entrepreneurs Raced for the Biggest Prize in Biology.* New York: Times Books, Henry Holt.

Wiener, Norbert, 1948. "A Rebellious Scientist after Two Years." *Bulletin of the Atomic Scientists* 4 (November): 338–339.

Wiener, Norbert, 1956. *I Am a Mathematician: The Later Life of a Prodigy.* Garden City, NY: Doubleday.

Wigner, Eugene P., ed., 1947. *Physical Science and Human Values.* Princeton, NJ: Princeton University Press.

Wigner, Eugene P., 1950. "The Limits of Science." *Proceedings of the American Philosophical Society* 94: 422–427.

Williams, D. L., 1947. *Planning of Research and Development.* New York: Wallace Clark.

Williamson, Merritt A., 1960. "The Art of Interviewing for Technical Positions." *Research/Development* (formerly *Industrial Laboratories*) 11, no. 1: 17–24.

Wilson, John W., 1985. *The New Venturers: Inside the High-Stakes World of Venture Capital.* Reading, MA: Addison-Wesley.

Wilson, E. Bright, Jr., 1952. *An Introduction to Scientific Research.* New York: McGraw-Hill.

Wilson, Mike, 1997. *The Difference Between God and Larry Ellison: Inside Oracle Corporation.* New York: William Morrow & Co.

Wilson, Robert E., 1949. "The Attitude of Management toward Research." *Chemical and Engineering News* 27, no. 5 (31 January): 274–277.

Wilson, Robert R., 1972. "My Fight against Team Research." In Holton 1972, pp. 468–479.

Wise, George, 1985. *Willis R. Whitney: General Electric, and the Origins of U.S. Industrial Research.* New York: Columbia University Press.

Withey, Stephen B., 1959. "Public Opinion about Science and Scientists." *Public Opinion Quarterly* 23: 382–388.

Work, Harold K., 1952. "The University's Role in Training Research Workers." In Hertz and Rubenstein 1952, pp. 126–140.

Yates, JoAnne, 1989. *Control Through Communication: The Rise of System in American Management.* Baltimore, MD: Johns Hopkins University Press.

Yeo, Richard R., 1981. "Scientific Method and the Image of Science, 1831–1890." Pp. 65–88 in Roy MacLeod and Peter Collins, eds., *The Parliament of Science: The*

British Association for the Advancement of Science, 1831–1931. Northwood, Middlesex: Science Reviews.

Yeo, Richard R., 1988. "Genius, Method and Morality: Images of Newton in Britain, 1760–1860." *Science in Context* 2: 257–284.

Yeo, Richard R., 1993. *Defining Science: William Whewell, Natural Knowledge, and Public Debate in Early Victorian Britain.* Cambridge: Cambridge University Press.

Yoffe, Emily, 1994. "Is Kary Mullis God? (Or Just the Big Kahuna?) Nobel Prize Winner's New Life." *Esquire* 122, no. 1 (July): 68–75.

[Youmans, E. L.?], 1885. "Editor's Table: Official Science at Washington." *Popular Science Monthly* 27, no. 6 (October): 844–847.

Zener, Clarence, 1957. "Planning of Research Work at Westinghouse Electric." *IRE Transactions on Engineering Management* EM-3, no. 4: 94–96.

Zilsel, Paul R., 1964. "The Mass Production of Knowledge." *Bulletin of the Atomic Scientists* 20, no. 4 (April): 28–29.

Zirkle, Conway, 1956. "Our Splintered Learning and the Status of Scientists." *Science* n.s. 121, no. 3146 (15 April): 513–519.

Zwemer, Raymund L., 1950. "Incentives from the Viewpoint of a Scientist." In Bush and Hattery 1950, pp. 75–78.

Zygmont, Jeffrey, 2001. *The VC Way: Investment Secrets from the Wizards of Venture Capital.* Cambridge, MA: Perseus.

Index

Page numbers in italics indicate figures.

academic scientists and culture: adaptation to industry from, 158–59, 364n105; approach to, 18; Big Science viewed from, 80–81; constraints on, 242–44, 253–54; entrepreneurship of, 210–13, 215–16; flexibility and freedom of, 139–40, 153, 161–62, 242–44, 362n89; as good society, 111; graduate students and, 240, 242–43, 247–49; individualism of, 174–75, 181–83, 259–61; industrial good life compared with, 252–53; industrial science as viewed by, 110–13, 128, 157–58, 175–76, 243–44; industry's convergence with, 231–32, 247–49, 384n68; multiple responsibilities of, 161, 216–17, 245–49, 364n111; nonteaching positions for, 385n73; openness as norm of, 147, 254, 255; organizational scientist as viewed by, 113–19, 121–24, 172–73, 178, 181–83; as percentage of all, 120; salaries of, 102–4, 105–9, 346n40, 347n52, 347nn56–57; specialization in, 88–89; tenure of, 245; on virtues attached to communal life of science, 127–28; "vulgar errors" in, 23, 47. *See also* universities; university-based research

Accel Partners, 273, 274, 275, 286

administration: general principles for, 351n100; research manager on, 131; sociological framework for studies of, 117–18. *See also* Big Science; bureaucracies

Administrative Science Quarterly (ASQ), 117–18, 131, 353n110

Advanced Research Projects Administration, 122, 353n113

agricultural research, 210–11

AIDS/HIV, 228

Alpha Chi Sigma, 185

Alvarez, Luis, 79, 80

American Association for the Advancement of Science, 78, 129–30, 322n7, 324n23

American Association of Scientific Workers, 68

American Chemical Society, 122

American Cyanamid, 95, 139–40, 152. *See also* Shepard, Norman

American Research and Development Corp., 398n137

American Viscose Corporation, Sylvania Division (research) of, 147, 148, 150, 349n77, 363n91

Andrews, Frank, 119

Andrews, Mrs. M. R., 372n79

angel investors: competition for, 390–91n39, 391n47, 391–92n50; on entrepreneurs' honesty and passion, 296–97, 396n121; on entrepreneurs' monetary investment, 293–94; entrepreneurs' relationships with, 279; on exit strategy awareness, 292; face-to-face meetings of, 301, 399n154; networking of, 304, 305–10; role of, 277; SEC requirements for, 390n35; use of term, 390n34. *See also* venture capitalists (VCs)

Anglo-Saxon heresy, 29

Anthony, Robert N., 123

anti-clericalism, 31

antinomies, presence of, 13–14

Antiquity: authorship and knowledge in, 8–9, 196–97; primitive knowledge of, 32–33; State use of expertise in, 34, 39

anti-Semitism, 67, 334n68

Apple Inc., 266

Appleton, Edward, 75

Applied Physics Laboratory, 167, 172

applied science. *See* basic vs. applied science

Archimedes, 32–33

Aristotle, 24, 32–33

Armour Research Foundation, 95

Arrowsmith (film), 63

Arrowsmith (novel by Lewis), 60–63, 169, 331–32n40

Arthur D. Little, Inc.: contractual work of, 95, 136; (dis)organization model of, 198, 374n106; founding of,

355n12; job offers of, 108; serendipity encouraged at, 203. *See also* Little, Arthur D.

Association of Scientific Workers (earlier, British National Association of Scientific Workers), 50

atomic bomb development: disseminating information on, 149; Einstein on, 64; interdisciplinary teamwork in, 180–81; is/ought distinction and, 75–76; moral responsibilities debated, 65–71; reflections on, 333n57; scientific vocation attention due to, 105–6; sociological view of, 118. *See also* Manhattan Project

Atomic Energy Commission, 104

atomic research facilities, 155

AT&T, 94, 95. *See also* Bell Labs (AT&T)

authority: limited sort of, 11–12; personal in, 2–4; of scientific knowledge, 13–14, 128; of truth-speakers, 6–9; use of term, 2; virtues distinguished from, 40–41. *See also* charisma and charismatic authority

authorship: in Antiquity, 8–9, 196–97; as fame-seeking, 35; of industrial scientists, 148–51, 360n67, 361n73; lack of pressure for, 326n57; of literature vs. scientific discovery, 8–9; organized science as undermining, 175; virtues linked to, 54

Babbage, Charles, 38, 42–43

Bacon, Francis, 32, 41, 54, 174

Baker, John R., 9, 171, 374–75n108

Baker, William O., 171

Ball, Carleton R., 371n70, 373n97

Band of Angels meeting, 390–91n39

Banks, Joseph, 38–39

Barber, Bernard, 121, 174, 321n3

Bardeen, John, 149, 150, 167

Barnes, Barry, 119

Barthes, Roland, 7–8, 33

Bartlett, Frederic C., 194–95

Barus, Carl, 133–34
baseball metaphor, in venture capital, 272–73, 387n8
basic vs. applied science: discourse on, 97–98, 161–62; expected research payback and, 145–46; "linear model" in, 343n19; NSF focus on (applied), 213; percentage of, in industry, 343n20; research uncertainties in, 132–35, 139–40, 144–45; teamwork vs. individualistic research and, 172–78; university's focus on (basic), 213, 344n21; U.S. Constitution and, 43–44
Bauman, Zygmunt, xv, 5
Bayh-Dole Act (Patent and Trademark Act Amendments, 1980), 139–40, 215, 241
Beck, Ulrich, xv
Beckman Instruments division, 150
Bell Labs (AT&T): basic research of, 134; budget of, 100; Depression-era personnel cutbacks of, 145; external ties of scientists of, 146–47; Feynman on early offer of, 380n28; flexibility and freedom of, 152, 160, 176–77; journalists' approval of, 370–71n66; planning of, 200; publication policy of, 149; research interests of, 94, 95; research payback of, 145–46; scientists of, 150; teamwork of, 179, 181. See also Fisk, James B.
Bello, Francis, 370–71n66
Bell Telephone Company, 112
Benchmark Capital, 274, 294
Benda, Julien, 56–57
Bendix, Reinhard, 121
Benson, Sidney W., 346n47
Bernal, J. D., 50–51, 142, 170, 368–69n41
Bernard, Claude, 6, 31–32
Bethe, Hans, 365–66n1
bias, 6–9. See also objectivity
Bichowsky, F. Russell, 153, 362n89
Big Science: criticism of, 80–87, 169–73, 339n120; democratic accountability

issues in, 88; features of, 166; organization forms and moral constitution of, 165–73, 351n95; scientists involved in, 87–88, 351n95; teamwork in, 16, 82–83; use of term, 365–66n1
Billings, John S., 322n7
biographers and biographical subjects, 8–9, 52–54, 329–30n19
biotechnology and biomedical research: convergence of university and industry in, 231–32, 247–49, 384n68; entrepreneurial initiatives in, 215–16, 223–29, 243–44, 249–50, 381n34; secrecy in, 254–55; training and mentoring in, 247–48. See also DNA research and technologies; entrepreneurship and entrepreneurial companies; venture capital
Birndorf, Howard, 263, 265–66, 302
Blodgett, George, 188
Blodgett, Katharine Burr, 188
Boger, Joshua, 281–82, 381n34
Bohr, Niels, 171
Booz, Allen & Hamilton, 138
Born, Max, 333n57
Boston University, 215
Box, Steven, 119
Boyer, Herbert W., 215–16
Boyle, Robert, 33, 35, 325n40
Brattain, Walter, 149, 150
Brewster, David, 52–53, 329–30n19
Breyer, Jim, 273, 275, 286
Bridgman, Percy W., 83, 324n31
Brinton, Daniel G., 324n23
Broad, William, 86
Bronk, Detlev, 76–77, 171
Bronowski, Jacob, 75
Bronson, Po, 294
Brougham, Henry, 8
Bryan, William Jennings, 62
Buffon, George-Louis Leclerc de (comte), 38–39
Bulletin of the Atomic Scientists, 65–66

bureaucracies: emergence of faceless, 128; ethical responsibilities and structure of, 336n85; as triumph of impersonal reason, 9–13. See also administration; Big Science

bureaucratic reports, 124

Burnham Cancer Institute, 383n54

Bush, Vannevar, 68, 135, 171

business organization: consequences of research uncertainties for, 132–35; research function and goals of, 230–31; scientists' location in scheme of, 364n105; teamwork presumed in structure of, 179; vertical integration in, 94

BusinessWeek magazine, 223, 224

Bute, Earl of (John Stuart), 38–39

Byers, Brook, 266, 288, 295, 302

Byster, Alfred (pseudonym), 246–47, 248

Caine Mutiny, The (film), 120

California Institute of Technology (Caltech), 236–38

calling: early modern science as, 34–46; fictional depiction of, 60–61; notions of individuals and, 32–33; Romantic conception of, 85; shift away from, 14–15, 44; Weber's concept of, 1, 17. See also career; vocation

Caltech (California Institute of Technology), 236–38

Cantillon, Richard, 377n1

capitalism: as cultural fault line, 195–97; entrepreneurship in, 266–67; rebellion against rational in, 10. See also entrepreneurship and entrepreneurial companies; venture capital

career: commercial terminology for, 79–80; criticism of science as, 173; de-moralization endemic to, 82; fraud in, 86–87; institutionalization of science as, 85; mid-century sensibilities about, 127; monetary rewards of, 209;

non-empirical evaluations of, 105–6; science as mass occupation, 173; selling notion of, 79–80; transformation of inquiry into, 14–15, 57–58. See also salaries and wages

Carlo, Dennis, 254

Carlyle, Thomas, 170

Carmichael, R. D., 367n26

Carothers, Wallace, 360n67

Carpenter, Edward, 40

Carr, E. H., 315n2

Carty, John J., 145–46

Carver, Emmett K., 371n69

Cavendish, Henry, 38–39

Celera Diagnostics, 243, 250, 266. See also White, Tom

Celera Genomics, 223, 225, 226, 266, 382n42

certainty, limits of, 30–31. See also normative uncertainty concept; uncertainties

Cetus Corporation: expansion of, 256; motivations of scientists at, 250, 262; patent rights of, 228–29; work rhythms of, 266

Chandler, Alfred D., 94

Chargaff, Erwin, 65, 84–85, 173

charisma and charismatic authority: approach to, 4–5; concept of, 3; entrepreneurship linked to, 266–67; Weber on, 3, 5, 91, 165, 208, 209, 266–67

Chemical and Engineering News (journal), 129

chemical industry: emergence of research in, 95; percentage of chemistry graduates in, 110; R&D budgets of, 100; recruitment criteria of, 185; scientists' flexibility and freedom in, 152–53; study of chemists in, 119. See also biotechnology and biomedical research

Cherwell, Lord (Frederick Lindemann), 333n57

China, People's Republic of, American fears of, 120

Christian gentility concept, 35–36

Christians, virtues monopolized by, 62

citizens and civic arena: experts as ordinary, 59–60; science integrated into, 15–16; scientists as, 60, 65, 68–69. See also economy of civic virtue; industrial scientists; organization man; virtues

Clark, Jim, 292, 397n128

clergy and priests: scientific interests of, 35–36, 42; terminology of, 72–80, 173

Clifford, W. K., 29

Cohen, Daniel, 381n34

Cohen, Stanley N., 215–16

Cold War: Big Science and moral equivalence presumptions during, 80–89; as cultural fault line, 195–97; defense of individualism in, 177; defense research scientists in, 74, 89–91; funds for industrial science in, 100–101; is/ought distinction in, 69–70; local-cosmopolitan distinction in context of, 118; science research funding in, 213–14; scientists in context of, 76–78, 101–2, 103–4; security concerns of, 67–68, 69; value conflict concept in, 114

collective and collectivism: criticism of, 168, 171, 175–76; individualism vs., 120–21, 169–71; merger of individual with, 196–97. See also teams and teamwork

Collins, Francis, 224

Columbia University: business school of, 353–54n114; Industrial Research Conferences of, 351–52n102; National Manpower Council of, 77–78, 104, 347n57

commerce: science enlisted in, 95–96; science integrated into, 41–43, 57–58, 85, 105–6; scientific career in terminology of, 79–80; values assigned to, 112–13, 114–17. See also business

organization; entrepreneurship and entrepreneurial companies; industry and industrial development; wealth and profit structure

common sense, 3–4, 10, 28

communications: conferences in, 205, 206, 207; online technologies of, 275; of research manager and scientists, 204–5

Communism, as cultural fault line, 195–97. See also Cold War

Compton, A. H., 90

Compton, Karl, 104

computer science research: entrepreneurial initiatives in, 236–38; industrial contracts for university, 239–41

Comte, Auguste, 29

Conant, James Bryant, 12, 71, 108–9, 362n84

Condon, Edward U., 67, 68, 145, 155, 334n72, 359n50. See also Westinghouse Research Laboratories

Condorcet, Marquis de (Marie-Jean Caritat), 36, 37

conflict of interest: allegations of, 216–17; in values of industrial vs. academic domain, 113–19

Congleton, William, 398n137

Conn, Robert C., 378–79n15

CONNECT: Financial Forum of, 390–91n39; panel on failed entrepreneurs held by, 293; "Springboard" meetings of, 277–80, 287, 391n45

conventionalism, defined, 27

Coolidge, William, 108, 152, 348–49n73

Copernicus, Nicolaus, 26, 35–36

Cordiner, Ralph, 171–72

Corleone, Luigi (pseudonym), 385n76

Cornell University, 117–18, 353–54n114

Cornwell, Ralph T., 349n77

Coser, Lewis A., 90–91, 341n154

Cotgrove, Stephen, 119

Cottrell, Frederick, 210, 212, 217

Coulomb, Charles Augustin, 39

creativity and invention: conflicting notions of, 176–77; context for, 248; individual nature of, 169–73; merger of individual and collective in, 196–97; planning juxtaposed to, 197–203; rational method of, 10; studied in industrial research, 122; in team composition, 183; in teamwork, 194

Crowther, J. G., 368–69n41

cultural nobility, 29

culture: fault lines in, 195–97; organization man as viewed in, 120–21; personal vs. impersonal boundary in, 7–8; scientific credibility and, 33–34. *See also* academic scientists and culture; industrial scientists and culture; late modern society and culture

Dalton, John, 38

Danielson, L. E., 122

Dartmouth University, 353–54n114

Darwin, Charles, 24, 25, 34, 41

David, Gary, 263–64, 384n68

Davis, Bob, 301

Davis, Tommy, 297

Dear, Peter, 26–27

de-magification: moral ordinariness in, 127; Weber on, 11, 25–26, 46

democratization of scientific knowledge, 28–29. *See also* citizens and civic arena

de-moralization of technical experts: academic culture's role in, 13–14; in Big Science, 80–81, 83–84; criticism of, 331n33; in current scientific enterprise, 89–91; as endemic to science as career, 82; limits of expertise and, 70–72; Whyte's outline of, 176–77. *See also* moral equivalence presumptions; Oppenheimer, J. Robert

Descartes, René, 8, 32, 35

Desk Set (film), 120

Dewey, John, 48

Digital Equipment Corporation, 289

disciplines: heterogeneity of, 229–31; human relations as, 122–23; industrial research as, 351–52n102; moral equivalence of scientist in, 21–23; sociology as, 321n2; utilitarian and self-cultivation arguments for, 59–60

disinterested inquiry: approach to, 18–20; independent means as allowing, 38–39; remuneration ignored in, 45–46, 103–4; as virtue, 57–59, 249–50

Distillation Products, Inc., 376n123

DNA Adventures Inc., *219*

DNA research and technologies: convergence of university and industry in, 231–32, 384n68; entrepreneurial initiatives in, 215–16, 223–29, 243–44, 381n34. *See also* biotechnology and biomedical research; Watson, James D.

DNAX Institute of Molecular and Cellular Biology, 244, 262–63

Doriot, Georges: background of, 394n98; as influence, 397n125; maxims of, 289–90, 291, 296; opinions summarized, 395n104

dot-com bust (2000), 274–75, 293, 392n58

Dow Chemical Corp., 95, 152, 353n113

Dragon's Den (television program), 390n34

Draper Fisher Jurvetson, 300–301

Dressler, Carole, 285

Drucker, Peter, 101, 182

DuBridge, Lee, 65–66, 80

due diligence: business academics on, 393n70; VCs on, 282–83, 284–85

Duhem, Pierre, 29, 30

DuPont Corp.: basic research of, 134; economic downturns faced by, 358–59n47; flexibility and freedom of, 152–53, 160; intellectual property policy of, 147; publication policy of, 148–49, 360n67; R&D budget of, 100; research

advisor of, 204; research interests of, 95; research uncertainties of, 138–39
Durkheim, Emile, 22, 319n33
Dynes, Robert C., 214, 320n45
Dyson, Freeman, 221, 337n101

early modern period: philosopher as morally superior in, 24–27, 47–48; science and metaphysics separated in, 28–33; science as calling in, 34–46; scientific entrepreneurs of, 210
Eastman, George, 136
Eastman Kodak Research Laboratory: accounting system of, 140–41, 141; basic research of, 134; conference system of, 205, 206, 207; as convergent laboratory, 193; first director of, 136, 137, 157; flexibility and freedom of, 152, 160; interdisciplinary teamwork of, 179, 180, 181; journalists' approval of, 370–71n66; planning of, 199, 202; publication policy of, 148, 151; research interests of, 94, 95; research uncertainties of, 132, 136–37, 139, 142; salaries of, 102–3, 108; teamwork success of, 194. See also Mees, C. E. Kenneth
eBay, 283, 294
economy of civic virtue: good society concept in, 195–97; industrial settings of, 16–17; interdisciplinarity and, 178–83, 180; mobilizing virtues and making knowledge in, 41, 183–90; moral constitution of Big Science in, 165–73; planning and scientific integrity juxtaposed in, 197–203; research manager role and, 203–8; teamwork models in, 190–95, 192; vices of organization man in, 173–78
Edison, Thomas, 94–95
education. See disciplines; universities
Ehrenpreis, Ira, 397n131
Ehrlich, Paul, 63
Einstein, Albert: American obsession with, 63–64; brain of, 7–8; on career

choice, 82; on defining truth, 30; as genius, 33; individualism of, 171, 173; on is/ought distinction, 11–12; mentioned, 129; reading of, 336n94; salary of, 103; on scientists as human, 49–50
Einstein complex, 79
Eisenhower, Dwight David, 167; on Big Science, 81–82; farewell address of, 81, 100, 166, 170, 338n119; on military-industrial complex, 81, 89, 91, 100, 170; science advisors of, 166, 167; on social sciences support, 77–78; speech writer of, 338–39n119; on technological revolution, 166
electrical industry: early research in, 94–95; R&D budgets in, 100. See also General Electric Research Laboratory
Electrical World (journal), 129
Eli Lilly Company, 216
Ellison, Larry, 266, 397n128
Ellul, Jacques, 76, 337n102
Emerson, Ralph Waldo, 7, 41, 55, 327n63, 367n26
Engineering Manpower Commission, 107
engineers: in Antiquity, 34, 39; as models for scientists, 30–31; schools for, 121–24; scientists distinguished from, 44. See also science and engineering workforce
Engstrom, Elmer W., 354n2
entrepreneurs and entrepreneurial scientists: backgrounds of, 377n2; business plans of, 280–81, 282, 287–89, 296, 298, 395n107; changes in knowledge production of, 264–65; definition of, 210; emergence of, 17, 209–11; familiarity and investor relationships of, 283, 285–89, 294–95, 299–300, 302; heterogeneity of, 229–31; Lemelson-MIT Prize for, 216; lifestyle choices of, 251–61; life-worlds of, 229–32; management ability of, 289–91; motivations of, 250–51, 261–64, 295,

entrepreneurs (*cont.*)
396–97n122, 397n128; personal investment by, 293–94; pitches to investors of, 277–82, 278, 286–87, 295–96, 390–91n39, 391n47, 391–92n50; portrayals of, 222–29, 224, 227; prudence and passion balanced by, 292–303, 396n121; role of, 211–13, 215–16, 377n1; vices of, 228

entrepreneurship and entrepreneurial companies: atmosphere and growth of, 256–58; criticism of, 216–17, 379n20; drug discovery in, 233–36; flexibility of, 258–60; goals and routines of, 265–67; heterogeneity of, 229–31; political economy of, 213–17; research contracts of, 95; secrecy of, 254–55; texture of life in, 251–61; training and mentoring in, 247–48; uncertainties of, 264–67; universities' support for, 214–16; VC criteria for investing in, 275, 282–89; VC firm's role in, 291–92, 298, 299–300; virtues of academia and industry in context of, 232–42. *See also* interest as performance; venture capitalists (VCs)

ethics: bureaucratic responsibilities and, 336n85; of pharmaceutical industry, 95; science distinguished from, 10–13; science of (Durkheim), 319n33. *See also* integrity

Ethyl Corporation, 185

Eustis, Warner, 363n98, 372n81

evolution, 62

Executive Suite (film), 120

expertise: limits of, 70–72; of ordinary citizens, 59–60; political power of, 81–82; sale and assimilation of, 56–57; of scientific witnesses, 45; scientist as human idea and, 48–52; State utilization of, 33–34, 39–41, 42. *See also* de-moralization of technical experts

Facebook, 275, 283

face-to-face meetings, 301–3, 399n154. *See also* familiarity; networking

Fairchild Semiconductors, 150, 266

falsificationism, defined, 28

familiarity: alleged loss of, 3; approach to, 4–5; in investor-entrepreneur relationships, 283, 285–89, 294–95, 299–300, 302. *See also* networking; personal, the

Faraday, Michael, 43

Fascism, as cultural fault line, 195–97

Fermi, Enrico, 90

Feuer, Lewis, 90, 217

Feynman, Richard P., *221*; on early job opportunity, 380n28; on naturalistic fallacy, 12–13; self-portrayal of, 83, 220–21

Fieser, Louis, 152–53

films and movies, 63, 120, 177

Fisher, John, 300–301

Fisk, James B., *167*; on external ties of scientists, 146–47; on organization models, 190, 191; on publication policy, 149; on pure vs. applied science, 98; on research payback, 146. *See also* Bell Labs (AT&T)

Fontenelle, Bernard le Bovier de, 36

Ford, Franklin L., 218

Ford, John, 63

Ford Foundation, 77–78

Fortune magazine: on flexibility and freedom, 152; on genius, 184; on location of scientists, 120; on organized research, 96; on salaries of scientists, 106, 107; on scientist as maverick, 182

Foucault, Michel, 22, 57, 70

Fouchy, Jean-Paul Grandjean de, 36

Franklin, Benjamin, 37, 41, 43

fraud exposés, 86–87

Fulbright, J. William, 72

"fusion of horizons" concept, 315n2

Gadamer, Hans-Georg, 315n2
Galilei, Galileo, 26, 173
Galton, Francis, 34, 44
gambling, corporate research decisions as, 271, 386n1
Gassendi, Pierre, 35–36
Gellhorn, Walter, 68
gender: recruitment considerations of, 185, 188; of venture capitalists, 285–86. *See also* women
Genentech, 215–16, 253, 297, 387n7
General Chemical Corp., 95
General Education for a Free Society (Conant), 12
General Electric Research Laboratory: accounting system of, 358n43; basic research of, 134; budget of, 100; conference system of, 205; economic downturns faced by, 145, 358–59n47; flexibility and freedom of, 152, 156, 160, 176–77; foreign-trained scientists of, 196; journalists' approval of, 370–71n66; publication policy of, 148, 151; recruitment criteria of, 185; research interests of, 94, 95; research payback of, 143; research uncertainties of, 139; salaries of, 108–9; women scientists of, 188, 372n79. *See also* Whitney, Willis R.
General Motors Research Laboratory, 137–38
genius: cult of, 83; as mechanical vs. inspired by God, 33; rejection of scientist as, 59–60; replacement of by mediocre researchers, 84–85; singularity of, 7–9; skepticism about, 34, 183–84; sociological arguments against, 22; teamwork as replacing, 196–97; as unnecessary in scientific life, 77; Watson as, 218, 220
Genset Corp., 296
gentlemen, industrial researchers as, 183
George, William H., 48, 319–20n35

Germany: drug patents of, 342n6; Nazification of science in, 51, 82; as scientific influence, xvi, 94, 172, 196
GI Bill of Rights, 101
Gibran, Khalil, 226
Giddens, Anthony, xv
Gilbert, Walter, 381n34
Gladstone, William E., 43
Glaser, Barney G., 352n105
Glass, Bentley, 80–81, 83–84
God and divine: death of, 56; fictional physician's prayer to, 62; identity of, 28–29; knowledge of nature inspired by, 33; truth of, vs. ultimate reality, 27–30
Goethe, Johann Wolfgang von, 39
Goffman, Erving, 296
Goldschmidt, Richard, 168
good society concept, 111, 195–97
Goodyear Corp., 95
Google, 194, 253, 276, 283
Gouldner, Alvin, 67, 118
government scientists: as human, 59; mobilization of, 128; as percentage of all, 120; salaries of, 106–7, 347n57. *See also* Manhattan Project
grant system: difficulties of, 242–44; federal sources in, 238, 239, 245, 246, 253, 390n37; insecurity of, 245–46
Gray, Stephen, 38
Gray Board, 69–70
Great Britain: individualism of science in, 172; organized science as viewed by Marxists in, 368–69n41; Society for Freedom in Science in, 170, 198; sociology of scientific knowledge in, 119; state self-interest and science in, 42–43
Greenberg, Daniel S., 87–88, 340–41n147. See also *Science* magazine
green start-ups, 295, 397n131
Gregory, Richard, 24
Greylock Management, 291, 397n125
Groves, Leslie, 334n68

Gudjonssen, Eidur (pseudonym), 247, 251, 253–54, 257–58
Gupta, Udayan, 395n104

Habermas, Jürgen, 81
Hagstrom, Warren O., 116–17, 174–75, 351n95, 351n98
Haldane, J. B. S., 25
Hales, Stephen, 35–36
Hall, Dorothy, 372n79
Hall, James, 39
Halley, Edmond, 35
Hamlin, Christopher, 45
Hamor, William Allen, 362n89
Hansen, William, 211
Hardy, G. H., 341n155
Harper's Magazine, 72–73, 177
Harris, Sir Arthur Travers ("Bomber"), 333n57
Harvard Business Review, 110
Harvard University: business school of, 123, 353–54n114; salaries of, 107; Vertex Pharmaceuticals' relationship with, 383n53; Watson's salary from, 218, 220
Hattery, Lowell H., 122–23
Hawicke, James (pseudonym), 242, 245–46, 384n71
Hawkins, Laurence A., 348–49n73, 360n61, 372n79
Healy, Bernadine, 225
Hebb, Malcolm, 139, 356–57n24
Heims, Steve, 66–67
Hermes Trismegistus, 32–33
Herschel, John, 24
Hertz, David Bendel, 122, 354n115
heterogeneity: of entrepreneurial scientists, 229–31; of good life, 251–61; of industrial science, 230–31; of late modern society, 14, 242–51; of motivations, 242–51
Hickman, K. C. D., 376n123
Highland Capital Partners, 301
Hill, A. V., 51
historical change, 14, 18–19

history, approach to, xiii–xiv
Hoffmann-LaRoche Corp., 228
Hogben, Lancelot, 368–69n41
Hood, Leroy, 266, 381n34
Hooke, Robert, 32, 35, 210
Hoover, Herbert, 342n5
Hoover Company, 156
Horsley Bridge Partners, 273
Hounshell, David A., 148–49, 153
House Un-American Activities Committee (HUAC), 68–69, 83, 118, 334n72, 335n76
Hower, Ralph, 123
Human Genome Project, 221
Human Genome Sciences, 381n35
Humboldt, Alexander von, 39
Hume, David, 10, 269, 318–19n27
Hutchins, Robert Maynard: on modern university, 378n10; on salaries, 103, 105–6, 346n40; on scientific specialization, 88–89
Huxley, T. H., 8–9, 28, 29
Hybritech Inc.: business plan of, 395n107; familiarity and, 288, 302; hiring of, 263–64; mentioned, 284; secrecy of, 254; work rhythms of, 265–66

identity and self-portrayal, 219, 221, 224, 227; of DNA researcher, 169, 218, 220–22; of entrepreneurial scientists, 222–29; of physicist, 83, 220–21; proper sense of, 46
impersonal, the: bureaucracies as triumph of, 9–13; cultural credibility of scientific, 33–34; insistence on, 6–9; 31–32; tension of personal with, 1–6
individualism: in academic experience, 174–75, 181–83, 259–61; communal conduct juxtaposed to, 173–74; conformity and collectivism vs., 120–21, 169–71; creative potential of, 169–73, 176, 177–78, 182–83; criticism of, 194, 373n97; defense of, 177; of science, 172;

sociological arguments against, 21–23; teamwork juxtaposed to, 82–83, 168–69, 259–61. *See also* creativity; genius

individuals/people: adaptation to industrial science of, 158–59, 364n105; as called and inspired, 32–33; creativity linked to, 169–73; credit for research of, 151; genius type of, 7–9; merger of collective with, 196–97; organization man compared with, 173–78; scientists as, 169–73, 182–83; tension of impersonal and, 1–6

inductive method, 32

Industrial Laboratories (journal), 122, 129, 200

Industrial Research Institute (IRI), 122, 123, 129, 130

industrial science: academic view of, 110–13, 128, 157–58, 175–76, 243–44; accounting system of, 138, 139, 140–42, *141*, 143–45, 358n43; budget of, 100–101, 107; business and engineering schools' views of, 121–24; convergent vs. divergent labs in, 192–93; examples of areas of, 94–95; facilities of, 154, *155*, 243, 363n91; German influence on, xvi, 94, 172, 196; heterogeneity of, 230–31; inquiry possibilities in, 19–20; intellectual property in, 139–40; interdiciplinarity in, 179–83, *180*; internal commentary on, 129–32; making knowledge in, 41, 183–90; management of, 121–22, 135–45, 156–59, 345–46n36; managers (directors) of, 203–8, 376–77n129; models for organization of, 190–95, *192*; national interest in, 94, 99–104, 342n5; naturalism and legend in discourse on, 160–64; planning in, 197–203; practical nature of, 250–51; preferences for jobs in, 109; problematics of outside views of, 99–100; project as focus of, 179–80, *180*; pros and cons of, 95–97;

pure vs. applied science distinction and, 97–98, 343n19, 343n20; research fellowships in, 145, 359n50; secrecy and publication policies of, 147–51, 361n73; sociological view of, 117–19; teamwork in, 158–59, 179–83, *180*, 364n105; temporal frame of, 152–56; testing and assay functions of, 135–36; uncertainties in, 132–35, 145–51, 156–59; university-based research compared with, 233–42. *See also* industry and industrial development

industrial scientists and culture: academic good life compared with, 252–53; in Cold War period, 76–78, 100–102, 103–4; company interests of, 158, 361n77; conflict of values presumed for, 113–19; constraints on, 145–46; depictions of, 120–21; emergence of, 94–99; flexibility and freedom of, 152–56, 160, 161–62, 176–77, 247, 253, 363n91; as gentlemen, 183; interdisciplinarity of, 178–83, *180*; internal commentary of, 129–32; local-cosmopolitan distinction for, 118; number of, 109–10; planning juxtaposed to integrity of, 197–203; professional ties of, 146–47, 148–49; psychological studies of, 119; publication of, 148–51, 360n67, 361n73; research managers' relationship with, 130–31; salaries of, 102–4, 105–9, 347n57; secrecy issues for, 116–17, 147–49; as social animal, 182–83; training and mentoring by, 247–48; virtues of, 119–21, 124–26, 184–90; women as, 185, 188, *188*, 372n79

industry and industrial development: approach to, 18–19; as home for inquiry, 157–58; as home of scientific research, 110; managerial idiom in, 121–22, 125; organized science compared with, 368n38; research managers' role in, 130–31, 135–45; research on, 121–24, 124, 351–52n102; science

industry (cont.)
integrated into, 57–58; as science leader, 194; scientific virtues and, 127–32; teamwork and practical problems of, 178–83, 180; tension between R&D and goals of, 114–17, 145–46, 163; uncertainties and consequences for, 132–35; universities' relationships with, 18–19, 212–13, 378n11; university's convergence with, 231–32, 247–49, 384n68. See also industrial science; industrial scientists and culture
initial public offerings (IPOs), 275, 276
innovation. See creativity
Institute for Genomic Research, The (TIGR), 223, 266, 381n35
Institute for Systems Biology, 266, 381n34
institutionalization: of is/ought distinction, 69–70; misconception of, 44; of moral equivalence presumptions, 52; of science as mass occupation, 85; of skepticism, 282–89. See also naturalistic fallacy
institutions: convergence of, 231–32, 247–49, 384n68; "ethos" of, 111–13; fault lines of, 254–55; free use of reason circumscribed by, 37–38; historical change in, 18–19; moral ordinariness and, 110–11; objectivity as, 23; as sites of virtue, 232–42; surveillance by, 85–86
instrumentalism, defined, 27
integrity: concept of, 13; of industrial scientists, 145–51, 185; scientific planning juxtaposed to, 197–203. See also ethics
Intel Corp., 283, 290
intellectual property: concept of, 139–40; entrepreneurial uses of, 210–13; industrial policies on, 147–48; Merton's view of, 112–13, 117, 350n84; secrecy and, 116–17; university policies on, 211–13. See also patents; technology transfer

interdisciplinarity, 178–83, 180
interest as performance: concept of, 277; entrepreneurs' pitches as, 277–82, 278, 286–87, 295–96, 390–91n39, 391n47, 391–92n50; networking at meeting and, 304, 305–10; prudence and passion balance in, 292–303, 396n121; VC investment criteria and, 282–86, 289–91
internal rate of return (IRR), 388n16
Invasion of the Body Snatchers (film), 120
invention. See creativity and invention
investment. See angel investors; venture capital; venture capitalists (VCs)
IPOs (initial public offerings), 275, 276
IRE [Institute of Radio Engineers] Transactions on Engineering Management, 129
IRI (Industrial Research Institute), 122, 123, 129, 130
IRR (internal rate of return), 388n16
is/ought distinction: Cold War institutionalization of, 69–70; Einstein on, 11–12; knowledge and knower in, 9–13; moral equivalence presumptions in, 83–84; in morality, 318–19n27; Oppenheimer on, 75–76; social responsibilities debate and, 71–72
Ivory Tower. See academic scientists and culture; universities

J. Craig Venter Institute, 226
J. H. Whitney & Co., 386–87n2
Jacobs, Irwin, 266
James, William, 29, 31
Jefferson, Thomas, 200
Jesus (biblical), 32–33
Jewett, Frank, 96, 145, 169
Jews, as scientists, 67, 334n68
Jobs, Steve, 266
Johns Hopkins University, 167, 172
Johnson & Johnson, 296
Jones, Mark Peter, 244, 321n47, 395n107
Jordan, David Starr, 29, 41

Journal of Chemical Education, 129
Journal of Industrial and Engineering Chemistry, 129
Judson, Horace Freeland, 86–87, 340n142

Kagle, Bob, 294
Kant, Immanuel, 7, 37–38
Kaplan, Norman, 119, 121, 163–64
Kekulé von Stradonitz, Friedrich August, 9
Kenney, Martin, 245
Kerr, Clark, 212, 231, 265
Kettering, Charles F. ("Boss"), 97, 137–38, 142, 147, 168
Kevles, Daniel, 58, 64
Kierkegaard, Søren, 8, 317–18n20
Killeffer, D. H., 143
Kipp, E. M., 345–46n36
Kleiner Perkins Caulfield & Byers (firm): familiarity and entry to, 215–16, 288; management fees of, 274; success/failure rate of, 273; VC funds of, 272, 295, 387n7
Knight, Frank H., 377n1
knowledge: apostolic handing on of, 32–33; as commodity (Weber), 47; conflicting notions of, 176–77; impossibility of moving from is to ought in, 11–12; mobilizing virtues in making of, 41, 183–90; monetary rewards of, 210; nature of, 23–27; the personal in making of, 4; personal vs. impersonal in, 1–8; as power *and* civic virtue, 41; production of, 169–70, 264–65, 366n3; profitless quest for, 45–46; religious justification for pursuing, 35–36; team vs. individual creation of, 168–73; teamwork in creation of, 179–83, *180*; ways of thinking about, xiv–xvi. *See also* expertise; technoscience and technoscientific knowledge; Truth and truth
knowledge economy, 230

Kodak Company. *See* Eastman Kodak Research Laboratory
Komisar, Randy, 282, 293, 294–95
Korean War: attitudes toward scientists in, 73–74, 77; mobilization of scientists in, 104; shortage of scientists in, 101–2
Kornberg, Arthur, 244, 246, 262–63
Kornhauser, William, 116–17, 163, 351n98
Kuhn, Thomas, 87, 132, 353n113
Kunze, Robert, 288

Lander, Eric, 381n34
Langlois, Richard, 266–67
Langmuir, Irving, 58, 108, 188, 202–3
Lapp, Ralph, 71, 81–82, 173, 338–39n119
Larrabee, Eric, 73
Lash, Scott, xv
late modern society and culture: absolutes and universals eliminated in, 56; academic common sense about, 3–4, 10; approach to, xv, 18; aspects of, 4–6; being part of the deal in, 302–3; descriptive to normative transit in, 10–11; entrepreneurial life-worlds in, 229–32; inspired genius concept in, 33; is/ought distinction in, 10–13; moral equivalence presumptions in, 47–48, 89–91; organization of science in, 165; organized research in, 95–97; personal reputation in, 311–12; research function and corporate goals in, 230–31; social organization and knowledge discourse in, 195–96; thematic links of past to, xvi–xvii. *See also* de-moralization of technical experts; entrepreneurship and entrepreneurial companies; heterogeneity; uncertainties; venture capitalists (VCs)
Latour, Bruno, 2–3, 389n31
Lavoisier, Antoine Laurent, 39, 174
Lawrence, Ernest O., 90
Leermakers, John, 139, 143, 199, 356–57n24

Leibniz, Gottfried Wilhelm, 39, 52–53
Lemelson-MIT Prize, 216
Levinson, Art, 253
Lewis, Sinclair, *Arrowsmith*, 60–63, 169, 331–32n40
Lewontin, Richard, 87
Life magazine, 68, 79, 218
life sciences, 83–85. *See also* biotechnology and biomedical research; DNA research and technologies; medical research
Likely, Wadsworth, 336–37n95
literature: on atomic bomb, 66; on genius and the personal, 7–9; on organization man, 120; on organized science, 169; on scientists, 60–63, 77; on venture capitalist, 276–77
Little, Arthur D., 41, 63, 96, 136. *See also* Arthur D. Little, Inc.
local-cosmopolitan distinction, 118, 352n105
Loche, Ken (pseudonym), 255
logical empiricism, 31
Lopez-Ibor, Juan, 336n87
loyalty oaths, 68
Lynd, Helen, 121
Lynd, Robert, 121
Lyons, Henry (pseudonym), 248–49, 252–53, 260–61

Macaulay, Thomas Babington, 54
Mach, Ernst, 29, 30
Mairan, Jean-Jacques Dortous de, 36
managerial ethos, 20
Manhattan Project: Feynman's role in, 221; Groves's concerns about, 334n68; interdisciplinary teamwork of, 180–81; legacy of, 167; moral responsibilities debated, 65–70; plutonium production for, 348n64; reflections on, 333n57; security concerns and, 68, 69; size and significance of, 64–65, 67; sociological view of, 118; targets for bombs of, 90

Man in the Gray Flannel Suit, The (film), 120
Marcson, Simon, 114–16, 117, 163, 350n87
Marcuvitz, Andy, 395n104
Margenau, Henry, 12
Martin, Miles J., 356–57n24
Martinis, Joanne, 264
Marvyn, Lee (pseudonym): on aesthetic of inquiry, 239–40, 241–42; complaints and resignation of, 238–39, 240–41, 383nn60–61; on good life and virtues, 252; on university teaching, 249
Marx, Jon (pseudonym): on good life and virtues, 253; on growth of company, 257; on motivations, 250, 261–62; on teamwork and individualism, 259–60
Marxism: opposition to, 170; organized science as viewed in, 368–69n41; scientists as workers in, 50–51
Massachusetts Institute of Technology (MIT): business school of, 353–54n114; journal of, 129–30; Radiation Laboratory of, 167; recruiting scientists from, 160–61; salaries of, 107, 108
Massachusetts Technology Transfer Center (MTTC), "Platform" program of, *278*, 297, 389n32
mathematicians: natural philosophers distinguished from, 27, 323n21; virtues of, 36–37
Matthew, Charles, 396n121
Mayo, Elton, 121, 123
McCance, Henry, 291
McCarran Senate Internal Security Subcommittee, 335n76
McCarthyism: as cultural fault line, 195–97; House Un-American Activities Committee (HUAC) and, 68–69, 83, 118, 334n72, 335n76
Mead, Margaret, 78
Mechanical Engineering (journal), 129
medical research: academics in, 246, 250; grant system in, 242, 245–46. *See*

also biotechnology and biomedical research; DNA research and technologies; life sciences; pharmaceutical research

Mees, C. E. Kenneth, *137*; on accounting system, 140–41, *141*, 143, 144; appointment of, 203; on autonomous research, 152; conference system of, 205, 206, 207; on disorganization, 198–200, 202, 265; on geniuses and teamwork, 183; as influence, 375n111; influences on, 157, 204, 363n101, 376n128; on interdisciplinary teamwork, 179, *180*; on laboratory management, 157, 363n101; on organization models, 191–93; publication policy of, 148, 151; on research director's job, 204–5, 206, 207; on research uncertainties, 132, 136–37, 139, 142; on salaries, 102–3, 108; successor to, 376n123; on uniqueness of photography research, 355–56n15; on universities as *not* home of scientific research, 364n111; writing collaboration with Baker, 374–75n108. *See also* Eastman Kodak Research Laboratory

Mellon Institute of Industrial Research, 95

Mencken, H. L., 48

Mendel, Gregor, 42

Merck Inc., 203

Mersenne, Marin, 35–36

Merton, Robert K.: on atomic bomb and social responsibility, 70–71; bias and limits of, 113–14, 131; journalists' criticisms of, 86–87; on local-cosmopolitan distinction, 118; mentioned, 121, 147, 163, 235, 360n66; on moral equivalence of scientist, 21–23, 47–48, 52; on moral ordinariness, 110–11; 1930s "thesis" of, 326n50; on responsibility and structure of labor, 336n85; on scientists as human, 48–49; on scientists' communal behavior, 173–74; on secrecy and

property rights, 112–13, 117, 350n84; on serendipity, 202

Metlay, Grischa, 212

Métraux, Rhoda, 78

Metzger, Nikolai (pseudonym): entrepreneurial satisfaction of, 250; on good life and virtues, 236–38, 241–42, 252; on motivations, 262

Midgley, Thomas, 179

Milgram, Stanley, 287

military: industrial science projects with, 135; as organizational model, 190; research management concerns of, 122–23. *See also* atomic bomb development; Manhattan Project

military-industrial-academic complex, 89

military-industrial complex: federal funds for R&D in, 100–101; use of term, 81, 89, 91, 100, 170

Mill, John Stuart, 377n1

Millennium Pharmaceuticals, Inc., 296

Millikan, Robert, 63, 64

minorities: as scientists, 189–90; as VCs, 285–86

MIT. *See* Massachusetts Institute of Technology (MIT)

modernity, xv, 4, 315n4. *See also* early modern period; late modern society and culture

Monsanto Company, 177

Moore, G. E., 10

Moore, Gordon, 150, 237, 283, 290

Moore, Jon, 247

Moos, Malcolm, 338–39n119

moral equivalence presumptions: in Cold War Big Science, 80–89; cultural credibility of, 33–34; in early twentieth century (1930s), 57–64; emergence of, 47–48; in late twentieth century, 47–48, 89–91; limits of expertise and, 70–72; in nineteenth century, 52–57; in post-WWII period, 72–80; revisited in context of organized science,

moral equivalence (*cont.*)
173–74; "scientists as human" in, 48–
52, 59, 61, 66, 71–72, 77, 79; Weber's
view of, 46; in WWII period, 21–23,
64–70
morality: atomic bomb development
and, 65–71; Big Science and, 165–
73, 351n95; Weber's view of science
and, 11, 12, 15, 20, 25, 46. *See also* de-
moralization of technical experts;
is/ought distinction; moral equivalence presumptions; moral ordinariness
moral ordinariness: concept of, 15, 16;
as democratic sensibility, 127; effects
of social structure and institutional
control on, 110–11; effects on science,
128; Oppenheimer's view of, 76; of
Watson, 218, 220
Morgan, Augustus de, 53–54, 329–30n19
Moritz, Michael, 276
Moses (biblical), 32–33
motivations: altruism as, 309, 312; in
choice of academia vs. industry, 232–
42; concept of a calling as, 1, 17; cynicism about, 48; of entrepreneurs and
entrepreneurial scientists, 250–51,
261–64, 295, 396–97n122, 397n128;
"free space" as, 263–64; fun and funds
as (or not), 217–29; heterogeneity
of, 242–51; lifestyles considered in,
251–61; money as, 57–60, 107, 250–51;
ordinary, universal type of, 76, 87;
social factors vs., 321n3; of venture
capitalists, 297–98
Motorola Inc., 203
Mozart, Wolfgang Amadeus, 9
Mullis, Kary, 226, 227, 228–29, 256
multiversity concept, 265
Muni, Paul, 63
Myrdal, Gunnar, 121

Nantucket Conference, 302
National Academy of Sciences, 106

National Cash Register Company, 156
National Defense Education Act, 102
National Institutes of Health (NIH), 225,
245, 246, 253
National Manpower Council, 77–78,
104, 347n57
National Opinion Research Center,
354nn119–20
National Research Council, 96, 154–55,
168
National Science Foundation (NSF):
applied research focus of, 213; career
choice study of, 78–79; congressional
hearings on, 104; establishment of, 122,
161; funding from, 238, 239, 245, 246
national security, 67–70. *See also* atomic
bomb development
National Venture Capital Association,
285
naturalistic fallacy: recognition of, 10–
12; revisited in 1960s, 12–13. *See also*
is/ought distinction
natural philosophy: divine guidance in,
33; early modern period of, 24–27,
47–48; mathematics distinguished
from, 27, 323n21; metaphysical aspects
of, 27–30; virtues linked to, 36–37
natural theological tradition, 24, 26
Nature, in sacred vs. secular domain,
23–27, 28
Nature magazine, 24, 45
NDAs (nondisclosure agreements), 272,
279
Needham, Joseph, 368–69n41
Nelson, Richard R., 356n22
Netscape, 292
networking: different accounts of, 310–
13; face-to-face activities in, 301–3,
399n154; familiarity in relationships
of, 283, 285–89, 294–95, 299–300, 302;
food in, 308–9; personal reputation
in, 311–12; at UCSD meeting, 304,
305–10; use of term, 277–78. *See also*
familiarity; personal, the

Neumann, John von, 66–67
Newfield, Christopher, 378n11
New Republic magazine, 61
Newton, Isaac: biographers on, 52–54, 329–30n19; mentioned, 171; on natural philosophy, 26, 27; self-importance of, 32–33
New Yorker magazine, 223
New York Times, 51, 63–64, 86
Nietzsche, Friedrich, 17, 55–56, 293
NIH (National Institutes of Health), 225, 245, 246, 253
Nobel Explosives Company, 205, 376n128
Nobel Prize: collective vs. individual recipients of, 170; funding issues and, 384–85n72
Nobel Prize, recipients of: Boyer and Cohen, 216; Einstein, 63; Feynman, 220; Kornberg, 244; Langmuir, 108; Mullis, 226, 228; Shockley, Bardeen, and Brattain (Bell Labs), 146, 149, *150*; Watson, 218, 220
Noble, David, 201–2, 375–76n121
nondisclosure agreements (NDAs), 272, 279
normative uncertainty concept, 5, 16, 157, 264. *See also* uncertainties
Northrop, F. S. C., 12
Noyce, Robert, 150, 266, 283, 290
NSF. *See* National Science Foundation (NSF)
Nutting, P. G., 168

Oakeshott, Michael, xiii–xiv, 386n1
objectivity: of entrepreneurial initiatives (or not), 217; personal eliminated in, 31–32; as social institution, 23; virtues necessary to, 15–16
oceanographic research, 233
Office of Naval Research (ONR), 122, 123
Office of Research Integrity, 86
Office of Scientific Research and Development (OSRD), 135

oil industry, 100, 138, 144–45
Olson, Ken (pseudonym), 394n95
O'Mair, Michael (pseudonym), 248, 255, 260
Omidyar, Pierre, 283, 294
Omidyar Network, 294
ONR (Office of Naval Research), 122, 123
On the Waterfront (film), 120
openness, as virtue, 147, 222, 254, 255, 260
open-source movement, 360–61n69
operationalism, defined, 27
Oppenheimer, J. Robert: on atomic bomb, 65, 66; hearings of, 68, 69, 118, 171; on is/ought distinction, 75–76; mentioned, 129; role of, 90; as specific intellectual, 70; Teller on, 71
Oracle Corp., 266
ordinariness, xv–xvi. *See also* moral ordinariness
O'Reilly, Sean (pseudonym): on company atmosphere, 256–57; context of, 241–42; on entrepreneurial companies, 245, 259; entrepreneurial satisfaction of, 249; on grant system, 242–43; research and positions of, 233–36; on teamwork, 260, 261
organization chart, 191, *192*, 198, 364n105, 374n106. *See also* organized research; teams and teamwork
Organization Man: as academic research topic, 113–19, 121–24, 235; appreciation of, 197; criticism of, 96–97; loss of virtue and, 119–21, 125–26; vices of, 173–78. *See also* industrial scientists
organized research: academic view of, 113–19, 121–24, 172–73, 178, 181–83; components of, 166–68; criticism of, 169–73; German influence on, xvi, 94, 172, 196; interdisciplinarity of, 178–83, *180*; managers (directors) of, 203–8, 376–77n129; models for, 190–95, *192*; planning of, 197–203; pros and cons of, 95–97, 128; significance of, 177–78; specific project steps in, 200, *201*.

organized research (*cont.*)
See also Big Science; industrial
science; organization chart; teams
and teamwork
Origin of Species, The (Darwin), 24, 34
Orth, Charles, 108, 123
OSRD (Office of Scientific Research and
Development), 135
Owens-Illinois Glass Company, 146

Packard, Vance, 120
Paris Academy of Sciences, 36
Parsons, Talcott, 117
Pascal, Blaise, 329–30n19
Pasteur, Louis, 63
Patent and Trademark Act Amendments
(Bayh-Dole Act, 1980), 139–40, 215,
241
patents: of academic research outcomes,
57–58, 211–13; of gene sequences, 221–
22; of German drugs, 342n6; industrial
management policies on, 147–48,
149–51; litigation over, 112; for PCR,
228. *See also* intellectual property;
technology transfer
Pauling, Linus, 339n120
PCR (polymerase chain reaction), 226,
228
Pearson, Karl, 29
Peirce, Charles Sanders, 58–59
Pelz, Donald, 119
Pennsylvania State University, 353–
54n114
People's Republic of China: fears of, 120
Perkins, Tom, 272, 284, 387n7
personal, the: elimination of, 6–9, 10, 31–
32; tension of impersonal with, 1–6.
See also familiarity; networking
Personnel (journal), 129–30
pharmaceutical research: convergence of
university and industry in, 232, 383n53;
entrepreneurial initiatives in, 233–
36, 245, 249; ethics of, 95; history of,
342n6; resources needed, 245, 384n70

phenomenalism, defined, 27
Phillips, Melba, 335n76
philosophers and philosophy: biographies of, 8; as heroes, 55–56; mathematicians distinguished from, 27,
323n21; as morally superior, 24–27,
47–48; scientist as *beau idéal* of, 44–45;
scientists and science distinguished
from, 9–13, 27–33; virtues of, 36–37.
See also natural philosophy
physicians and surgeons: in Antiquity,
39; fictional depiction of, 60–63, 224–
25; as researchers, 246–47; scientists
distinguished from, 44
physicists: Big Science and nostalgia
of, 82–83; disinterested research of,
58; Einstein's link to, 64; moral responsibilities debated, 65–71; particle
research of, 167–68. *See also* atomic
bomb development
physics: crisis in, 30; Feynman's approach
to, 220; popularizing of, 61
Pinker, Steven, xv
Pixar Animation Studios, 266
planning: emergence of, 166–68; rebellion against rational, 10; scientific integrity in context of, 197–
203; of specific project steps, 200,
201
"Platform" program (MTTC), 278, 297,
389n32
Plutarch, 36
Poincaré, Henri, 10–11, 29, 30, 319n28
Polanyi, Michael, 49, 147, 170, 197–98
political economy, of scientific entrepreneurship, 213–17
polymerase chain reaction (PCR), 226,
228
Popper, Karl, 23, 28
Popular Science Monthly, 58–59
Porter, Roy, 326n57
positivism, defined, 27
postmodernity, 5
Powers, Phillip N., 348n63

pragmatism: application of, 29; definition of, 28; as philosophy of science, 31

President's Science Advisory Committee, 166, 167

President's Scientific Research Board, 106, 354nn119–20

Price, Derek de Solla, 9, 365–66n1

PricewaterhouseCoopers, 287

Priestley, Joseph, 24, 26

Princeton University, 107

prisca sapientia (and *prisca theologi*) traditions, 32

Private Securities Litigation Reform Act (1995), 290

probabilism, defined, 28

production: of knowledge, 169–70, 264–65, 366n3; planning of, 199; of plutonium, 348n64; research vs., 135

professional and trade associations: conferences of, 130; industrial scientists in, 146–47, 148–49; in recruitment criteria, 185; research contracts of, 95

professionalization of science: mediocrity in, 84; misconception of, 42, 44; motivations in context of, 49; standards in, 58

professional journals: industrial scientists' publication in, 148–51, 361n73; internal commentary on industrial science in, 129–32

professional technicians, use of term, 175

project request form, example of, 201

psychologists: on industrial scientists, 119; on teamwork, 194–95

public attitudes: on disloyalty of scientists, 68; on power of scientists, 65–66, 72; on scientists as odd or eccentric, 77–80

Public Health Service, 122

Pupin, Michael, 63

pure science. See basic vs. applied science

Puritanism, 326n50

QUALCOMM, 266

quantum theory, 30

Rabi, I. I., 72, 345n33

Rabinow, Paul: on academics' responsibilities, 245; focus of, 6; on knowledge production, 264; on modernity, 4; on motivations, 250, 262

radar development, 180, 369n50

Radin, Amy, 272

Rainwater, Lee, 119

Rand, Ayn, 120

RAND Corporation, 138

rationalism and reason: bureaucracies as triumph of impersonal, 9–13; free use of, 37–38; methodology of, 32; rebellion against, 10

Rauschloeb, Rachel von, 219

Ray, John, 35–36

RCA, 131, 149

R/D: Research Development (journal), 129

Reagan, Ronald, 215

realization ratio (RR), 388n16

Rebel without a Cause (film), 120

recruitment: benefits offered in, 163–64; examples of, 108–9, 128, 160–61; half-time offer in, 348–49n73; problems in, 76–80, 217; recruit's terms of acceptance, 137–38; sample letter of reference in, 186–87; social virtues considered, 184–85, 186–87; of women and minorities, 188–90

Reese, Charles L., 138–39

Reich, Leonard S., 108, 151

Reichenbach, Hans, 31

Renan, Ernest, 7

Renouvier, Charles Bernard, 330n28

research, defined, 139

Research Corporation, 210, 212

Research Management (journal), 129

research manager (director): as buffer to uncertainties, 207–8; communications of, 204–5; conference system

research manager (cont.)
of, 205, 206, 207; hybrid nature of, 376–77n129; models of, 203–4
Research Technology Management (journal), 129
Revenue Act (1936), 100
Rhodes, Joan (pseudonym), 249–50, 258–59
Riesman, David, 79, 120
Rintoul, William, 376n128
Robinson, Edward G., 63
Robinson, Robert J., 393n70
Rock, Arthur, 283, 290, 394n99
Rockefeller Foundation, 78
Roethlisberger, F. J., 121, 123
Roizen, Heidi, 285
Romanticism, genius in, 33
Root, Elihu, 168
Rosenberg, Charles E., 87
Roth, Philip, 9
Route 128, 216
routinization of science, 82, 84
Rowland, Henry, 29, 63
Royal Navy, 35
Royal Society of London, 35
Royston, Colette, 288
Royston, Ivor: baseball metaphor of, 272–73; entrepreneurial pitch of, 288; on leaving university, 243–44, 384n66; on post-bubble opportunities, 293; on VC investment criteria, 284, 291
Rubenstein, Arthur H., 354n115

safe harbor statement, 290, 395n101
Saintsbury, George, 322n7
salaries and wages: of academic scientists, 102–4, 105–9, 346n40, 347n52, 347nn56–57; argument for increase in, 76–77; argument for living wage, 105; in Big Science, 83–84; in early modern period, 34–35, 38–39; of Einstein, 103; expectations of, 209, 217–18, 220, 228; expected indifference to, 45–46, 103–4; of government scientists, 106–7,

347n57; independent means vs. necessity for, 38–39; of industrial scientists, 102–4, 105–9, 108–9, 347n57; limits of, 38, 112; as motivation, 57–60, 107, 250–51
San Diego Enterprise Partners, 378–79n15
SBIR (Small Business Innovation Research) program grants, 390n37
Schering-Plough Corp., 263
Schley, George, 57–58
Schmidt, Benno, 272, 386–87n2
scholarly life, 37–38, 41
Schumann, André (pseudonym), 243, 250, 253–55, 256
Schumpeter, Joseph, 266, 377n1
science: as adventure, 79; art compared with, 7–9; calls for moratorium on developments in, 73, 336n93; changes affecting identity of, 99–100; as common sense, 28; conditions for doing good, 235; as corruptible, 81–82; death of, 176, 368–69n41; definitions of, 43, 270–71; deflationary concept in, 88–89; ethics distinguished from, 10–13; of ethics (Durkheim), 319n33; as impersonal, 6–9; legitimacy of, 40; metaphysics distinguished from, 27–33; as practical vs. abstract, 43–44; professionalization of, 42, 44, 49, 58, 84; routinization of, 82, 84; as "sacred cow," 73–74; sanctity of, 24–25; unity or disunity of, 71–72, 336n87; wartime mobilization of, 197–99. *See also* basic vs. applied science; industrial science
Science and Engineering (journal), 218
science and engineering workforce: clock discipline for (or not), 154–56; Cold War needs for, 76–77; Depression-era cutbacks in, 145; as "professional technicians," 175; shortage of, 68, 74, 101–2, 103–4; women and minorities in, 189–90, 285–86. *See*

also recruitment; salaries and wages; scientists

Science magazine: on *Arrowsmith* (Lewis), 61; on Association of Scientific Workers, 50; establishment of, 129–30; on recruitment criteria, 185; on scientist-on-the-make, 87–88; on university mores, 182; on wages, 105

Science Service (organization), 336–37n95

scientific and control instruments industry, 100

scientific expert witnessing, 45

scientific life and community: attractions of, 77–78; collective methodical discipline in, 34; collective voice of, 32; communal virtue of, 45; cosmopolitanism of, 68–69; moral responsibilities debated in, 65–71; skepticism about, 73–74, 77; sociological study of, 21–23; virtues attached to, 127–28. *See also* calling; career; science and engineering workforce; scientists; vocation

Scientific Method: capitalization of, 324–25n38; elimination of personal in, 6–9, 31–32; growing acceptance of, 33–34, 39–41; nonscientific imitation of, 90; Weber's view of, 45–46. *See also* objectivity; utility, rhetoric of

Scientific Monthly (journal), 129–30

scientific naturalists, 25, 40, 42–43

scientific role, defined, 26

scientists: approach to, xv–xvi; changes affecting identity of, 99–100; Cold War libel against, 67–69; competition of, 218, 220–22; conflict presumed between academic and industrial, 113–19; deflationary concept and, 88–89; engineer and technician as models for, 30–31; heterogeneity of, 229–31; personality traits of, 184–85, *186–87*; philosophers distinguished from, 9–13; post-WWII survey of, 109;

socialization of, 174; "stockpiling" and commodification of, 101, 112, 345n33; term for, 322n7; witch-hunt against, 68–69, 118; work satisfaction survey of, 354nn119–20. *See also* academic scientists and culture; entrepreneurs and entrepreneurial scientists; government scientists; industrial scientists; moral equivalence presumptions; recruitment; virtues

scientists, characterizations of: as businessmen, 61; in fiction, 60–63; as heroes, xv, 63, 79, 87; as holy men, 64; as human, 48–52, 59, 61, 66, 71–72, 77, 79; as individuals, 169–73, 182–83; as invisible vs. genius, 6–9; as mad, warped, 73; as mavericks, 182; as no longer intellectuals, 90–91; as normal, 80–89; as priests of power, 72–80; in self-portrayals, 218, *219*, 220–22, *221*; as workers, 375–76n121

Scopes trial, 62

Scripps, E. W., 336–37n95

Scripps Research Institution, 383n54

Scripture, E. W., 29

Seaborg, Glenn T., 77, 105–6, 347n52

secrecy and property rights: of entrepreneurial companies, 254–55; industrial management policies on, 147–51, 361n73; industrial scientists' adaptation to, 116–17; as institutional fault line, 254–55; Merton's view of, 112–13, 117, 350n84; in software industry, 360–61n69; in Springboard meetings, 279–80. *See also* intellectual property; patents; professional journals

Secrist, Horace A., 364n105

secularization: anti-metaphysical notions linked to, 31; of inspired genius, 33; of Nature and scientific inquiry, 23–27, 28

Securities and Exchange Commission (SEC), 290, 390n35, 395n101

semiconductor industry, 288–89
Sequoia Capital (firm), 274, 276, 287
Seragen (company), 215
serendipity, 202–3
serial entrepreneur type, 292
Shapley, Harlow, 68
Shearer, Kevin, 250
Shell Development Company, 142
Shepard, Herbert A., 118, 122
Shepard, Norman, 139–40, 204–5. See
   also American Cyanamid
Shockley, William, 149, 150
Shockley Semiconductor Laboratory
   (Beckman Instruments division), 150
Sidgwick, Henry, 10
Silicon Graphics, 292
Silicon Valley: biometrics company in,
   236–38; competition to have most
   money in, 397n128; dot-com bust and,
   274–75, 293, 392n58; entrepreneurial
   initiatives of, 216; fun and funds in,
   222, 294; idealistic entrepreneurs of,
   295; legendary deals in, 283. See also
   specific companies
Simpson, O. J., 228
Sinclair, Upton, 103
Six Sigma (Motorola), 203
Sloan, Alfred, 137
Slosson, Edwin, 61
Small Business Innovation Research
   (SBIR) program grants, 390n37
Small Business Technology Transfer
   (STTR) program, 390n37
small-world phenomenon, 287–88
Smiles, Samuel, 34
Smith, Adam, 36–37
Smith, John Kenly, Jr., 148–49
Smith, William, 38
Smyth, Henry DeWolf, 345n33
Snow, C. P., 30, 65, 91, 300, 336n87,
   341n155
social and moral responsibilities, debates
   on, 65–71. See also moral equivalence
   presumptions

social constructionism, 340n142
social constructiveness concept, 194–95
social ethic concept, 175–76
socialism, 10. See also collective and collectivism
social sciences: attitudes toward scientists studied in, 78–79; cultural
   fault lines in, 195–97, 254–55, 312–13;
   government support for, 77–78; managerial, practical research in, 121–24,
   125, 376–77n129; practical past concept and, xiv; science as too far ahead
   of, 73
Society for Freedom in Science, 170, 198,
   374–75n108
sociology: on conflict of values for industrial scientists, 113–19; criticism
   of, 119; growth of discipline, 321n2;
   on individualism, 21–23; on industrial
   science, 111–13; managerial literature compared with, 131–32; research
   methodologies of, 350n87, 351n98;
   on science and State power, 105–6;
   structural-functionalist tradition of,
   117–18; on teamwork and organized
   science, 174–75; value conflict concept in, 113–19; Whyte's references to,
   121
Socony-Vacuum Oil Company, 177
Socrates, 32–33
Söderqvist, Thomas, 330n23
software industry: entrepreneurial initiatives in, 236–38, 250; secrecy in,
   360–61n69
Sorcerer II (yacht), 226
Soviet Union: fears of, 120; Sputnik
   launch of, 79, 101–2, 345n35
specialization, 88–89, 258
specific intellectual, idea of, 57, 70
Spencer, Herbert, 29
Sperry Gyroscope Company, 211
Springfield, Diane (pseudonym), 243,
   255–56
Sputnik, 79, 101–2, 345n35

Standard Oil companies: accounting system of, 141; conferences of, 130; research interests of, 95, 140; research management concerns of, 122; research uncertainties of, 142; serendipity encouraged at, 203

Standen, Anthony, 72–73, 336n94

Stanford University: business school of, 353–54n114; entrepreneurial initiatives of, 215–16; intellectual property policy of, 211

start-ups. *See* entrepreneurship and entrepreneurial companies

State and State power: expertise used by, 33–34, 39–41, 42; increased WWI-era access to, 65; industrial science and, 94, 99–104, 342n5; science integrated into structures of, 41–43, 57–58, 69–70, 105–6; wages paid in early modern period, 34–35. *See also* Big Science

Steele, Lowell W., 361n77, 363–64n103

Steelman Report: origins of, 354n119; on planning, 198, 199, 200–201; on research uncertainty, 133; on scientists' motivations, 109; vocabulary of "strain" and "stress" in, 354n120

Steenbock, Harry, 210–11

Stefánsson, Kári, 381n34

Stensrud, Bill, 293

Stent, Gunther, 85–86

Stevenson, Earl, 169, 369n50

Stine, Charles, 358–59n47

Strauss, Anselm, 119

structural virtues of scientific community: concept of, 15–16; venues of, 19–20

STTR (Small Business Technology Transfer) program, 390n37

Superconducting Supercollider, 167–68

Swanson, Robert A., 215–16

Swift, Jonathan, 170

Sylvania Division (American Viscose Corp.), 147, 148, 150, 349n77, 363n91

syndication issues, 298–99

Synthetic Genomics, Inc., 226

Szilard, Leo, 64, 334n68

Tandem Computers, 387n7

Taylor, Frederick W., 123, 138, 351n100

Tead, Ordway, 351n100

teams and teamwork: accommodation to, 158–59, 364n105; criticism of, 169–71, 174–78; defense of, 193–94; emergence of, 166–68; flexibility of, 181, 258–61; good society in context of, 195–97; interdisciplinarity of, 178–83, *180*; leadership's tension with, 166; models of, 190–95, *192*; practical commentary on, 178–79; sensibility of, 16–17; virtues mobilized in, 41, 183–90; women and minorities in context of, 189–90

technocracy: concept of, 81; fear of, 338–39n119; "priesthood" of, 173

technological revolution: criticism of, 169–73; organizational components of, 166–68

Technology Partners, 397n131

Technology Review (journal), 129–30, 236

technology transfer: academic's reaction to, 240–41; legislation on, 139–40, 215, 241; universities' policies on, 214–16; university centers for, 277, *278*, 389n32. *See also* entrepreneurship; intellectual property; patents

technoscience and technoscientific knowledge: approach to, 1–6, 18; concept of, 2–3, 270–71; cultural authority of, 13–14; future of, 17–18; moral heterogeneity of, 242–51; personal dimension of, 3–4; social constructiveness of, 195; uncertainties of, 269–70; unfamiliarity of, 289–91. *See also* charisma and charismatic authority; familiarity; personal, the

Teitelman, Robert, 295

Teller, Edward, 64, 71, 74, 149, 180

Tennessee Eastman, 376n123

Terman, Lewis M., 68–69
Thatcher, Margaret, 19, 327n69
Thiel, Peter, 283
Thomas, Charles Allen, 348n64
Thomas Aquinas, 32–33
Thoreau, Henry David, 61
Thorpe, Charles, 316–17n10, 383n56
TIGR (The Institute for Genomic Research), 223, 266, 381n35
Time magazine, 79, 106–7, 217, 223
Tocqueville, Alexis de, 43, 196
Tolstoy, Leo, 11, 40, 319n32
transistor development, 146, 149–51, *150*, 179, 181
Trikon Technologies, 378–79n15
Truman, Harry, 65–66
Truth and truth: absolute, 29; biographers' duty to, 53; commerce as influence on, 217; death of, 56; as God's vs. ultimate reality, 27–30; journalist's perspective on, 340–41n147; limits of, 30–31; organization vs., 165; role of freedom in, 82; salaries and, 103; search for two kinds of, 10, 319n28; as solitary vs. collective, 16; as solitary vs. social, 170–71
truth-speakers, 6–9
Turner, Stephen, 5
Tuve, Merle A., 169, 172
Tyndall, John, 29, 40

Ulam, Stanislaw, 66
uncertainties: acceleration of, xv; accounting system in context of, 138, 139, 140–42, *141*, 143–45, 358n43; in basic vs. applied research, 132–35, 139–40; concept of, 132–35; different levels of, 269–70; entrepreneurial life and, 264–67; expectations of, 132, 136–39, 142; familiarity as counter to, 301; practical approach to, 178–79; research managers' focus on, 162–64, 207–8; research organization and management in context of,

156–59; tension between business goals and, 145–46, 163; VCs' attempts to mitigate, 297–99, 301–3; in venture capital, 276
universities: business and engineering schools of, 121–24; business and management schools of, 122–23, 353–54n114; entrepreneurial initiatives of, 214–16, 237, 238–39; heterogeneity of, 229–31; as home for inquiry, 19–20, 45–46, 158, 249–50, 364n111; as home of intellectuals and science, 91, 128, 212, 238–39, 357n25; HUAC allegations and, 334n72, 335n76; Hutchins's vision of, 378n10; industrial and labor relations schools of, 122–23, 353–54n114; industry's convergence with, 231–32, 247–49, 384n68; industry's relationships with, 18–19, 212–13, 378n11; intellectual property of, 139–40; intellectual property policies of, 211–13; knowledge production of, 264–65; limits of, 160–61; power structures of, 365n123; professor-student relationships in, 174–75, 181–83; as sites of virtue, 232–42; specialization in, 88–89; technology transfer policies of, 214–16; utilitarian and self-cultivation arguments for science in, 59–60; values assigned to, 111–13, 114–17. *See also* academic scientists and culture; university-based research
university-based research: demoralization of technical experts in, 13–14; discoveries of, 57–58; divergent laboratories in, 193; funding for, 213–14, 222–23, 346n47; industrial contracts and funds for, 239–41, 346n47; industrial science compared with, 160–64, 233–42; interdisciplinary organized science compared with, 172–73, 178, 181–83; pure vs. applied science distinction and, 97,

344n21; uncertainties in, 132–35, 139–40. *See also* academic scientists and culture; universities

University of California: Human Subjects Committee of, 280, 383n56; Regulation no. 4 of, 212–13; salaries of, 347n52

University of California, San Diego: entrepreneurial initiatives of, 214; networking at meeting at, *304*, 305–10; "Springboard" meetings of, 277–80, 287, 391n45

University of California, San Francisco, 215–16

University of Chicago, 105–6, 353–54n114

University of Cincinnati, 95

University of Denver, 354nn119–20

University of Illinois, 353–54n114

University of Massachusetts. *See* Massachusetts Technology Transfer Center (MTTC)

University of Pennsylvania, 353–54n114

University of Wisconsin, 66

Updike, John, 8

Urey, Harold, 97

U.S. Air Force, Personnel and Training Research Center of, 122, 353n113

U.S. Army, 122

U.S. Coast and Geodetic Survey, 58–59

U.S. Congress: attitudes toward scientists of, 72; hearings concerning science, 68, 69, 86, 104; House Un-American Activities Committee of, 68–69, 83, 118, 334n72, 335n76; McCarran Senate Internal Security Sub-Committee of, 335n76; science support and management considered, 68–69, 72

U.S. Constitution, 43–44

U.S. Department of Defense, 122, 353n113

U.S. Department of Labor, 106

U.S. Forest Service, 59

U.S. Naval Observatory, 59

U.S. Navy, Office of Naval Research of, 122, 123

U.S. Steel Corp., 94

utilitarianism, 43

utility, rhetoric of: components of, 39–43; in placing science in universities, 60

values: of industrial vs. academic science, 111–13, 114–17. *See also* virtues

Van Osnabrugge, Mark, 393n70

Varian, Russell, 211

Varian, Sigurd, 211

VCs. *See* venture capitalists (VCs)

Veblen, Thorstein, 6, 11, 45–46, 102, 315n4

Venter, J. Craig, *224*; companies of, 223, 381n35, 382n42; Mullis compared with, 228; portrayal of, 223–26; work rhythms of, 266

venture capital: coining of term, 386–87n2; concept of, 271–72; firm structure in, 272–73; impact of dot-com bust on, 274–75, 276; investment criteria in, 275, 282–86; limits of professionalism in, 392n62; measures of fund success in, 388n16; success/failure rate of, 272–74; time frame of, 299–300, 395–96n108; 2006 figures for, 388n19; uncertainty as key feature of, 276; Web 2.0 interests of, 275–76

venture capitalists (VCs): backgrounds of, 214, 215; baseball metaphor of, 272–73, 387n8; criteria of, 275, 282–89; criticism of, 297–99, 302, 398n139; decision making of, 17–18; entrepreneur's investment expected by, 293–94; on entrepreneur's management ability, 289–91; entrepreneurs' pitches to, 277–82, *278*, 286–87, 295–96, 390–91n39, 391n47, 391–92n50; on entrepreneurs' prudence and passion, 292–303, 396n121; on exit strategy awareness, 292; face-to-face meetings

venture capitalists (*cont.*)
  of, 301–3, 399n154; familiarity and
  networking of, 283, 285–89, 294–
  95, 299–300, 302, *304*, 305–10; focus
  of, 270–71; how-to literature on,
  276–77; institutionalized skepticism
  and, 282–89; management fees of,
  274; maxims and proverbs of, 289,
  290, 291, 392n54; number of, 271; as
  performing future in present, 276–
  82; self-perceptions of, 297, 300;
  special-purpose public rhetoric of,
  295; wash-out rate of, 274–75. *See
  also* angel investors; entrepreneurship and entrepreneurial companies;
  technoscience and technoscientific
  knowledge
Vertex Pharmaceuticals, Inc., 247, 383n53
virtues: academia vs. industry as sites of,
  115, 232–42; Adam Smith on, 36–37;
  authority and expertise distinguished
  from, 40–41; of *beau idéal*–type scientist, 44–45; biographical views of,
  52–53; Christian monopoly of, 62;
  conflicting notions of, 176–77; demise
  of, 90–91; differences allowed in,
  183–84; differing characterizations of,
  52–54, 329–30n19; of entrepreneurs
  looking for VC, 289–303; gendered
  nature of, 189–90; industrialization
  and, 127–32; of industrial scientists
  (or not), 119–21, 124–26, 162–63; interdisciplinarity and, 178–83, *180*; loss
  of, 119–21; mobilization of, 41, 183–90;
  moral heterogeneity of, 242–51; organization as threat to, 165; personal vs.
  impersonal in, 1–6; reasons to dismiss,
  6–9, 31–32; in recruitment criteria, 184–
  85, *186–87*, 188–90; research uncertainty as context of, 135; residual type
  of, 75; in venture capitalist world, 270
virtues, specific: aesthetic of inquiry,
  239–40, 241–42; altruism, 309, 312;
  asceticism, 55–56, 105; disinterest,

57–59, 249–50; hard work and persistence, 33–34; honesty, 86, 296–97;
  independence and autonomy, 67–68,
  113, 238, 240, 242, 243, 245, 252–53;
  intuition, 284; likability, 300–301; list
  of, 63, 184–85; loyalty, 67–68, 102, 118;
  moral heroism, 63; normality, 59–60,
  87; openness, 147, 222, 254, 255, 260;
  out-of-the-box thinking, 288–89;
  poverty, 222–23; prudence and passion balance, 292–303, 396n121; social,
  184–90; team spirit, 196; tolerance,
  51. *See also* economy of civic virtue;
  good society; integrity; moral equivalence presumptions; structural virtues
  of scientific community; teams and
  teamwork
vita activa and *vita contemplativa*, 41, 65,
  103
vocation: definition of, vi; entrepreneurship encompassed in, 250–51; fictional
  depiction of, 62; is/ought distinction
  in, 11; lost sense of, 75; mid-century
  sensibilities about, 127–29; non-
  empirical evaluations of, 105–6; of
  shopkeeper vs. scientist, 46; specialization in, 258; theological rhetoric of,
  63; of Venter, 223–24; virtues assigned
  to institutions in view of, 233–34; Weber's concept of, 86, 93, 250–51. *See also*
  calling; career

Wade, Nicholas, 86
Waite, Charles, 397n125
Walpole, Horace, 202
War Emergency Board of Plant Patholo-
  gists, 168
WARF (Wisconsin Alumni Research
  Foundation), 210–11
Warner, Lloyd, 121
Waterman, Alan T., 74
Watson, David Lindsay, 48–49
Watson, James D., *219*; *Arrowsmith*
  as influence on, 331–32n40; DNA

discovery of, 9; on patents, 221–22; self-portrayal of, 169, 218, 220–22; Venter compared with, 225–26

Watson, Thomas, 394n95

"way we live now": academic characterizations of, 3–4, 10; antinomies and, 13–14; approach to, xiii–xiv; "community" in, 311; diverse accounts of, 14, 312–13; folk consciousness about, 306–7; identity of science and scientist in, 99–100; personal reputation in, 311–12; personal vs. impersonal in, 1–6; pre-WWII criticisms of, 49–50

wealth and profit structure, 45–46. See also commerce; salaries and wages; venture capital

weapons of mass destruction, 65–70, 74. See also atomic bomb development

Web 2.0, 275

Weber, Max: on "challenges of the day," 305; on charisma and charismatic authority, 3, 5, 91, 165, 208, 209, 266–67; on concept of a calling, 1, 17; mentioned, 50; on modernity, 128; on science and morality, 11, 12, 15, 20, 25, 46; on science as vocation, 86, 93, 250–51; on scientific discovery, 45–46, 127, 269; on scientist's task, 21; on specialization in vocation, 258; on teachers and sale of knowledge, 47; Tolstoy's influence on, 40; Whyte's references to, 121

WebMD, 292

Weinberg, Alvin, 167; on Big Science, 81–82, 166; on teams and knowledge production, 169–70, 366n3

Weiss, Paul, 84

Weisskopf, Victor F., 331–32n40

Werth, Barry, 247, 252, 281

Western Cartridge Company, 140, 143–44

Westinghouse Research Laboratories: accounting system of, 355n14; basic research of, 134; facilities of, 155;

research director of, 162–63, 207; research fellowships of, 145, 359n50; research interests of, 95; teamwork success at, 194. See also Condon, Edward U.

Whewell, William, 322n7

White, Tom, 266, 279–80

Whitehead, Alfred North, 10, 343n15

Whitney, Jock, 386–87n2

Whitney, Willis R.: on flexibility and freedom, 156; foreign-trained scientists under, 196; management tactics of, 157, 363n101; Mees on, 157, 204, 363n101; publication policy of, 151; recruitment by, 108–9, 348–49n73, 372n79; on research payback, 143, 358n43; on serendipity, 202. See also General Electric Research Laboratory

Whyte, William H., Jr.: mentioned, 183, 184, 235, 259; on organization man, 120–21, 125–26; response to, 353n110; on teamwork, 16, 175–77

Wiener, Norbert, 82

Wigner, Eugene, 64, 80

Wilde, Oscar, 9

Wilkins, John, 35–36

Wilson, E. Bright, Jr., 153

Wilson, R. R., 331–32n40

Wilson, Robert E., 141, 142, 170

Wired magazine, 225, 226

Wisconsin Alumni Research Foundation (WARF), 210–11

Wittgenstein, Ludwig, 11, 317–18n20

Wollaeger, Tim, 284

women: as scientists, 80, 185, 188, 188, 189–90, 372n79; stereotypes of, 190, 372n81; as VCs, 285–86

World War I: as "chemists' war," 64; drug patents seized in, 342n6; science mobilization in, 64, 168; technological teamwork in, 195

World War II: legacy of science mobilization in, 167–68; moral equivalence

World War II (*cont.*)
  presumptions in, 21–23, 64–70; as
  "physicists' war," 64. *See also* atomic
  bomb development; Manhattan
  Project
Wyant, Peg, 286
Wylie, Philip, 120
Wythes, Paul, 387n8

Yates, JoAnne, 138
Yellowlees, Heather (pseudonym), 245–
  46, 384n71
Yeo, Richard R., 329n18

Zilsel, Paul R., 368n38
Zirkle, Conway, 67–68
Zuckerberg, Mark, 275, 283